ANNUAL REVIEW OF NUCLEAR SCIENCE

EDITORIAL COMMITTEE (1973)

ANNUAL REVIEW OF NUCLEAR SCIENCE

EMILIO SEGRE, *Editor*
University of California, Berkeley

J. ROBB GROVER, *Associate Editor*
Brookhaven National Laboratory

H. PIERRE NOYES, *Associate Editor*
Stanford University

VOLUME 23

1973

ANNUAL REVIEWS INC.   4139 EL CAMINO WAY   PALO ALTO, CALIFORNIA 94306

ANNUAL REVIEWS INC.
Palo Alto, California, USA

International Standard Book Number 0-8243-1523-5
Library of Congress Catalog Card Number 53-995

Assistant Editor	Joanne Cuthbertson
Indexers	Mary Glass
	Susan Tinker
Compositor	Typesetting Services, Ltd,
	Glasgow, Scotland

PRINTED AND BOUND IN THE UNITED STATES OF AMERICA

CONTENTS

CERENKOV COUNTER TECHNIQUE IN HIGH-ENERGY PHYSICS

J. Litt and R. Meunier[1]

CERN, Geneva, Switzerland

CONTENTS

1 INTRODUCTION

Following the discovery of Cerenkov radiation in 1934 by Vavilov (1) and Cerenkov (1) and the subsequent theoretical explanation by Tamm & Frank (2) in 1937, a great deal of effort has been put into the investigation of the properties of this phenomenon and its application. Several review articles have appeared (3) that cover the whole subject of Cerenkov radiation in great detail. The aim of the present article is to supplement the previous reviews by placing emphasis on the recent developments in the Cerenkov counter technique as applied to the physics research at high-energy accelerators. The applications to cosmic ray research and to the measurement of total deposited energy, such as in shower detectors, will not be reviewed (3).

[1] Visitor from CEN, Saclay, France.

1

Tamm & Frank (2) developed a classical theory to account for Cerenkov radiation based on the idea that a charged particle moving in a transparent medium with a uniform velocity larger than the velocity of light in the medium emits a radiation along a conical wavefront. The angle of emission θ of the radiation of a given wavelength λ is related to the relative velocity $\beta(=v/c)$ of the particle and the refractive index n of the medium by the relation

$$\cos \theta(\lambda) = 1/\beta n(\lambda) \qquad\qquad\qquad 1.$$

The theory also produced the spectral distribution and hence the intensity of the radiation. These predictions were confirmed experimentally by Cerenkov (4), and there followed a quantum-mechanical theory of the effect by Ginsburg (5) in 1940.

Relation 1 can be rewritten as

$$\cos \theta(\lambda) = v_{light}(\lambda)/v_{particle} \qquad\qquad\qquad 2.$$

which shows that the angle of emission of the light is directly related to the ratio of the velocity of the light in the radiator to the particle velocity. This fact explains the usefulness of Cerenkov radiation in the field of elementary particle physics. The optical measurement of the emission angle of the light together with the knowledge or measurement of the velocity of light in the same medium affords the means of determining the velocity of a particle and its direction within the accuracy of optical techniques.

For a high-energy charged particle, the measurement of its relative velocity β together with a measurement of its momentum p is practically the only way to determine the mass m of the particle, which is deduced from the relation

$$m = p/\beta\gamma \quad\text{with}\quad \gamma = (1-\beta^2)^{-\frac{1}{2}} \quad\text{and}\quad c = 1 \qquad\qquad 3.$$

By using the Cerenkov effect, the velocity β can be measured to an accuracy in the range of 10^{-2} to 10^{-7}, with an accuracy in m for a single particle given by

$$dm/m = \gamma^2(d\beta/\beta) + (dp/p) \qquad\qquad\qquad 4.$$

A Cerenkov counter consists of a transparent medium (gaseous, liquid, or solid) in which the radiation is emitted, and an associated electronic detector. The research at high-energy particle accelerators uses several types of Cerenkov counters which differ in their mechanical arrangements and quality of the optical systems. From relation 1 it is seen that the radiation is not produced unless $\beta n(\lambda) > 1$ (i.e. the Cerenkov angle must be greater than zero). The possibility of detecting particles with velocities above a given value is exploited in the *threshold Cerenkov counter*. An instrument which detects the radiation over a small range of angle at a nominal value θ is known as a *differential Cerenkov counter*.

The Cerenkov effect is the result of an electromagnetic interaction that produces light which can be detected by photomultipliers. Cerenkov detectors, which are standard equipment in high-energy physics experiments, provide a nondestructive method for measuring the mass of individual charged particles.

2   GENERAL PROBLEMS ASSOCIATED WITH THE DETECTION OF CERENKOV RADIATION

2.1   The Small Light Yield

The number of Cerenkov photons N_γ emitted per unit length of a radiator of length L in the Tamm-Frank theory is given by

$$dN_\gamma/dL = 2\pi\alpha Z^2 \int_{\lambda_2}^{\lambda_1} [1 - (1/\beta^2 n^2)](d\lambda/\lambda^2) \qquad 5.$$

where α is the fine structure constant ($\sim 1/137$), Ze is the charge of the particle, and λ_1, λ_2 define the limits of the spectral range of the detected radiation. Relation 5 shows that most of the Cerenkov light appears at small values of λ.

Due to the Z^2 dependence in equation 5, a particle with large charge can produce sufficient Cerenkov light intensity to be recorded on photographic emulsion (6). However for singly charged particles the light yield is quite small: for example, the energy loss by Cerenkov radiation in the wavelength bandwidth from 200 to 600 nm is about 1 keV, which is about 100 times less than the energy loss by ionization. It is therefore most important in a Cerenkov detector to minimize the effect of ionization which can cause some radiators to scintillate, and to detect the Cerenkov radiation over as wide a bandwidth as possible.

Thus, many efforts have been made to improve the light transmission and extend the sensitivity of the photodetector into the ultraviolet region where the light yield is greater. At long wavelengths the detection by the counter is limited simultaneously by the low-energy threshold of the photocathodes and the reduced Cerenkov photon emission as given in equation 5. There is not much to be gained in this region of the spectrum.

In relation 1 the refractive index $n(\lambda)$ and Cerenkov angle $\theta(\lambda)$ are functions of the wavelength λ. Extending the wavelength bandwidth leads to a dispersion $d\theta/d\lambda$ of the Cerenkov light that is detrimental to the precision of the measurement of θ, and hence β. The dispersion can be made smaller by *reducing* the bandwidth using optical filters. However, this is possible only when the flux of particles can compensate for the reduced light yield in a given radiator thickness. For single-particle detection one can either reduce the dispersion by using a smaller value of θ or apply achromatic correction to the optics.

2.2   Quantum Efficiency and Noise

Only the best quality photomultipliers or image intensifiers can be considered for use in single-particle detectors. The detection efficiency of a given counter is obtained by folding the quantum efficiency of the photodetector with the optical transmission and the Cerenkov light spectrum.

As a rule, only the specifications of the photocathode efficiency are available from the manufacturers. However, the efficiency for collecting the photoelectrons at the anode depends upon the wavelength of the light, especially at the level for detecting single photoelectrons (7). High-speed photomultipliers collect the photoelectrons by focusing them onto a small-size first dynode, which is elaborately calculated to minimize transit time differences. Photoelectrons that are emitted with initial energies exceeding a level of about 0.8 eV are likely to miss the first dynode. This occurs for wavelengths shorter than about 450 nm, which is the region of interest for Cerenkov light detection. This effect depends strongly on the magnitude of the accelerating field in the region between the photocathode and the first dynode, although raising the collecting field also increases the noise level of the detector.

This reduction of collection efficiency does not continue at wavelengths shorter than 300 nm because a larger fraction of the photoelectrons is emitted with very low velocities (see Duteil et al, 8); however in this region the detection of Cerenkov light is restricted by the transmission through the material of the counter optics and the entrance window of the photodetector. By using ultraviolet-transmitting materials such as fused silica, the detection of the Cerenkov light can be accomplished out to wavelengths of about 200 nm.

The effective detection efficiency of a photodetector as used in a Cerenkov counter with the associated electronic equipment can be measured if one defines a wavelength bandwidth that would be transmitted by a standard set of optics. It is convenient to define the standard optics as those that would be used in the simplest threshold Cerenkov counter, consisting of (a) one front-aluminized mirror, especially coated for reflection at ultraviolet wavelengths and containing a protective-interference layer of $\lambda/2$ thickness of MgF_2 at 350 nm, and (b) one exit window of ultraviolet-transmitting fused silica which is antireflection coated with a layer of $\lambda/4$ thickness of MgF_2 at 350 nm. These details are not so restrictive, but do introduce the necessary transmission attenuation that would be present in any threshold or differential Cerenkov counter. With this standard optics it is possible to experimentally measure (8) and compare the performances of different types of photomultipliers.

For a Cerenkov counter of length L and Cerenkov angle θ, the number of photoelectrons N produced by the passage of a singly charged high-energy particle is given by

$$N = AL\theta^2 \qquad\qquad 6.$$

where A characterizes the photodetector taking into account the Cerenkov light spectrum and the transmission of the standard optics. Values of the parameter A have been measured in several laboratories (8) and are in good agreement. The best photomultipliers available today, having a fused silica entrance window and a bi-alkali (K-Cs-Sb) photocathode, have a value of the parameter A from about 100–150 cm^{-1}. Photomultipliers having a photocathode of lower quantum efficiency and a glass entrance window provide a value of A from about 50–60 cm^{-1}.

In a focusing type of Cerenkov counter the light flash associated with a single particle is isochronous for a particle traveling along the counter axis because the time taken by the light is equal to the time of flight of the particle. For particles at a distance from the axis, because the wavefront of the Cerenkov light is conical, the duration of the flash of light is typically a few picoseconds. Hence it is the transit time-spread in the photomultiplier which is the main contributor to the coincidence resolving time of the Cerenkov detector; this time can be of the order of a few nanoseconds.

The noise pulses in a photomultiplier, mainly arising from single photoelectrons, are indistinguishable from true signals since the threshold for the detection electronics is generally set below this level. Therefore for the differential type of Cerenkov counter the light output is in most cases shared between several photomultipliers, and the output signals are placed in a coincidence arrangement. In this manner the accidental count rate can be reduced to an extremely low level at the expense of reducing the overall counting efficiency of the detector.

2.3   Choice of Cerenkov Angle and Radiator

Since the Cerenkov light is emitted at a given angle θ, it appears to an observer to come from a circle positioned at infinity. Focusing optics can be used to concentrate this light onto a ring image of radius r in the focal plane of the system. The radius r is given by $r = f \tan \theta$, where f is the focal length of the optics. The basic principle of the differential Cerenkov counter is to detect these ring images using diaphragms placed in front of photodetectors. The counter detects the radiation

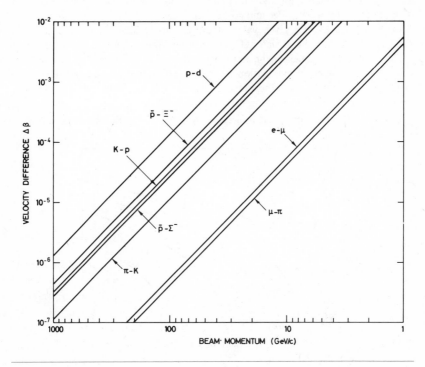

Figure 1   Velocity difference between pairs of particles as a function of the beam momentum.

emitted at a given angle θ and within an angular range of $\Delta\theta$. The velocity resolution $\Delta\beta/\beta$ achieved by such a detector is obtained by differentiating equation 1, and is given by

$$\Delta\beta/\beta = \tan\theta(\lambda)\Delta\theta \qquad\qquad 7.$$

In a beam of particles of momentum p, the velocity difference between two particles of mass m_0 and m_1 is given by

$$(\Delta\beta/\beta)_{m_0 m_1} = (m_1^2 - m_0^2)/2p^2 \qquad\qquad 8.$$

The velocity differences between different pairs of particles as a function of the beam momentum are indicated in Figure 1. The velocity resolution of a differential counter at the highest beam momentum is normally designed to be *two or three times smaller* than the velocity difference between the particles m_0 and m_1.

From equation 7 the velocity resolution of a differential counter is proportional to θ and, providing that the optics is well designed, can be improved by reducing $\Delta\theta$. The light output from the counter, from equation 6, is proportional to θ^2. Thus, the final choice of angle will be a compromise between a required velocity resolution and angular acceptance for a given application, and a minimum tolerable amount of light output for a given size and cost of the device.

In order to achieve a given value of the Cerenkov angle, a suitable material must be chosen for the radiator. Figure 2 shows the variation of the Cerenkov angle for different values of the refractive index n as a function of the particle velocity β, calculated from equation 1. Transparent materials are available for use in Cerenkov counters over the range of n from 1.0 to about 1.8. The different regions of n are obtained by using various materials:

1. $1.0 \lesssim n \lesssim 1.13$ is provided by gases below their critical points (9), and by a

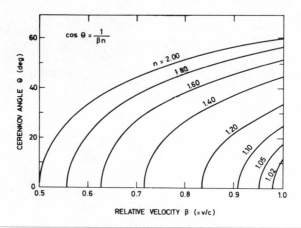

Figure 2 Variation of the angle θ of emission of Cerenkov radiation for different values of the refractive index n as a function of particle velocity β.

few liquified gases such as $n = 1.11$ for liquid hydrogen and $n = 1.13$ for liquid deuterium.

2. $1.13 \lesssim n \lesssim 1.35$ is provided by gases above their critical points.

3. $1.28 \lesssim n \lesssim 1.33$ is provided by fluorocarbons of quite high density that have high transparency in the ultraviolet and yet low dispersion.

4. $n \gtrsim 1.33$ is provided by many liquids including light hydrocarbons and also water and glycerine mixtures.

5. $n \gtrsim 1.46$ is achieved by many solids such as fused silica, plastics, and glasses.

An inhomogeneous medium can be used in a Cerenkov counter, provided that the scale of the inhomogeneity is small compared to the wavelength of the Cerenkov radiation. The use of a mixture of very fine silica powder and air has been suggested (10) which, by varying the compaction factor, can produce average values of the refractive index in the range from 1.05–1.20.

There are four cases of the application of Cerenkov light to particle velocity determination that can be distinguished:

1. Threshold Cerenkov counter using a solid or liquid radiator: The threshold velocity $\beta_0 = 1/n$ (corresponding to $\theta = 0$) can be quite low. For example, using Plexiglas with $n = 1.50$ produces a threshold value $\beta_0 \gtrsim 0.67$ or $\gamma_0 \gtrsim 1.3$.

2. Threshold gas Cerenkov counter: To obtain high values of n it is often necessary to use extremely high pressures. Hence a reasonable upper limit to the available range of n could be to use ethylene gas at 50 atm, which produces a value $n = 1.13$. This corresponds to a threshold velocity $\beta_0 \gtrsim 0.88$ or $\gamma_0 \gtrsim 2.15$.

3. Differential Cerenkov counter using a solid or liquid radiator: Since this type of counter uses a relatively large value of θ, the velocity resolution cannot be made extremely small by reducing $\Delta\theta$ since the angular acceptance becomes too small for the counter to be useful. The lowest index liquid which is commonly used is FC75 which, for $\Delta\beta/\beta \lesssim 5 \times 10^{-4}$ and $\Delta\theta \sim 10^{-3}$ rad, corresponds to a value $n = 1.28$. Thus the upper limit of β becomes equivalent to particle separation of typically the doublet π, K in beams of up to about 5 GeV/c. The lower limit to β corresponds to the largest n value, which is about 1.58 for styrene or the liquid DC704. Hence this type of counter is approximately restricted to values of $\beta \gtrsim 0.63$ or $\gamma \gtrsim 1.3$.

4. Differential gas Cerenkov counter: These counters have an application in beams up to a few hundred GeV/c, but the lower limit is set by the requirement of using a gas at a reasonable pressure, which results in a limit of approximately $\beta \gtrsim 0.8$ or $\gamma \gtrsim 1.60$.

A gas radiator provides a simple means of changing the refractive index merely by altering the operating pressure. The change of index is given by the Lorentz-Lorenz law:

$$(n^2 - 1)/(n^2 + 2) = (R/M)\rho \qquad\qquad 9.$$

where R is the molecular refractivity, M is the molecular weight, and ρ is the gas density. For pressures that are not too high, equation 9 can be approximated to high accuracy by

$$(n-1) = (n_0 - 1)P \qquad\qquad 10.$$

where n_0 is the refractive index of the gas for a given wavelength of light, at the temperature of the counter, and at a pressure of 1 atm. The pressure P is in atmospheres.

2.4   Attainment of Prescribed Refractive Index

Sufficient precision is required in the value of the refractive index of the radiator to achieve a given velocity resolution $\Delta\beta/\beta$. For low-velocity particles there is no problem with the liquid or solid radiator whose refractive index can be measured to a few parts in 10^4 or 10^5. The only necessary precaution is taking care of any temperature effects. As the particle velocity increases, so does the required precision of the index measurement. At high energies gases are used as the radiator and knowledge of the refractive index becomes more difficult. The index depends upon the gas density, as given in equation 10, which in turn depends upon the temperature and pressure according to the equation of state. The refractive index can be calculated from a measurement of the gas pressure and temperature, but such a procedure can be seriously affected by any impurity in the gas. Other methods have been used for measuring the refractive index, such as having an oscillating circuit containing a capacitor which is immersed in the gas, although such methods require a refractometer for absolute calibration.

For accurate work the refractive index should be continuously monitored using an interferometer, which directly measures the quantity $(n-1)$ by counting the number of fringes. In this way temperature, pressure, and purity controls are not necessary. This method also leads to an *absolute* determination of the value of β provided that the Cerenkov angle θ is known, as occurs in an achromatic counter.

If one aims at, say, 10 increments of the refractive index setting between the pion and kaon peaks in a mass spectrum for a beam of momentum p, the required accuracy $\Delta n/n$ is given by

$$\Delta n/n \approx \tfrac{1}{10}(\beta_\pi - \beta_K)/\beta_\pi \approx (m_K^2 - m_\pi^2)/20p^2 \qquad\qquad 11.$$

where m_π, m_K are the masses of the pion and kaon, respectively. Hence for

$$p = 200 \text{ GeV}/c \qquad \Delta n/n = 2.8 \times 10^{-7}$$
$$p = 300 \text{ GeV}/c \qquad \Delta n/n = 1.2 \times 10^{-7}$$
$$p = 400 \text{ GeV}/c \qquad \Delta n/n = 7.0 \times 10^{-8}$$

With a Rayleigh type of refractometer, an accuracy $\Delta n/n$ of better than 10^{-7} can be achieved.

2.5   Optical Dispersion in the Radiator

In equation 1 the Cerenkov angle θ is given as a function of the wavelength λ of the light. Hence there is a spread (dispersion) of the Cerenkov angle due to the variation of the refractive index of the radiator as a function of λ. The amount of chromatic dispersion $\Delta\theta_{\text{DISP}}$ is given by differentiating equation 1.

$$\Delta\theta_{DISP} = \Delta n/[n(\lambda)\tan\theta(\lambda)] \qquad 12.$$

where Δn is the change in refractive index over the range of wavelengths. If one defines an average wavelength λ_2 (which is the mean of the distribution of detected photoelectrons versus wavelength), and wavelengths λ_1, λ_3, which are the means of the distributions on either side of the average wavelength, then equation 12 can be rewritten as:

$$\Delta\theta_{DISP} = [n(\lambda_2)-1]/[n(\lambda_2)v\tan\theta(\lambda_2)] \qquad 13.$$

where

$$v = [n(\lambda_2)-1]/[n(\lambda_1)-n(\lambda_3)] \qquad 14.$$

The parameter v characterizes the optical dispersion in the radiator and is a quantity having the same definition as the Abbe number for glasses but for the wavelength of interest in Cerenkov detectors. Table 1 contains values of the parameter v and refractive index for some commonly used gas radiators. The wavelengths chosen are those that are compatible with fused silica optics and S13 or S133 photocathode spectral response. These values merely illustrate the variation of these parameters for typical gases. An exhaustive list of the physical quantities of radiator materials can be found in the review articles by Jelley (3) and Zrelov (3).

2.6 Optical Aberrations

Typical optical systems for a threshold and differential Cerenkov counter are illustrated in Figure 3. In the threshold counter, the Cerenkov light is focused by

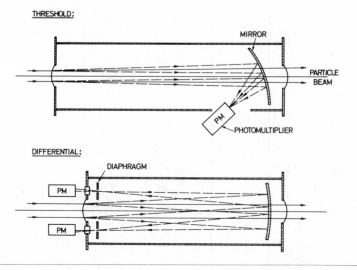

Figure 3 Typical optical systems used in a threshold and a differential Cerenkov counter.

Table 1[a] Values of the refractive index, optical dispersion (as defined in equation 14), and multiple scattering factor (equation 19) for some common gas radiators. The wavelengths chosen are those compatible with fused silica optics and S13 or S133 photocathode spectral response.

Gas	Refractive index $(n_0 - 1)10^6$ at 1 atm, 20°C and wavelengths			Optical dispersion ν	Multiple scattering factor $\left[\dfrac{\rho}{(n_0 - 1)X_0}\right]^{-\frac{1}{2}}$ cm$^{-\frac{1}{2}}$
	$\lambda_1 = 280$ nm	$\lambda_2 = 350$ nm	$\lambda_3 = 440$ nm		
He	33.27	32.90	32.67	54.5	0.24
Ne	64.07	63.37	62.85	52.2	0.67
H$_2$	140.6	135.3	132.0	15.7	0.10
N$_2$	294.8	287.0	282.0	22.5	0.33
CH$_4$	447.8	430.3	419.7	15.3	0.19
CO$_2$	447.1	433.3	427.9	19.5	0.32
SF$_6$	739.9	727.3	719.4	35.5	0.56

[a] The source of optical data was Landolt-Börnstein, 1962, 6 Auflage, 2 Band, 8 Teil (Springer Verlag). It has been interpolated by the authors.

a mirror onto a single photomultiplier. In the case of the differential counter the reflected ring of Cerenkov light is focused onto an adjustable annular diaphragm. The light passing through the aperture of the diaphragm is detected by a number of photomultipliers equispaced around the annulus. The simplest way to collect the light is by using a lens or a spherical mirror; however a spherical mirror is preferred since it produces less optical aberration.

A spherical mirror of radius of curvature R and focal length $f(=R/2)$ focuses the cone of light in the differential counter into a ring image of radius r, given by

$$r = f \tan \theta \qquad\qquad 15.$$

The total radial spread Δr of the ring image is then determined by the spherical and coma aberrations which, up to third order, are given by:

$$\Delta r = -\tfrac{1}{8}(d/f)^3 + \tfrac{1}{8}(d/f)^2 \theta \qquad\qquad 16.$$

The first term in equation 16 is the spherical error and the second term is due to coma, where d is the useful diameter of the mirror, f is the focal length of the mirror, and θ is the Cerenkov angle. Thus, the angular broadening $\Delta\theta_{\text{OPT}}$ of the ring image due to the optical aberrations is given by

$$\Delta\theta_{\text{OPT}} = \Delta r/f \qquad\qquad 17.$$

This simple manner of focusing the light is satisfactory in many cases, but to reach the best performance in velocity resolution and angular acceptance the focusing system must be optically corrected. Details of these corrections are given in section 4.

2.7   Energy Loss, Scattering, and Diffraction Effects

As the particle traverses the radiator it loses energy, mainly by the process of ionization, which will alter the value of the velocity β and hence the value of the Cerenkov angle. Multiple Coulomb scattering of the particle causes a broadening of the Cerenkov angle.

The ionization loss, which is about $2 \text{ MeV} \times \text{g}^{-1} \times \text{cm}^{-2}$ for most materials, can in some applications (11) be corrected by varying the refractive index of the radiator along the length of the counter in such a way as to keep the Cerenkov angle constant. This effect soon becomes negligible as the energy increases.

The angular spread $\Delta\theta_{\text{MSC}}$ (root mean squared projected angle) due to multiple scattering of the particle of momentum p can be expressed (12) as

$$\Delta\theta_{\text{MSC}} = (E'/p\beta)t^{\frac{1}{2}} \qquad\qquad 18.$$

where E' is a constant (≈ 15 MeV) and t is the length of the radiator in the units of its radiation length. The broadening of the Cerenkov angle has been studied by Dedrick (13) for the case of particles undergoing multiple scattering. Within the approximations of his solution it is concluded that the broadening of the radiation pattern is less than that calculated for the particle trajectory. Thus it is reasonable to use equation 18 as the broadening of the Cerenkov angle.

For a gas radiator of length L, equation 18 can be rewritten as

$$\Delta\theta_{\text{MSC}} = (E'/p\beta)[\rho L(n-1)/\{(n_0-1)X_0\}]^{\frac{1}{2}} \qquad\qquad 19.$$

where ρ is the gas density, X_0 is the radiation length, and the quantities n, n_0 are defined in equation 10. Values of the quantity $[\rho/\{(n_0-1)X_0\}]^{\frac{1}{2}}$ are included in Table 1 for some commonly used gases. The multiple scattering error, as given by equation 19, is determined by the nature of the radiator and does not depend upon the optical configuration of the counter.

All Cerenkov counters require approximately the same thickness of radiator, expressed in radiation lengths, for a given photoelectron yield irrespective of the design Cerenkov angle. (This is in the approximation that $\gamma\theta > 1$, as shown in Ref. 14.) Since multiple scattering and the magnetic bending of high-energy particles scale like $1/p$, they have the same relative effect on the smearing of the momentum determination of a particle in a given beam-line configuration as the momentum increases.

For the design of Cerenkov counters that require extremely high velocity resolution light diffraction effects eventually give a limit. The broadening of the Cerenkov cone $\Delta\theta_{DIFF}$ by diffraction is approximately given by:

$$\Delta\theta_{DIFF} \sim \lambda/(L \sin\theta) \qquad\qquad 20.$$

For example, for a 5-m differential gas Cerenkov counter operating at $\theta = 25$ mrad and $\lambda = 350$ nm, the broadening $\Delta\theta_{DIFF} \sim 3 \times 10^{-6}$ rad. If this were the largest contribution to the image spread, the effect on the velocity resolution of the counter would be $\Delta\beta/\beta = \tan\theta\,\Delta\theta_{DIFF} \sim 8 \times 10^{-8}$.

2.8   Phase Space Acceptance

In a focusing counter, for example in differential counters and for some threshold counters, the Cerenkov light is directed onto the photocathode via an *imaging* optics and will be detected only if the light passes through the aperture of a diaphragm. For differential counters the aperture is usually an annular slit subtending an angle $\Delta\theta_0$, and for threshold counters can be defined by a single circular aperture placed in front of the photodetector.

The collection of light by the photocathode is therefore simultaneously affected by the value of the angle of emission of the Cerenkov light and the particle divergence with respect to the optical axis of the counter. Thus, all focusing counters accept only particles that are within a limited region of phase space. This reflects the fact that particle identification is achieved through the geometric analysis of the light distribution, uniquely linked to the geometric properties of the particle trajectory.

The acceptance A_c of a focusing Cerenkov counter can be defined as the product of the Liouville invariant (i.e. the beam emittance) and the momentum acceptance such that:

$$A_c = kS_c\,\Delta\Omega_c\,\Delta p \qquad\qquad 21.$$

where S_c is the sensitive area of the photodetector, Δp is the momentum band within which particles collinear with the counter axis can be detected, and k is a factor resulting from the integration. The quantity $\Delta\Omega_c$ is the solid angle subtended by the particles of the central velocity or momentum which are detected such that

$$\theta_0 \Delta\Omega_c = (\pi/4)\Delta\theta_0^2 \qquad\qquad 22.$$

The acceptance A_b of a beam transport system is a constant of motion for a particle not undergoing acceleration. If there is a Cerenkov counter at some position in this beam the acceptance can be evaluated at any point along the beam, such as at the target position where

$$A_b = S_t \Delta\Omega_t \Delta p = A_c \qquad\qquad 23.$$

In this relation, the subscript t denotes the values at the target position and k is a numerical factor which has been shown (15) to be equal to 1/3.

The acceptance of an identified beam that uses a Cerenkov counter is defined by the requirements for mass separation, which thus fixes the value of $\Delta\theta_0$. The largest value of $\Delta\theta_0$ that can be used to detect particles of mass m, when particles of mass $(m+\Delta m)$ are producing light just at the edge of the diaphragm, is given by

$$(\Delta\theta_0/2)\tan\theta = (\Delta\beta/\beta) = (1/\gamma^2)(\Delta m/m) \qquad\qquad 24.$$

and therefore

$$\Delta\theta_0 \leqq (2\Delta m/m\gamma^2 \tan\theta) \qquad\qquad 25.$$

A more detailed study of matching a differential Cerenkov counter into a beam design (15) has shown that

$$\Delta\theta_0 \leqq (2\Delta m f/m\gamma^2 \tan\theta) \qquad\qquad 26.$$

where f is a function limited to values in the range from 0 to 1 depending upon parameters such as target size, collimator size, matrix elements of the beam transport system, beam momentum, mass of the particles under consideration, and the quantity $\tan\theta$ of the differential Cerenkov counter.

The number of secondary particles N_{sec} produced in a target by N_{inc} incident particles is given by

$$N_{sec} = N_{inc}(d^2 N/d\Omega\,dp)\Delta\Omega_b\, p(\Delta m/m)g \qquad\qquad 27.$$

where $\Delta\Omega_b$ is the solid angle of the beam and g is an efficiency function with a value between 0 and 1.

The study of the above functions f and g is the basis of the design of a beam system containing particle identification. Such a study has been applied to a charged hyperon beam at CERN, and has shown that a highly efficient identified beam can be obtained. It results that the target size is very important, and because the effective target size is many times larger than its physical size, due to the aberrations of beam optics, the correction of these aberrations is essential.

Two optimum beam configurations are possible. One version can be without a momentum slit, which is therefore a short beam (or a spectrometer), and requires for matching that the angular dispersion of the beam at the counter position be equal to $1/(\gamma^2 \tan\theta)$. The second version includes a momentum slit, which is a general purpose high-resolution beam, and requires that the angular dispersion at the counter location be zero.

By scaling the length of a differential Cerenkov counter and the value of the Cerenkov angle (14), it is possible to make a beam design that will maintain a good efficiency of particle identification over the energy range from hundreds of MeV up to hundreds of GeV.

3 THRESHOLD CERENKOV COUNTER

The threshold Cerenkov counter detects particles that have a velocity sufficient to produce Cerenkov light in the radiator. The threshold velocity β_0 is defined as that velocity corresponding to a Cerenkov angle $\theta = 0$, that is $\beta_0 = 1/n$ from equation 1. In practice a finite value of the Cerenkov angle is required before the recording efficiency of the photodetector reaches an acceptable value.

To obtain a good detection efficiency it is essential to optimize the circuitry of the photomultiplier and the subsequent discriminator in order to detect single photoelectrons. Due to the statistical fluctuations in the emission of an average number N of photoelectrons from the photocathode, one can define an electronic detection efficiency ε for a counter using a single photomultiplier as:

$$\varepsilon = 1 - \exp(-N)$$

The values of ε are close to unity, i.e. for $N = 4.5$ photoelectrons the quantity $(1 - \varepsilon)$ equals 10^{-2}, and for $N = 6.9$ then $(1 - \varepsilon)$ equals 10^{-3}.

The resolution of a threshold counter is determined by the shape of its efficiency curve near threshold. The slope of this curve is affected by several factors such as the dispersion of the radiator, spread in beam momentum, statistical fluctuation in the number of photoelectrons emitted at the photocathode, and the quality of the counter optics. In general, the resolution can be improved by increasing the photoelectron yield of the counter, that is, by increasing the length of the radiator and maximizing the efficiency for the detection of the Cerenkov photons.

Some Cerenkov detectors with solid or liquid radiators are built to detect the light reaching the boundaries of the radiator at the critical angle. Since values of the critical angles are quite large, the Cerenkov light intensity is also large and hence these counters have a substantial output signal on which to set the threshold. Counters of this type are described in the review articles of Ref. 3.

An important factor in the design of a threshold counter is minimization of the effect of spurious counts due to photomultiplier noise pulses and delta-rays produced in the material of the counter. For counters operating near to the threshold velocity it is often necessary to use at least two counters, and to use the output signals in coincidence. This reduces the effect of photomultiplier noise pulses and, to some extent, the effect of delta-rays, but at the cost of a lower electronic detection efficiency. The number of delta-rays with energies above the threshold can be calculated (12) and, in most cases, is at the level of 10^{-3} of the rate of desired particles.

There have been many ways in which the Cerenkov light has been focused onto the photocathode, such as by using a cylindrical mirror placed around the radiator

(16, 17), light funnels (18), spherical mirrors (19), parabolic mirrors (17), ellipsoidal mirrors (20), or even a Fresnel lens (21).

There have been numerous examples of the threshold detector. In unseparated beams of pions, kaons, and protons, for example, two or three threshold counters can be used to separate the different types of particles. One counter can be set to detect only pions, and a second to detect pions and kaons. Thus it is possible to identify each type of particle in the beam with simple electronics logic. At higher energies, as the thresholds for different particles in a given beam come closer together, it becomes more difficult to use the threshold technique. The effect of spurious signals from noise pulses and from delta-rays make the threshold counter technique unsuitable for some applications.

In recent years the application of threshold Cerenkov detectors has been moving towards higher energies. Only gas radiator threshold counters can be considered in the multi-GeV energy region. For relativistic particles the Cerenkov relation of equation 1 can be rewritten as

$$n(\lambda) - 1 = (1/2)\theta^2(\lambda) + 1 - \beta \qquad 28.$$

In equation 8 the velocity difference $\Delta\beta_{m_0 m_1}$ between two particles of masses m_0 and m_1 is given in terms of the beam momentum p. Hence if a threshold counter detects the lighter particle of mass m_0 at an average Cerenkov angle $\langle\theta\rangle$, and the heavier particle is at threshold ($\theta = 0$), from equations 8 and 28 it results that:

$$\langle\theta\rangle^2 = 2\Delta\beta_{m_0 m_1} = [(m_1^2 - m_0^2)/p^2] \qquad 29.$$

assuming that the dispersion in the radiator is small compared to the value of $\langle\theta\rangle$.

From equations 13 and 29 the dispersion in the gas is given by

$$\Delta\theta_{\text{DISP}} = (\langle\theta\rangle/2v)[1 + (1/\gamma_0^2 \langle\theta\rangle^2)] \qquad 30.$$

and hence,

$$\Delta\theta_{\text{DISP}}/\langle\theta\rangle = (1/2v)[m_1^2/(m_1^2 - m_0^2)] \qquad 31.$$

The dispersion does not depend upon p, but is only a function of the masses of the particles. In Table 2 the dispersions have been calculated using equation 31 for some common gases and several particle separations. The values are always small, typically a few percent of the value of $\langle\theta\rangle$, which would not affect significantly the counting rate near threshold.

If one neglects the error due to dispersion for the threshold gas Cerenkov counter, the design parameters can be derived from equations 6 and 29 to be given by

$$\langle\theta\rangle = (N/AL)^{\frac{1}{2}} \lesssim [(m_1^2 - m_0^2)^{\frac{1}{2}}]/p$$
$$L \gtrsim (N/A)[p^2/(m_1^2 - m_0^2)] \qquad 32.$$
$$\Delta\beta/\beta_{\text{LIMIT}} = (\langle\theta\rangle^2/2) = N/2AL$$

Figure 4 shows a focusing threshold gas Cerenkov counter (22) in operation at the Serpukhov accelerator for secondary-beam particle identification. The Cerenkov

Table 2 Values of the chromatic angular dispersion calculated using equation 31. The values of the optical dispersion parameter ν required for these calculations have been taken from Table 1.

Particle separation		Mass difference $(m_1^2 - m_0^2)$ GeV^2	Chromatic dispersion $\Delta\theta_{DISP}/\langle\theta\rangle$		
m_0	m_1		Hydrogen	Nitrogen	Helium
e	μ	0.0112	0.032	0.022	0.009
μ	π	0.0083	0.034	0.024	0.010
π	K	0.224	0.035	0.024	0.010
K	p	0.636	0.044	0.031	0.013
p	$\overline{\Sigma}$	0.553	0.083	0.057	0.024
$\overline{\Sigma}$	$\overline{\Xi}$	0.312	0.175	0.111	0.051
$\overline{\Xi}$	$\overline{\Omega}$	1.051	0.085	0.059	0.024

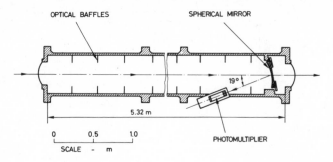

Figure 4 A focusing threshold Cerenkov counter (22) in operation at the Serpukhov accelerator.

light is focused by an inclined spherical mirror of focal length 60 cm onto a single photomultiplier. The spectral range of the detected light is from 180–600 nm. The counter is filled with helium gas and operates at a pressure close to atmospheric for the detection of 50-GeV/c pions. The curves of electronic detection efficiency versus pressure for different values of the photomultiplier high voltage are shown in Figure 5. The velocity resolution, which is defined to correspond to a change in counting efficiency from 0 to 0.63 [i.e. $(1-\varepsilon) = 1/e$], is calculated to be 6.5×10^{-6}. By summing the outputs of two similar threshold counters it was possible to achieve a velocity resolution of 3.6×10^{-6}. The background event rate was less than 3×10^{-4}, and the efficiency on the plateau was better than $(1-\varepsilon) = 6 \times 10^{-7}$.

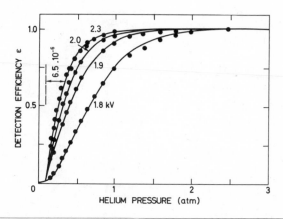

Figure 5 Pressure curves for the threshold counter of Figure 4 in a beam of 50-GeV/c negative pions. The electronic detection efficiency is shown for different values of the photomultiplier high voltage. The velocity resolution is calculated to be 6.5×10^{-6}, corresponding to a change in the efficiency from 0 to 0.63.

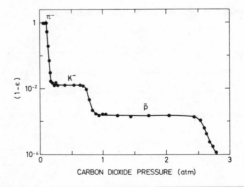

CARBON DIOXIDE PRESSURE (atm)

Figure 6   Pressure curve for the threshold counter of Figure 4 in an unseparated 20-GeV/c negative beam. The variation of the parameter $(1 - \varepsilon)$ is shown as a function of gas pressure. The background level above the antiproton region is about 10^{-6}.

A typical pressure curve for the operation of this counter with a carbon dioxide gas filling and used in a 20-GeV/c unseparated negative beam is shown in Figure 6. The background level above the pressure region for detecting antiprotons is $\sim 10^{-6}$.

There have been several examples of threshold counters detecting particles over a large angular acceptance (23), mostly for the identification of particles emitted from a production target. The light-collection problems are usually simplified by dividing up the acceptance area into several independent counters, consisting of separate

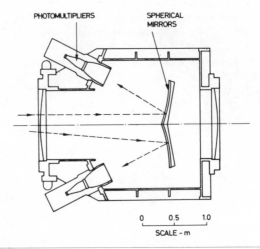

PHOTOMULTIPLIERS   SPHERICAL
MIRRORS

0   0.5   1.0

SCALE – m

Figure 7   A section through a wide-aperture threshold Cerenkov hodoscope of Williams et al (23). Each of eight spherical mirrors focuses the light onto a separate photomultiplier.

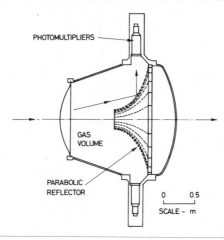

Figure 8   A section through the threshold Cerenkov counter of Ashford et al (23). The particles enter from the left and the Cerenkov light produced by electrons is directed out radially by the reflector onto 24 photomultipliers.

mirrors and photomultipliers. A recent example of this type of counter has been reported by Williams et al (23) and is shown in Figure 7. Each of eight spherical mirrors (62.5 by 62.5 cm) focuses the Cerenkov light onto a separate photomultiplier. Such a counter can be used as eight separate detectors, or the output may be electronically added. This counter has been used to separate K^+K^- states from $\pi^+\pi^-$ in an experiment using a 15-GeV pion beam at SLAC.

A large threshold Cerenkov counter has been used by Ashford et al (23) for detecting electrons in experiments of K_{e3} decays and hyperon leptonic decays. A section through the counter is shown in Figure 8. The Cerenkov light emitted in the gas volume is reflected outwards by a sectional parabolic mirror onto a ring of 24 photomultipliers. The detection of pions by this counter was measured to be less than about 0.02%.

4 DIFFERENTIAL CERENKOV COUNTER

The differential Cerenkov counter selects particles of a given velocity by detecting their Cerenkov radiation at a fixed angle. As mentioned in section 2.3, the velocity resolution is given by $\Delta\beta/\beta = \tan\theta\,\Delta\theta$, where $\Delta\theta$ is the total angular acceptance of the counter. The design of the differential counter can be separated into two main classes depending upon the nature of the radiator; counters using solid or liquid radiators are thus distinguished from counters using a gas radiator.

4.1   Differential Counter Using a Solid or Liquid Radiator

For low-energy particles having $\gamma < 5$, the required refractive index at threshold must be greater than $n = 1.02$, which is inconveniently in the range of compressed gases. As a consequence of the low value of γ, only a modest velocity resolution of the order of 10^{-2} to 10^{-3} is needed to achieve mass separation of the particles. Thus a solid or liquid material can be considered as the radiator for a differential counter.

The first optical systems for Cerenkov counters were proposed by Getting (24) and discussed in detail in the review by Jelley (3). Several focusing counters have been made using solid or liquid radiators (25). The early counters not only suffered from a poor velocity resolution, but also had a fairly small sensitive area. There followed several designs for counters that could operate in wide particle beams by means of focusing the Cerenkov light (26).

Solid or liquid radiators are available over a range of refractive index, although they cannot be tuned as easily as by varying the pressure of a gas. Therefore, attempts have been made to build an optical system of variable focal length to focus the Cerenkov light emerging from the radiator onto an annular diaphragm. This provides a means of tuning the velocity setting around the value of the refractive index of the radiator.

For some differential counters operating with a small angular spread in an experiment where multiple scattering and energy loss effects are quite small, often the main contribution to the resolution of the counter is the effect of the dispersion in the radiator. This situation can be improved by compensating for the dispersion by means of an achromatic optical system. The general conditions for achromatic optics in Cerenkov detectors were derived by Frank (27) using refraction and reflection at the interface between different media.

An optically corrected and partly achromatic differential Cerenkov counter using a liquid radiator was built by Meunier et al (28) in 1958. Since then about five

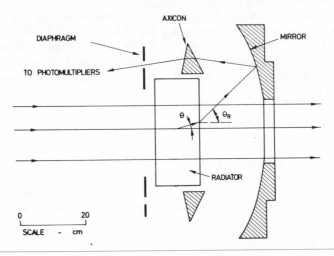

Figure 9   Optics of an achromatic liquid differential counter (DISC) built at CERN (28, 29).

different designs based upon the same principle have been used in experiments at CERN (11, 29). Their design incorporates an exchangeable liquid radiator cell covering a range of refractive index of about 5% (from $n \approx 1.55$ with high-index liquids, to n from 1.33–1.40 with water-glycerine mixtures, to $n \approx 1.28$ using low-index liquids such as FC75). This type of detector has come to be known as a DISC Cerenkov counter (differential isochronous self-collimating).

The principle of the design for the largest liquid DISC counter is shown in Figure 9. The liquid radiator FC75 is contained in a thin cell having a fused silica window. The Cerenkov light is focused by a toroidal mirror of 50-cm radius and passes through a ring-shaped prism onto the aperture of a diaphragm. The light passing through the aperture of the diaphragm is detected by four photomultipliers in coincidence. The ring prism is an axicon lens, that is, it has two conical faces that are described by rotating straight lines around a common optical axis. The axicon not only deflects the Cerenkov light but also compensates for the spread in the cone of light due to dispersion in the radiator.

The light emerging from the exit of the radiator travels at some angle θ_R, which is the refracted Cerenkov angle and not the true Cerenkov angle θ in the radiator. Hence, there exists a relation between θ_R and β given by

$$\tan \theta = \beta \sin \theta_R \qquad\qquad 33.$$

and hence, from the Cerenkov relation 1, becomes:

$$\sin^2 \theta_R = n^2 - (1/\beta^2) \qquad\qquad 34.$$

The optical aberrations in this type of counter have to be carefully minimized. The requirements of the optics are that for a single displacement of the axicon lens along the axis: (a) the focal length is altered to cover a range of θ_R, (b) the focal length remains fixed, (c) geometric aberrations are corrected, and (d) chromatic aberrations are corrected. These requirements have been met in the design of the toroidal mirror and axicon lens. The necessary optical precision can be obtained using a mirror and lens made of plastic.

The counter shown in Figure 9 focuses a small angular interval within the range from $40.5° < \theta_R < 50.6°$, and is tuned to a given velocity by moving the position of the axicon relative to the mirror. Thus, the counter can operate over a range of β values:

i.e. using FC75 $0.91 < \beta < 0.97$
water $0.85 < \beta < 0.94$
pentane $0.83 < \beta < 0.88$

The range covered indicates that these counters are used at beam momenta generally below about 5 GeV/c. Another design has been made for the detection of low mass particles (i.e. electrons, muons, and pions) where the β values close to unity are encompassed using an angular range from $54.9 < \theta_R < 60.3$.

The velocity resolution for this type of counter, by differentiating equation 34, is given by

$$d\beta = \beta^3 \sin \theta_R \cos \theta_R \, d\theta_R \qquad\qquad 35.$$

The range of angular acceptance $d\theta_R$ is chosen by adjusting a variable aperture diaphragm. With an angular setting of $d\theta_R \approx 10$ mrad, the velocity resolution becomes approximately $\Delta\beta \sim 5 \times 10^{-3}$.

For detecting low-energy particles, the energy loss in the radiator produces a significant deterioration in the performance of the DISC counter. This effect was corrected in the multicell DISC counter of Fischer et al (11) in which each of five cells was filled with a liquid of slightly different refractive index n_i so that the refracted angle θ_R was kept constant according to the relation:

$$n_i^2 = \sin^2 \theta_R + (1/\beta_i^2) \qquad 36.$$

For these DISC counters a typical optical resolution $(d\theta_R)$ has been about 1 mrad for the smallest diaphragm aperture size of 0.2 mm. Hence, if we neglect multiple scattering effects and assume that the axicon produces perfect chromatic correction, this type of counter can produce velocity resolutions approaching $\Delta\beta$ values of $\sim 5 \times 10^{-4}$.

In the design of a differential Cerenkov counter using solid or liquid radiators there are several important parameters to be chosen:

1. The velocity resolution with no chromatic correction is, from equation 34, given by

$$\Delta\beta/\Delta\lambda = -\beta^3 n(\Delta n/\Delta\lambda) \qquad 37.$$

This should be as small as possible, and should not depend upon other geometric factors (such as the range of θ) in a well-chosen configuration.

2. The largest divergence angle should be chosen for a given velocity resolution. The Cerenkov angle θ and refracted angle θ_R are related in equation 33, which can be rewritten as

$$\Delta\theta = (\Delta\beta/\beta^2)[n^2 - (1/\beta^2)]^{-\frac{1}{2}} \qquad 38.$$

Hence the angular acceptance depends only upon β and n, and not on design parameters of the counter such as θ.

3. The effect of multiple scattering on the velocity resolution from equations 6, 7, and 18 can be expressed as

$$\Delta\beta/\beta = \tan\theta\,\Delta\theta_{\text{MSC}} = (15n/p)(N/AX)^{\frac{1}{2}} \qquad 39.$$

which is proportional to the refractive index n.

Hence, from the above three points the optimization of the counter design does not depend directly upon the range of the angle θ_R, but only upon the values of β, n, and Δn. The best choice for a liquid will have the smallest value of n and Δn. Thus the design of the counter begins with the choice of the radiator, and then the required velocity range determines the range of θ_R of the optics.

The fluorocarbons are commonly used in these counters due to their low dispersion and transparency at ultraviolet wavelengths. The main disadvantages of these liquids are the high density (multiple scattering) and immiscibility with other liquids.

In some applications (30) a liquid with a low index of refraction, and hence low multiple scattering, is used, even though this involves some technical problems in building the counter. In this respect liquid hydrogen ($n = 1.11$) and liquid deuterium ($n = 1.13$) have been used as radiators. The increase in angular acceptance of a liquid hydrogen counter as compared with using FC75 would, from equation 39, be a factor of 1.7.

At the upper momentum limit for the liquid or solid radiator differential Cerenkov counter, at say 5 GeV/c, the velocity resolution $\Delta\beta \approx 5 \times 10^{-3}$ can be typical. If the counter is filled with liquid FC75 an angular acceptance at the highest resolution of about 6 mrad can be used—which is about the size of the divergence of typical secondary beams at these momenta. This type of counter is not suitable for application at higher energies.

4.2 Differential Gas Counter

To separate pions and kaons in a high-energy beam of, say, 100 GeV/c requires a velocity resolution of about 10^{-5}. Such a resolution can be obtained with a differential gas Cerenkov counter. At these high energies the effects of multiple scattering and energy loss become less important, and the main contribution to the velocity resolution is due to the dispersion in the radiator.

The differential gas counter contains a spherical lens that focuses the cone of Cerenkov light onto an adjustable annular diaphragm. The light passing through the aperture is detected by a number of photomultipliers in coincidence. Several differential gas Cerenkov counters have been built (31) with velocity resolutions of about 10^{-4}. Even better resolutions can be achieved by using small values of θ and $\Delta\theta$.

The chromatic spread of the Cerenkov light $\Delta\theta_{\text{DISP}}$ due to dispersion is given by

$$\Delta\theta_{\text{DISP}} = (\theta/2v)[1 + (1/\gamma^2\theta^2)] \qquad\qquad 40.$$

where v is defined as in equation 14. If we assume that this is the largest error in the image, the effect of the dispersion error on the velocity resolution is given by

$$\Delta\beta/\beta = \tan\theta\,\Delta\theta_{\text{DISP}} \approx (\theta^2/2v) + (1/2v\gamma^2) \qquad\qquad 41.$$

Using equation 8, the maximum value of the Cerenkov angle is thus defined by

$$\theta^2/v \lesssim [(1/p^2)(m_1^2 - m_0^2)] - (1/v\gamma^2) \qquad\qquad 42.$$

In this approximation, the design formulae for the differential gas Cerenkov counter simplify to:

$$\theta \lesssim \frac{1}{p}[v(m_1^2 - m_0^2) - m_i^2]^{\frac{1}{2}}$$

$$L \gtrsim \frac{N}{A}\left[\frac{p^2}{v(m_1^2 - m_0^2) - m_i^2}\right] \qquad\qquad 43.$$

$$(\Delta\beta/\beta)_{\text{LIMIT}} = \theta^2/2v$$

where the subscript i refers to the particle (0 or 1) being detected by the counter.

The ultimate resolution using this type of counter is achieved by using the least

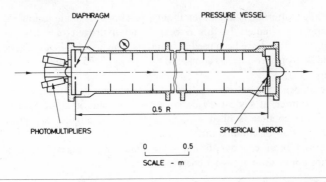

Figure 10   A differential gas Cerenkov counter in use at Serpukhov (32).

dispersive gas, helium. For example, Figure 10 shows a differential gas counter used in the secondary beams at Serpukhov (32). There are two basic designs: one is 5-m long with a Cerenkov angle of 23 mrad, and the second is 10-m long with θ equal to 12 mrad. A velocity resolution of $\Delta\beta < 10^{-5}$ has been achieved by using a helium radiator without correcting for dispersion in the gas. Figure 11

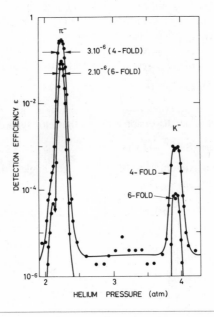

Figure 11   A typical pressure curve for a 10-m gas differential counter, of the type shown in Figure 10, when used in a 45-GeV/c unseparated beam. Results are shown for a 4- and 6-fold coincidence arrangement of the photomultipliers.

shows a typical pressure curve for the 10-m counter when used in a 45-GeV/c beam of pions and kaons. The variation of the detection efficiency is shown when the outputs of the 6 photomultipliers are placed into 6- and 4-fold coincidence arrangements. A background level of better than 1 in 10^6 can be routinely obtained with these counters.

4.3   Optically Corrected Gas Differential Counter

It is possible to design optics for the differential gas counter that will correct for the dispersion error by a factor of about 15. This achromatic gas counter (DISC) makes it possible to envisage resolutions in the region of $\Delta\beta \approx 10^{-6}$ to 10^{-7}.

In the differential counter the length of the radiator approximately defines the focal length of the optics. For an acceptable level of detection efficiency and good background rejection (see section 6.1 for further details) at least 24 photoelectrons should be detected by the photomultipliers. Thus, with the length and Cerenkov angle known, it remains to define the required amount of chromatic correction and the range of adjustment of the correction needed for a particular application. Although the correction optics are an added complication, they do provide a counter that can operate over a wide range of operating beam momenta and detect particles over a large mass range. For certain applications, such as in hyperon physics where the constraint on the counter length is severe, it is necessary to use a specially designed achromatic counter.

The required amount of chromatic correction, as given in equation 40, is a function of the velocity of the particle. This variation can be achieved by moving the correction element along the axis of the counter. There is a fairly wide choice of optical configurations that can achieve the correction. The choice is dictated by factors such as complexity of manufacture, number of lens elements, and availability of optical materials.

For a Cerenkov counter the optics must be ultraviolet-transmitting, which excludes all optical glasses except fused silica. There are also some monocrystals which can be used such as NaCl, KCl, CaF_2, and LiF. However, only NaCl and KCl have a suitable dispersion compared with that of fused silica, and are available in large sizes (up to 350-mm diameter) at a reasonable cost.

Chromatic correction implies that refracting optics will be needed. One can contemplate a design with a single element of fused silica which, in conjunction with a mirror, must have a positive power. However, this combination cannot be tuned to correct over a variable range of chromatism, since the velocity setting of the counter would be affected simultaneously by the corrector lens position and the gas pressure. This solution was possible for the liquid DISC counter (28) because the refractive index in this instance could not be varied. The large Cerenkov angle and the use of an axicon lens in the liquid DISC counter allows the focus to be maintained at a fixed position within an accuracy acceptable for this counter at energies up to about 5 GeV.

The required accuracy for the gas DISC counter is more than two orders of magnitude more precise than for the liquid DISC. The chromatic correction has to be applied in a way that is proportional to $(n-1)$. It is desirable to have a counter in which the tuning of the velocity setting is independent of the setting of the chromatic corrector. Also, the chromatic corrector should not affect to first order the focal length of the optics for the mean wavelength, irrespective of its longitudinal position, so that the diaphragm can be located at a fixed position. Thus a corrector of zero power is required. One must also incorporate into the design of the corrector the possibility of minimizing longitudinal chromatic aberrations, spherical aberrations, coma, and astigmatism.

The design variables are the curvature, thickness, and positions of all the lenses. The optimization of the design has been performed through a detailed study of ray-tracing for a range of particle velocities, wavelength bandwidths, and corrector positions. In this way an error matrix is developed which must be carefully studied in relation to the effect of these residual errors on the performance of a specific detector.

One type of design that has been shown to be amenable to satisfactory correction over large variations in the design Cerenkov angle from 20 to 120 mrad consists essentially of four elements:

1. A mirror, which is aspheric for counters with small values of Cerenkov angle, and of the Mangin-type for larger angles.

2. A single fused silica lens which, together with the mirror, corrects most of the geometric aberrations. This element also transforms the optics into a telecentric system in which the entrance pupil is at infinity, thus allowing for the stability of the Cerenkov angle even in the presence of some defocusing.

3. A chromatic corrector element (which is a doublet, or preferably a triplet, of fused silica and sodium chloride), which cancels the variation of chromatism by displacement along the counter axis. This element does not affect the velocity setting of the counter.

4. Field lenses of fused silica, used as exit windows, which transfer the light passing through the diaphragm aperture onto the photocathode area of the photomultipliers.

A gas DISC counter was built by Duteil et al (33) with the optics shown in Figure 12. The dispersion was corrected by an achromatic doublet of fused silica and sodium chloride. The Cerenkov angle was 44 mrad and a radiator of 2 m

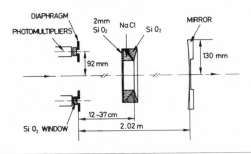

Figure 12   Optics of a gas DISC counter built by Duteil et al (33).

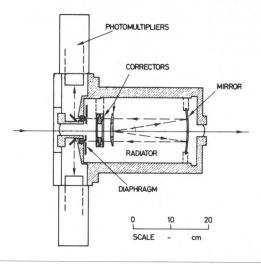

Figure 13   A gas DISC counter used in the 20-GeV charged hyperon beam at CERN (34).

nitrogen (or carbon dioxide) gas was used. A velocity resolution of better than 10^{-5} was achieved with this counter.

An optically corrected gas differential counter operating (34) in the charged hyperon beam at the CERN proton synchrotron is shown in Figure 13. The application of this counter to hyperon physics demanded a high velocity resolution since the Σ^- and Ξ^- particles are close in mass, and the length had to be minimized because the Σ^- half-life is only ~ 57 cm at 20 GeV. Two counters were built with a limiting velocity resolution of 5×10^{-5} covering the range of β from 0.993 to 1.000. The short length was achieved by using a Cerenkov angle of 120 mrad, providing sufficient light in a counter external length of 41 cm.

At this large Cerenkov angle, without optical correction, the geometric aberration would limit the velocity resolution to 2×10^{-4}. The chromatic dispersion alone would limit the velocity resolution to 2×10^{-4} at $\beta = 1.000$, and 4×10^{-4} at $\beta = 0.993$. Since these aberrations are of similar importance, the optics were designed to simultaneously minimize both effects. This was achieved with a design incorporating 6 lenses into a 3-element system, and using spherical surfaces. A Mangin-type mirror was used, consisting of a rear-surfaced aluminized negative meniscus and a single positive meniscus lens. This component removes most of the geometric aberration. A chromatic corrector triplet, which moves along the counter axis to cover the range of β, corrects for the dispersion in the gas radiator. An amount of longitudinal chromatism produced by the chromatic triplet is canceled by introducing a similar amount, but with opposite sign, in the mirror. By a careful evaluation of the correction optics it has been possible to reduce the velocity resolution down to the order of $\Delta\beta \sim 5 \times 10^{-5}$. A typical pressure curve for this DISC counter is shown in Figure 14.

An optically corrected counter is presently under construction for use in the high-energy beams at the NAL and CERN SPS multihundred GeV accelerators. This counter is 5-m long, contains helium gas, and uses a Cerenkov angle of

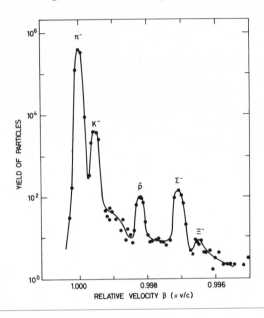

Figure 14　A pressure curve for the DISC counter shown in Figure 13.

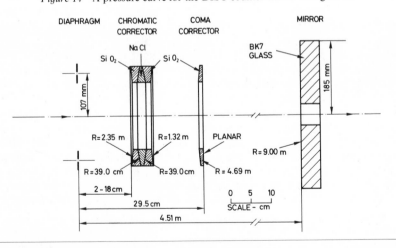

Figure 15　Optics for a gas DISC counter which is to be used in the multihundred GeV beams at NAL and the CERN SPS.

24.5 mrad. The optics are shown in Figure 15. This counter, which is of the type B outlined in Ref. 35, is designed for a velocity resolution of about 4×10^{-7}, which should be capable of distinguishing between pions and kaons in unseparated beams up to about 250 GeV/c. The focusing errors in this counter are about five times larger than the diffraction limit.

It is not so simple to present design equations for the optically corrected differential counters since the correction involves the simultaneous minimization of several kinds of aberrations. However, by studying the several counters already in operation it appears that the limiting velocity resolution is approximately equivalent to having a diaphragm aperture setting that is close to $\Delta r \sim 0.1$ mm ($\pm 50\%$). Hence from this *approximate experimental* fact, the limiting velocity resolution can be expressed as

$$(\Delta\beta/\beta)_{\text{LIMIT}} = \tan\theta \, \Delta\theta_{\text{LIMIT}} \approx \theta(\Delta r/L) \qquad\qquad 44.$$

where L is the counter length. Using equations 6, 8, and 44, it is possible to derive an approximate set of design equations for the optically corrected gas differential counter:

$$\theta \lesssim [N(m_1^2 - m_0^2)/(2A\Delta rp^2)]^{\frac{1}{3}}$$
$$L \gtrsim [N/A]^{\frac{1}{3}}[2\Delta rp^2)/(m_1^2 - m_0^2)]^{\frac{2}{3}} \qquad\qquad 45.$$
$$(\Delta\beta/\beta)_{\text{LIMIT}} = (A\theta^3\Delta r/N)$$

5 COMPARISON OF THRESHOLD AND DIFFERENTIAL CERENKOV COUNTERS

In sections 3 and 4 the design parameters of the threshold and differential Cerenkov counters have been expressed in the form of equations. A comparison of these parameters for counters at high energies, that is for gas counters, is shown in a graphical form (14) in Figures 16–18. The parameters have been calculated using equations 32, 43, and 45. Included in two of these figures are the parameters of four Cerenkov counters presently in operation (32, 34, 36) and three counters currently being designed for use at the CERN SPS (35). The parameters of these selected counters are given in Table 3.

For the curves representing the differential Cerenkov counter, typical values have been used for the ratio (N/A), as defined in equation 6. The curves for the threshold counter are shown for electronic detection inefficiencies $(1 - \varepsilon)$ at the level of 10^{-2} to 10^{-6}. The broken regions in the curves denote that the required gas pressures are in excess of 30 atm for helium or nitrogen, and in excess of 20 atm for the liquefying gas SF_6. The scale on the abscissa for two of the figures is drawn to show the limiting momentum scale for several pairs of particles. The *limiting momentum* is that value at which the velocity resolution of the counter is *equal to* the velocity difference between the two particles. In general a Cerenkov counter should not be used at a momentum as high as the limiting value.

Figure 16 shows the maximum values of the Cerenkov angle for the threshold, differential, and DISC types of counter. From equations 32, 43, and 45 the maximum

Table 3 Design parameters for some Cerenkov counters already in operation in high-energy beams or presently in the design stage.

Type	Counter	Ref.	Length m	θ mrad	Gas filling
Differential	IHEP (1)	} 32	5	23	} He, N_2
	IHEP (2)		10	12	
	CERN (hyperon)	34	0.3	120	SF_6
	CERN	33	2	44	} CO_2
	IHEP	36	2.5	45	
DISC	Future designs:				
	Type A	} 35	2	45	SF_6, Fr-13
	Type B		5.5	24.5	} He
	Type C		10	20	

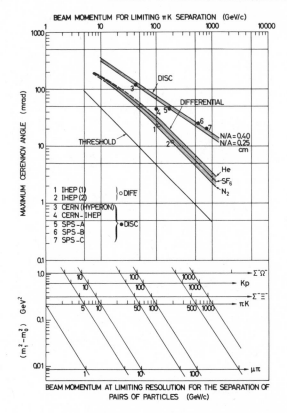

Figure 16   Maximum values of the Cerenkov angle for threshold, differential, and DISC types of gas counters. The abscissa corresponds to values of the beam momenta at which the limiting velocity resolution of each counter is equal to the velocity difference between pairs of particles. The seven Cerenkov counters included in this figure are listed and referenced in Table 3.

values of the angles for these three types of counter scale with the limiting momentum as p^{-1}, p^{-1}, and $p^{-\frac{2}{3}}$, respectively.

The variation of the minimum lengths of the three types of Cerenkov counters is illustrated in Figure 17. The minimum lengths scale with limiting momentum as p^2, p^2, and $p^{4/3}$ for the threshold, differential, and DISC counters, respectively. If we assume a constant value of (N/A), the minimum lengths of the threshold and differential counters are approximately related by

$$L_{\text{THRESH}} \approx \nu L_{\text{DIFF}} \qquad\qquad 46.$$

Thus, even if we allow for a smaller value of N for the threshold counter, the

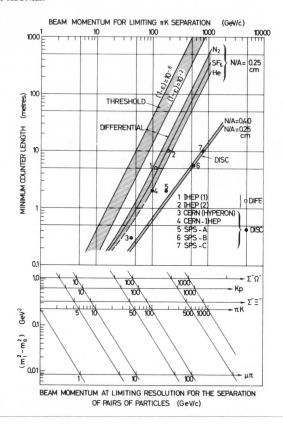

Figure 17 Variation of the minimum lengths of threshold, differential, and DISC Cerenkov counters as a function of the beam momentum for the limiting separation of different pairs of particles.

length of the threshold counter will always be several times the length of a differential counter.

The limiting velocity resolution $(\Delta\beta/\beta)_{\text{LIMIT}}$ is shown as a function of the Cerenkov angle θ in Figure 18 for the three types of counters. From the design equations, the limiting resolutions for the threshold, differential, and DISC counters scale as $\langle\theta\rangle^2$, θ^2, and θ^3, respectively.

In summary, the gas threshold Cerenkov is the least expensive counter to build as it only requires an average-quality optical mirror, a single photomultiplier, and simple electronic configuration. The counter vessel need only withstand working pressures from vacuum to a few atmospheres, and only a simple refractive index measurement is needed. However, for a given rejection the threshold counter becomes extremely long as the beam energy increases, and therefore these counters have a poorer rejection than the differential counters. There can be counting rate problems, especially in a veto counter, since all charged particles above threshold (wanted and unwanted) will produce light.

The differential and DISC counters are more precise instruments and, especially the latter, can be designed for the ultimate in velocity resolution at high energies. These counters are relatively more expensive than the threshold counter since they require a higher mechanical precision, especially in the diaphragm, and a high-quality optical mirror. The multicoincidence photomultiplier arrangement demands a substantial amount of fast electronic logic, and the fine velocity resolution requires a precision interferometer for the refractive index measurement. Special optics are needed to correct for the geometric and chromatic aberrations in the DISC counter.

The differential and DISC counters provide a positive signature of particles falling within the accepted velocity band. These detectors are not primarily designed for 100% efficiency, which would be difficult to obtain with multicoincidence logic, but to achieve a very good rejection of particles outside the acceptance in velocity and angle defined for the counter.

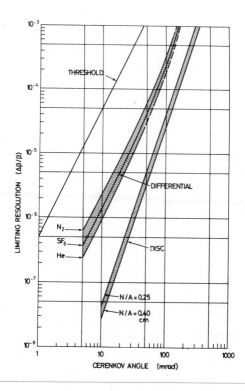

Figure 18 The limiting velocity resolution as a function of the Cerenkov angle for threshold, differential, and DISC types of gas counters.

Threshold counters are generally built to have a good detection efficiency. This is especially important for detecting a particle that is not the lightest mass in the beam. In this case one threshold counter must be used as a veto signal to reject the lighter particles.

At their limiting resolutions the differential and DISC counters have a more restricted angular acceptance than the threshold counter. However, the limiting resolution is needed only in *exceptional cases* where one is prepared to trade acceptance for resolution.

6 PATTERN RECOGNITION IN A FOCUSING CERENKOV COUNTER

6.1 Ring Image Detector

An objective lens using spherical optics will transform Cerenkov light rays emitted at an angle $\theta(\lambda)$ along a particle trajectory into a ring image of radius r in the focal plane of the lens, as shown in Figure 19, such that

$$r = f \tan \theta(\lambda) \qquad\qquad 47.$$

where f is the focal length of the lens.

The image may suffer from optical aberrations which, if required, can be corrected by using refracting optics such that the quantity $dr/d\lambda$ becomes zero. The Cerenkov photons emitted by the passage of the charged particle will be distributed around the ring image which will then be transformed into a smaller number of photoelectrons by the photodetector. From the geometry of the light dots in the focal plane the velocity and direction of the particle can be obtained. There is therefore a problem of pattern recognition. The radius of the circle and its center must be found for an image defined by random points with a radial distribution which is not necessarily Gaussian.

There have been many attempts to record the dots of light directly onto a photographic plate, and such a method has been used when working with a well-focused beam of monoenergetic particles (37). To detect light from single events it has been necessary to use an image intensifier (38). The best and most recent example is by Giese et al (38) where a velocity resolution of $\Delta\beta = 6 \times 10^{-7}$ has been observed

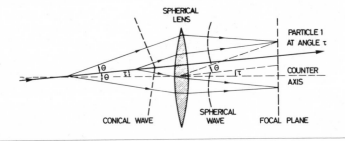

Figure 19   Principle of ring-focusing of Cerenkov light using a spherical lens.

with a systematic error of $\pm 2 \times 10^{-7}$. They used a four-stage cascade image intensifier and recorded the output screen image using photographic film. With a radiator of 10-m helium gas and using $\theta = 11.5$ mrad they obtained a mean value of 1.8 spots of light for each particle. From equation 6 this is equivalent to having an A parameter of 13.6 cm^{-1}, which is quite low when compared with photomultiplier detection but is due to their restricted wavelength bandwidth from 360 to 480 nm at half maximum.

The pattern recognition of a circle requires at least three spots of light, and for this case the fit is always perfect. Giese et al selected events in which at least four spots were evenly spaced around the circumference of the ring, thus reducing the fraction of good events to 5%. The selection of events in this manner leads to the above value of the velocity resolution. This is clearly an off-line procedure and is rather slow.

A simple alternate method is to detect the light that passes through an annular diaphragm placed in front of a photodetector. For detection by a single photo-detector the level of discrimination must be higher than a few photoelectrons so as to reduce the number of spurious noise signals, and there is no guarantee that the circle of light does not extend into the opaque part of the diaphragm. This method can produce results for particles that are collinear with the axis, but will never achieve a very low efficiency for particles that are slightly offset in velocity or collinearity (i.e. it will provide a poor rejection of unwanted events).

An improvement is collection of the light falling outside the annular aperture and using this signal in anticoincidence (39). This method suffers from the difficulty that if the anticoincidence threshold is set very low, even the tail of the light distribution that passes through the diaphragm aperture can trigger the anti-coincidence channel and thus reduce the efficiency and the effective aperture of the diaphragm. When analyzing less-abundant particles (such as kaons) close to abundant particles (pions), the anticoincidence channel might become overloaded, thus reducing the count rate of the wanted particles. In fact, the anticoincidence method has never achieved a very good quality of rejection of unwanted particles.

A more useful method is to fragment the annual diaphragm into several sectors (q), each connected to a photomultiplier. It can be required that all sectors (or a fraction of the sectors) give an output signal before an event is selected (see Huq & Hutchinson, 26). With all q photomultipliers in coincidence, the accidental background rate due to spurious noise pulses soon becomes negligible as q increases, even when the photodetectors are set to fire on single photoelectrons. Also the q outputs can be grouped in parallel, i.e. the outputs from p adjacent photomultipliers are passed into an *or* circuit, and the outputs from q/p *or* circuits are arranged in coincidence. These different configurations have different efficiencies ε given for example by

$$q\text{-fold} \quad \varepsilon \quad = \quad [1 - \exp(-N/q)]^q$$
$$q/p\text{-fold} \quad \varepsilon(q,p) = \quad [1 - \exp(-Np/q)]^{q/p} \qquad 48.$$

where N is the average total number of photoelectrons produced. If an on-line record is made of the counts in these different channels it is possible to continuously extract the value of ε, and hence N. The background and rejection can also be obtained from the comparison of $\varepsilon(q,p)$ in the region of the mass spectrum between the particle peaks.

The efficiency for the q-fold coincidence requirements (i.e. assuming $p = 1$) is shown in Table 4 as a function of the value N. These efficiencies are generally at the level of 70 to 90%. A higher efficiency can be obtained by using a longer counter, but this is not a recommended choice in view of the fast rising cost and the increased amount of material in the beam.

Table 4   Electronic detection efficiencies for different values of the q-fold coincidence and for values of the average number of photoelectrons N, calculated using equation 48 assuming $p = 1$.

Efficiencies	N = 12	N = 24	N = 48
q = 3 fold	0.946	0.999	1.000
4	0.815	0.990	1.000
6	0.418	0.855	0.998
8	0.133	0.665	0.985

The slope of the velocity resolution curve increases as the value of q is increased. For $q > 4$, a q-fold coincidence signal implies that the light ring must be completely included in the diaphragm aperture. It is only this configuration that makes the counter *self-collimating*, that is, insensitive to particles having a divergence greater than $\Delta\theta/2$.

The expected response of a differential or DISC counter versus the number of coincidences q, the total number of photoelectrons N, and the diaphragm angular aperture D (expressed in units of the width of the light spread $\langle \Delta\theta \rangle$) has been computed (15), and a sample of the results is shown in Figure 20. A Poisson distribution is assumed for the distribution of N and for the radial distribution of the intensity of the ring image. Although this assumption is a reasonable approximation of the spread due to multiple scattering and geometric aberrations, it is not strictly valid for the chromatic spread in an uncorrected differential counter, nor is it true for the residual chromatic spectrum in a DISC counter. However, the results of the analysis do provide some conclusions that are in agreement with the experience of running these detectors:

1. The slope of the sides of the peaks is governed by the magnitude of the angular spread $\langle \Delta\theta \rangle$ of the light distribution, and is practically independent of q.

2. The width of the curves depends strongly upon q.

3. The detection efficiency on the peak shows a developing plateau only when the diaphragm angular aperture D is set at more than twice the angular spread $\langle \Delta\theta \rangle$ of the light.

4. It is rather misleading to quote the velocity resolution as the width at half-height of the detection efficiency curves. An ideal Cerenkov counter would present

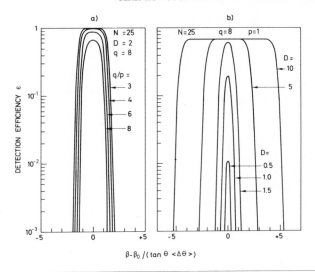

Figure 20   Calculated response curves for a differential or DISC Cerenkov counter as a function of the number of inputs (q/p) to the coincidence circuit and angular aperture of the diaphragm setting (D). The abscissa is a velocity-difference scale expressed in units of the velocity resolution of the counter. N is the number of photoelectrons detected by the total number (q) of photomultipliers.

a rectangular response curve, with a flat top, and would be capable of rejecting particles to the side even with a large width of the curve.

From curves similar to those shown in Figure 20 it is apparent that a good compromise for the working conditions is setting the diaphragm at two or three times the value of $\langle \Delta\theta \rangle$, and choosing the total number of the groupings (q/p) of the photomultipliers to be greater than about 6. This compromise ensures that there is a small plateau region on the top of the efficiency curve, with an efficiency as given by equation 48, and that there is no sizable degradation of the mass resolving power of the counter. If the value of q is 6 or 8, from Table 4 one can conclude that a value of N of about 24 is recommended to obtain an acceptable counting efficiency.

Actual experimental results can be illustrated by the pressure curve of the IHEP 5-m differential counter (32). Figure 21 shows the counting efficiency curves for two settings of the diaphragm aperture as a function of the number (q/p) of inputs into the coincidence arrangement.

6.2   Spot Image Detector

Another possibility for electronic pattern recognition is focusing the Cerenkov light from a single particle to form a single dot of light, a technique which has been applied since the start of the development of Cerenkov counters (24, 40). The focusing conditions for a dot are such that for a narrow particle beam they are

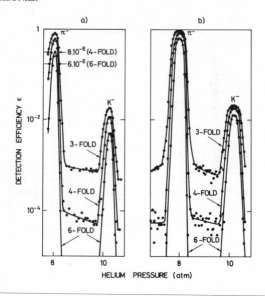

Figure 21   Pressure curves for the IHEP 5-m differential Cerenkov counter (32) illustrating quality of detection, and rejection of unwanted events, in an unseparated 40-GeV/c negative beam. The detection efficiency is shown as a function of the number of inputs into the coincidence arrangement for the two diaphragm slit settings of (*a*) 1.6 mm and (*b*) 3.2 mm.

approximately obtained through the use of conical, cylindrical, or more complicated revolution surfaces, all having a singular point on the axis. This focusing, although approximate, was sufficient for the precision required from liquid or solid radiators with particles of about 100 MeV. For particles outside the velocity acceptance band the spot enlarges and falls outside the region defined by a circular diaphragm.

The spot-focusing idea has been recently revived in a computer study of a proposed

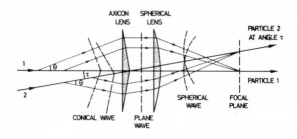

Figure 22   The principle of the spot-focusing of Cerenkov light using an axiconic lens to produce a plane wavefront, followed by a spherical lens to form the final point image in the focal plane.

counter design by Benot et al (41). The principle of the counter optics is illustrated in Figure 22. The conical wavefront of the Cerenkov light is converted into a plane wavefront using an axicon lens, and the final image is formed by a spherical lens. The counter can detect *multiparticle* production as long as the position of the target coincides with the position of the apex of the axicon (or a virtual image of the axicon). The Cerenkov light from a particle of velocity β_0 is focused to a spot, whereas light from a particle of velocity different from β_0 produces a ring image. Benot et al have studied a possible design for this type of detector which, assuming a production target of 1-mm radius, is capable of providing the angular coordinates of multiparticles to ± 0.1 mrad and the velocities (in the range of $70 < \gamma < 400$) to $\Delta\beta \approx 2 \times 10^{-6}$.

The spot-focusing idea offers the possibility of simultaneously detecting the particle velocity and direction. This is a clear advantage over the crude pattern recognition afforded by the single annular diaphragm of a differential Cerenkov counter.

6.3   Multiplexed Cerenkov Counter

The restricted phase-space acceptance of a differential counter with a single diaphragm is a drawback in the measurement of particles coming from an extended phase space (large target size). One way to obtain more data on each event is to multiplex the output of the Cerenkov counter.

A six-channel Cerenkov counter (42) using a specially zoned mirror has been used in a precision measurement of the pion-proton total cross section (43). As shown in Figure 23, the Cerenkov light is focused by a spherical mirror into a ring image at the focal plane. A special spherical mirror is placed in the focal plane to focus the light from each of the six zones onto pairs of photomultipliers connected in coincidence. A particle traveling along the optical axis of the counter, and producing Cerenkov light at an angle θ, will form a ring image at the focal plane of radius $r = f \tan\theta$, and centered on the point 0. If a particle enters the counter at an angle α with respect to the optical axis, the center of the ring image is displaced to the

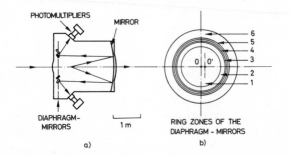

Figure 23 A six-channel Cerenkov counter built by Ivanov et al (42) in which Cerenkov light is focused onto special diaphragm-mirrors in the focal plane of the counter optics. Using a suitable electronic logic between the light outputs from each ring zone of the diaphragm-mirrors it is possible to determine the angle of each incoming particle.

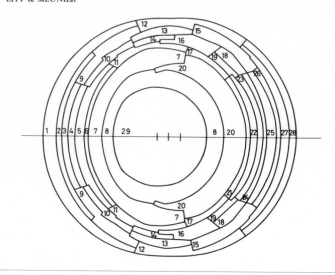

Figure 24   A special mirror proposed for the mass sensitive image-dissecting Cerenkov counter of Roberts (45) planned for use in the charged hyperon beam at NAL.

point $0'$ by an amount equal to $f \tan \alpha$. Hence by using an electronic logic (in coincidence and veto arrangement) of the light outputs detected from each of the six zones it was possible to define the scattering angles α in the experiment.

A more elaborate multiplexed Cerenkov counter with segmented mirror has been proposed by Soroko (44) at Dubna. Also Roberts (45) has designed a mass sensitive image-dissecting Cerenkov counter for the 150-GeV/c charged hyperon beam at NAL. As proposed, the mirror is divided into several segments which are adjusted to focus the Cerenkov light onto 29 photomultipliers. The complexity of the mirror is shown in Figure 24. By a suitable electronic logic, this counter is designed to produce specific trigger signals for a range of particle masses (up to the deuteron mass). A multichannel Cerenkov device is being constructed at Serpukhov (46) in which a diaphragm having variable radius and aperture size is placed in front of each of 12 photomultipliers.

The possibility introduced by special photomultipliers with crossed magnetic and electric fields (47) has been incorporated into the design of a multichannel counter by Vishnevsky et al (48). These photomultipliers provide a determination of the point of emission of the photoelectrons from the output signal delay, and can therefore replace an aperture-diaphragm arrangement.

7   CONCLUSIONS

With the advent of the new accelerator at NAL, physics instrumentation has entered into the multihundred GeV energy range. The possibilities for different types of particle detectors and identifiers are quite restricted in this new energy range, and will consist mainly of the following devices:

1. Scintillation counters, which provide fast timing (down to ~ 0.1 nsec) and track position information to the accuracy of a few millimeters.

2. Wire chambers, which have a poorer timing accuracy than scintillators but provide accurate position information [down to ~ 0.1 mm for high pressure wire chambers (49) or drift chambers (50)].

3. Total absorption counters or nuclear detectors based upon the emission of scintillation or Cerenkov light in large-size radiators. Unfortunately these detectors make a destructive measurement on the particles but can produce an energy determination approaching the accuracy of a few percent (51).

4. Ionization detectors based upon the measurement of the relativistic rise of the energy loss by ionization. This technique is in an early stage of development (52).

5. Transition radiation detectors, which are also in the developmental stage (53), are particularly suited for the detection of high γ particles. Thus they will be useful in detecting low-mass particles.

6. Cerenkov counters, which can provide velocity and angular coordinates of charged particles over a wide range of masses and beam energies. The effect is well known, and counters can be designed with confidence. Due to the large statistical fluctuation associated with a low light yield, an *intensity measurement* of Cerenkov radiation provides a rather poor result for single particles and is valid only for a large group of particles. However, the usefulness of the Cerenkov detector is that an *angular measurement* of the light, which is not subject to the effect of statistical fluctuations, can produce a high accuracy even for single particles.

The limit for the application of the Cerenkov technique at high energies will be established by the ability to resolve the velocity of the particle with an accuracy that is going approximately as $1/p^2$. In the differential Cerenkov counter the optical aberrations will set this limit, whereas the DISC counter will be limited eventually by the diffraction of light. It appears that practical limits for the application of Cerenkov counters may be reached within the energy range afforded by the new multihundred GeV accelerators at NAL and CERN.

Literature Cited

1. Cerenkov, P. A. 1934. *Dokl. Akad. Nauk SSSR* 2:451
   Vavilov, S. I. 1934. *Dokl. Akad. Nauk SSSR* 2:457; Ibid 1956. *Sobranie Sochinenii (collected works)* 4:356. Moscow: Izdatel'stvo Akad. Nauk SSSR
2. Tamm, I. E., Frank, I. M. 1937. *Dokl. Akad. Nauk SSSR* 14:107
3. Jelley, J. V. 1958. *Cerenkov Radiation And Its Applications.* London: Pergamon
   Hutchinson, G. W. 1960. *Progr. Nucl. Phys.* 8:197–236

Zrelov, V. P. 1968. *Cerenkov Radiation In High-Energy Physics.* Moscow: Atomizdat. 2 vols. (English transl.: 1970. *Israel Program Sci. Transl.*)
Denisov, S. P. 1971. Advances in Cerenkov counters. *IHEP Rep.* 71/47
4. Cerenkov, P. A. 1944. *Tr. Fiz. Inst. Akad. Nauk SSSR* 2:4
5. Ginsburg, V. L. 1940. *Zh. Eksp. Teor. Fiz.* 10(6):589
6. Pinsky, L. S., Eandi, R. D., Osborne, W. Z., Rushing, R. B. 1971. *XII Int. Conf. Cosmic Rays, Tasmania* 4:1630–34

7. Lakes, R. S., Poultney, S. K. 1971. *Appl. Opt.* 10:797–800
8. Duteil, P., Garron, J. P., Meunier, R., Stroot, J. P., Spighel, M. 1968. *CERN Rep.* 68/14
   Yovanovitch, D. D. et al 1971. *Nucl. Instrum. Methods* 94:477–80
9. Vovenko, A. S. et al 1964. *Sov. Phys. Usp.* 6(6):794–824
10. Linney, A., Peters, B. 1972. *Nucl. Instrum. Methods* 100:545–47
11. Fischer, J., Leontic, B., Lundby, A., Meunier, R., Stroot, J. P. 1959. *Phys. Rev. Lett.* 3:349–50
12. Rossi, B., Greisen, K. 1941. *Rev. Mod. Phys.* 13:240
13. Dedrick, K. G. 1952. *Phys. Rev.* 87:891–96
14. Benot, M., Litt, J., Meunier, R. 1972. *Nucl. Instrum. Methods* 105:431–44
15. Meunier, R. Characteristics of charged particle beams for mass identification using a differential Cerenkov counter. In preparation
16. Bayatyan, G. L., Zel'dovich, O. Ya., Landsberg, L. G. 1964. *Instrum. Exp. Tech.* 7:805–10
17. Vivargent, M., Von Dardel, G., Mermod, R., Weber, G., Winter, K. 1963. *Nucl. Instrum. Methods* 22:165–68
18. Hinterberger, H., Winston, R. 1966. *Rev. Sci. Instrum.* 37:1094–95; Ibid 1968. 39:1217–18
19. Denisov, S. P. et al 1970. *Nucl. Instrum. Methods* 85:101–7
20. Dobbs, J. MacG., McFarlane, W. K., Yount, D. 1967. *Nucl. Instrum. Methods* 50:237–41
21. Swanson, R. J., Masek, G. E. 1961. *Rev. Sci. Instrum.* 32:212
22. Denisov, S. P. et al 1970. *Nucl. Instrum. Methods* 85:101–7
23. Hinterberger, H. et al 1970. *Rev. Sci. Instrum.* 41:413–18
    Aubert, J. J. et al 1970. *Rev. Sci. Instrum.* 87:79–85
    Williams, H. H., Kilert, A., Leith, D. W. G. S. 1972. *Nucl. Instrum. Methods* 105:483–91
    Ashford, V. et al 1972. *Nucl. Instrum. Methods* 98:215–20
    Cence, R. J. et al 1973. *Nucl. Instrum. Methods* 108:113–18
24. Getting, I. A. 1947. *Phys. Rev.* 71:123–24
25. Chamberlain, O., Segrè, E., Wiegand, C., Ypsilantis, T. 1955. *Phys. Rev.* 100:947–50
    Huq, M. 1958. *Nucl. Instrum. Methods* 2:342–47
26. Leontic, B. 1957. *CERN Rep.* 59/14
    Von Dardel, G. 1960. *Proc. Int. Conf. Instrum. High Energ. Phys., Berkeley,* 166–71
    Huq, M., Hutchinson, G. W. 1959. *Nucl. Instrum. Methods* 4:30–35
27. Frank, I. M. 1956. *Usp. Fiz. Nauk* 58(1):111
28. Meunier, R., Stroot, J. P., Leontic, B., Lundby, A., Duteil, P. 1962. *Nucl. Instrum. Methods* 17:1–19, 20–30
29. Dowell, J. D. et al 1962. *Phys. Lett.* 1:53–56
    Gilly, L. et al 1964. *Phys. Lett.* 11:244–48
    Gilly, L. et al 1965. *Phys. Lett.* 19:335–38
    Davies, J. D. et al 1967. *Phys. Rev. Lett.* 18:62–64
    Davies, J. D. et al 1968. *Nuovo Cimento A* 54:608–19
    Binon, F. et al *Nucl. Phys. B* 33:42–60
30. Ayres, D. S. et al 1969. *Nucl. Instrum. Methods* 70:13–19
31. Mermod, R., Winter, K., Weber, G., Von Dardel, G. 1960. *Proc. Int. Conf. Instrum. High Energ. Phys., Berkeley,* 172–73
    Kycia, T. F., Jenkins, E. W. 1961. *Proc. Conf. Nucl. Electr., Belgrade* 1:63–70
32. Denisov, S. P. et al 1971. *Nucl. Instrum. Methods* 92:77–80
33. Duteil, P., Gilly, L., Meunier, R., Stroot, J. P., Spighel, M. 1964. *Rev. Sci. Instrum.* 35:1523–24
34. Badier, J. et al 1972. *Phys. Lett. B* 39:414–18
35. Meunier, R. 1970. *NAL Summer Study Rep.* SS/170, 85–108
36. Bushnin, Yu. B. et al 1971. *Instrum. Exp. Tech.* 14:67
37. Zrelov, V. P. 1963. *Instrum. Exp. Tech.,* 6:215–19
    Zrelov, V. P., Pavlovic, P., Sulek, P. 1972. *Nucl. Instrum. Methods* 105:109–16
38. Roberts, A. 1960. *Nucl. Instrum. Methods* 9:55
    Butslov, M. M., Medvedev, M. N., Chuvilo, I. V., Sheshunov, V. M. 1963. *Nucl. Instrum. Methods* 20:263–66
    Reynolds, G. T., Waters, J. R., Poultney, S. K. 1963. *Nucl. Instrum. Methods* 20:267–70
    Binnie, D. M., Jane, M. R., Newth, J. A., Potter, D. C., Walters, J. 1963. *Nucl. Instrum. Methods* 21:81–84
    Benwell, R. 1964. *Nucl. Instrum. Methods* 28:269–73
    Iredale, P., Hinder, G. W., Parham, A. G.,

Ryden, D. J. 1966. *Advan. Electron. Electron Phys.* B 22:801–12

Giese, R., Gildemeister, O., Paul, W., Schuster, G. 1970. *Nucl. Instrum. Methods* 88:83–92

39. Kycia, T. F., Jenkins, E. W. See Ref. 31 Lindenbaum, S. J., Love, M., Ozaki, S., Russell, J., Yuan, L. C. L. 1963. *Nucl. Instrum. Methods* 20:256–60

40. Marshall, J. 1952. *Phys. Rev.* 86:685–93; Ibid 1954. *Ann. Rev. Nucl. Sci.* 4:141–56
    Ozaki, S., Russell, J. J., Sacharidis, E. J. 1965. *Nucl. Instrum. Methods* 35:301–8

41. Benot, M., Howie, J. M., Litt, J., Meunier, R. 1973. A spot-focusing Cerenkov counter for the detection of multiparticle events at high energies. Unpublished

42. Ivanov, V. I., Moroz, N. S., Radomanov, V. B., Stavinsky, V. S., Zubarev, V. N. 1970. *JINR Rep. E13/5459, Dubna*

43. Giordenesku, N. et al 1970. *JINR Rep. P1/5460, Dubna*

44. Soroko, L. M. 1971. *JINR Rep. P13/6019, Dubna*

45. Roberts, A. 1972. *Nucl. Instrum. Methods* 99:589–98

46. Vorontsov, V. L., Goldberg, G. R., Denisov, S. P., Kudrjashov, A. M., Lifshits, E. M. 1968. *Bull. Izobretatelja, N36*

47. Aleksandrova, N. F. et al 1969. *IHEP Rep. 69/22, Serpukhov*

48. Vishnevsky, N. K. et al 1970. *Proc. Int. Conf. Instrum. High Energ. Phys., Dubna* 1:305–20

49. Willis, W. J., Hungerbuehler, V., Tanenbaum, W., Winters, I. J. 1971. *Nucl. Instrum. Methods* 91:33–36

50. Charpak, G., Rahm, D., Steiner, H. 1970. *Nucl. Instrum. Methods* 80:13–34

Charpak, G. 1970. *Proc. Int. Conf. Instrum. High Energ. Phys., Dubna* 1:217–50
Charpak, G., Sauli, F., Duinker, W. 1973. High accuracy drift chambers and their use in strong magnetic fields. Unpublished

51. Hughes, E. B., Hofstadter, R., Lakin, W. L., Sick, I. 1969. *Nucl. Instrum. Methods* 75:130–36
    Hayashi, M., Homma, S., Kajiura, N., Kondo, T., Yamada, S. 1971. *Nucl. Instrum. Methods* 94:297–99
    Engler, J. et al 1973. *Nucl. Instrum. Methods* 106:189–200

52. Landau, L. D. 1944. *J. Phys.* 6:201–5
    Vavilov, P. V. 1957. *Sov. Phys. JETP* 5:749–51
    Blunck, O., Westphal, K. 1951. *Z. Phys.* 130:641–49
    Dimčovski, Z., Favier, J., Charpak, G., Amato, G. 1971. *Nucl. Instrum. Methods* 94:151–55

53. Garibian, G. M. 1960. *Sov. Phys. JETP* 37(10):372–76
    Garibian, G. M. 1970. *Proc. Int. Conf. Instrum. High Energ. Phys., Dubna* 2:509–29
    Alikhanian, A. I., Avakina, K. M., Garibian, G. M., Lorikian, M. P., Shikhliarov, K. K. 1970. *Phys. Rev. Lett.* 25:635–39
    Wang, C. L., Dell, G. F. Jr., Uto, H., Yuan, L. C. L. 1972. *Phys. Rev. Lett.* 29:814–17
    Bamberger, A., Dell, G. F. Jr., Uto, H., Yuan, L. C. L., Alley, P. W. 1973. *Phys. Lett.* B 43:153–56
    Alikhanian, A. I., Kankanian, S. A., Oganessian, A. G., Tamanian, A. G. 1973. *Phys. Rev. Lett.* 30:109–11
    Harris, F. et al 1973. *Nucl. Instrum. Methods* 107:413–22

PHOTON-EXCITED ENERGY-   ✖ 5534
DISPERSIVE X-RAY FLUORESCENCE
ANALYSIS FOR TRACE ELEMENTS[1]

F. S. Goulding and J. M. Jaklevic
Lawrence Berkeley Laboratory, University of California, Berkeley, California

CONTENTS

INTRODUCTION

Like other methods devised to measure quantities of interest in basic physics, X-ray fluorescence spectroscopy is now becoming a tool of the analytical chemist. Semiconductor detectors that have contributed much of the experimental data on nuclear energy levels are proving equally valuable in measuring fluorescent X rays, and are so providing a versatile tool for rapid multielement analysis of many types of sample.

X-ray fluorescence analysis, as shown in Figure 1, involves creation of vacancies in atomic shells by primary radiation, and observation of the characteristic X rays emitted when the vacancies fill from outer shells. Charged particles or photons may be employed as the primary radiation. This paper will mainly be concerned with photon excitation, although some comparisons will be made with particle

[1] This work was performed under the auspices of the United States Atomic Energy Commission.

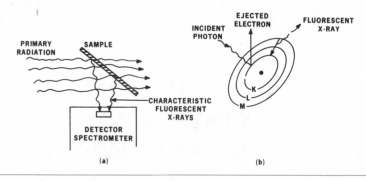

Figure 1   Illustration of the principles of an X-ray fluorescence analyzer.

excitation. The emitted X rays may be analyzed using either "wavelength dispersive" Bragg-crystal spectrometers or "energy dispersive" semiconductor detector spectrometers. While Bragg spectrometers offer the better energy resolution over most of the energy range of interest, several factors make their use somewhat inconvenient and expensive for multielement analysis of specimens. Consequently, they will not be discussed in this paper.

We further restrict the paper by dealing only with trace element analysis. This type of analysis places the greatest demands on the instrument but, in return, has potential for many practical applications. Since trace elements are important in biological systems, both as essential nutrients and as hazardous substances, medical and environmental monitoring applications are obvious. Trace element distributions can also provide information on the origins and history of objects and materials, suggesting important applications in archeology, geology, and criminology. Restricting attention to trace element analysis neglects a wide range of applications where elements are present at high concentrations, such as the analysis of alloys and minerals. It also neglects the fruitful combination of scanning electron microscopes or microprobes with X-ray spectrometers to analyze spatial distributions of elements in samples. Many of these high-concentration problems have been discussed (1–4).

In the following discussion, samples are assumed to consist of an organic matrix containing trace elements ranging in concentration from less than 1 ppm to 1000 ppm by weight. Unless otherwise stated, a carbon matrix is assumed, since carbon represents a reasonably good average of the atomic numbers of common elements in organic samples.

PHOTON-INDUCED FLUORESCENCE—PHYSICAL MECHANISMS

The ejection of inner-shell electrons by the photoelectric interaction of an incident photon requires that the photon energy exceed the binding energy of the appropriate shell. Figure 2 shows the binding energies and dominant X-ray energies used in X-ray fluorescence analysis. X-ray fluorescence analyzers generally measure the effects of K-shell vacancies in the light elements ($Z < 55$), and L-shell vacancies in the heavy elements. It is also possible to employ the K-shell of heavy elements, but the difficulty of producing high-energy primary radiation, combined with excessive backgrounds at high energies, has led to the use of the L-shells instead.

It is also necessary, insofar as possible, to employ almost monochromatic primary radiation. Low-energy components in the primary radiation (such as in the bremsstrahlung produced by a conventional X-ray tube) scatter from the sample to produce considerable background that conceals the spectral lines of the elements of interest. Methods of producing nearly monochromatic primary radiation will be discussed later; the following discussion will assume monochromatic primary radiation of energy E_I.

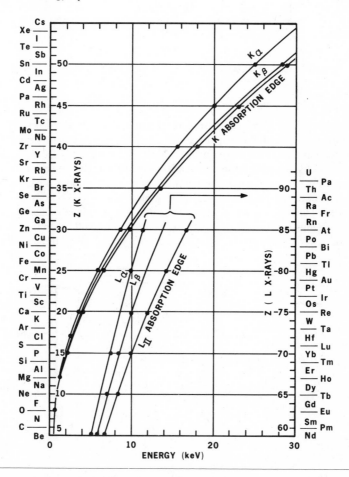

Figure 2   Binding energies and dominant X-ray energies for the elements.

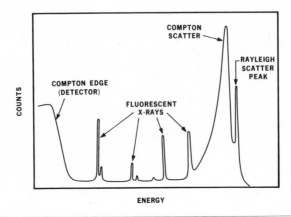

Figure 3   Idealized spectrum observed with an X-ray fluorescence analyzer.

Production of Characteristic Spectral Lines

Figure 3 shows an idealized version of the spectrum produced by an X-ray fluorescence analyzer. Its principal features are the two high-energy peaks, caused by Compton and Rayleigh scattering of incident photons from the sample into the detector, the characteristic fluorescent lines of elements in the sample superimposed on a generally flat background through the whole spectrum, and a plateau in the extremely low-energy region caused by Compton scatter of photons out of the detector leaving only low-energy electrons in it to produce signals. In trace element analysis, the integrated counts in the scatter peaks far exceed those in the rest of the spectrum—over 90% of the total counts are often contained in these peaks. Also

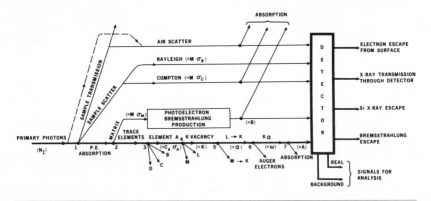

Figure 4   Diagram of the mechanisms involved in producing fluorescent X rays and background.

low-energy Compton scatter from the detector can be neglected for practical purposes since it occurs below the energy range of interest.

The flow diagram (Figure 4) shows the physical processes involved in a fluorescence spectrometer, and will be referred to throughout the following discussion. The diagram emphasizes the production and observation of K_α X rays that are important for the analysis for low-Z elements; a similar drawing could be made stressing production of L_α (and L_β) X rays used to analyze for high-Z elements. Nodes are numbered where competing processes occur that reduce the yield of the required fluorescent X rays. The processes occurring at these nodes are:

NODE 1   For thin samples, such as are generally used for trace element analysis, most of the N_I incident photons are *transmitted* through the sample. They may be scattered by air surrounding the sample, or may excite fluorescence in the elements in the air to produce unwanted X rays in the detector. A vacuum or helium atmosphere is often employed around the sample to essentially eliminate this type of background. *Scatter* of the incident X rays by the carbon matrix of the sample also produces X rays that reach the detector. Some of the incident X rays are photoelectrically *absorbed* in the sample.

NODE 2   Photoelectric absorption occurs primarily in the *matrix,* producing photo-electrons with energy near that of the incident radiation which cause a bremsstrahlung radiation background in the detector. A few of the incident photons interact photoelectrically with *trace elements* in the sample.

NODE 3   We focus attention on a particular element A present among several trace elements (B, C, D, etc). Since the elements are assumed to be present only in trace quantities, and as the matrix contains only low-Z elements, interelement effects such as the absorption of the X rays from one element in a second element to produce its fluorescent X rays can be neglected.

NODE 4   Photoelectric excitation may create vacancies in any of the atomic shells. As we are concerned with only K-shell vacancies, vacancy production in the L, M, etc shells constitutes a loss. The requirement for conservation of momentum in the interaction favors absorption in the innermost shell that is energetically possible, so, if the incident energy exceeds the K-shell binding energy, most of the interactions will create K-shell vacancies. The fraction K of interactions producing K-shell vacancies in an element is indicated by the size of the absorption jump that occurs in the elemental X-ray absorption curve when the incident energy just exceeds the K-shell binding energy. The absorption jump ratio changes from 7.2 for arsenic ($Z = 33$) to 9.9 for argon ($Z = 18$); therefore, for this range of elements, over 80% of the vacancies are created in the K-shell when the incident energy exceeds its binding energy. Photoelectric absorption above the K-shell absorption edge varies approximately as $E_I^{-3.5}$, so the optimum condition for creating K-shell vacancies requires a primary energy just greater than the K-shell binding energy for the element. Since this condition cannot be achieved simultaneously for all elements in multielement analysis, convenient incident X-ray energies are used, each permitting reasonably efficient K-vacancy production in a range of elements. Typical primary radiations are the K X rays of copper (for low-Z elements), of molybdenum (for the K X rays of medium-Z elements, and for the L X rays of heavy elements), and of a rare earth (for the K X rays of rather high-Z elements).

NODE 5   Refilling of a K-shell vacancy may occur from any outer shell and is accompanied by either X-ray or Auger-electron emission. The K_α X rays of interest arise from $L \rightarrow K$ transitions, while K_β X rays (for the most part) result from $M \rightarrow K$ transitions. We will assume that K_α X rays constitute a fraction α of the total X rays emitted. The value of α is generally greater than 0.8.

NODE 6   If Auger-electron emission occurs, no X rays are emitted. The ratio of X-ray photon emission to total vacancy-filling is the fluorescent yield ω, which varies from a low value (< 0.1) for low-Z elements to a relatively high value (> 0.5) for medium-Z elements such as arsenic. A similar range applies to the L-shell fluorescent yield of the heavy elements.

NODE 7   Absorption of the fluorescent X rays in the sample itself, in the air path (if air is present) to the detector system, or in entry windows to the detector will reduce the K X-ray rate seen by the detector. For most purposes, absorption in the sample itself is the dominant contributor, as helium can be used to surround the sample when necessary, and only a very thin (~ 25-μ) beryllium window exists in the detector system. Absorption of fluorescent X rays in the sample can be made insignificant except at low energies (< 5 keV) by using thin samples (< 50 mg/cm^2).

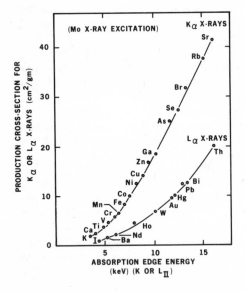

Figure 5 K_α and L_α production cross sections. Mo ($K_\alpha + K_\beta$) incident radiation is assumed.

The final number of K_α X rays of element A reaching the detector is $N_I C_A \sigma_A K_A \alpha_A \omega_{K_A} A_A G$, where the subscripts A denote that the parameters for the various processes are determined for element A, and where the geometry factor G represents the probability of an X ray emitted from the sample reaching the detector.

Data on the various parameters involved in these processes are available from both theoretical and experimental studies. The compilation of X-ray cross sections by McMaster et al (5) provides comprehensive data on photoelectric cross sections as a function of incident energy, and also on the K-edge jump ratios for the elements. Fluorescent yield data are summarized and discussed in (6) and (7), while the K_α/K_β and L_α/L_β ratios are discussed in (8). By combining these data, cross sections for K_α and L_α production (i.e. $\sigma K \alpha \omega$) can be computed with the results shown in Figure 5 taken from the paper by Giauque et al (9). This particular set of curves is computed for Mo $(K_\alpha + K_\beta)$ exciting radiation. Table 1 shows how well these

Table 1 Relative yields for several elements

Line	Calculated	Measured
Cr K_α	0.381	0.370 ± 0.011
Mn K_α	0.450	0.435 ± 0.003
Fe K_α	0.587	0.559 ± 0.003
Ni K_α	0.884	0.882 ± 0.007
Cu K_α	1.000	1.000 ± 0.015
As K_α	1.653	1.660 ± 0.083
Se K_α	1.776	1.753 ± 0.057
Pb L_α	0.804	0.774 ± 0.019

computed values agree with relative yields for thin standards of several elements as measured on a standard X-ray fluorescence analyzer (9) using copper as the reference standard. The agreement is better than $\pm 5\%$ for all elements measured. This is adequate for most analytical purposes.

Figure 6 shows the relative production of X rays for various elements using three different primary radiations ($K_\alpha + K_\beta$ of Cu, Mo, and Tb). As can be seen, correct choice of the excitation increases the yield for a given element by a large factor. For example, the yield for chromium ($Z = 22$) is ten times higher when using Cu excitation as compared with Mo excitation. The curves in Figure 6 include corrections for absorption of X rays in the 25-μ Be window common in semiconductor detector spectrometers and also for the roll-off in efficiency for a 3-mm thick Si detector at high energies. Correcting for absorption of X rays by the sample is a more difficult problem as it requires knowledge of the distribution of trace elements through it. Assuming a uniform distribution, the method described by Giauque et al (9) can be used, but this is not applicable to air particulate filters where the particles are distributed nonuniformly through the filter medium. Studies of typical calibrated

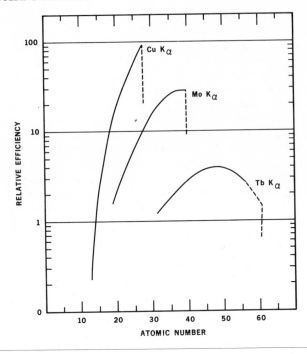

Figure 6 Relative yield of K_γ X rays of the elements for three different exciting radiation energies. The curves are corrected for absorption in the 25-μ Be window to the detector, and for the fall-off in efficiency of a 3-mm thick silicon detector at higher energies.

samples are essential in these cases to determine valid approximate absorption corrections. Fortunately, such corrections are only necessary for the low-Z elements. For very low Z elements even absorption within single particles can cause loss of accuracy.

Scattering Processes

Scattering from the sample matrix is important in photon-excited X-ray fluorescence since it produces most of the photons observed by the detector. Rayleigh (coherent) and Compton (incoherent) scattering are the dominant effects causing a pair of peaks at the high-energy end of the detector spectrum. Since Rayleigh-scattered photons have the same energy as the photons incident on the sample, the spectral peak corresponding to this process is at energy E_I, and has a width characteristic of the spectrometer resolution. Compton-scattered photons, on the other hand, have less energy than E_I, the loss in energy depending on the scattering angle. Therefore, the Compton-scatter peak is at a lower energy than E_I, its width reflecting both the spectrometer resolution and the distribution of scattering angles.

In Compton scattering, the loss of energy ΔE between incident and photons scattered at an angle θ is given by:

$$\Delta E/E_I = (1 - \cos\theta)E_I/m_0 c^2 \qquad\qquad 1.$$

where $m_0 c^2$ is the rest energy of the electron. Even for $180°$ backscatter, the energy loss is only 8% at 20 keV, while the loss falls to 4% at a more typical angle used in a fluorescence analyzer ($90°$). Choice of E_I is made on the basis that the energy of singly Compton-scattered photons falls well above the range of the fluorescent X rays of the trace elements of interest.

Multiple scattering is rare and its effect is restricted to a low-energy tail on the Compton-scatter peak. Therefore, no interference is produced by scattered photons unless processes occur to degrade signals into the energy range where interesting spectral lines occur. On the other hand, since 90% or more of the detected photons are due to scatter processes, the electronic system in the spectrometer must handle a high rate of these high-energy events to accumulate within a reasonable time sufficient numbers of counts in the spectral lines of interest to achieve useful accuracy.

Some reduction in the rate of Compton-scattered photons observed by the detector can be achieved by using a $90°$ angle between the incident and detected photons since the spatial distribution of photons exhibits a minimum in the $90°$ direction (10). Typically, the observed rate in the $90°$ direction is less than half the rate expected on the basis of an isotropic distribution of Compton-scattered photons.

While the cross section for Rayleigh scattering in carbon at 10 keV is similar to that for Compton scattering, Rayleigh scattering in low-Z elements necessarily involves only shallow glancing-angle collisions. Therefore, most of the Rayleigh-scattered photons travel in the same direction as the incident photons, and the number reaching the detector is much smaller than that of Compton-scattered photons.

Background-Producing Processes

The ability of a spectrometer to observe spectral lines produced by trace elements is largely determined by fluctuations in the background existing in the energy range where these lines are produced. Achieving the minimum possible background levels is therefore essential in a spectrometer designed for trace element analysis. Figure 4 shows a number of processes that contribute to the background shown in the idealized spectrum of Figure 3. Identifiable physical processes are:

(a) Production of bremsstrahlung radiation by photoelectrons in the sample itself. These electrons arise mainly from matrix atoms. Photons of all energies smaller than E_I are produced, so those reaching the detector contribute to background through the whole energy region used for analysis.

(b) Escape of photoelectrons from the front surface of the detector. Since the photoelectrons result mostly from scattered photons from the sample, the escape electron rate is directly proportional to the scatter rate from the sample. Since the electrons produce some ionization in the detector prior to leaving its surface, they cause signals through the whole amplitude range of interest.

(c) Bremsstrahlung photons produced by photoelectrons in the detector may escape from the detector, removing their energy from the detector signals. As most of the photoelectrons are produced by scattered photons from the sample, the amplitude of the bremsstrahlung escape background is directly proportional to the number of scattered photons reaching the detector.

(d) Escape of fluorescent silicon X rays from the detector may also occur. Their low energy (1.74 keV) makes the escape of these X rays improbable, but tiny satellite peaks are produced 1.74 keV below strong spectral features. These do not produce a general background, so they will be neglected in the following discussion.

(e) Any detector process causing loss of ionization signal will produce background. As the vast majority of the photons reaching the detector are scattered from the sample, these will be the main source of degraded signals.

Processes (a), (b), and (c) may be treated as physical processes basic to the method. Process (e), however, is technological in nature, depending on the quality of silicon and the design of the detector. On the other hand, processes (b), (c), and (e) are all directly attributable to scattered photons from the sample, while process (a) is unrelated. In making comparisons, it is convenient to express all backgrounds in terms of a fixed number of Compton-scattered photons observed by the detector.

Accurate calculation of the backgrounds due to the basic physical processes (a), (b), and (c) is impossible because of the tortuous paths of electrons at low energies, but approximate estimates have been made (11) based on certain simplifying assumptions. The contributions to background of incomplete charge collection in detectors have also been studied in some detail (12), and the effects of surface

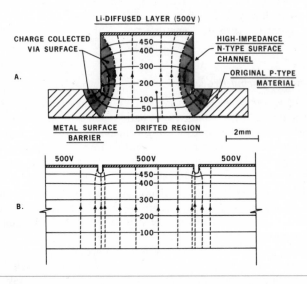

Figure 7 (a) Field distortions causing poor charge collection in a simple detector. (b) Field configuration in a guard-ring detector.

states on the sides of conventional detectors have been demonstrated. Figure 7a shows one of the standard lithium-drifted silicon detector configurations used for several years in nuclear and X-ray spectroscopy. Conduction in the n-type surface states extending down the sides of the detector causes a distortion of the equipotentials, normally expected to be parallel to the faces of the detector, with the resulting change as shown in the electric field lines. Consequently, electrons produced by photon interactions in the shaded regions of the detector are collected in surface layers and are trapped there. Since production of the correct output signal depends on complete transit of holes and electrons across the detector to the two electrodes, only partial signals are produced by events in the shaded regions. These constitute a large fraction of the total volume of the small-area detectors used for X-ray spectrometers, so many of the photons scattered from the sample produce degraded signals. In some cases, nearly 50% of the signals from a simple detector of this type are degraded by this process.

Two approaches have been adopted to reduce this intolerable source of background. The first is to collimate incoming X rays to a small region at the center of a detector. This is fairly effective, but results in much reduced efficiency as compared with a noncollimated detector. Also, fluorescent X rays produced from the collimator material may complicate observed spectra. A more satisfactory method employs the guard-ring detector configuration shown in Figure 7b, where the boundaries of the central detector are defined by internal electric field lines, thereby achieving internal collimation in the detector. In the arrangement shown in Figure 8a, the same potential is applied to the central region and the outside

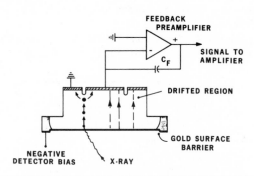

Figure 8a   Simple guard-ring detector circuit.

ring, but signals are only taken from the central region. The removal of surface effects produces a reduction in background that may be as large as a factor of ten, as compared with the simple detector. A further reduction of background is achieved by accepting only those signals in the central region not accompanied by signals in the surrounding guard ring as shown in Figure 8b. (In this figure, a second guard ring is employed to reduce surface noise effects in signals from the inner ring.) This "guard-ring reject" method eliminates those events whose charge

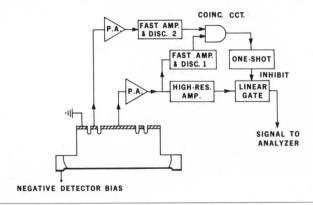

Figure 8b   Double guard-ring reject circuit.

collection is shared by the guard ring and central region. The result is a further reduction in background by a factor between 3 and 5.

Figure 9 shows the striking improvement in background achieved by using the guard-ring reject system as compared with a conventional detector when analyzing a dried blood serum specimen. The general form of the spectrum is similar to the idealized spectrum of Figure 3, but the excessive background due to surface collection in the simple detector is quite apparent. The statistical fluctuations in the higher background clearly cause a higher detection sensitivity limit than in the case of the guard-ring reject system.

Figure 10 shows the results of background calculations (11) and also experimental measurements on simple detector, guard-ring detector, and guard-ring reject systems.

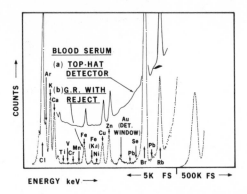

Figure 9   An example of the relative performance of a simple detector and a guard-ring reject detector. The sample is lyophilized blood serum, and Mo $(K_\alpha + K_\beta)$ excitation was used.

The calculated backgrounds are expressed in terms of the number of background counts accumulated in the full width at half maximum (200 eV) of a spectral peak while accumulating 10^6 counts in the scatter peaks. The values for sample bremsstrahlung, expressed in these terms, depend on an assumption concerning the spatial distribution of Compton-scattered photons from the sample—for the purpose of illustration, Figure 10 assumes an isotropic distribution. Figure 10 also shows the calculated yields for K_α X rays at the detector from S, Ti, Fe, Zn, and Br, and the yield of L_α X rays from Pb, all present at a level of 1 ppm by weight in an organic matrix. These are also expressed in terms of the number of counts in the K_α peak for 10^6 counts in the scatter peaks, assuming an isotropic spatial distribution of Compton-scattered photons from the sample. On the other hand, the experimental data on detector systems, and on the electron and bremsstrahlung escape from the detector (calculated), are all expressed in terms of 10^6 actual counts in the scatter peaks as seen by the detector. Quantitative interpretation of the data of Figure 10 must take into account these differences in presentation for the various curves, and

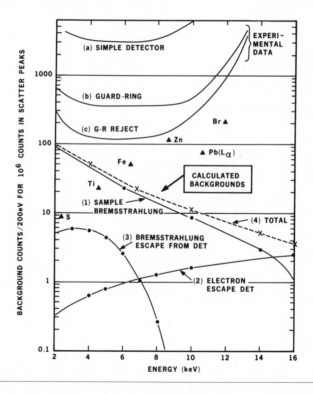

Figure 10 Experimental and theoretical backgrounds based on an organic matrix and Mo excitation. Also shown are the yields for the K_α (or L_α) peaks of several elements present at the 1 ppm by weight level.

must also recognize the approximate nature of the bremsstrahlung calculations, which probably overestimate the background level. Despite these reservations, Figure 10 provides a useful picture of the background factors.

The large improvement in experimental performance achieved by using the guard-ring reject method is illustrated by Figure 10, but it is clear that even this method falls well short of the theoretical performance capability. A 10:1 improvement in background can be anticipated over most of the energy range. Some of the background observed in the experimental results can be attributed to contributions from scattered low-energy incident radiation (i.e. the primary radiation is not perfectly monochromatic), and also to charge-trapping in microscopic poor regions in the detector. Further work is needed to clarify these points.

Detection Sensitivity

The capability of an X-ray fluorescence analyzer to detect traces of an element is determined by the yield of counts in the spectral peak corresponding to that element relative to statistical fluctuations in the background in the spectral region of the peak. The detection limit is commonly defined, somewhat arbitrarily, as the amount of an element producing a number of counts equal to three times the variance of the background contained in the full width at half maximum of a spectral peak. Using a value of 200 eV as a typical peak width (as in Figure 10), a background level of N counts/200 eV results in a detection limit of 3 $N^{1/2}$ counts in a spectral peak.

The detection limit for various elements can readily be determined with the aid of Figure 10. Thus, since 1 ppm of Zn produces 100 counts, while the background of a guard-ring reject system in the region of the Zn peak is about 150 counts/200 eV, the detection limit for Zn, assuming that 10^6 counts are accumulated in the scatter peaks, is approximately 0.3 ppm. At a total counting rate of 5000 cps, this result can be achieved in 200 sec. If a counting rate of 20,000 cps can be achieved, the same sensitivity is achieved in only 50 sec, or the detection limit can be improved by a factor of two if the same 200-sec counting time is employed. These limits could be improved nearly by a factor of three if the theoretical values of background could be achieved. Similar calculations can be performed for other elements—as can be seen from the relative ratios of background-to-peak counts for the various elements in Figure 10, Zn is in the most sensitive energy region when using Mo X-ray excitation of the sample. Nevertheless, the sensitivities vary by less than a factor of two over the range of elements from Fe to Br, and also for the L_α X rays of Pb.

CALIBRATION

The smoothly varying and calculable behavior of the relative yield curves of Figure 6 lead to simple calibration of an X-ray fluorescence analyzer. It is only necessary to determine an efficiency factor (mainly geometrical) for the overall excitation-detection process to obtain absolute concentration data for the trace elements in a sample. This factor can be determined by measuring a thin "intensity standard" containing a known weight of one element. Unknown samples can then be quantitatively analyzed by first comparing the peak areas for spectral lines produced by these samples with the peak area produced by the intensity standard, and then by applying the appropriate energy-dependent relative-efficiency factor (see Figure 6). Automatic computer analysis of X-ray spectra by this method has been used to provide quantitative analysis for up to 30 elements (13).

The calibration procedure is complicated to some degree by interference between spectral lines of one element with those of another. Two good examples are the K_α arsenic and the L_α lead lines, which occur at the same energy (within the spectrometer resolution), or the iron K_β and cobalt K_α lines where a similar situation occurs. Further problems in spectral analysis are caused by weak satellite peaks produced by the escape of silicon K_α X rays from the detector. Iron K_α X rays (which are very strong in the analysis of many specimens) produce a weak Si-escape line roughly equal in energy to the K_α line of vanadium. These complications may be largely overcome by careful choice of the spectral analysis and stripping sequence, and by complex spectral analysis computer programs. However, an inevitable loss in sensitivity results for an element whose spectral features are obscured by strong lines from other elements. This is one area where the better energy resolution of the Bragg-crystal spectrometer proves valuable. Fortunately, serious interference effects occur for very few elements and, for most purposes, the elements in question are not very important ones.

ANALYSIS SPECTROMETER DESIGN

A spectrometer system consists of the primary radiaton source, the electronic system including the detector, and the data-processing system. The following discussion will be restricted to the first two items. Spectral analysis using computers is a separate topic considered in other papers (13).

Primary Radiation Source

The primary radiation source must provide photons of sufficient energy to excite the elements of interest at a rate high enough to accumulate the necessary number of counts in a short time. Experimental tests show that analysis of elements present at levels of 1 ppm or less in a few minutes requires source activities in the 10-Ci range even if efficient geometries are used. A further requirement is that the source should produce virtually no photons in the energy range where trace element lines of interest may occur. If this requirement is not met, such photons scattered from the sample will constitute an undesirable background.

Radioactive sources can be used to provide monochromatic radiation, are small so that good geometries can be realized, and require no associated equipment or maintenance. However, the large activities required for fast analysis present difficulties, and, furthermore, the range of suitable sources is limited, so problems arise in providing optimum excitation energies. Suitable sources include ^{125}I, which has a half-life of 60 days and emits 35-keV γ rays and Te K_α and K_β X rays (~ 28 and 31 keV), or ^{109}Cd, which emits Ag X rays (~ 22 and 25 keV) and

88-keV γ rays, and has a half-life of 470 days. The 60-keV γ ray of ^{241}Am is often employed for analysis of elements requiring higher excitation energies.

Better detection sensitivities can generally be achieved when using radioactive sources by exciting secondary fluorescers whose characteristic X rays irradiate the sample. The secondary fluorescer material is selected to produce optimum excitation of the elements of interest with minimum scatter from the sample. One example of the ring-source fluorescer geometry (14) is shown in Figure 11.

Difficulties in obtaining the required activities and fears over large-scale dispersal of such sources have led to increased use of X-ray tubes for sample excitation. They can easily provide the necessary photon flux, and the X rays can be turned on only when required and be controlled in intensity. Also the anode voltage of the X-ray tube can be adjusted to determine the maximum X-ray energy emitted by the tube. Bremsstrahlung radiation generated by a conventional reflection-geometry X-ray tube is not suitable for use as primary radiation in an X-ray fluorescence analyzer since its scatter from the sample causes excessive backgrounds. Two basic methods have been used (15) to develop nearly monochromatic primary radiation. The first employs a thin transmission-anode tube in which the anode material is chosen so that its characteristic X rays are of the desired energy, and the anode potential is adjusted to enhance the ratio of characteristic X rays to bremsstrahlung generated by the tube. The X rays which pass *through* the thin anode are then employed. The bremsstrahlung radiation generated at the anode is largely filtered out by absorption in the anode material, while the fluorescent X rays are largely transmitted by the anode material since their energy is just below that of the K-absorption edge. Almost the same result can be achieved by filtering the X rays from a conventional reflection-geometry X-ray tube whose anode material and operating voltage are chosen to emphasize the characteristic anode X rays. The second basic method uses a secondary fluorescer to convert the bremsstrahlung and the characteristic

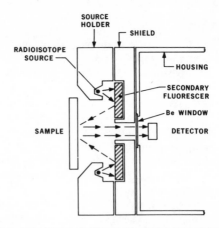

Figure 11   Ring source—secondary fluorescer geometry.

anode material X rays from a conventional reflection-geometry X-ray tube into the characteristic radiation of the fluorescer material. This is analogous to the ring-source fluorescer arrangement of Figure 11. One such geometry employed in an analyzer designed to analyze particulate material on filters (13) is shown in Figure 12. The secondary fluorescer (target) and the voltage on the X-ray tube are changed to provide different energies of radiation incident on the sample. Using this tight geometry, 20 W of anode dissipation is sufficient to permit analysis, in 10 min, for elements present in the sample at well below the 1 ppm level.

Electronic System

The evolution since 1960 of electronics for semiconductor detector spectrometers has been directly responsible for the resurgence in interest in X-ray fluorescence analysis. It is appropriate therefore to devote a portion of this paper to the important developments that have led to the present types of spectrometer.

A primary requirement in an X-ray spectrometer is its ability to separate the spectral lines due to adjacent elements in the periodic table. Semiconductor detector spectrometers have the inherent advantage over Bragg-crystal spectrometers of providing simultaneous analysis over a broad energy range, but statistical fluctuations in the conversion process from energy to signal in the detector and noise in amplifiers limit the resolution of these spectrometers. Low-noise amplifiers developed in recent years provide adequate energy resolution for low-energy X-ray analysis to become practical, since resolution of the K X-ray lines due to all the elements above

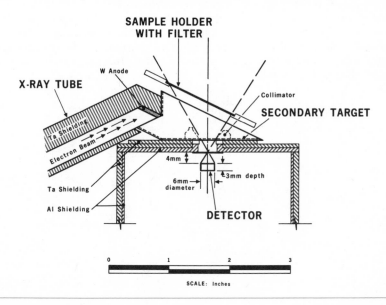

Figure 12 One type of X-ray tube-sample-detector assembly used in an analyzer for air particulates deposited on filters.

carbon in the periodic table can be achieved. As discussed in the previous section, interference problems between spectral lines would be reduced if better energy resolution could be achieved, but these problems are manageable for most types of analysis.

A further aspect of the spectrometer design that is extremely important in trace element fluorescence analysis is the ability of the spectrometer to maintain excellent energy resolution at high counting rates. While accumulating statistically meaningful numbers of counts in the peaks due to trace elements in a sample, large numbers of counts are inevitably accumulated in the backscatter peaks (for photon excitation) or in the low-energy bremsstrahlung background (for particle excitation). This implies that the total counting rate must be high if analysis is to be performed in short counting times.

Figure 13 shows schematically the overall detector-electronic system used in an X-ray spectrometer. A lithium-drifted silicon detector, normaly 3–5 mm thick and 0.5–1 cm in diameter is cooled to liquid nitrogen temperature, primarily to reduce its leakage current (resulting from thermally generated carriers) to a very low value ($\sim 10^{-15}$A). Holes and electrons, produced by absorption of photons, are collected by applying an electric field across the detector (voltage ~ 500 V), and the resulting signal current flows into a field-effect transistor (FET) that forms the first stage of a charge-sensitive preamplifier. The FET is also cooled to a low temperature (about 130°K) chosen to optimize the signal/noise ratio of the spectrometer.

Step-function signals from the preamplifier are amplified and shaped in a main amplifier unit before being fed to a pulse-height analyzer (or computer). The pulse shaper in the amplifier is a critical part of the system, being selected on the basis of optimizing the signal/noise ratio. Where the input circuit leakage currents are small, as in the case of X-ray spectrometers, the ideal shaper produces quite long pulses—commonly tens of microseconds in duration. These long times limit the maximum rate at which pulses can be processed to the 10 kc/sec range.

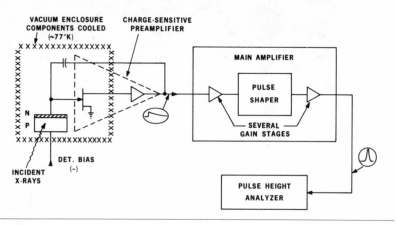

Figure 13   The overall detector-electronics system used in an X-ray fluorescence analyzer.

Energy Resolution

Energy resolution is determined partly by statistical fluctuations in the charge-production process in the detector and partly by electronic noise resulting from fluctuations in leakage currents and in currents in amplifying elements. Fluctuations in charge trapping in the detector can also contribute, but selection of detector material exhibiting low trapping generally makes this contribution negligible for low-energy detector applications. Since the *absolute* value of the signal spread due to charge-production statistics decreases as the photon energy decreases, while electronic noise factors are independent of energy, the relative importance of the two contributions depends on the energy of photons being measured. In modern X-ray spectrometers, electronic noise is the major contributor at energies below about 5 keV, but charge-production statistics dominate at higher energies.

The basic charge-production mechanism in the detector starts with the photon interacting photoelectrically with an electron (usually a K-shell electron) in the silicon. The ejected photoelectron causes a shower of holes and electrons, successive generations in the shower representing lower and lower energies. Finally, a large number of holes and electrons are produced with more than thermal energies, but less energy than that is required to produce further hole-electron pairs. The excess energy of these holes and electrons is dissipated by exciting vibrational modes in the silicon crystal lattice; this energy is therefore lost to the ionization signal. The loss of energy in the end products of the shower varies statistically from one incoming photon to the next, so the output signals reflect these fluctuations. Details of this process have been studied theoretically (16–18) and a number of accurate experimental measurements of the resulting contributions to energy resolution have been made (19, 20). The detector contribution to resolution is given by:

$$\Delta E_{(FWHM)} = 2.35 \, (FE\varepsilon)^{1/2} \qquad\qquad 2.$$

where $\Delta E_{(FWHM)} =$ contribution to full width at half maximum resolution, $E =$ energy of photons, $\varepsilon =$ average energy required to produce a hole-electron pair, and $F =$ Fano factor, representing the statistics of energy loss to vibrational processes. For a silicon detector, ε is approximately 3.8 eV at 77°K, and F is approximately 0.12. The detector statistical contribution to resolution is shown in Figure 14 as the limiting resolution with zero electronic noise.

Two types of electronic noise are produced. The first results from fluctuations in currents in the preamplifier input circuit, which consists of discrete electrons flowing into the input capacity (detector plus circuit), therefore producing small voltage step functions at the input. This type of noise may be called "step noise." The other type of noise is caused by statistical fluctuations in the electron (or hole) flow through the input FET. As these electronic charges are not integrated in the input capacity of the preamplifier, their basic shape as passed on to the main amplifier is a delta function, the duration of each impulse being representative of the transit time of the carrier through the FET. This noise can be called "delta noise." These two types of noise, together with the signal, are processed by the pulse shaper in the main amplifier.

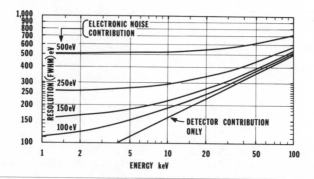

Figure 14   Resolution vs energy for typical X-ray spectrometers.

The pulse shaper commonly used in modern spectrometers develops a Gaussian pulse shape. Although sophisticated circuits are used to develop this pulse, their behavior can be represented by a single RC differentiator and multiple RC integrators. This pulse shape approaches the optimum insofar as signal-noise ratio is concerned and, being almost symmetrical about its peak, gives a narrow total width that reduces pulse pile-up problems at high rates. Choice of the peaking time depends on the ratio of step-noise and delta-noise components, and on the need for good performance at high count rates. Small peaking times (τ) cause increasing delta noise (noise $\propto \tau^{-1/2}$), but reduce pile-up problems at high rates. Long peaking times reduce delta noise, increase step noise (noise $\propto \tau^{1/2}$), and increase pile-up problems. A typical X-ray spectrometer exhibits the noise performance (expressed in terms of equivalent FWHM spread in the energy absorbed in a silicon detector) shown in Figure 15 as the peaking time of the Gaussian pulse shaper is varied. In

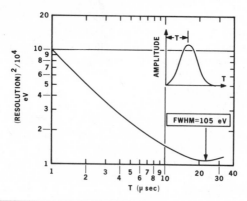

Figure 15   Electronic noise—expressed in terms of equivalent energy resolution (Si detector) plotted as a function of the peaking time of the Gaussian pulse shaper used in a spectrometer.

addition to the basic noise factors already mentioned, microphonic noise becomes important at long peaking times.

The effects of electronic noise, combined with those of charge-production statistics in the detector, cause the variation of resolution with energy illustrated in Figure 14. The best low-capacity detector spectrometers exhibit noise contributions < 100-eV FWHM when used with shapers having long peaking times (~ 50 μsec). A more typical system for trace element analysis exhibits a noise contribution near 130 eV with a shaper peaking time of about 15 μsec.

Counting-Rate Performance

Accumulation of a large number of counts is necessary if adequate statistics are to be obtained in the background beneath weak peaks to permit their detection. The speed of analysis is therefore directly determined by the ability of the spectrometer to analyze photons at a high rate, providing that the intensity of primary radiation source is adequate. In practice, analysis in a few minutes for trace elements present at levels below 1 ppm requires total detector counting rates up to 20 kc/sec. Since long pulses (~ 30 μsec total width) must be employed to give good energy resolution, the high-rate requirement imposes a very serious design problem in the signal-processing electronics.

A major counting-rate problem arises in the conventional charge sensitive preamplifier configuration shown in Figure 16a. In this type of preamplifier, charge pulses through the detector flow from the feedback capacitor C_F producing small positive steps at the output. The resistor R_F across C_F provides a path for the

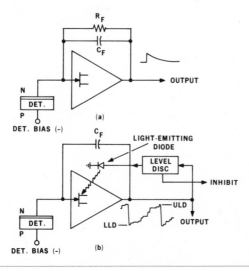

Figure 16 (*a*) A conventional charge-sensitive preamplifier using a resistor to provide the charge leak across the feedback capacitor. (*b*) Pulsed-light feedback method of discharging the feedback capacitor.

charge on C_F to slowly leak away. Typical values of C_F and R_F in low-energy systems might be 0.2 pF and 10,000 MΩ giving a decay time-constant of 2 msec. As is well known (21), only a single differentiator must be present in a system to be used at high counting rates so as to avoid the effect of overshoots on pulses modulating later pulses. Since the main pulse shaper contains one differentiator (~ 3 μsec RC), pole-zero cancellation methods are generally employed to effectively remove the second differentiation produced by the charge decay from C_F via R_F. This would be quite satisfactory if the feedback circuit $R_F C_F$ behaved as a single time constant, but high valued resistors, such as R_F, exhibit large changes in their effective resistive value at high frequencies. Therefore the pole-zero cancellation is ineffective, and each signal pulse is followed by residual overshoots that interfere with later pulses.

This circuit also suffers from the effects of stray capacity and noise produced by the feedback resistor R_F. These degrade resolution to a degree that is intolerable in low-energy X-ray systems. The pulsed-light feedback system (22) shown in Figure 16b was developed to overcome the resolution degradation and counting-rate problems of the conventional circuit. In this system, signal pulses from the detector are integrated by the feedback capacitor C_F resulting in a buildup of voltage at the output. When the output excursion reaches an upper limit set by the output-level discriminator, the light-emitting diode is turned on. Its light couples into the drain-to-gate diode of the FET which, acting as a photodiode, produces a current into the gate circuit that rapidly discharges C_F. When the output voltage drops to its lower limit the light-emitting diode is turned off. The counting system is gated off during the brief recharge period, and for a short time later, to take account any short-term aftereffects. To achieve this mode of operation, FETs are removed from the standard commercial

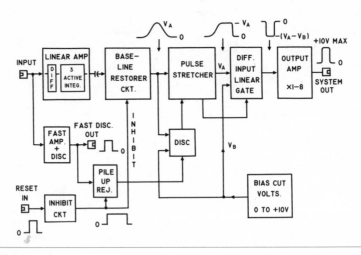

Figure 17 A block diagram showing the analogue-signal processing elements used in a modern X-ray spectrometer.

package and mounted together with the light-emitting diode in a low-loss package. This also removes the lossy material which surrounds the gate lead in the conventional commercial FET package, thereby removing another serious noise contribution.

These measures have produced the excellent resolutions of modern X-ray spectrometers, and have made operation at high counting rates possible. However, very sophisticated analogue-signal processing operations are also performed. The block diagram of Figure 17 illustrates the complexity of electronics used to process signals in a typical system. One element is a pile-up rejector that permits only those pulses uncontaminated by others to be passed on to the pulse-height analysis system. As the input counting rate is increased, this circuit rejects an increasing fraction of the pulses so that the output rate peaks at a certain input rate and then falls off as the input rate increases. However well a system is designed to avoid degradation in resolution at high-input counting rates, the pile-up effect causes a limitation in output rate, and thereby on the sensitivity of an X-ray fluorescence analyzer. To partially overcome this limitation, a pulsed X-ray tube excitation system has been devised (23) that operates as shown in Figure 18. The X-ray tube is turned on initially, but

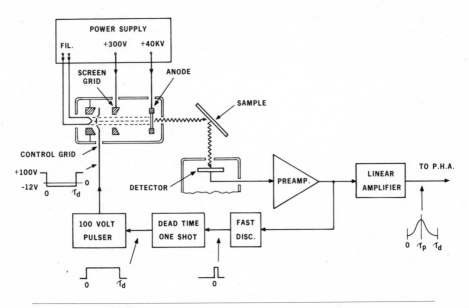

Figure 18   Diagram of a pulsed X-ray tube excitation scheme to reduce pile-up losses.

is turned off immediately as X ray is sensed in the amplifier system. The signal pulse is then processed before the X-ray tube is turned on again. If the X-ray tube pulse current is high, the signal output assumes the appearance of almost regular pulses with no pile-up occurring. The improvement in output rate capabilities is

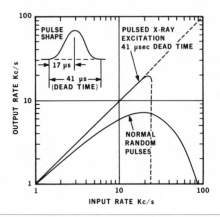

Figure 19   Illustration of the improvement in output counting rate achieved using the pulsed X-ray tube method.

shown in Figure 19, where the decline in output rate at high input rates exhibited by the conventional system is contrasted with the steady increase in output rate exhibited by the pulsed X-ray tube system until the full counting rate capability of the system ($=1$/pulse width) is realized.

EXPERIMENTAL RESULTS

As discussed in the introduction, multielement analysis at trace levels has a wide variety of applications to environmental and medical problems. A few illustrative examples are discussed here.

Monitoring of the elemental composition of particulates in air represents an ideal application. The sample is prepared by drawing a known amount of air through a membrane (e.g. Millipore) filter, and is analyzed on an X-ray fluorescence analyzer. A particle-size separator may be used ahead of the filter if the elemental distribution as a function of particle size is to be determined. Sample collection times ranging from 1 to 24 hr are commonly employed, depending on the information required on time variations in elemental composition. Typical air-flow rates through the filter (~ 8 cm^2 in area) are about 1 liter/sec—about twice the breathing rate of an adult. This means that about 1 μg of particulate material will deposit per cm^2 of the filter in 2 hr when the atmosphere contains 1 μg/m^3 of particles.

Figures 20 and 21 show the spectra obtained in a 5-min analysis of two filters exposed in Berkeley, California for two contiguous 4-hr periods from 2:30 to 6:30 AM, and from 6:30 to 10:30 AM, under moderate atmospheric conditions. The concentrations of several elements are seen to increase radically as a result of the industrial and traffic activity after 6:30 AM. Table 2 shows the concentrations of several elements in the two time periods. These concentrations were computed by a small on-line computer analyzing and stripping the spectra shown

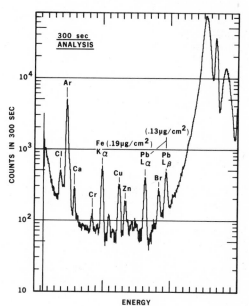

Figure 20 Spectrum obtained in a 5-min count on an air filter exposed in Berkeley for 4 hr at night.

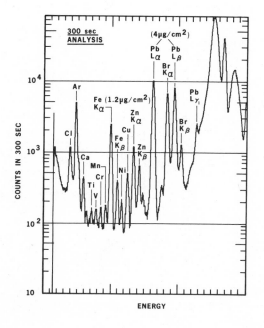

Figure 21 Similar to Figure 20, but the filter was exposed in the morning rush hour.

Table 2   Concentration (ng/m³ in air) of elements in particulates in Berkeley in night and day periods

Element	2:30 to 6:30 AM	6:30 to 10:30 AM
Pb	70	2400
Fe	115	750
Ca	75	300
Cl	500	1350
Br	19	770
Ti	0	50
V	0	50
Cr	0	30
Mn	0	40
Ni	0	38
Zn	0	180
Cu	0	30

in Figures 20 and 21. The speed of analysis, combined with a sensitivity of less than 5 ng of most elements per cm² of the filter, make this method a powerful tool in such studies.

Another far-reaching application is the analysis of biological specimens. One example is the analysis of blood to diagnose lead poisoning in infants (24). Figure 22 shows a spectrum obtained in such an analysis. The blood sample was lyophilized (freeze-dried), pulverized, and pressed into a tablet about 1 inch in diameter and 0.5-mm thick. The total volume of whole blood used is less than 1 ml.

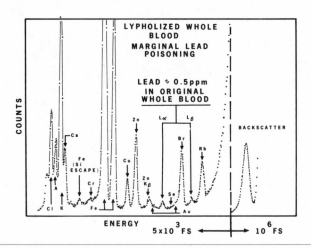

Figure 22   Spectrum obtained on a lyophilized whole blood sample containing a rather high lead concentration.

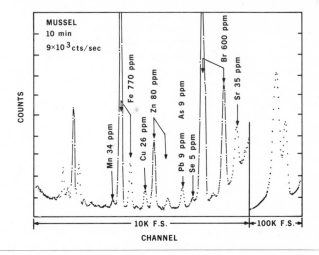

Figure 23   Spectrum from a dried Mussel sample.

The concentration achieved by the drying process is about a factor of five. The particular sample analyzed in Figure 22 contained 0.5 ppm of lead in the original blood, which represents a level at which clinical symptoms of poisoning begin to appear.

Analysis of another type of specimen is shown in Figure 23. This spectrum was obtained on a dried specimen of mussel from the nothern end of San Francisco Bay. The ability of such species to concentrate elements from the water in which they live provides a useful tool to study the sources and distribution of metals in large water bodies. Research studies based on this method are already providing information on topics of environmental interest.

COMPARISON WITH PARTICLE EXCITATION

A number of accelerator laboratories have studied the possible use of their accelerators for trace element analysis (25–29). At first sight, charged-particle excitation appears to offer a potential advantage over photon excitation since the incident radiation has quite a different character from the measured radiation, so less background interference might be expected. In practice this advantage is not realized, except for a few very special types of sample, since charged particles produce electrons in the sample that, in turn, produce bremsstrahlung radiation causing a photon background in the detector.

Since photon backgrounds set the ultimate physical limit to sensitivity in spectrometers using either type of excitation, the comparative performances depend on the relative numbers of photons and their energy distributions. When using charged-particle excitation, bremsstrahlung photons reaching the detector are primarily of low energy, whereas the sample-scattered photons in a photon-excited system are of much higher energy. Therefore, particle-excited fluorescence analysis is less critical of the behavior of the detector in producing degraded signals. On the other hand, whereas charged particles produce very intense bremsstrahlung due to their highly preferential excitation of the low-Z matrix atoms which constitute the matrix of organic samples, photon excitation preferentially excites those elements whose binding energy is just below the incident energy. While direct comparisons in sensitivity are difficult to make [though they have been attempted experimentally for some types of sample (30)], the preferential excitation of trace elements by photons makes photon excitation more sensitive than particle excitation for organic samples. Very short range charged particles can be employed with advantage to excite elements on the surface of a specimen while not producing background from the whole sample. They can therefore be a very useful tool in analysis of surfaces or deposits on surfaces (31).

To establish a physical basis for comparing the sensitivities that can be achieved by the two methods, it is essential to generate data for particle excitation similar to that shown in Figure 10 for photon excitation. Results of such calculations (11), based on the binary encounter model of Garcia (32), are shown in Figures 24 and 25 for the cases of 30-MeV α-particle and 4-MeV proton excitation. The energy distribution of bremsstrahlung photons per 200-eV interval is shown, assuming 10^{12} particles incident on a 1 mg/cm^2 sample. Also shown are the K X-ray yields for several elements (S, Ti, Fe, Zn, and Br), assuming that each element is present at the 1 ppm by weight level.

The data of Figures 24 and 25 can be used to deduce detection sensitivity limits for the various elements, as discussed in connection with Figure 10 for photon excitation. For a total of 1 μC of beam incident on a 1 mg/cm^2 sample, and assuming that the 0.25 cm^2 detector is positioned 10 cm from the sample, the following detection sensitivities are derived for zinc:

(a) 30-MeV α-excitation limit = 20 ppm   (3σ limit)

(b) 4-MeV proton limit = 14 ppm   (3σ limit)

These results compare very unfavorably with the 0.3 ppm sensitivity derived for photon excitation earlier in this paper—and even that limit may be improved if theoretical backgrounds can be achieved in photon-excited spectrometers. Increasing the integrated beam charge improves the detection sensitivity but the very large improvement necessary to equal the performance of photon excitation cannot be realized due to excessive low-energy counting rates and target heating. Even the use of a low-energy absorber to reduce the total detector counting rate fails to achieve the desired result.

Lower energy particles such as 2-MeV protons produce much lower levels of high-energy bremsstrahlung, and therefore offer the potential for improved detection sensitivity. However, target heating limits the value of these particles and, furthermore, the range of such particles is very limited, causing nonuniform irradiation of specimens of reasonable thickness (e.g. a few mg/cm^2). This feature can be advantageous if surface layers are the interesting regions of a specimen, but is not desirable for general trace element analysis problems.

Figure 24 Calculated brems-strahlung background and K_α X-ray yields for various elements produced by 30-MeV α-particles. The bremsstrahlung data is based on a target thickness of 1 mg/cm^2 and an integrated particle flux of 10^{12} particles. X-ray yields are based on the elements being present at a level of 1 ppm in the sample (i.e. 1 ng/cm^2 of each element).

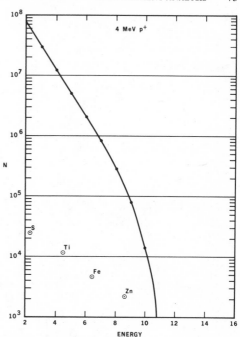

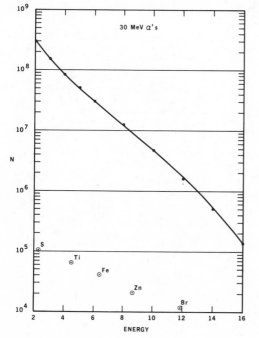

Figure 25 Similar to Figure 24, but 4-MeV protons provide the excitation.

On the whole, there seems little point to the use of charged particles produced by accelerators for trace element analysis. The high cost of accelerator facilities could only be justified if sensitivities better than photon excitation could be achieved. This appears not to be the case.

Literature Cited

1. Liebhafsky, H. A., Pfeiffer, H. G., Winslow, E. H., Zemany, P. D. 1960. *X-Ray Absorption and Emission in Analytical Chemistry,* 179–91. New York: Wiley
2. Birks, L. S. 1969. *X-Ray Spectrochemical Analysis.* New York: Wiley—Interscience
3. Russ, J. C. 1970. *Energy Dispersion X-Ray Analysis: X-Ray and Electron Microprobe Analysis.* Philadelphia: Am. Soc. Testing Materials
4. *Seventh National Conference on Electron Probe Analysis 1972.* Pittsburgh, Pa: Electron Probe Analysis Soc. Am.
5. McMaster, W. H., Kerr Del Grande, N., Mallet, J. H., Hubbell, J. H. 1970. *Lawrence Livermore Lab. Rep. UCRL-50174*
6. Bambynek, W. et al 1972. *Rev. Mod. Phys.* 44: 716–813
7. Burhop, E. H. S., Asaad, W. N. 1972. *Advan. At. Mol. Phys.* 8: 163–284
8. Hansen, J. S., Freund, H. U., Fink, R. W. 1970. *Nucl. Phys. A* 142: 604–10
9. Giauque, R. D., Goulding, F. S., Jaklevic, J. M., Pehl, R. H. 1973. *Anal. Chem.* 45: 671–81
10. Evans, R. D. 1955. *The Atomic Nucleus,* p. 683. New York: McGraw
11. Goulding, F. S., Jaklevic, J. M. 1973. *Lawrence Berkeley Lab. Rep. LBL-1742*
12. Goulding, F. S., Jaklevic, J. M., Jarrett, B. V., Landis, D. A. 1972. *Advan. X-Ray Anal.* 15: 470–82
13. Goulding, F. S. et al 1973. *Lawrence Berkeley Lab. Rep. LBL-1743*
14. Giauque, R. D. 1968. *Anal. Chem.* 40: 2075–77
15. Jaklevic, J. M., Giauque, R. D., Malone, D. F., Searles, W. L. 1972. *Advan. X-Ray Anal.* 15: 266–75
16. Alkhazov, G. D., Komar, A. P., Vorobev, A. A. 1967. *Nucl. Instrum. Methods* 48: 1–12
17. Van Roosbroeck, W. 1965. *Phys. Rev. A* 139: 1702
18. Klein, C. A. 1968. *IEEE Trans. Nucl. Sci. NS-15, No. 3,* 214–23
19. Eberhardt, J. E. 1970. *Nucl. Instrum. Methods* 80: 291–92
20. Pehl, R. H., Goulding, F. S. 1970. *Nucl. Instrum. Methods* 81: 329–30
21. Chase, R. L. 1961. *Nuclear Pulse Spectrometry,* Chap. 2. New York: McGraw
22. Landis, D. A., Goulding, F. S., Jaklevic, J. M. 1970. *Nucl. Instrum. Methods* 87: 211–16
23. Jaklevic, J. M., Landis, D. A., Goulding, G. S. 1972. *IEEE Trans. Nucl. Sci. NS-19, No. 3,* 392–97
24. Chisolm, J. J. 1971. *Sci. Am.* 224: 15–23
25. Johansson, T. B., Akselsson, R., Johansson, S. A. E. 1970. *Nucl. Instrum. Methods* 84: 141–43
26. Watson, R. L., Sjurseth, J. R., Howard, R. W. 1971. *Nucl. Instrum. Methods* 93: 69–76
27. Flocchini, R. G., Feeney, P. J., Sommerville, R. J., Cahill, T. A. 1972. *Nucl. Instrum. Methods* 100: 397–402
28. Rudolph, H., Kliwer, J. K., Kraushaar, J. J., Ristenen, R. A., Smythe, W. R. 1972. *Proc. 18th Ann. Instrum. Soc. Am. Anal. Instr. Symp,* p. 151
29. Johansson, T. B., Akselsson, R., Johansson, S. A. E. 1972. *Advan. X-Ray Anal.* 15: 373–75
30. Cooper, J. A. 1973. *Nucl. Instrum. Methods* 106: 525–38
31. Musket, R. G., Bauer, W. 1972. *Appl. Phys. Lett.* 20: 411–13
32. Garcia, J. D. 1969. *Phys. Rev.* 177 (No. 1): 223–29

✻ 5535

FLASH-TUBE HODOSCOPE CHAMBERS

Marcello Conversi
Istituto di Fisica dell'Università di Roma, Sezione di Roma dell'I.N.F.N., Rome, Italy

Gennaro Brosco
Istituto di Fisica dell'Università di Roma, Rome, Italy

CONTENTS

1 INTRODUCTION

1.1 *Scope of Present Review*

Progress in high energy and particle physics has been considerably influenced, in the last 10 or 15 years, by the development of a new family of particle detectors, which collectively may be called "Electrically Pulsed Chambers" (EPC). In our definition an EPC is any detector of particle tracks—typically a spark chamber

(1–2)—normally insensitive, but sensitized for a short time by the application of an impulsive electric field. The "hodoscope chamber" made of electrodeless flash tubes (3) is the oldest member of this family. Like all other EPCs developed later, the basic idea underlying its working principle is that the probability of finding a free electric charge at a given instant in a macroscopic quantity of a gas is negligible. But in spite of the close similarity (1–2) with the ordinary optical spark chamber (see Figure 1), the hodoscope chamber (HC) has a number of characteristic features that commend it for certain applications to nuclear and high energy physics, as discussed later on in this article. In particular, the individuality of the hodoscope chamber's elements (flash tubes), the possibility of digitizing the information by means of simple external pickup probes (4) or of elaborating it by means of iconoscopic vidicon-type systems (5), the adaptability to different geometries and to cover large collecting areas, and the low cost for huge sensitive volumes are some of the reasons for the renewed interest (5, 6) in the possible applications of this first EPC to experiments with particle accelerators. As a matter of fact such a possibility was pointed out 18 years ago (3), but the actual application to such experiments has been hindered by the long "memory time" of the original device as compared to the memory time of the spark chamber developed a few years later. Thus only very recently (7) large hodoscope chambers, supplied with an ac clearing field to shorten the memory time of their flash tubes (8, 9) (see section 4), have been applied to experiments carried out at a particle accelerator.

On the other hand, ever since the first developments (10–21a) following its

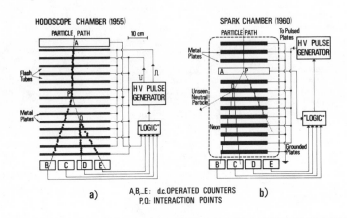

Figure 1 (*a*) Reproduction of the first hodoscope chamber to scale (3). Its working principle is also shown: a selected mode of coincidence among external counters triggers the high voltage pulse generator that sensitizes the detector. Tracks of ionizing particles are seen as a sequence of flashes (black spots in the figure) from the neon tubes traversed by them. (*b*) Sketch, from (1), showing the principle of operation of the spark chamber. Tracks of ionizing particles are now seen as a sequence of sparks (vertical lines in the figure) along the particle paths.

introduction in 1955, the HC technique has been used extensively to fulfill the need of large area, low cost detectors of particle tracks for modern cosmic ray research. Here only low radiation background is present and high repetition rates are not usually required; under these circumstances the "slow" ordinary neon-filled flash tube can be employed satisfactorily. Its long memory and recovery times have both been exploited to the advantage of some specific experiments (22, 23).

As discussed later on in this article (section 4), one of the most attractive features of the HC is its capacity as a detector of many-particle events. This explains why applications of the new technique to high energy cosmic ray showers were initially envisaged (14) and made (17, 20, 21). To illustrate such a capacity we reproduce in Figure 2 two rather impressive events recorded in a recent investigation (24, 25) in this field of physics: (a) The detection of a single-cored shower containing 1.8×10^5 particles (recorded by a 31-m^2 unshielded hodoscope containing nearly 180,000 small flash tubes); and (b) The detection of a many-burst event induced by shower hadrons (recorded by a system of 14-m^2 shielded hodoscope chambers) (see also section 4.2).

Although the most recent developments and applications of the flash-tube chamber have been recently reviewed (26), no complete review of the subject has appeared. The recent development of the "plastic chamber" (27), an interesting extension of the HC technique discussed in the last section of this article, adds interest to this subject.

1.2   Origin and Principle of Operation of the Hodoscope Chamber

The starting point for the development of the new technique was the observation of Adriano Gozzini late in 1954 at Pisa that a neon bulb kept in the dark does not glow when exposed to the high power impulsive electromagnetic field from a magnetron. It is well known, on the other hand, that a neon bulb does glow when exposed to the much weaker but uninterrupted electromagnetic field of a low power FR transmitter. Since the 1-MW power pulses from the magnetron lasted about 1 μsec and occurred with a repetition rate of the order of 1 per sec, the conclusion was that no electrons were present in general inside the neon bulb at the time of the magnetron pulse. The reason is that the electrons freed in neon by impinging ionizing particles (cosmic rays or radioactive radiation) diffuse quickly to the glass walls of the bulb, where they stick and become ineffective (28).

If this interpretation is correct, then it should be possible to "see" the track of any ionizing particle traversing a large number of neon tubes by subjecting the latter to an intense impulsive electric field immediately after the traversal. Such a possibility was tested successfully on March 25, 1955 on 1.5-cm diameter soda glass tubes filled with pure argon at about 1/2 atmospheric pressure and placed in a parallel plate condenser. With the help of a radar modulator technique two pulses of opposite polarity (of a maximum amplitude of 20 kV, see section 2.4 and Figure 7) were applied to the condenser plates to produce an impulsive electric field of up to 10 kV/cm immediately after the coincidence between the pulses of two dc operated counters, one placed above and the other below the condenser. We then built the particular chamber reproduced in Figure 1a, filled with 0.65-cm diameter tubes containing pure neon at 350 torr, and operated with external counters in a logic circuitry allowing the selection of various simple types of trigger.

The basic mechanism of flash-tube operation can be specified as follows. When a charged particle traverses a tube it ionizes the gas along its track. If a high voltage

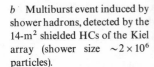

a Single-cored electron shower containing 1.8×10^5 particles, recorded by a system of HCs of 31-m^2 collecting area.

b Multiburst event induced by shower hadrons, detected by the 14-m^2 shielded HCs of the Kiel array (shower size $\sim 2 \times 10^6$ particles).

Figure 2 Examples of very high energy events obtained by the cosmic ray group at Kiel. See section 4.2 for further details.

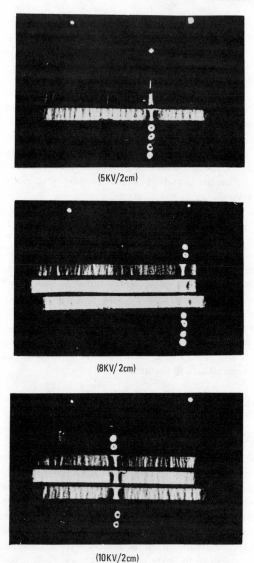

(5KV/2cm)

(8KV/2cm)

(10KV/2cm)

Figure 3 Photographs of the discharge in a flash tube observed sideways by Fukui and Miyamoto in 1957. The path of the primary ionizing particle was identified by the flashes of other tubes placed perpendicular to the former one. In all photographs, taken at different values of the field strength, the discharge appears as a bright column at the place of particle traversal. This suggested "the possibility of developing a new instrument which can detect the path of the particle more precisely" (12).

"triggering pulse" is applied soon enough, one or more free electrons are still present in the gas and may produce a local streamer under the accelerating effect of the electric field. Photon emission then causes the discharge to be propagated in a short time along the full length of the tube. This discharge results in a bright flash when the gas mixture, its pressure, and the applied high voltage pulse are chosen conveniently.

The principle of "selective trigger," illustrated in Figure 1 and common to all EPCs, confers on these types of detector a great stability of operation and allows one to pick up rare events out of a background. The most relevant feature of the new technique—a feature that is considerably improved in the spark chamber and streamer chamber—is the association of good space and time resolutions. It is this association that makes the operation of these chambers successful in the presence of a severe background.

Independently of, but somewhat later than the Pisa work, A. A. Tyapkin applied the above-mentioned principle (29) to pulse a hodoscope of discharge counters for use with particle accelerators, and H. P. Henning (30) applied it to enhance the sparks in a parallel plate counter system. The principle was also applied, again at Pisa (31), and at Harwell (32), to trigger parallel plate spark counters; and by Fukui & Miyamoto (33) in Japan in their "discharge chamber." This can well be regarded as a particular type of spark chamber, developed from the work carried out by the same authors (12) on the hodoscope chamber. More precisely, Fukui & Miyamoto were the first to investigate the discharge mechanism in a flash tube and to observe that streamers occur preferentially at the place where the primary particle passes. This is clearly shown by the bright columns in their photographs (12) reproduced in Figure 3.

2 GENERAL PROPERTIES OF ELECTRODELESS FLASH TUBES

2.1 Introduction

Even though the working principle of the hodoscope chamber is very simple from a conceptual point of view, the detailed mechanism of the discharge in the flash tubes is complex and not yet fully understood. There are several effects, related partly to the nature of the filling gas and partly to the high resistivity of the glass, which are difficult to interpret quantitatively. In the following section we shall show that some of the relevant experimental features of the flash tubes are amenable to theoretical treatment. In the present section, however, we give merely a qualitative discussion of the requirements for good flash-tube behavior and of the parameters influencing this behavior while presenting some of the most recent and significant improvements. The central part of this section is dedicated to a brief description of the methods for extracting the information from a hodoscope chamber and for investigating the flash-tube performance in the laboratory.

Generally speaking, the requirements for a satisfactory tube performance are: 1. high flashing efficiency after traversal by ionizing particles, 2. relatively short sensitive and recovery times, 3. low rate of spurious flashing, 4. good spatial resolution, 5. stability of operation and long lifetime, 6. large amount of output light for photographic purposes, and 7. reduction or cancellation of the long-lasting effects observed by several groups and presumably due to a deposit of electric charge on the high resistivity wall of the tube as discussed later (section 2.5).

Concerning point 2 above, we have already mentioned in section 1.1 that for some specific applications to cosmic ray physics—where in practice there are no problems of background radiation—a long sensitive time may be desirable. This is indeed the case of some large-scale flash-tube installations, where a limited number of flash tubes, out of a total of many thousands, are used "actively" to select a complex event that is then recorded by beams of a fast shutter camera (23). Also, the long recovery time of the ordinary flash tubes is sometimes useful, for example when reignition is used to obtain a greater amount of light from the same tubes (7, 22).

The actual flash-tube performance depends on a number of items such as: (*a*) tube geometry; (*b*) type of glass; (*c*) treatment of the glass tube previous to filling; (*d*) nature and pressure of the filling gas; (*e*) strength and time dependence of the applied electric field; and (*f*) temperature and relative humidity, both influencing the glass resistivity and thereby causing some long-lasting effects that will be defined and discussed later.

Most of the flash tubes employed in actual experiments have an internal diameter from a few mm to 2 cm and lengths up to 2 m, are usually made of soda glass 0.5- to 1-mm thick, and are filled with neon, or with a neon-helium mixture, at pressures ranging from about 0.5 to 3 atm. The high voltage (HV) pulse has rise times of typically 20–200 nsec and it decays exponentially with decay constants of 1–10 μsec. Field strengths from a few to several kV/cm may be used. The HV pulse is applied after a time θ from the passage through the tube of the particle to be recorded. This time in general cannot be made shorter than ~ 100 nsec, even though methods for very fast triggering (~ 10 nsec) had been envisaged and developed on a laboratory scale early in 1956 (31). As we shall see, improvements in flash-tube performance have been achieved recently through appropriate choices of the glass, gas filling mixture, and externally applied electric fields.

With the recent improvements mentioned above and to be discussed later, both the sensitive time, t_s, and the recovery time, t_r, of the flash tube have been reduced considerably with respect to the old values. These quantities will now be defined.

2.2 Some Definitions

The flash tube *sensitive time*, t_s, can be regarded as the memory time of the passage of a particle, in that it is essentially the time required for the relatively few electrons freed by the primary particle (typically some tens) to disappear via attachment to the glass walls or to impurities possibly present in the noble gas mixture filling the tube.

For a quantitative definition of t_s one needs to know the *efficiency vs delay curve*, $\eta(\theta)$; i.e. the curve that gives the flash-tube detection efficiency, η, as a function of the delay, θ, between the passage of the recorded particle and the instant of application of the HV pulse. Obviously $\eta(\theta)$ is a decreasing function of θ. If $\eta(\theta)$

were a step function at time θ, t_s would equal θ. Since this is not the case, a convenient definition of t_s may be given by the condition[1]

$$\eta(\theta = t_s) = \tfrac{1}{2}\eta(\theta = 0) \qquad\qquad 1.$$

Needless to say, in order to derive from the curve $\eta(\theta)$ the correct value of t_s, one needs to subtract the *spurious* and the *random flashes*. Spurious flashes are here defined as those occurring without a charged particle traversing the tube. The *probability of spurious flashes*, P_{sf}, is an important parameter of the flash tube and the performance of the latter should not be considered satisfactory if P_{sf} is in excess of 0.01.

Random flashes are here defined as those due to background ionizing particles traversing the tubes just before or during the application of the HV pulse. In fact if the latter has an "effective width" τ_e and if a given tube of sensitive time t_s is exposed to a background rate of b particles per sec, the *probability of a random flash* is $P_{rf} = 1 - \exp\{-b(t_s + \tau_e)\}$. Obviously a short sensitive time t_s allows the flash-tube chamber to operate in the presence of background rates as large as compatible with the condition $b(t_s + \tau_e) \ll 1$. However, like the case of the spark chamber or of any other EPC, t_s cannot be shorter than the minimum time (typically 100–200 nsec) required to produce the HV triggering pulse.

If a tube has flashed due to the passage of a particle, and a second HV pulse is repeated after a time T_r, there is a probability $P_{af}(T_r)$ that the tube flashes again even if no correlated particle is traversing it. Of course this *probability of after flashing*, or *probability of reignition*, is a decreasing function of T_r, which tends to $P_{sf} + P_{rf}$ for large enough values of T_r. This allows us to define a *recovery time*, t_r, by the condition

$$P_{af}(T_r > t_r) = \sim P_{sf} + P_{rf} \qquad\qquad 2.$$

With this definition t_r can be considered the memory time of the flash, and it is somehow related to the long disappearance time of the very many electrons liberated during the tube discharge.

There is indeed some arbitrariness in calling this quantity the "recovery time" since, as already mentioned, effects have been observed in flash-tube behavior that are considerably longer than t_r. Nevertheless we shall maintain this definition, also because recent methods have been found by which these long-lasting effects are drastically reduced, as we shall see in section 2.5.

2.3   Extraction of the Information from Hodoscope Chambers

The light emitted by ordinary flash tubes is large enough for photographic recording and the optical method for extracting the information from a system of hodoscope chambers has been extensively exploited in the past. The photographic method has been employed for example in the first magnetic spectrographs that used flash tubes (17, 34–37), in the early investigations of extensive air showers

---

[1] The alternative definition $t_s = \displaystyle\int_0^\infty \eta(\theta)\,d\theta$ is, however, preferred for a more accurate evaluation of the probability of random flashes defined below.

(38–40), and in the study of neutrino interactions at great underground depths (41).

However, as the experiments have become larger and more sophisticated, the problems of photography and film scanning have increased considerably. Thus, like the case of many "second generation" experiments with particle accelerators, there has been a trend, in the last years, to digitize the flash tube and read the experimental data directly into a computer "on-line."

In 1961 Coxell et al (22) suggested a method of selecting extensive air shower events by means of photomultipliers, Bacon & Nash (42) have used flash tubes containing sealed electrodes, and Reines (23) has described a method utilizing phototransistors on individual tubes.

More recently Ayre & Thompson (4) have developed a method that is particularly simple, reliable, and inexpensive. This method offers interesting possibilities for future cosmic ray experiments and has been adopted in MARS (43), the 300-ton "Magnetic Automated Research Spectrograph" operating at Durham since 1971, briefly described in section 4.5.

The method, as illustrated in Figure 4, consists merely in placing a simple probe (e.g. a screw with a washer at its head) against the front of the flash tube that is kept a few cm out of the metal plates between which the high voltage pulse is applied. A screening shield protects the probe against the HV pulse applied to the "hot" electrode. Thus only a small pulse is induced on the probe by the hot electrode, whereas a large pulse is obtained whenever the flash tube discharges if the probe is grounded through a resistor R. The polarity of the detected pulse is the same as the polarity of the appplied HV pulse and has a very similar shape when observed on a large output impedence ($R = 10$ MΩ). The amplitude of the detected pulse does not depend critically on the geometrical arrangement and decreases slowly with decreasing R, being typically 200–400 Volt for $R = 10$ MΩ. The pulse length decreases more rapidly with decreasing R.

By a suitable circuit the pulse from the probe can be reduced and shaped as desired. The rather large fluctuations of the pulse heights can be accommodated by a suitable choice of the discrimination voltage of the stage into which the pulses are fed. The signal to noise ratio is very large if the probe is in touch with the tube and the latter sticks a few cm out of the metal plates between which the HV pulse is applied.

Ayre & Thompson have also made an attempt to predict some of their experimental observations. They have assumed that the gas in the tubes away from the

Figure 4 The digitization method of Ayre and Thompson.

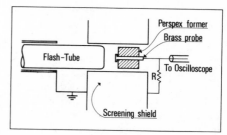

plates becomes a conductor when the tube discharges, due for example to photo-ionization or distortion of the electric field related to the grounded shield. They have furthermore assumed that a uniform charge density exists in the end of the tube. Neglecting the effect of the probe in distorting the field, they have calculated the potential at various axial distances from the end of the tube. Agreement with experimental observations can be considered satisfactory on account of the simplified theoretical treatment outlined above.

The probe can be realized in a slightly different way that allows one to retain fully the optical information from the front side of the flash tubes. It has been shown, in fact, that a small piece of copper glued on the opposite side of the flash tube and grounded through a resistor R as before yields a pulse of quite similar characteristics (44).

An interesting application of the digitization method has been made recently by Ayre et al (45) to measure the time delay in arrival of the digitization pulses corresponding to different longitudinal positions of the particle traversal. The velocity of propagation of these pulses has thus been determined to be 3.6×10^6 m/sec for flash tubes of 1.5-cm diameter, 2-m long, filled with neon at 600 torr; and 6.7×10^6 m/sec for tubes of 0.55-cm diameter, 2-m long, filled at 2.4 atm. The results of this investigation show the possibility of locating the initiating particle of the flash tube along the tube axis, by timing techniques, to approximately ± 10 cm, which corresponds to a relative error of $\pm 5\%$ for the 2-m tubes.

2.4   Methods for the Investigation of Flash-Tube Properties

Flash-tube performance has been investigated in the past using cosmic rays as triggering particles. Evans & Baker (6) were the first to use a ^{106}Ru source (which emits β rays energetic enough to penetrate a thin scintillation counter and the glass of the tube wall) and studied the flash-tube behavior with the help of the external probe of Ayre & Thompson (4).

Recently this method has been adopted in other laboratories. As an example we show in Figures 5 and 6 the experimental setup employed by the Rome group (9) for a systematic investigation of the flash-tube properties made in view of their

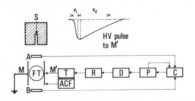

Figure 5   Experimental setup used by the Rome group to investigate flash-tube perfor-mance by means of β rays from a radioactive source. S: Radioactive source (^{106}Ru). A, B: Scintillators. FT: Sample flash tube. M, M': Metal plates. C: Coincidence circuit. D: Delay. P: Paralyzing circuit. R: Repetitor. T: Trigger circuit supplying to M' high voltage pulses of rise time $\tau_r = 50$ nsec and exponential decay time $\tau_d = \sim 5$ μsec. ACF: Circuit providing the alternating clearing field.

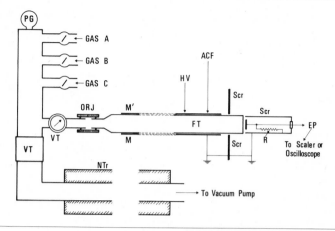

Figure 6   Vacuum-filling equipment, and system of electric signal pick-up probe, utilized in conjunction with the setup of Figure 5. *ACF*: Alternate clearing field. *EP*: Electric probe. *FT*: Flash-tube sample. *GAS A, B, C*: Three selected kinds of gas. *HV*: High voltage pulse. *M, M'*: Metal plates. *NTr*: Liquid nitrogen trap. *ORJ*: "O-ring" junction. *PG*: Pressure gauge. *R*: Output resistance of *EP* (usually 100 kΩ). *Scr*: Grounded metal screening. *VT*: Vacuum tap.

possible application to experiments with particle accelerators. The triggering counting rate, corresponding to the coincidences AB between the pulses of the two thin plastic scintillators A and B, could be in this case as large as $\sim 5/\text{sec}$. The minimum overall time delay, θ_0, between instant of particle traversal and the application of the HV pulse was ~ 150 nsec.

The setup, conceived to investigate the performance of flash-tube samples of different glasses, fillings, and dimensions, allows one to derive quickly the efficiency vs delay curve, $\eta(\theta)$, the curve relative to the reignition probability, $P_{af}(T_r)$, and the probability of spurious flashing, P_{sf}, or of random flashes, P_{rf}.

The sample flash tube, FT, is placed in a parallel plate condenser MM'. An HV pulse is supplied to M' at the occurrence of any coincidence AB. By the circuit D an artificial delay, T_d, can be added so as to extend the curve $\eta(\theta)$ to large values of the total delay $\theta = \theta_0 + T_d$. A "paralyzing" circuit, P, prevents the coincidence circuit C from giving any other output pulse before a time T_p has elapsed. The paralyzing time T_p can be adjusted so that the repetition rate of the triggering HV pulses vary between about 5/sec and 1/min. The "repetition" circuit R transforms each single input pulse into a sequence of n ($1 \leqq n \leqq 10$) equidistant pulses. By changing the time distance T_r of any two consecutive pulses in this sequence, the recovery time t_r of the sample can be determined by setting $n > 1$, so that the tube flashes only once for $T_r > t_r$ and more times for $T_r < t_r$.

An alternating square wave voltage can be supplied to the parallel plate condenser MM' with the purpose of studying the effect of an ac electric field of given frequency and amplitude on flash-tube behavior (see section 2.6). One tube end is used to observe the flashes and/or to pick up the corresponding induced electric signals by means of the external probe EP. The other end of the tube is connected to the vacuum and filling system of Figure 6 through an "O-ring" junction *ORJ* that allows easy installation and dismounting of the sample tubes.

Flash-tube performance depends on the characteristics of the triggering HV pulse, which should be therefore specified, as a rule, when presenting experimental results on flash-tube behavior. In the model setup we have just illustrated the HV pulse had the shape sketched in Figure 5, with a rise time of 50 nsec and an exponential decay time of approximately 5 μsec.

Square pulses of well-defined time width should be preferred to the exponential pulses commonly in use, obtained by discharging a condenser through a thyratron or a spark gap. In the original hodoscope chamber (3) square pulses of an effective voltage as large as 40 kV were obtained from an 8-kV power supply by the simple method reproduced in Figure 7.

More recently HV pulses of various wave forms have been studied, particularly at Case Western Reserve University (46), in order to obtain the best flash-tube performance. As explained in detail in the following subsection, bipolar pulses have been found to reduce long-lasting effects.

2.5   Long-Lasting Effects

Many authors have reported the existence of long-lasting effects that give the flash tube a sensitive time dependent on the rate at which the tube flashes. This phenomenon was first observed at Durham (47) and was later investigated at Case Western University by Crouch (46), who correctly interpreted the decrease in flash-tube efficiency at comparatively high flashing rates in terms of a "built-in clearing

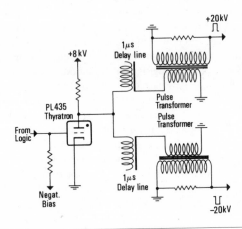

Figure 7   The triggering system in the original 1955 hodoscope chamber, providing two square high voltage pulses of opposite polarity (maximum amplitude 2×20 kV, time width 2 μsec with 150-nsec rise time).

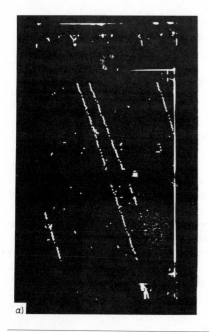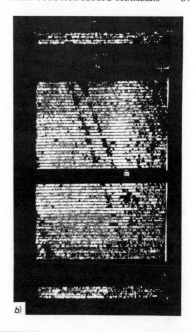

a)   b)

Figure 8   A striking example of the effect of the remnant built-in clearing field on the sensitive time of ordinary ("slow") flash tubes when operated with a large delay time (40 μsec) between occurrence of physical event and application of HV pulse.

field." It appears that charges remaining on the glass walls of the flash tubes after the HV pulse has been applied produce a remnant clearing field occurring on the wake of the HV pulse. More precisely the many electrons and positive ions produced in the discharge move in opposite directions during the time of application of the HV pulse. Even if the motion of the positive ions is slow, a fraction of these many ions may reach the tube wall opposite the side reached by the electrons, thus producing the above-mentioned internal clearing field.

This interpretation is supported by various experimental observations contributed by a number of groups (46–51). In particular it has been shown at Nottingham (49) that the observed effects cannot be due to polarization of the glass and are more likely to be caused by a deposit of electric charges on the tube walls as explained above.

A rather impressive example of reduction of efficiency due to the remnant clearing field in ordinary flash tubes has been obtained by Ashton et al (50) and is exhibited in Figure 8. Event (*a*) preceded event (*b*) by 38 sec. Both events were recorded with nearly rectangular HV pulses of 2.7 kV/cm, of 14-μsec width, applied with a delay time as large as 40 μsec. Four charged particles penetrated the chamber in event (*a*) and were followed 38 sec later by a large shower [event (*b*)] that deposited ionization in most of the tubes. It will be noticed that the effect of the remnant clearing field during the 40 μsec preceding the application of the HV pulse in event (b) is so large that even tubes adjacent to those that had flashed in event (a) have been cleared.

The existence of these long-lasting effects hinders the possibility of using flash-tube chambers in experiments with particle accelerators where high repetition rates are often required. A substantial reduction of these effects has been obtained recently, however, by two different methods: (a) by using commercial glass of resistivity two orders of magnitude lower than that of the soda glass employed in ordinary flash tubes; (b) by applying bipolar HV pulses to the flash-tube chamber.

It has been known ever since the first developments of the hodoscope chamber technique (12, 13) that the type of glass can influence considerably the flash-tube performance. One probable reason for this influence is the glass resistivity that determines the average time required for the charges deposited on the tube walls to disappear. Some systematic research on the effects of the glass resistivity has been carried out in Rome (9, 52) and has confirmed the prediction that low resistivity glass should be used. This is illustrated in Figure 9 where the efficiency curves, obtained at a triggering repetition rate of 0.25/sec, are reported for flash tubes made of three types of commercial glass of different resistivities. All sample tubes, of 1-cm internal diameter, were filled with the same gas mixture (30% Ne, 70% He) at the same pressure (380 torr). The faster decrease observed for the flash tubes made of G20 and 16III glass, with respect to the case of the 16B glass flash tube, may be interpreted in terms of the higher resistivity of the former glasses and of the higher clearing fields that are consequently present inside the tubes after a given time from the previous discharge.

The other method for reducing the long-lasting effects in flash tubes has been adopted recently by Bemporad et al (51) who have extended the earlier investigation by Crouch (46), having in mind possible applications to experiments with high energy particle accelerators. Bemporad et al confirm that the sensitive time t_s of their

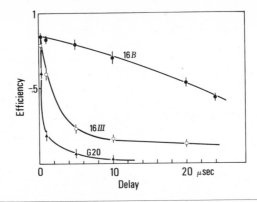

Figure 9 Efficiency vs delay curves for flash tubes made of three different types of Jena glass, showing effect of the glass resistivity on flash-tube sensitive time.

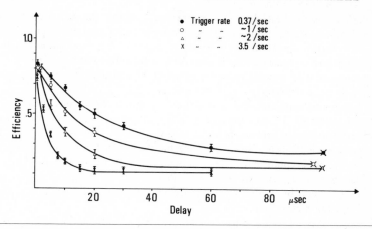

Figure 10   Efficiency vs delay curves for the same flash tubes, obtained at different trigger rates, v, showing effect of v on t_s.

1-cm diameter flash tubes, made of glass Jena G20 and filled with a 30/70 Ne/He mixture at 360 torr, clearly depends on the flashing repetition rate as shown in Figure 10. The authors show next that instabilities in flash-tube performance occur as a consequence of variations of the internal clearing field from one to another measurement. They show, in fact, that an external ac field (square wave, 50 Hz) simulates the same effects, and may be used also, of course, to control the flash-tube sensitive time, as done in other laboratories (8, 9). Finally the authors show

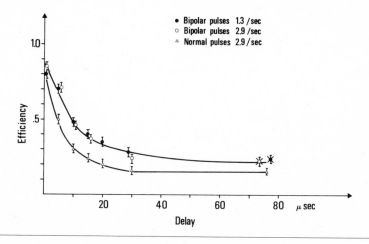

Figure 11   Efficiency vs delay curves for the same flash tubes, obtained with ordinary and bipolar HV triggering pulses, showing cancellation of long-lasting effects by means of bipolar pulses.

that the long-lasting effects due to the charge deposit on the tube walls are drastically reduced by employing a bipolar HV triggering pulse of convenient pulse shape. Best performance is achieved with HV pulses made of two exponential signals of opposite polarity, separated by 15 μsec each, with a 4-μsec decay constant. The effectiveness of these bipolar pulses in reducing the above-mentioned long-lasting effects is quite clear from the results reported in Figure 11.

By a computer simulated model for diffusion and drift of electrons in flash tubes, Holroyd & Breare (53) have estimated the strength of the remnant electric field as a function of the mean interval between tube discharges. They have obtained the curve of Figure 12 using known data for the drift velocity of electrons in neon (54), taking into account that there is a useful region in the tube (S in Figure 13) for a flash to occur (see section 3.4). Points derived from an approximate analytical method by Ashton et al (50) are also reported in the figure and found to fall near the curve.

A second type of long-lasting clearing field effect, with time constants of the order of days, has been reported by Holroyd (55) for the flash tubes of high resistivity glass currently in use at Durham for cosmic ray investigation. The effect consists again in a reduction of the flash-tube efficiency measured with large delay times (typically 50 μsec) after the flash tubes have been subjected to a large number of flashes (some 10,000 per tube). For applied HV exponential pulses of 40-μsec decay constant, the observed reduction in efficiency is appreciable only for field strengths in excess of about 6 kV/cm. The effect does not occur when the HV pulses are applied without the tubes being discharged.

Possible mechanisms for this long-clearing field effect have been suggested: polarization of the glass by the high fields that appear across it when the tube discharges, or electrons that are trapped in the glass surface and are therefore

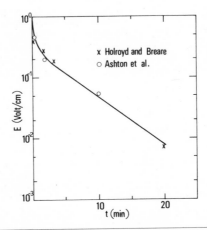

Figure 12   Time dependence of the remnant clearing field, built-in on the wake of previous discharges in a flash tube of the type commonly used for cosmic ray research.

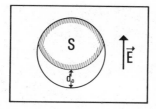

Figure 13   Only electrons present in the region S can produce the tube discharge under the effect of the field **E**, since the path available for the associated avalanches is $\geq d_0$ and avalanche-streamer transitions may occur (see section 3).

unable to move to neutralize positive ions on other regions of the tube. At any rate the effect is of no importance at short time delays of the triggering pulses and it appears to be strongly reduced or eliminated when bipolar ringing pulses are applied.

2.6   Space-Time Resolution

The limited space resolution of the hodoscope chamber is obviously related to the thickness of the flash-tube walls, which can hardly be less than 0.25 mm, and especially to the tube diameter, which under ordinary conditions cannot be made smaller than a few millimeters. Flash tubes of diameter as small as 3 mm, filled with commercial neon at a pressure as large as 3 atm (18), have been made and operated normally. Space resolution down into the millimeter region has been achieved recently (56) for glass tubes 5-cm long and filled with a neon-helium mixture at atmospheric pressure, by applying the impulsive electric field longitudinally rather than transversally with respect to the tube axes. Examples of particle tracks of 400-MeV/c momentum from a beam of the CERN proton-synchrotron are shown in Figure 14. The advantage of the longitudinal triggering mode is that it leaves enough space for the avalanche to transform into a streamer (see Figure 13 and section 3) independently of the tube diameter. An obvious limitation, however, concerns the tube length which cannot be much greater than 5 cm if unduly high values of the HV pulse amplitude are to be avoided.

On the other hand the early investigation by Ashton and co-workers (16) on staggered stacks of flash tubes of the standard types developed at Durham led to the conclusion that track location can be achieved with an accuracy comparable to that obtained in a cloud chamber, i.e. about 1 mm.

The accuracy of track location also was investigated theoretically by Bull et al (57); using a computer method for analysis of flash-tube data (58) they concluded that, experimentally, the error of location does not decrease as rapidly with increasing number of layers as expected from the calculation. They also concluded that high pressure tubes are more efficient, give greater accuracy of track location, and that there is no advantage in increasing the vertical separation of the flash-tube layers used to detect cosmic ray particles.

As already mentioned, the long memory time of the ordinary flash tubes has been exploited to advantage in some cosmic ray experiments. A long memory time

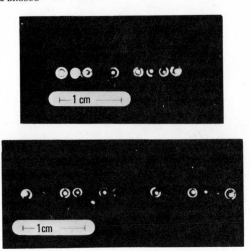

Figure 14   Tracks of charged particles recorded through the flashes of 1-mm diameter tubes operated with an impulsive electric field parallel to the tube axes.

is objectionable, however, for most experiments that are to be carried out with particle accelerators. On account of the renewed interest in this possible application of the hodoscope chamber technique, efforts have been made recently in various laboratories to reduce the flash-tube sensitive and recovery times.

Reduction of the flash-tube sensitive time, t_s, down into the microsecond region has been achieved either by applying alternate clearing fields to the chamber (8, 9) or by adding small amounts of impurities characterized by a high electron attachment probability (9, 51, 59) to the noble gas mixture.

The necessity of using ac rather than dc clearing fields as in ordinary spark chambers is related to the presence of electric charges deposited from previous discharges on the internal walls of the tube. The ineffectiveness of dc clearing fields can be explained, in fact, as due to a movement and rearrangement of the charges on the high resistivity glass surface, which causes the nearly complete backing off of the external dc field. This interpretation, first suggested by Breare and collaborators (8), is not invalidated by the small decrease in efficiency observed (60) for field strengths that would have reduced the efficiency of a spark chamber to virtually zero.

The effect of an ac clearing field (ACF) is demonstrated experimentally by the curves reported in the Figures 15 and 16. The curves of Figure 15 were obtained at a low triggering rate, with HV pulses of 4.2 kV/cm peak field decaying exponentially with a 36-μsec constant. Since a sinusoidal ACF was applied, a long tail is present in each curve due to the trigger particles that pass through the system at the zero or low field part of the ac cycle. The long tail is no longer present, in fact, when nearly square waves of alternating polarity are used, as in the case of Figure 16.

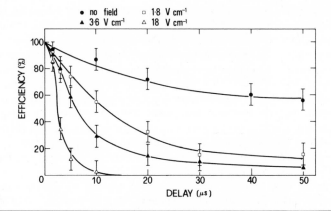

Figure 15 Effect of a sinusoidal electric clearing field (ACF) on the efficiency-delay curve of a flash tube.

Notice that an ACF of 1 kHz was applied in this case. However, no appreciable dependence on the ACF frequency was observed by the Rome group (9) over the frequency interval explored, from 50 Hz to 1 kHz. This is not surprising because essentially the frequency v_c of the ACF must only fulfill the very loose condition $t_s < 1/v_c < T_b$, where T_b is the time required to "back off" the externally applied dc fields through the movement and rearrangement of the electric charges on the internal walls of the flash tube. This time is of course an increasing function of the surface resistivity of the glass and of the tube capacitance, but can hardly be less than some milliseconds. Chaney et al (8) have investigated this point also changing the temperature, and therefore the resistivity, of the glass. At a temperature of 100°C, where the resistivity has fallen to 5×10^9 ohm·cm, a frequency of 10 Hz still has a small effect. This substantially agrees with the above considerations.

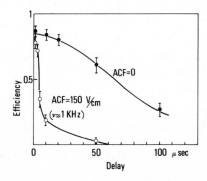

Figure 16 Effect of a nearly square wave ACF on flash-tube sensitive time.

Chaney et al (8) also point out that the shortening of the hodoscope chamber sensitive time is obtained for rms values of the ACF that correspond to values of the dc field shortening the spark chamber sensitive time to the same extent (61).

Reduction of the flash-tube sensitive time t_s by an ACF should be preferred to reduction obtained by adding small amounts of electronegative impurities to the noble gas mixture. This is because the former method involves an "external action" on the flash-tube behavior. However, adding impurities is more effective in reducing the flash-tube recovery time as we shall see in section 2.8.

Sensitive times shorter than 1 μsec were reported by Barsanti et al (11) early in 1956 for tubes filled with commercial neon that contained unknown impurities. Recently it has been shown (9) that small but controllable amounts of SF_6 impurities (0.75×10^{-6}) added to a 30% neon-70% helium mixture reduce the sensitive time of 1-cm diameter flash tubes to about 2 μsec. The stability of operation has been tested subjecting the flash tube to 1.6 million flashes. The efficiency curve obtained after this treatment is slightly shifted with respect to the "old" curve, as shown in Figure 17. The small observed increase in t_s (from about 2 μsec to 3 μsec) may

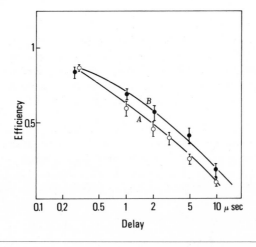

Figure 17   *A*: Efficiency-delay curve for a flash tube of glass Jena 16III containing a small contamination of SF_6. *B*: Same curve repeated after subjecting the tube to 1.6 million flashes.

be due to some extent to SF_6 decomposition caused by the many discharges to which the flash tube has been subjected. Further confidence in the stability of performance is derived from not finding any substantial change in the flash-tube characteristic curves after several months of inoperation. Tests were made, however, on only a few sample tubes.

The effect of impurities is apparent also in Figure 18. Efficiency curves for the same tube differ considerably depending on the method of evacuation previous to

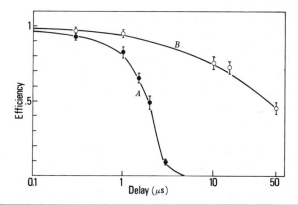

Figure 18   Efficiency-delay curve for tubes of the same type. *A*: Evacuated with a rotative pump at room temperature. *B*: Evacuated with a diffusion pump while kept for 2–3 hr at a temperature of 200–300°C.

filling. Tubes evacuated with a rotative pump at room temperature exhibit much shorter sensitive times than those evacuated at high temperature (200–300°C) with a diffusion pump. Since the light intensity of the flashes is much smaller for the former tubes, and since it has been observed at Durham that small amounts of water vapor cause a decrease in light intensity, it is also plausible that the main impurity present in the former tubes is water.

2.7   The Efficiency-Delay Curve

The efficiency vs delay curve $\eta(\theta)$, which defines the sensitive time t_s (section 2.2), is an important experimental feature most useful for studying the flash-tube properties and the mechanism of flash-tube discharge.

The effects produced on the curve $\eta(\theta)$ by changing field strength, pulse shape, gas pressure, and flashing repetition rate, or by adding suitable impurities to improve flash-tube performance, have been investigated extensively by many authors.

In general, layers of several tubes were exposed to cosmic rays to derive the curve $\eta(\theta)$. Thus it is convenient (18) to define two efficiencies: (*a*) the *layer efficiency*, η_L, i.e. the ratio of the number of single flashes observed in a layer to the number of particles having passed through that layer; (*b*) the *internal efficiency*, η, derived from η_L by multiplying by the ratio ρ of the separation of the tube centers to the internal diameter of the tubes. It is assumed of course that only the particles traversing the gas of a tube can cause a flash in that tube. If N is the observed number of flashes, the internal efficiency is given with its statistical uncertainty by (18) $\eta = \rho\eta_L\{1 \pm [(1-\eta_L)/N]^{1/2}\}$.

The first systematic investigation on the efficiency-decay curve under various experimental conditions was carried out early in 1956 by Barsanti et al (11), mainly on flash tubes of 6.5-mm internal diameter (0.3-mm thick walls) filled with spectroscopically pure neon, or with neon plus 0.2% of argon, at a pressure that in most cases was 350 torr. The chamber was in all cases triggered by HV rectangular pulses. The results of this investigation can be summarized as follows:

1. The flash-tube efficiency is nearly independent of pulse rise time in the explored range of values, from about 0.1 to 1 μsec. This result has not been confirmed by later work of other authors (62) who have found a linear decrease of η with increase in rise time, but under rather different experimental conditions.

2. The efficiency increases with pulse width and reaches a plateau value at pulse widths of about 4 μsec for tubes filled with pure neon and for a field strength of 8 kV/cm.

3. The efficiency increases with field strength and reaches a plateau value at about 7 kV/cm for tubes filled only with neon and for a pulse width of 2.2 μsec.

4. The plateau value is reached at a much smaller field strength (about 3 kV/cm) for tubes filled with Ne + 0.2% Ar, due evidently to the so-called Penning effect (63, 63a) to be discussed in section 3. This effect is remarkable in that the pulse width can also be reduced to somewhat less than 1 μsec without affecting the flash-tube efficiency, as it is seen from the curves reproduced in Figure 19.

5. The efficiency-delay curve indicates that the sensitive time is about 6 μsec for tubes of 6.5-mm diameter containing only neon at 350 torr.

6. In the case of tubes containing Ne + 0.2% Ar the curve giving the efficiency vs gas pressure at a constant field strength of 8 kV/cm exhibits a broad maximum at a pressure of about 350 torr.

This last result has not been confirmed by the subsequent experiments carried out at Durham (13) on tubes that were filled only with neon. The neon in this latter case was of a commercial type, made of 98% neon, 2% helium, and small quantities of other elements. Tests made by Barsanti et al (11) on tubes filled with commercial neon of unknown composition showed a quick decrease in the efficiency-delay curve corresponding, as already mentioned, to sensitive times shorter than 1 μsec. Such short sensitive times are of course unnecessary and may be also undesirable for some cosmic ray experiments; but these early results suggested the possibility of using the

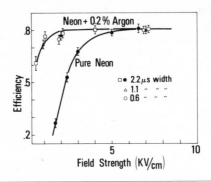

Figure 19 Efficiency-delay curves for pure neon and neon +0.2% argon, obtained in Pisa in 1956 and clearly showing the Penning effect.

hodoscope chamber technique in the presence of a severe background radiation (3), and even of placing the chamber in a beam of some 10^5 particles per sec, as it was done in fact five years later with the spark chamber.

The systematic work carried out at Durham (13) by Wolfendale and collaborators early in 1957 proved that the requirements for high flash-tube efficiency are not rigorous, that commercial gas can be used to fill tubes previously evacuated with a rotative pump, and that time delays of several microseconds in the application of the HV pulse can still be tolerated. Special credit must be given to the Durham group for the subsequent development of the standard types of flash tubes later employed on a large scale by a number of experimental teams for cosmic ray research.

2.8 Flash-Tube Recovery Time

The recovery time, t_r, of the first flash tubes and of those built on a large scale a few years later for use in cosmic ray experiments was in the range 0.1 to 1 sec. These long recovery times limit the field of application of the hodoscope chamber technique to experiments not requiring fast data acquisition. For most experiments with particle accelerators this is a serious limitation. A factor greater than 10 in reduction of t_r, as recently obtained (9), makes it feasible to employ flash-tube chambers in many such experiments.

Reduction of t_r by means of an ACF is illustrated in Figure 20. The reduction factor (about 3) is not as large as it is for the sensitive time (see section 2.6). A larger reduction factor is obtained by adding small amounts of electronegative impurities such as SF_6 (9), CO_2 (51), and O_2 (59).

If the information from the hodoscope chamber is recorded photographically, then it is important that the impurities added to the noble gas mixture do not absorb any relevant fraction of the emitted light. This is definitely known to be the

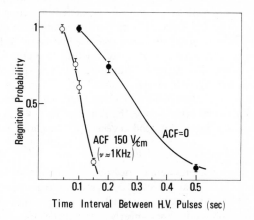

Figure 20   Reduction of flash-tube recovery time by means of an external ac clearing field.

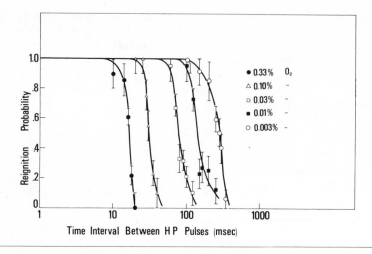

Reignition Probability

Time Interval Between H P Pulses (msec)

- 0.33% O₂
△ 0.10% "
□ 0.03% "
■ 0.01% "
○ 0.003% "

Figure 21   Reduction of flash-tube recovery time by oxygen impurities.

case with SF_6 (64). Recently Breare and collaborators have shown that oxygen at concentrations not exceeding 0.1% can also be adopted as a non-light-absorbing impurity to reduce the flash-tube sensitive time (59). The effect depends on the percent concentration of oxygen impurity as one can see from the curves reported in Figure 21.

Recovery times of about 10 msec have been obtained recently by Bemporad and collaborators adding 0.5% of CO_2 impurity to flash tubes filled with neon (51). Also, recovery times of some 10 msec had been found by the Rome group (9) for flash tubes contaminated with SF_6 in the way explained in section 2.6, even after subjecting them to 1.6 million flashes. The performance stability of this type of flash tube has been studied extensively on a single sample and found to be satisfactory, with HV exponential pulses lasting a few μsec, for field strengths not in excess of ~ 6 kV/cm. The recovery time decreases with decreasing field strength, but it is not affected apparently by any type of clearing field (the sensitive time also is not affected). These tests should be extended to a large stock of such flash tubes before employing the latter in real experiments. The Rome group has also found that the flash-tube recovery time does not depend in general on the glass resistivity.

2.9   Optical Properties of Flash Tubes

Extensive optical measurements were carried out at Durham (22) early in 1961 on soda glass flash tubes of internal diameter 1.5 cm, wall thickness 1 mm, and of either 115-cm or 27-cm length, filled with commercial neon at a pressure of 650 torr. The triggering HV pulse had in some cases a nearly rectangular shape of 3-μsec duration.

The form of the discharge was studied with a method similar to that of Fukui & Miyamoto (12) but the results of the two groups differ, the bright discharge columns of Figure 3 not being found by the Durham group. The discrepancy, possibly due to differences in pulse characteristics, is somewhat reduced by the observation that in the measurements with a radioactive source at Durham the discharge seems to be brighter in the region of the source.

Coxell et al (22) have studied in particular the variation in intensity from one pulse to the next for the same tube, the variation in intensity from one tube to the next, the variation of the light output with field strength, the variation of integrated intensity with the number of discharged tubes, the polar diagram, the after-flashing characteristics, and the profile of the light pulse, and have also estimated the number of photons emitted in the light pulse. The results of these measurements can be summarized as follows:

1. From the measurements of the intensity of the light emitted by the same tubes it is concluded that the discharge mechanism is essentially reproducible and that quantitative use of light intensity measurements should be possible.

2. Variations of intensity among tubes of the same type may be expected due to variations in the "quality" of the end window from which the light is emitted and to differences in gas filling. A standard deviation of 17% was obtained from observations on tubes belonging to six different groups, each one composed of 60 tubes filled at the same time. Presumably these variations would be reduced by using greater care in the glass blowing and filling technique.

3. The light output is found to increase with increasing field strength in the way shown in Figure 22.

4. Under suitable conditions of field strength and pulse shape, the total intensity of the light emitted by n flash tubes, as viewed by a photomultiplier system, is a nearly linear function of the number n. These results, exhibited in Figure 23, led the authors to suggest the complete elimination of the photographic recording of the flashes for particular applications, and recording in its place the pulse heights from one or more photomultipliers. Alternatively, the photomultiplier method could be used to make a selection of events to be recorded photographically, the camera shutter being operated by the selection device and a second HV pulse being applied within the recovery time of the tubes, to insure their reignition. These possibilities are hindered by the sharply peaked polar diagrams of the tubes (point 5 below). Covering the front of the tubes with translucent material considerably reduces this difficulty.

5. The polar diagrams in the plane perpendicular to the direction of the electric field are as shown in Figure 24. A lower limit to the expected variation also has been calculated by the authors disregarding the filamentary character of the discharge, assuming a uniform cylinder of emitting gas, and neglecting reflections and self-absorption. The intensity I_α along the direction α with respect to the tube axis is then given by the formula

$$I = I_0(4r/3\pi l)\operatorname{ctg}\alpha \qquad\qquad 3.$$

where r is the radius of the tube and $2l$ the length between the electrodes. Reflections are probably mainly responsible for the much greater intensity observed experimentally. This intensity can be increased further by painting the outside of the tube white.

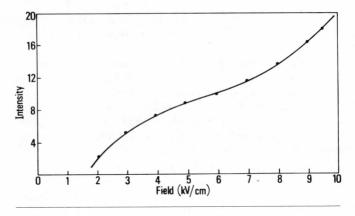

Figure 22   Dependence of light intensity on field strength.

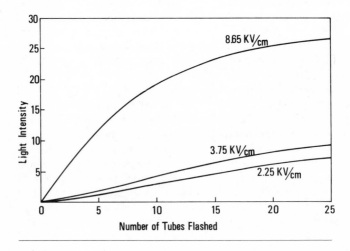

Figure 23   Total light intensity of *n* flash tubes as a function of *n*.

6. "After-flashing," or reignition, gives an array of tubes a "memory" which the authors suggest could be used as follows: with the help of photomultipliers one selects a chosen pattern of tubes flashed under the triggering HV pulse in order to open a fast camera shutter. Then a second HV pulse occurring before the recovery time of the tubes allows one to record the selected events. The reignition probability curves obtained for two types of neon tubes differing in the inner diameter (1.5 cm and 0.5 cm) and in the filling pressure (600 torr and 2.3 atm, respectively) yield recovery times of the order of 100 msec ($\sim$ 200 msec and $\sim$ 70 msec, respectively).

7. Figure 25 shows the variation of intensity of the flash with time (i.e. the "profile" of the light pulse) and the shape of the corresponding triggering HV pulse. The

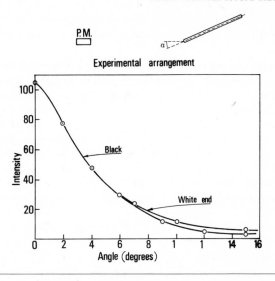

Figure 24   Polar diagram of the light emitted from a flash tube 2-m long.

intensity rises rapidly to a maximum value that depends on the field strength, and then falls with a time constant of about 5 μsec, while the HV pulse, of a nearly rectangular shape, has a 3-μsec duration. The fall of the light pulse is seen to have an interruption at the instant of cessation of the HV pulse. The lengthening of the light pulse may be caused by the metastable states of neon that have half-lives of 7.5 μsec at 1 atm (65). According to Coxell et al the second peak in the light profile may be due to the fact that when the applied voltage is removed, electrons, which were held on the glass by the pulse, return into the gas and cause an increase of photon emission.

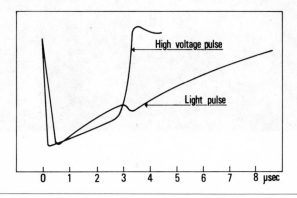

Figure 25   Time variation of the light intensity in a flash.

8. The estimated number, n_γ, of photons per cm^3, emitted after the application of a pulse of 7 kV/cm to a tube containing neon at 600 torr, is $n_\gamma = 2 \times 10^{10}$ photons/cm^3.

The light emitted by a flash tube has a spectrum that obviously depends on the filling gas. Small amounts of impurities may greatly influence this spectrum and the light intensity. For example it has been found (9) that the light intensity of a flash tube evacuated at room temperature with a rotative pump, and then filled with a 30/70 Ne/He mixture at 380 torr, is a factor three smaller than that of the same flash tube evacuated with a diffusion pump while kept at a temperature of 200–300°C.

Figure 26 shows the spectrum of the light emitted from such a flash tube. The light is concentrated in a narrow band (from 5.852 Å to about 6.400 Å). Hodoscope chambers with flash tubes of this type can also operate in the presence of some "white background" if appropriate filters are inserted before the camera that records the flashes (7).

In practice, due to the polar diagram, which is strongly peaked around the axis for long flash tubes, an accurate alignment of the latter is required (7) to take photographs from a large distance (say 20 m).

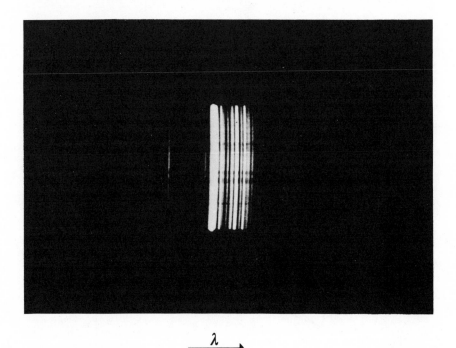

$\lambda \longrightarrow$

Figure 26 Spectrum of the light emitted from a flash tube filled with 70% helium and 30% neon at a pressure of 380 torr.

3   THEORY OF FLASH-TUBE OPERATION

3.1   Introduction

As stated in section 2.1, the detailed mechanism of flash-tube discharge is so complex that at present there is no theory that can interpret quantitatively all experimental data. Nevertheless some of the main experimental features are amenable to theoretical treatment as we will show in this section.

This particularly applies to the efficiency-delay curve that has been discussed qualitatively in section 2.7. A theoretical expression for this curve was first derived in 1955 (10) taking into account the basic process, which is the diffusion of the electrons to the glass walls where they stick and become ineffective (28). However the derivation was based on assumptions concerning the diffusion and discharge mechanism too crude to allow adequate comparison of the theoretical formula with experimental data. A theory with a correct treatment of the electron diffusion was later developed by Lloyd (66); however, he did not include the effect of possible electronegative impurities present in the noble gas filling the tube.

Further theoretical work concerning the flash-tube operation has been carried out more recently at Nottingham (67) and in Rome (68). In what follows we shall briefly review all of this theoretical work and whenever possible the predictions will be compared with the experimental facts that have been discussed only qualitatively thus far.

3.2   The Flashing Mechanism

Let $n(x)$ be the number of "primary electrons" liberated in the gas by an ionizing particle that traverses a flash tube of radius r at a distance x from its axis. Under the action of the applied impulsive electric field each of these electrons is accelerated and may cause an avalanche to occur through subsequent ionization processes. The number N_a of "secondary electrons" present in the avalanche after the latter has advanced over a path of length d is given by (69)

$$N_a = \exp(\alpha d) \qquad\qquad 4.$$

where α is the first Townsend coefficient, which depends on the strength of the electric field E. For a given field strength it can be shown (see section 3.4) that only for large enough values of d (70) and therefore of N, does the avalanche transform into a streamer (67). Under these circumstances, the photons emitted from the streamer may cause new avalanches and streamers in other regions of the flash tube via photoionization processes. Thus the discharge propagates along the full length of the tube and a flash does occur. The detailed mechanism by which this flash is produced, as well as the intensity and the spectral distribution of the emitted light, depend of course on a number of parameters, primarily on the nature and pressure of the filling gas.

3.3   Disappearance of Primary Electrons

Let m_0 be the average number of electron-ion pairs, per cm, per atm, that the primary particle generates in the gas. As an example we may recall that for a minimum ionizing particle in neon (71) $m_0 = 33.6/\text{cm} \times \text{atm}$. Then if p is the pressure (in atm) of the gas filling the tube, the number $n(x)$ of primary electrons is given by:

$$n(x) = 2pm_0(r^2 - x^2)^{1/2} \qquad\qquad 5.$$

After a time t this number is reduced to a number $n(t, x) \leq n(x) \equiv n(0, x)$, because some of the electrons may have disappeared through one of the following processes: (a) recombination with the positive ions; (b) attachment to electronegative impurities; (c) diffusion to the glass walls where the electrons stick and cannot come back into the gas, not even under the action of relatively strong electric fields (28); (d) drifting to the glass wall under the action of any electric field that may be present.

Process (a) can be neglected because of the smallness of m_0 and the recombination coefficient ($\sim 10^{-7}$ cm^3/sec in neon). [See Loeb (69), p. 561.]

Process (b) can be neglected only if the tubes have been accurately evacuated and then filled with noble gases of sufficiently high purity. Even small amounts of particular impurities may cause a quick reduction of the number of primary electrons. For example in oxygen the attachment probability is $b = 1.8 \times 10^{-4}$ (72) and the mean free path of thermal electrons—which have a mean quadratic velocity $u = 1.17 \times 10^7$ cm/sec—is $\lambda_0 = 0.6 \times 10^{-4}$ cm (73), at atmospheric pressure. Using these data one sees that if only process (b) were present, the probability $p(t)$ of electron disappearance at time t in a noble gas with the impurity at a percent concentration c, namely

$$p(t) \equiv \frac{n(t, x)}{n(0, x)} = \exp(-bcut/\lambda_0) \qquad\qquad 6.$$

for an oxygen concentration of 0.1% would drop below 50% after about 20 μsec.

In writing the previous formula we have assumed that the electrons captured by the electronegative impurity have a negligible detachment probability. This assumption is essentially correct for electric field peak values, E_0, that are not exceedingly high. For example in oxygen E/p should not exceed 40 V/cm·torr [see Raether (70), p. 183], or about 30 kV/cm at atmospheric pressure.

Actually, due to processes (c) and (d), the number $n(t, x)$ decreases faster than predicted by equation 6 with increasing the time t.

Assuming for the time being that no electric fields are present, so that there is no contribution from process (d), the effect of the diffusion to the tube wall can be taken into account following Lloyd's theoretical treatment (66). If D_- is the electron diffusion coefficient in the noble gas filling the tube [$D_- = \sim 400$ cm^{2}sec^{-1} in neon at atmospheric pressure (73)] equation 6 should then be replaced by the formula

$$p(t, x) = [\exp(-bcut/\lambda_0)] \left\{ \sum_\beta \frac{\exp(-\beta^2 D_- t/r^2) \int J_0(\beta^2/r) \, dy}{\beta J_1(\beta)(r^2 - x^2)^{1/2}} \right\} \qquad\qquad 7.$$

in which J_0 and J_1 are the Bessel functions of order 0 and 1, respectively, and β represents the roots of the equation $J_0(x) = 0 (\beta = 2.4, 5.5, 8.6, \ldots)$ (74).

To a first approximation (which can be shown to be good for $x/r < 0.5$) only the first term in the sum appearing between braces in the last factor of equation 7 can be retained. Then one finds

$$p(t, x) = \exp\left\{-\left[3bc/\lambda\lambda_0 + (\beta_1/r)^2\right]D_- t\right\}$$ 8.

since, as it is well known from the diffusion theory, D_- can be expressed in terms of the electron mean free path in the (noble) gas, λ, and the velocity, u, by the formula

$$D_- = \lambda u/3$$ 9.

For a tube of radius $r = 0.5$ cm filled with neon at atmospheric pressure, where $\lambda = 1.03 \times 10^{-4}$ cm (73), one sees that $p(t, x)$ drops below 50% after a time $t \geqq t_{1/2} = 15.7$ μsec. Notice that in the absence of impurities ($c = 0$) the time

$$t_{1/2} = \frac{r^2 \cdot \ln 2}{\beta_1^2 \cdot D_-}$$ 10.

increases quadratically with the tube radius and would be about 75 μsec for the numerical values given above.

Thus far we have assumed that D_- is the diffusion coefficient for electrons in thermal equilibrium with the gas. Actually electrons are liberated in the gas with energies much greater than the thermal energy and are slowed in times that sometimes cannot be neglected. Since D_- increases with the electron energy, the corrected formula of the number of residual electrons at the time t should lead to somewhat shorter sensitive times. These corrections have been estimated by Lloyd (66) and Brosco (68). Similar calculations have been made also by Schneider (75) for the spark chamber.

3.4   The Avalanche-Streamer Transition

As mentioned in section 3.2, for the flash to occur in a tube subject to a field of strength E after it has been traversed by a charged particle, it is necessary that the number N_a of electrons in the initial avalanches be large enough (70). Now it is generally accepted as an empirical law, verified over a wide range of values of E and of the gas pressure p, that for an avalanche to transform into a streamer the number N_a must exceed about e^{20} (70, 76); that is, the avalanche must advance over a path greater than:

$$d_0 = 20/\alpha$$ 11.

The above condition for the advance of the avalanche is also obtained by requiring that the electric field associated with the space charge in the avalanche be comparable or greater than the external field.

This condition implies that the only electrons that may cause a flash to occur are those having a distance greater than d_0 from the tube walls at the instant of application of the HV pulse. These electrons are those that occupy the region S in the transversal section of the tube shown in Figure 13. It will be assumed that d_0, and therefore S, does not depend on the presence of small amounts of electronegative impurities possibly present in the noble gas mixture.

An exact calculation of the number of electrons present in the region S, taking into account all important parameters, is quite complex. If, however, a uniform electron distribution over the section of the tube is assumed,[2] then the fraction f of electrons which can cause a flash to occur is given by (53, 68)

$$f = S/\pi r^2 = 1 - (2\phi + \sin 2\phi)/\pi \qquad \qquad 12.$$

where

$$\phi = \arcsin (d_0/2r) \qquad \qquad 13.$$

The values of f that can be derived from this formula are in good agreement with those obtained from the experimental curves (62). For example for a flash tube of 0.5-cm diameter, filled with neon at atmospheric pressure and triggered with an impulsive field of 10 kV/cm, one finds $f = 0.28$ to be compared with an experimental value of about 0.25 (62).

For pressures in excess of about 1 atm the discharge mechanism becomes quite different from the one outlined above and the theoretical evaluation of the fraction f is no longer so simple. More precisely Hampson & Rastin (67) have proved recently what had been conjectured earlier by Lloyd (66), namely that at these higher pressures the electrons, which are free inside the gas when the HV pulse is applied, are no longer capable of directly producing the flash through the avalanche-streamer mechanism, and that the flashes still observed at these higher pressures are produced via photoionization of the glass.

3.5 Derivation of the Efficiency-Delay Curve

Using the expressions 5, 7 (or 8), and 12 for $n(x)$, $p(t, x)$, and f, respectively, the efficiency $\eta(x)$ for a particle traversing the tube at a distance x from its axis is immediately obtained if a Poisson distribution is assumed:

$$\eta(x,t) = 1 - \exp\left[-n(x) \times p(t,x) \times f\right] \qquad \qquad 14.$$

This last assumption can be retained even though it is not rigorously verified (53). Then the average of $\eta(x,t)$ over the tube radius

$$\eta(t) = \int_0^r \eta(x, t)\, dx \qquad \qquad 15.$$

yields the internal efficiency of the flash tube as a function of t, i.e. the efficiency-delay curve. The sensitive time t_s derived from equation 14 with the approximate expression (8) for $p(t, x)$, is:

$$t_s = \frac{\ln\left\{(2rpm_0 f/\ln 2)[1 - (x/r)^2]^{1/2}\right\}}{[3bc/\lambda\lambda_0 + (\beta_1/r)^2]D_-} \qquad \qquad 16.$$

The dependence of t_s on x can be eliminated by taking its average value over x:

$$\bar{t}_s = \frac{\ln (2Arpm_0 f/\ln 2)}{[3bc/\lambda\lambda_0 + (\beta_1/r)^2]D_-} \qquad \qquad 17.$$

[2] We discuss this assumption at the end of section 3.5.

In this formula A is a geometrical factor which under the assumption that $2rpm_0f/\ln 2 \gg 1$ can be evaluated by computing the average of $\ln \{2rpm_0f[1-(x/r)^2]^{1/2}/\ln 2\}$. The result is $A = 0.74$.

The curves derived from equations 14 and 15, as well as the sensitive times derived from equation 17, are sometimes in excellent agreement with the corresponding experimental data. This is for instance the case of the experimental curves $\eta(\theta)$ reported in (62) for tubes of 5-mm diameter filled at 1 atm with neon and a 0.1% of oxygen (field strength $\simeq 10$ kV/cm). In other cases, however, the agreement is not good and the measured sensitive times are smaller than predicted by the theory. For example Coxell & Wolfendale (18) report a sensitive time of 12 μsec for a tube of 6-mm diameter filled with pure neon at 600 torr, under a field of 6.3 kV/cm. The corresponding theoretical value derived from equation 17 is instead ~ 25 μsec, i.e. about twice as large. There are several possible explanations for similar discrepancies: first, the possible presence of small amounts of electronegative impurities in gases that have not been subjected to a high accuracy analysis; second, the finite rise time of the applied HV pulse, because of which f may be even considerably smaller than predicted; finally, the long-lasting effects discussed in section 2.5.

The assumption of a uniform electron distribution inside the tube, upon which the derivation of equation 12 has been based, is certainly not correct for times very small in comparison with the average time

$$t_d = r^2/2D_-$$ 18.

required for the primary electrons to diffuse over a length equal to the tube radius r. On the other hand, for $t \ll t_d$ the electrons are still on the trajectory of the primary particle and it is easily shown that in equation 14 f should then be replaced by

$$f(x) = 1 - d_0/2r[1-(x/r)^2]^{1/2}; \quad \{x/r \leq [1-(d_0/2r)^2]^{1/2}\}$$ 19.

This yields average values of f about 30% larger than those obtained from equation 12, with little effect, however, on the expected value of the sensitive time, since f appears in a logarithmic expression in equation 17.

Actually f also depends in a complicated way on the characteristics of the HV pulse (in particular on the pulse rise time) so that it may sometimes be convenient to consider it as an unknown to be determined experimentally.

When $t \ll t_d$ and no impurities are present in the gas, the average efficiency is given with good approximation by the simple formula

$$\eta(t \ll t_d) = [1-(1/2rm_0+d_0/2r)^2]^{1/2}$$ 20.

which yields values of η very close to those obtained from equations 14 and 15 for $t = 0$. Notice that equation 20 gives correctly $\eta = 1$ in the limit in which the mean free path for ionization, $1/m_0$, and the path d_0, for an avalanche-streamer transition to occur, are small compared to the tube diameter.

As we have shown in section 2.7 (Figure 19), adding a small percentage of argon (0.2%) to the neon produces a remarkable change in the efficiency-delay curve and, in particular, a decrease in the minimum field strength required to operate the flash tube efficiently. This effect (63) is explained by the fact that during the discharge neon atoms are excited to a metastable state of 16.55 eV (77). Since the ionization potential of argon is 15.76 eV (77), new electrons may be produced through the reaction $Ne^* + A \rightarrow Ne + A^+ + e^-$, thus increasing the avalanche and therefore the Townsend coefficient α. By the numerical values of α available [see Loeb (69), p. 695] for pure Ne and for Ne with added A in small quantities, corresponding values of field strengths are obtained which do explain also quantitatively the 1956 Pisa experimental data reported in section 2.7.

3.6   Evaluation of the Recovery Time

Immediately after a flash has occurred, large numbers of positive and negative charges are present inside the tube, with initial densities $\rho_+(0) = \rho_-(0) \equiv \rho_0 \approx 10^9$ cm^{-3}. This number is roughly estimated on the basis of the number of electrons present in a streamer, which is roughly $e^{20} \approx 5 \times 10^8$ (see section 3.4), and on the average number of streamers concurring to the flash, which is about 2.5 per cm (78).

In the following treatment we neglect fluctuation effects, and cannot therefore derive a realistic expression of the reignition probability. An evaluation of the flash-tube recovery time is possible, however, as we shall show.

The processes by which the many electrons present in the gas after the discharge can disappear are the same as listed in section 3.3. But of course disappearance by recombination with the positive ions can no longer be neglected.

We assume that when reaching the tube walls all positive ions pick up electrons from the glass and thus become neutralized. We shall neglect the effects of the remnant builtup clearing field discussed in section 2 and, for the time being, assume that no electronegative impurities are present in the gas.

Then the two equations that take into account both diffusion to the tube walls and recombination for the electrons and the positive ions can be written as follows (68):

$$\frac{\partial}{\partial t} \int_0^x \rho_\pm \cdot x \cdot dx = xD_\pm \frac{\partial}{\partial x}\rho_\pm - \delta \int_0^x \rho_\pm \rho_\mp x \cdot dx$$
$$+ 4\pi \rho_\pm \, eK_\pm \int_0^x [\rho_\mp - \rho_\pm] x \cdot dx \qquad \text{21.}$$

where $\rho_+(x,t)$ and $\rho_-(x,t)$ are the charge densities at the time t at a distance x from the tube axis, D_+ and D_- are the diffusion coefficients for positive ions and electrons respectively, δ is the electron-ion recombination coefficient, K_+ is the ion mobility, K_- the electron mobility, and e is the electron charge. The last term in equation 21 describes the effect of the radial electric field arising from ambipolar diffusion (79). The latter occurs as far as the number of positive ions is sufficiently large to produce a radial field which counteracts the diffusion of the electrons to the walls.

On the basis of the above equations Brosco (68) has shown that the time T_1 during which ambipolar diffusion occurs in a flash tube of radius r is given by the approximate expression

$$T_1 = \frac{r^2}{12D_+} \ln \frac{1 + 2dD_+/r}{1 + 12D_+/\delta\rho_0 r^2}$$ 22.

where $d = 4\pi(e^2/KT) \sim 0.7 \times 10^{-4}$ cm at room temperature ($T = 300°C$). He has shown, furthermore, that an additional time

$$T_2 = \frac{r^2}{6D_-} \ln \frac{6Vf}{dr^2 \ln(10/9)}$$

must elapse in order to reduce the reignition probability of a tube of volume V below 10%. For the numerical values $D_- = 400$ cm^2/sec, $D_+ = 0.16$ cm^2/sec (80), $\delta = 10^{-6}$ cm^3/sec (81), one finds always $T_2 \ll T_1$ (for instance $T_1 = 0.1$ sec, $T_2 = 0.5$ msec for $2r = 0.5$ cm, and $V = 20$ cm^3); so that the recovery time t_r is given approximately by T_1, equation 22.

The values of t_r given by equation 22 are in good agreement with those measured for flash tubes filled with pure noble gases.

If electronegative impurities are present and dominate the electron disappearance after the tube discharge, then the recovery time should be

$$t_r = \ln(\rho_0 V \lambda_0 f/bcu)$$ 23.

where b is the attachment probability to the impurity present in the volume V of the gas with a percent concentration c, u is the thermal velocity of the electrons, and λ_0 is the electron mean free path in the impurity at atmospheric pressure.

The values of t_r derived from equation 23 are considerably smaller than those measured. The reason for this discrepancy is not yet understood.

4   LARGE FLASH-TUBE INSTALLATIONS AND SPECIAL HODOSCOPE CHAMBERS

4.1   Introduction

It is beyond the scope of the present review to give even a brief account of the extensive work carried out in cosmic ray physics with the help of the hodoscope chamber technique. Nevertheless the review would be incomplete without mentioning the large-scale apparata that have been in use successfully or are at present in operation.

A brief presentation of some of the most important flash-tube installations and a few examples of applications to high energy physics are given in this section, which ends with a description of some recent developments on nonconventional hodoscope chambers.

4.2   The Hodoscope Chamber as a Shower Detector

Especially because of the individuality of the flash tubes the hodoscope chamber technique is particularly suitable to the detection of multiparticle events. This is why applications to the study of electron-photon showers were envisaged (14) and made (17, 20, 21, 38) since the early stage of its development.

Today a full exploitation of the possibilities offered by the technique in investigations of high energy cosmic ray showers is attained in a number of large-scale apparata installed in various laboratories. We quote, just as an example, the large system of hodoscope chambers in operation at Kiel (24, 25, 82) since 1965, which has produced several results of physical interest (83, 84). Two remarkable examples of recorded events have been shown in Figure 2. The apparatus consists of a 31-m^2 hodoscope chamber system installed at Kiel for investigating the electron component of air shower cores, and two other hodoscopes, of 14-m^2 area each, shielded by thick absorbing material and operated at two different altitudes (sea level and 2860 m above sea level) for measuring the hadron component of air showers. The 31-m^2 unshielded hodoscope alone contains nearly 180,000 small flash tubes of inner volume approximately spherical, of 1-cm diameter. These small flash tubes are filled with neon at 600 torr and have an internal efficiency of about 95%. They are placed with a density of about 5700 tubes/m^2 between two electrodes, one of which is transparent. At the occurrence of an air shower, recorded as a coincidence among the pulses of a system of scintillation counters, an HV pulse triggers the chamber system. The tubes that have been traversed by charged particles then flash and are recorded photographically through the transparent electrode. A similar technique had been developed by S. Fukui (38) in his early investigation on cosmic ray showers.

Other large flash-tube installations for the investigation of extensive air showers (EAS) have been developed at Nottingham (85). Recent measurements have also been performed with the hodoscope chamber technique (86) on the longitudinal development of the nuclear-electromagnetic cascades with energies in excess of 200 GeV.

The response of the hodoscope chamber as a detector of lower energy electrons or photons has been investigated recently (26, 68) by exposing a system of 15 small bigap hodoscope chambers (30 cm × 30 cm) to electrons in the energy range of 40 to 500 MeV. The chambers contained flash tubes 30-cm long, of 1-cm diameter. An ac clearing field of about 70 V ptp, 50 Hz, was applied to reduce random flashes due to background radiation from the machine to a tolerable rate. Thin Pb plates were placed between the chambers to achieve ∼1 radiation length per chamber. The transition curves obtained in this way were compared with those expected on the basis of the elaborated calculations of Crawford & Messel (87) at the same primary electron energy. The following conclusions were reached:

1. Once the chamber system has been calibrated, it allows one to measure the energy E_e of the primary electron with a percent error $\varepsilon < 25\%$ for $E_e > 50$ MeV. The error ε decreases approximately as $1/(E_e)^{1/2}$ with increasing E_e.

2. The energy determination of the primary electron (or photon) can be based merely on the total number of tubes flashed in the system, even at energies as low as 50 MeV.

3. The accuracy in the energy determination of the primary electron or photon is better than that achieved with similar spark chamber systems (88, 89), especially for energies $E_e > 200$ MeV where many-track events are involved. The accuracy is comparable to that reported for the "limited-current spark chambers" (90) but it is

not as good as that obtainable with a "sandwich" of plastic scintillators and thin Pb plates (91), or with a liquid Xe bubble chamber (92–94).

4.3   Application of Hodoscope Chambers to Quark Searches

A systematic search for $e/3$ quarks in extensive cosmic ray air showers (EAS) has been conducted by Ashton et al at Durham (95, 96) since 1966, using in an ingenious way the characteristic flash-tube efficiency-delay curve, $\eta(\theta)$. The basic idea is that the sensitivity to the ionization can be varied at will by varying the time delay θ. The latter can be adjusted, therefore, so that fast fractionally charged particles (Figure 27) are recorded with smaller efficiency than ordinary charged

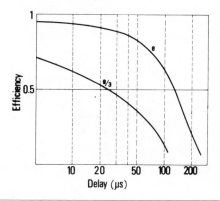

Figure 27   Efficiency vs delay time for same flash tube when traversed by an ordinary minimum ionizing particle (curve e) and a hypothetical $e/3$ quark.

minimum-ionizing particles. Inspection of the track recorded by hodoscope chambers containing many layers of flash tubes can then be used, statistically, to determine whether the recorded particle is fractionally charged. Figure 28 shows the installation. An example of a recorded EAS event is given in Figure 29.

The advantages of this method are that a large detector can be built easily, a high rate of events can be achieved, and comparatively high density shower events can be resolved. The drawbacks are the limited spatial resolution and the necessity of many layers of tubes to obtain reliable estimates of the ionization density.

Another search for quarks in which some thousand 2-m long flash tubes were employed as a track detector has been carried out recently by Crouch, Mori & Smith (97) who have set an upper limit of $2.2 \times 10^{-6}/\text{sec} \cdot \text{m}^2 \cdot \text{ster}$ at 90% cl on the vertical intensity of relativistic $2e/3$ cosmic ray quarks.

4.4   Hodoscope Chambers for Neutrino Physics

A large-scale system of hodoscope chambers containing about 50,000 2-m (1.85-cm outer diameter) flash tubes has been developed by the Case-Wits-Irvine collaboration and employed for the neutrino experimental program undertaken in a 2-mile deep

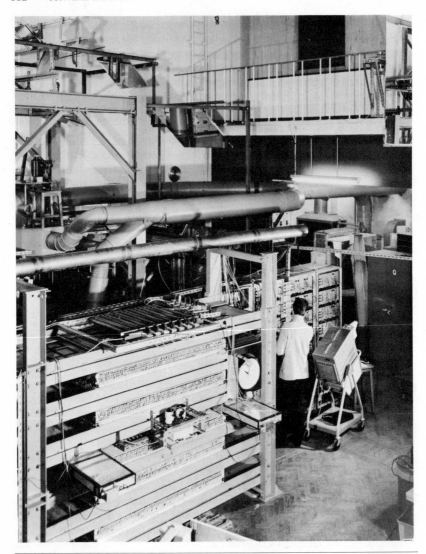

Figure 28   Hodoscope chamber installation at Durham to search for cosmic ray $e/3$ quarks.

laboratory in the East Rand Proprietary, east of Johannesburg. This is one of the largest existing flash-tube installations and is producing a number of results in the field of high energy neutrino physics (98–100). As mentioned in a previous section, part of the flash tubes are used "actively" in the installation to select the events to be recorded.

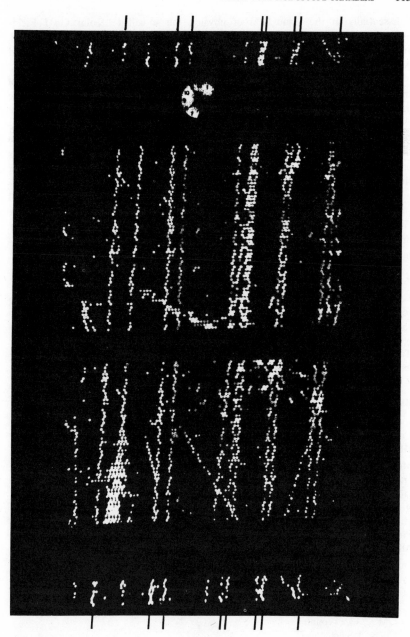

Figure 29   Example of EAS event obtained with the installation of Figure 28.

Essentially in the same field of physics, the so-called "Kolar Gold Fields experiment," which is the result of a collaboration between groups from England, India, and Japan (101), has been in operation for several years in the Kolar Gold Mines, Southern India, at a depth of 2.316 m of rock.

Search for the interaction products of cosmic ray neutrinos, measurements of the intensity of atmospheric cosmic ray muons deep underground, and possible astrophysical interpretations of the results are varied aspects of this large-scale experiment.

It is perhaps worthwhile to point out that the hodoscope chamber technique might have been applied as well to "first generation" experiments on neutrinos with particle accelerators even using the "slow" standard flash tubes. This is because the large detectors required in these experiments to identify the neutrino interactions are separated by thick shieldings from the machine and are, therefore, well protected from the machine background radiation. On the other hand, hodoscope chambers of very large volumes are easily constructed and particularly suitable for distinguishing single muon tracks from many track events associated with electromagnetic showers.

4.5 Flash-Tube Magnetic Spectrographs

Large flash-tube magnetic spectrographs capable of detecting momenta in the TeV/c region have been developed recently and applied, in particular, to obtain cosmic ray muon spectra under various experimental conditions.

Table 1 Figures of merit for cosmic ray spectrographs constructed in the period 1950–1962, as reported in (34)

Exp. Team	Collecting Power (cm^2 sterad)	mdm (GeV/c)	Detectors
Caro et al (1951)	0.69	90	GM[a]
Glaser et al (1950)	0.082	50	GM, CC[b]
Manchester Spectrograph			
Hyams et al (1950)			
Owen & Wilson (1955)	0.93	31	GM
Rodgers (1957)			
Holmes et al (1961)	0.93	356	GM, CC
Pine et al (1959)	7.9	260	GM, CC
Allkofer (1960)	0.39	18	SC[c]
Durham Spectrograph			
Counters	8.0	26.5	GM
Flash tubes	8.0	660	GM, FT[d]

[a] GM = Geiger counter.
[b] CC = Cloud chamber.
[c] SC = Spark counter.
[d] FT = Flash tube.

The characteristics of the first Durham flash-tube spectrograph, completed more than 10 years ago, are compared in Table 1 [taken from (34)] with those of other cosmic ray spectrographs based on different techniques. The figures of merit are the maximum detectable momentum (mdm) and the "collecting power" (measured in $cm^2 \cdot ster$) that are seen to be the largest for the Durham spectrograph.

Maximum detectable momenta in the TeV/c region were attained with the neon-hodoscope magnetic spectrograph of the Nagoya group (102) with which cosmic ray muon spectra and the charge ratio at large zenith angles have been investigated in particular.

A value of mdm as large as 5.85 GeV/c has been reported for MARS (103), the Durham "Magnetic Automated Research Spectrograph" reproduced in Figure 30.

This 300-ton vertical spectrograph uses ordinary neon flash tubes for particle trajectory defining purposes. The data, taken out of the tubes by means of Ayre-Thompson probes, are fed directly into an on-line computer. The ratio of the rms scattering angle to the magnetic deflection is 10%. The integral $\int Bdl$ has a value of nearly 8.1×10^6 gauss $\cdot$ cm. A momentum selector is incorporated in the spectrograph, with a low momentum cutoff of about 200 GeV/c. It is claimed that there is no rejection of events due to burst production in the iron blocks or air showers incident upon the apparatus from this device.

Each of the two sides of the instrument has an acceptance of (408 ± 2) cm^2 for particles of infinite momentum. Traversal of cosmic ray muons through each side occurs at a rate of more than 800 per hour.

The construction of this spectrograph was motivated by the desire to clarify hitherto unsolved questions concerning the momentum spectrum and the charge ratio of high energy cosmic ray muons. The results of preliminary investigations in this field have been reported recently at the Hobart Conference (104).

We mention finally the investigations on the high energy muon flux that do utilize hodoscope chambers but with no magnetic facilities. As an example we quote the work of Higashi et al (105) by which the muon energy spectrum in the vertical direction has been determined indirectly up to energies as large as 4 TeV/c.

4.6   Plastic Chambers

We have already mentioned some types of nonconventional flash tubes such as those operated with longitudinal electric field (56) or the so-called "spark tube" (42). The latter is essentially a neon glass tube with sealed electrodes, conceived with the idea of combining the advantages of automatic electronic data transfer with the retention of visual inspection. Other types of nonconventional flash tubes, most with sealed electrodes, have been tested in various laboratories (26) more or less successfully. Nevertheless they did not find wide application in actual experiments, preference being given, as a rule, to "external" means of extracting the information from electrodeless flash tubes.

Work on the "plastic hodoscope chamber" (27), at present being conducted in Rome, will be reported now in some detail, as this detector appears particularly promising for potential wide-range applications to nuclear, cosmic ray, and particle physics.

Basically, the plastic chamber consists of a large number of small "detecting elements," which are thin tubes of plastic material, of cross section either circular or rectangular, through which a noble gas mixture flows. Tests have been made on a number of plastic materials: various types of polyvinyl, polyethylene, nylon, teflon, moplen (a type of polypropylene of the "Montedison," Italy).

Figure 30   A view of MARS, the Durham "Magnetic Automated Research Spectrograph," characterized by a high collecting power and a maximum detectable momentum in excess of 5.800 GeV/c.

Obviously a large number of solutions can fulfill any geometrical requirement since these materials are produced in a variety of different shapes and are often flexible, or can be bent at a convenient temperature. We shall illustrate here, as examples, two types of plastic chambers quite similar to those being built in Rome for which we tested sample elements.

The cylindrical chamber shown in Figure 31 has been described elsewhere (27). It can be mounted directly on the vacuum chamber of a machine like the Frascati

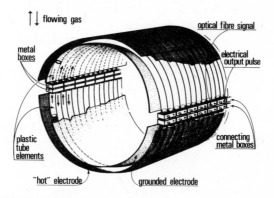

Figure 31 Simplified representation of a cylindrical plastic chamber (not to scale) with possibility of extracting the information by means of either electrical probes or optical fibers. The "hot electrode," which receives the HV pulse sensitizing the tube elements, is made of wire gauze in this example, to minimize Coulomb scattering.

storage ring, Adone, to cover a nearly 4π solid angle around the crossing region of the colliding beams. It consists of two semicylinders, each of which contains a double layer of semicircular black nylon tubes of 1-cm inner diameter. A thin metal plate (a wire gauze in the figure) is placed between the two layers to receive the triggering HV pulses which sensitize the chamber. The semicircular tubes that have been traversed by ionizing particles are then discharged. As in the case of the ordinary glass flash tubes the luminous discharge propagates along the complete length of the tube element but it is interrupted at the ends, where the tube elements are inserted into connecting boxes (metal boxes as illustrated in Figure 31). Figure 31 also shows how the gas flows from one tube element to the next.

Figure 32 gives a schematic view of a modular parallelepiped bigap moplen chamber. The modulus (Figure 32a) is made up of two honeycomb structures, in each of which the detecting elements are thin wall moplen tubes of rectangular cross section (3.5 mm × 5 mm in Figure 32). The two opposite large-area surfaces of each structure can be covered with a conductive varnish, or thin conductive mylar foils, one of which is grounded and the other one representing the "inner hot electrode" to which the HV pulse is applied. In this sample the gas flows "in parallel" through the rectangular tubes of the modulus, following a path determined

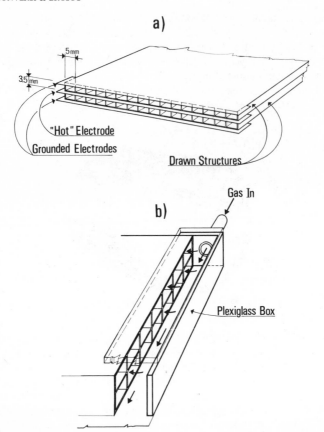

Figure 32   Schematic views of a modular bigap plastic chamber, made of thin wall honeycomb structures of moplen.

by lateral plexiglass boxes (Figure 32*b*) which also allow visual inspection, or photographic recording, if desired.

On the other hand, if the chamber has to be used as an automatic detector, then the same lateral boxes in which the gas flows can become the support of electrical probes or of small light pipes as shown in Figure 31. In fact, as in the case of Figure 31 or of any other type of plastic chamber, the binary information corresponding to the "flashing" or "not flashing" state of the tube elements can be extracted from each element by two different methods: (*a*) by a probe that is in touch with the plasma formed in the flashing tube and thereby receives an electric signal of ~ 100 V that last ~ 3 μsec when grounded through a resistor of ~ 5 kΩ; (*b*) by a low-cost monofilament optical fiber, which receives from the flashing tube an optical signal of intensity sufficient to be recorded by a vidicon-type system.

Elaboration of the information extracted by the first method involves techniques currently employed, e.g. with proportional wire chambers. A somewhat more elegant and well-balanced solution might be achieved, however, with the second method. Such a solution avoids resorting to a large number of integrated circuits and connecting cables. Moreover it fits well large apparata, because the space resolution of the chamber-optical-fiber system is in the mm range and the demagnification of a large sensitive area (several square meters) down to the dimensions of the vidicon cell reduces this figure to one that is just appropriate to the vidicon resolution.

Many elements are needed of course to make a detector of large collecting area, the number of elements obviously increasing with decreasing tube radius or, more generally, the transversal linear dimensions, l, of the detecting elements. Tests have been made, for example, on samples of the above-mentioned honeycomb moplen structures with $l = 3.5$ mm, and with a 30/70 Ne/He mixture at atmospheric pressure flowing through them. The detecting efficiency is in this case nearly 90% over a wide range of field strengths, provided that the time delay in the application of the HV pulse does not exceed a few 100 nsec. The dimension along the direction of the applied field presumably can be reduced down to about 1 mm if the pressure p of the noble gas mixture that flows through the tube elements is large enough. The condition $lp > (1/3)$ cm·atm should insure a high detecting efficiency over a large range of values for l (and $p > 1/3\,l$ atm).

The sensitive time of the plastic chamber depends on the particular type of material and in general also on the rate of gas flow. In most cases it can be adjusted at about 1 μsec for quite reasonable values of the latter rate.

The experimentation on various samples of different plastic materials is being carried out with a ^{106}Ru radioactive source and the countercoincidence system illustrated in Figure 5. Typical conditions of operation are as follows: (a) pulsed electric field raising in ~ 50 nsec to a peak value of 7–9 kV/cm, lasting about 100 nsec, and then decaying in some 10 nsec; (b) commercial gas mixture of Neon (30%) and Helium (70%) at atmospheric pressure flowing at a variable rate in the sensitive elements; (c) maximum repetition rate of the HV triggering pulses: ~ 10 pulses/sec.

The recovery time t_r of the sensitive elements depends again on the type of plastic and the field strength. Values of t_r from 50 msec to a few hundred msec have been measured.

No long-lasting effects due to charge deposit on the tube walls have been observed. In particular the efficiency vs delay curve of the plastic elements does not appear to depend on the flashing repetition rate as in the case of the electrodeless glass flash tubes.

The use of the moplen honeycomb structures which are inexpensive, commercially available on a variety of dimensions, and characterized by a very small amount of low-Z material now appears particularly promising. Coulomb scattering can be reduced to a minimum in this case, also because—as already mentioned—the metallic electrodes used to trigger the chamber can now be replaced by thin layers of conductive varnish on the outer surfaces of the plastic structures. The latter, further-more, are characterized by a minimum amount of "dead space," unlike the chambers based on the use of circular tubes.

Ease of construction and operation, great flexibility, adaptability to any desired geometry, small amount of dead material, and the possibility of automatic track recording while retaining the visual inspection are the main features which make the plastic chamber presently appear a promising instrument for future use in nuclear physics, cosmic ray research, and elementary particle physics.

Literature Cited

1. O'Neill, G. K. 1962. *Sci. Am.* 207:37
2. Davis, D. G. 1963. *Endeavour* 22:19
3. Conversi, M., Gozzini, A. 1955. *Nuovo Cimento* 2:189
4. Ayre, C. A., Thompson, M. G. 1969. *Nucl. Instrum. Methods* 69:106; Ibid 1970. *Acta Phys. Hung. Suppl.* 4:541
5. Conversi, M., Giannoli, G., Spillantini, P. 1972. *Nuovo Cimento Lett.* 3:483
6. Evans, W. M., Baker, J. C. 1971. *Rutherford High Energy Lab. Int. Rep. RHEL/M/H/4*
7. Ceradini, G. et al 1973. Unpublished
8. Chaney, J. E., El Disouki, W., Breare, J. M. 1973. *Nuovo Cimento Lett.* 6:339
9. Brosco, G., Conversi, M., Giovannini, M. 1973. *Nucl. Instrum. Methods.* In press
10. Conversi, M., Focardi, S., Franzinetti, C., Gozzini, A., Murtas, G. P. 1955. *Nuovo Cimento Suppl.* 4:234
11. Barsanti, G., Conversi, M., Focardi, S., Rubbia, C., Torelli, G. 1956. *Proc. CERN Symp. High Energy Accel. Pion Phys. (Geneva)* 2:55
12. Fukui, S., Miyamoto, S. 1957. *Inst. Nucl. Study (Univ. Tokyo) Tech. Rep. INS-TCA-10*
13. Gardener, M., Kisdnasamy, S., Rössle, E., Wolfendale, A. W. 1957. *Proc. Phys. Soc. London B* 70:687
14. Bertanza, L., Bruin, M. W., Rinaldi Fornaca, G., Murtas, G. P. 1957. *Proc. Int. Conf. Mesons Recently Discovered Particles (Padua-Venice)*, 13–43
15. Fukui, S., Miyamoto, S. 1958. *Inst. Nucl. Study (Univ. Tokyo) Tech. Rep. INS-TCA-11* (In Japanese)
16. Ashton, F., Kisdnasamy, S., Wolfendale, A. W. 1958. *Nuovo Cimento* 8:615
17. Ashton, F., Nash, W. F., Wolfendale, A. W. 1959. *Proc. Roy. Soc. A* 253:163
18. Coxell, E., Wolfendale, A. W. 1960. *Proc. Phys. Soc. London* 75:378
19. Ashton, F. et al 1960. *Nature* 185:364
20. Nichimura, H., Miyamoto, S. 1960. *Uchusen-Kenkyn* 5:43 (In Japanese)
21. Ashton, F. et al 1960. *Proc. Moscow Cosmic Ray Conf.* 1:302
21a. Fukui, S. et al 1960. *Progr. Theor. Phys. Suppl.* 16:1; Hasegawa, H. et al 1961. *Rep. INSJ-41*; Oda, M., Tanaka, Y. 1962. *J. Phys. Soc. Japan A* 3, 17:282
22. Coxell, M., Meyer, M. A., Seull, P. S., Wolfendale, A. W. 1961. *Nuovo Cimento Suppl.* 21:7
23. Reines, F. 1967. *Proc. Roy. Soc. A* 391:125
24. Böhm, E., Nagano, M., Van Staa, R., Trümper, J. 1971. *12th Int. Conf. Cosmic Rays, Hobart* 4:1438
25. Samorski, M. 1973. *Nucl. Instrum. Methods* 108:285
26. Conversi, M. 1973. *Riv. Nuovo Cimento.* In press
27. Conversi, M. 1973. *Nature (Phys. Sci.)* 241:160
28. Massey, H. S. W., Burhop, E. H. S. 1952. *Electronic and Ionic Impact Phenomena*, p. 321. London: Oxford Univ.
29. Tyapkin, A. A. 1956. *Prib. Tekh. Eksp.* 3:51
30. Henning, P. G. 1957. *Atomkernenergie* 2:81
31. Focardi, S., Rubbia, C., Torelli, G. 1957. *Nuovo Cimento* 5:275
32. Cranshaw, T. E., De Beer, J. E. 1957. *Nuovo Cimento* 5:1107
33. Fukui, S., Miyamoto, S. 1959. *Nuovo Cimento* 11:113; See also: Ref. 78
34. Brooke, G. et al 1962. *Proc. Phys. Soc. London* 80:674
35. Bull, R. M., Nash, W. F., Rastin, B. C. 1965. *Nuovo Cimento* 40:348
36. Aurela, A. M., Wolfendale, A. W. 1967. *Ann. Acad. Sci. Fenn. A* VI:227
37. Palmer, N. S., Nash, W. F. 1968. *Can. J. Phys.* 46:313
38. Fukui, S. 1961. *J. Phys. Soc. Jap.* 16:604
39. Bagge, E. et al 1965. *Proc. Int. Conf. Cosmic Rays, London* 2:738
40. Earnshaw, J. C. et al 1967. *Proc. Phys. Soc. London* 90:91
41. Achar, V. C. et al 1965. *Phys. Lett.* 18:196
42. Bacon, D. F., Nash, W. F. 1965. *Nucl. Instrum. Methods* 37:43
43. Ayre, C. A. et al 1969. *Proc. 11th Int. Conf. Cosmic Rays, Budapest* 4:547; Ibid 1972. *Nucl. Instrum. Methods* 102:19, 29
44. Bogazzi, A. 1970. Thesis, Univ. Rome. Unpublished

45. Ayre, C. A., Thompson, M. G., Whalley, M. R., Young, E. C. M. 1972. *Nucl. Instrum. Methods* 103:49
46. Crouch, H. F. 1969. *Case Western Univ. Internal Rep. UCI-10 P 19–20.* Unpublished
47. Pickersgill, D. R. 1968. *Univ. Durham Internal Rep.* Unpublished
48. Holroyd, F. W., Breare, J. M. 1971. *Proc. Int. Conf. Cosmic Rays, Hobart* 4:1538
49. Ferguson, H., Rastin, B. C. 1971. *Nucl. Instrum. Methods* 96:405
50. Ashton, F., Breare, F., Holroyd, J. M., Tsuji, K., Wolfendale, A. W. 1971. *Nuovo Cimento Lett.* 2:707
51. Bemporad, C., Calvetti, M., Costantini, F., Guidi, C., Lariccia, P. 1973. *Proc. Int. Conf. Instrum. High Energy Phys., Frascati.* In press
52. Giannoli, G. 1969. Thesis, Univ. Rome. Unpublished
53. Holroyd, F. W., Breare, J. M. 1972. *Nucl. Instrum. Methods* 100:277
54. Pack, J. L., Phelps, A. V. 1961. *Phys. Rev.* 121:798
55. Holroyd, F. W. 1971. PhD thesis, Univ. Durham. Unpublished
56. Breskin, A., Charpak, G. 1973. *Nucl. Instrum. Methods* 108:27
57. Bull, R. M., Coates, D. W., Nash, W. F., Rastin, B. C. 1962. *Nuovo Cimento Suppl.* 23:39
58. Bull, R. M., Coates, D. W., Nash, W. F., Rastin, B. C. 1962. *Nuovo Cimento Suppl.* 23:28
59. Breare, J. M. et al 1973. *Proc. Int. Conf. Instrum. High Energy Phys., Frascati.* In press
60. Coxell, H. 1961. PhD thesis, Univ. Durham. Unpublished
61. Meyer, M. A. 1963. *Nucl. Instrum. Methods* 23:277
62. Hampson, H. F., Rastin, B. C. 1971. *Nucl. Instrum. Methods* 95:337
63. Kruithof, A. A., Penning, F. M. 1937. *Physica* 4:430
63a. After the introduction of the "microwave discharge chamber" (Fukui, S., Hayakawa, S. 1960. *J. Phys. Soc. Japan* 15:532) it was shown that (Fukui, S. et al 1962. *J. Phys. Soc. Japan* 17:250), and explained why (Tsukishima, T. 1963. *J. Phys. Soc. Japan* 18:558) the "optimum" concentration of argon in neon (about 0.2% for d.c. fields) increases with decreasing the duration of the HV pulse.
64. Eckardt, V. 1971. *Nuovo Cimento Lett.* 1:997
65. Jesse, W. P., Sadauskis, J. 1955. *Phys. Rev.* 100:1755
66. Lloyd, J. L. 1960. *Proc. Phys. Soc. London* 75:387
67. Hampson, H. F., Rastin, B. C. 1971. *Nucl. Instrum. Methods* 96:197
68. Brosco, G., 1972. Thesis, Univ. Rome. Unpublished
69. Loeb, L. B. 1955. *Basic Processes of Gaseous Electronics.* Berkeley: Univ. California
70. Raether, H. 1964. *Electron Avalanches and Breakdown in Gases,* p. 125. London: Butterworth
71. Eyeions, D. A., Owen, B. G., Price, B. T., Wilson, J. G. 1955. *Proc. Phys. Soc. London A* 68:793
72. Healy, R. H., Reed, J. W. 1941. *The Behaviour of Slow Electrons in Gases.* Sydney: Amalgamated Wireless Press. 85 pp.
73. Laporte, M. 1933. *Les Phénomènes Elémentaires de la Décharge Electrique dans les Gaz (Gaz Rares).* Paris: Presses Universitaires de France. 44 pp.
74. Jahnke, E., Emde, F. 1945. *Tables of Functions.* New York: Dover. 166 pp.
75. Schneider, F. 1963. *Internal CERN Rep. AR/Int. GS/63-7*
76. Zeleny, J. 1942. *J. Appl. Phys.* 13:444
77. Moore, C. E. 1949. *Atomic Energy Levels.* Washington, DC: Nat. Bur. Stand.
78. Fukui, S., Miyamoto, S. 1961. *J. Phys. Soc. Jap.* 16:2574
79. Loeb, L. B. 1939. *Fundamental Processes of Electrical Discharge in Gases,* p. 586. New York: Wiley
80. Tyndall, A. M. 1938. *The Mobility of Positive Ions in Gases.* London: Cambridge
81. Biondi, M. A., Brown, S. C. 1949. *Phys. Rev.* 76:1697
82. Samorski, M. 1972. *Internal Rep. EAS 11-72, Univ. Kiel*
83. Böhm, E. et al 1967. *Can. J. Phys.* 46:S41
84. Böhm, E., Nagano, M., van Staa, R., Trümper, J. 1971. *Proc. 12th Int. Conf. Cosmic Rays, Hobart* 4:1438
85. Blake, P. R., Ferguson, H., Nash, W. F. 1972. *J. Phys. A* 5:L125; Ibid *J. Phys. Soc. Jap.* 33:1197
86. Gillespie, C. R. et al 1971. *Proc. 12th Int. Conf. Cosmic Rays, Hobart* 4:1282
87. Crawford, D. F., Messel, H. 1970. *Electron-Photon Shower Distribution Curves for Lead, Copper and Air Absorbers.* Oxford: Pergamon
88. Cronin, J. W. et al 1962. *Rev. Sci. Instrum.* 33:946
89. Schneider, K. 1965. PhD thesis, MIT
90. Kajikawa, R. 1963. *J. Phys. Soc. Jap.* 18:1365

91. Barbiellini, G., Borgia, B., Conversi, M., Santonico, R. 1968. *Atti Accad. Naz. Lincei* 44:233
92. Slovinski, B., Strugalski, Z. S., Janowska, B. 1968. *Dubna Rep. JINR, P1:3919*
93. Ogrzevalski, Z. 1969. *Dubna Rep., JINR P1:4959*
94. Strugalski, Z. S. 1972. *Dubna Rep., JINR P13:6191*
95. Ashton, F. et al 1967. *10th Int. Conf. Cosmic Rays, Calgary;* Ibid 1968. *J. Phys. A* 1:569
96. Ashton, F. et al 1971. *J. Phys. A* 4:895
97. Crouch, M. F., Mori, K., Smith, G. M. 1972. *Phys. Rev. D* 5:2667
98. Reines, F. 1969. *Neutrino Meeting, CERN, Rep. 69-28,* p. 103
99. Meyer, B. S. et al 1970. *Phys. Rev. D* 1:2229; Reines, F. et al 1972. *Proc. Neutrino '72 Europhys. Conf. Balatonfhüred* 2:199
100. Reines, F. et al 1971. *Phys. Rev. D* 4:80; Sandie, W. G. et al 1970. *Proc. 6th Interamerical Seminar Cosmic Rays, La Paz;* Chen, H. H. et al 1971. *12th Int. Conf. Cosmic Rays, Hobart* 4:1494-95
101. Menon, M. G. K. et al 1967. *Proc. Roy. Soc. A* 301:137; Krishnaswamy, M. R. et al 1971. *Proc. Roy. Soc. A* 323:489, 511; Krishnaswamy, M. R. et al 1971. *Proc. 12th Int. Conf. Cosmic Rays, Hobart* 4:547; Krishnaswamy, S. et al 1969. *Neutrino Meeting, CERN Rep. 69-28,* p. 81; Osborne, J. L., Wolfendale, A. W., Young, E. C. M. 1972. *Proc. Neutrino '72 Europhys. Conf., Balatonfhüred,* p. 223
102. Kamiga, Y., Kawaguchi, S., Iida, S. 1971. *Proc. 12th Int. Conf. Cosmic Rays, Hobart* 4:1534
103. Ayre, C. A. et al 1972. *Nucl. Instrum. Methods* 102:10, 29
104. Ayre, C. A. et al 1971. *Proc. 12th Int. Conf. Cosmic Rays, Hobart* 4:1309, 1364, 1447, 1458
105. Higashi, S., Kitamura, T., Watase, Y., Oda, M., Tanaka, Y. 1964. *Nuovo Cimento* 32:1

SYMMETRIES IN NUCLEI[1] ˟ 5536

K. T. Hecht

Department of Physics, University of Michigan, Ann Arbor, Michigan

CONTENTS

INTRODUCTION

A good case can be made that nontrivial continuous groups found their first important application in nuclear physics when Wigner (1), motivated by the approximate spin-isospin independence of the nuclear interaction, introduced the group SU(4) to the study of the spectra of light nuclei. Although Racah in his famous series of contributions (2, 3) initially laid down his fundamental group theoretical program for spectroscopy with atomic spectra in mind, his seniority quantum number has been connected with a good symmetry primarily in nuclear spectra through the unitary symplectic groups $Sp(2j+1)$ where the goodness of this symmetry is related to the pairing property of the nuclear interaction. Finally, Elliott (4) showed that collective effects, in particular the rotational character of nuclear spectra, can be related to the symmetry group SU(3). The symplectic symmetries on the one hand and SU(3), together with SU(4), on the other form the bases for two opposite extreme nuclear coupling schemes insofar as a pairing interaction is diagonal in the first and a quadrupole-quadrupole interaction in the second. Although Elliott's original application of the group SU(3) makes use of a generating procedure and an intrinsic many-particle wave function that is closely related to the aligned coupling scheme of Bohr & Mottelson (5), the goodness of SU(3) symmetry breaks down in heavy deformed nuclei due to strong spin orbit

[1] Supported in part by the National Science Foundation.

coupling. A symmetry group that gives a simple basis for rotational spectra of heavy deformed nuclei or that is simply related to the aligned coupling scheme in these nuclei has not yet been discovered. It is also known that the symplectic symmetries are good primarily in configurations of identical nucleons, and no alternate, comparably good symmetry classification for configurations of both neutrons and protons has been found for states with isospin $T \ll T_{\max}$. For these reasons the search for new symmetry groups in nuclei continues. Secondly, although continuous groups have for many years played an important role in our understanding of nuclear spectra, their real status in most cases is not very clear. The question to what extent an approximate symmetry is broken or preserved is often not easy to answer in a complicated nuclear configuration. A systematic study of the goodness of many nuclear symmetries has only been undertaken recently. In addition, attempts are now being made to give quantitative measures of symmetry breaking so that the question of the potential usefulness of a symmetry can be answered before the fact (before a detailed decomposition of many n-particle wave functions into irreducible representations of a given symmetry group has been carried out).

This article will deal with the symmetries associated with the finite-dimensional vector spaces of the nuclear shell model,[2] not with the more fundamental symmetries associated with the conservation of nucleon number, parity, total angular momentum, or isospin, for example, except insofar as these are related to the former. The symmetries to be considered are therefore all related to the unitary groups in g dimensions $U(g)$ and their subgroups, where g is determined by the degeneracies or near degeneracies associated with the single particle levels of the nuclear shell model—for example, $g = (2j+1)$ or $2(2j+1)$ for an isolated single particle level, or $q = \frac{1}{2}(N+1)(N+2)$ for the Nth major oscillator shell. In reviewing recent developments in this field, progress in three main areas will be stressed. The first involves the search for new useful symmetry groups, which will be discussed together with a brief description of some of the well-known symmetries in section 1. In some cases the search for new symmetry groups has led merely to alternate, but nonetheless very useful ways of looking at well-known symmetries. This is true in particular of the so-called quasispin and quasiparticle groups, themselves special cases of noninvariance groups, which have led to new calculational techniques for exploiting the underlying conventional symmetries. The second topic to be discussed involves the question of the goodness of nuclear symmetries (section 2). Since extensive shell model calculations have now been carried out for nuclei in several regions of the periodic table, it is possible in specific examples to measure the content of favored irreducible representations of certain groups for shell model wave functions spanning some relatively large vector spaces. For the most part, however, shell model calculations are carried out in highly truncated vector spaces that presuppose the goodness of some symmetry, often on tenuous ground; and other techniques have to be developed to give a reliable measure of the goodness of symmetries. In several

---

[2] The recent work of Weaver & Biedenharn (69) and Ogura (70) on the application of infinite-dimensional irreducible representations of the group $SL(3, R)$ to the problem of nuclear collective motion should be mentioned as first steps in a brand new application of the theory of noncompact groups to problems of nuclear spectroscopy.

cases the realistic effective interactions used in shell model calculations have now been completely classified as to their irreducible tensor character under the higher symmetry groups, making it possible to compare the strengths of the symmetry-breaking and symmetry-preserving terms (although it is not completely clear how these strengths are best weighted to give the most reliable measure of symmetry breaking). The most promising recent approach is due to J. B. French and collaborators who have applied spectral distribution methods to the study of the goodness of group symmetries in nuclei. The method involves the calculation of low moments of the Hamiltonian averaged over the states of an irreducible representation. The first moments give the centroids of the various irreducible representations, while the second moments are related to the spectral widths of the irreducible representations. The symmetry-breaking contributions to the widths, when compared with the centroid separations, give perhaps the best measure of symmetry breaking. Finally, even in cases where the validity of certain symmetries is highly approximate, recent mathematical progress, largely through the calculation of Wigner and Racah coefficients for the higher symmetry groups, have provided us with new calculational tools that can facilitate nuclear structure calculations. This will be illustrated with some specific examples in section 3.

1 SYMMETRIES IN THE VECTOR SPACES OF THE NUCLEAR SHELL MODEL: DESCRIPTION

1a   Unitary Groups and Space Symmetry

The discussion of nuclear symmetries must begin with the unitary groups in g dimensions $U(g)$. It is convenient to construct the many-particle states in terms of nucleon (Fermion) creation operators a_i^+ where $i = nljm_jm_t$ or $i = nlm_lm_sm_t$ $(= \mu m_sm_t)$ and may encompass one or several single particle levels. An n-particle function is then given by

$$\sum_i C_{i_1i_2\ldots i_n} a_{i_1}^+ a_{i_2}^+ \ldots a_{i_n}^+ |0\rangle$$

where the "vacuum" is usually a closed shell configuration and the Cs are some kind of coupling coefficients. Alternately, if it is advantageous to split the total wave function into a product of space and spin-isospin functions, for example, it may be convenient to construct the n-particle wave functions in terms of Boson creation operators in the separate spaces. Making the association between the space part of the single particle function and the Boson operator $\alpha_{\mu k}^+$

$$\psi_\mu(\mathbf{r}_k) \leftrightarrow \alpha_{\mu k}^+ |0\rangle$$

where $\mu = nlm_l$, and $k = 1,\ldots,n$ is the particle index, the n-particle space function can be written

$$\sum_\mu C_{\mu_1\mu_2\ldots\mu_n} \alpha_{\mu_11}^+ \alpha_{\mu_22}^+ \cdots \alpha_{\mu_nn}^+ |0\rangle$$

To some degree of approximation the Hamiltonian will be invariant if the a_i^+ (or

$\alpha_{\mu k}^+$) are subjected to a unitary transformation in a g-dimensional subspace of the space of single particle states; e.g.

$$a_{\mu m_s m_t}^+ \to a_{\mu m_s m_t}^{+\,\prime} = \sum_v U_{\mu v}\, a_{v m_s m_t}^+$$

for all $m_s m_t$. Alternately

$$\alpha_{\mu k}^+ \to \alpha_{\mu k}^{+\,\prime} = \sum_v U_{\mu v}\, \alpha_{v k}^+$$

for all $k = 1, \ldots, n$. It is usually sufficient to consider infinitesimal transformations $U_{\mu v} = \delta_{\mu v} + i\varepsilon_{\mu v}$, where the matrix ε is infinitesimal and Hermitean. In the space of the n-particle states the infinitesimal unitary transformations can then be generated by the operator

$$1 + i \sum_{\mu v} \varepsilon_{\mu v} A_{\mu v}$$

where

$$A_{\mu v} = \sum_{m_s m_t} a_{\mu m_s m_t}^+ a_{v m_s m_t} \quad \text{or} \quad A_{\mu v} = \sum_k \alpha_{\mu k}^+ \alpha_{v k}$$

in our example are the infinitesimal generators of the group $U(g)$. [Here the annihilation operators are given by $\mathbf{a} = (\mathbf{a}^+)^+$ or $\boldsymbol{\alpha} = (\boldsymbol{\alpha}^+)^+$.] With $\mu \neq v$ the operator $A_{\mu v} = E_{\mu v}$ is the shift operator that shifts a particle out of state v into state μ without changing the quantum numbers $m_s m_t$ (or alternately the particle index k). The operators with $\mu = v$, $A_{\mu\mu} = H_\mu$, form a set of g-commuting operators. They are the number operators whose eigenvalues w_μ count the number of particles in state μ. The full set of eigenvalues $(w_1 w_2 \ldots w_g)$ give the so-called weight of the state. If $U(g)$ is a good symmetry it will of course be important to choose the coupling coefficients (the Cs in the above equations) so that the n-particle functions transform according to specific irreducible representations of $U(g)$ (decomposition of the direct product of n fundamental IRs into IRs of $U(g)$; we shall abbreviate irreducible representation by IR throughout). The IRs of $U(g)$ are characterized by the highest weights, that is the g positive integers $[f_1 f_2 \ldots f_g]$ where $f_1 = $ maximum possible value of $w_1 = $ maximum possible number of particles in state 1, and $f_2 = $ maximum possible value of w_2 with the restriction that state 1 is already occupied by the maximum possible number of particles, etc for $f_3, \ldots f_g$. The numbers f_i with $\sum f_i = n$ and $f_1 \geq f_2 \geq f_3 \geq \ldots$ form a partition of n and are related to the permutation symmetry of the n-particle wave functions (in particular the possibility of symmetrizing simultaneously in one group of f_1 particles, a second group of f_2 particles,..., a gth group of f_g particles, but with an underlying antisymmetry built in that prevents further symmetrization). It is common to picture the IR labels by a g-rowed Young tableau with f_i squares in the ith row. In the most general case a set of $\frac{1}{2}g(g-1)$ additional quantum numbers are needed to fully specify the n-particle states in a given IR. If these quantum numbers can be put into simple correspondence with IR labels of a chain of subgroups of $U(g)$, the state labeling problem is solved in a mathematically tractable manner. In the applications to nuclear spectroscopy it will usually be vital that the chain of subgroups contain the rotation group in three dimensions so that the relevant angular momentum is a good quantum number. The physically relevant subgroup chain is therefore of the form $U(g) \supset \ldots \supset \ldots \supset R(3)$ [with $R(3)$ replaced by $SU(2)$ in the case of $\frac{1}{2}$-integral angular momenta, although we shall sometimes not be careful to make the distinction, particularly when the emphasis is on the Lie algebras of the group]. In the ideal case it would be possible to find a number of subgroups sandwiched in between $U(g)$ and $R(3)$ in this chain, so that the set of Casimir invariants of the group chain furnish a sufficient set of commuting operators. In practice this is possible only in very special cases, usually only when the f_i are severely restricted. A well-known example is that of a shell of identical nucleons ($f_i \leq 1$) with $j \leq 7/2$ for which the quantum numbers n, v (seniority), JM_J associated with the group chain $U(2j+1) \supset Sp(2j+1) \supset SU_J(2)$ give a complete classification scheme. For almost all of the interesting symmetries in nuclear spectroscopy, however, there is a state labeling problem since the physically relevant group chains do not in general furnish us with a sufficient number of commuting operators. From the strictly mathematical point of view there is a canonical subgroup chain for the group $U(g)$ that gives a complete solution to the state labeling problem, the chain studied by Gel'fand (6).

$$U(g) \supset U(g-1) \supset U(g-2) \supset \ldots \supset U(2) \supset U(1)$$

where the subgroup $U(g-1)$ is generated by the subset of the infinitesimal operators, $A_{\mu v}$, with $\mu, v = 1, \ldots, g-1$; that is, the group of unitary transformations that do not involve the particles in the state labeled g. In a nuclear shell the removal from the basis of the particles in the gth quantum state with a particular value for the magnetic quantum number $m(m_l$ or $m_j, \ldots)$ of course automatically precludes the possibility that state vectors labeled by the IRs of $U(g-1), \ldots$ contained in a given IR of $U(g)$ can be eigenstates of the angular momentum operator.

In recent years much progress has been made in the detailed study of the Gel'fand basis. In particular, the work of Biedenharn, Louck, and collaborators (7–9) on the Casimir invariants and the Wigner and Racah coefficients of $U(g)$ would perhaps have reduced most of the problems of nuclear spectroscopy to fairly trivial calculations if only the Gel'fand basis were physically relevant for nuclear spectroscopy. Some authors, e.g. Moshinsky & Devi (10), do make use of the Gel'fand basis directly, but merely at the expense of transferring the difficulties associated with the state labeling problem to the problem of calculating the transformation coefficients from the Gel'fand basis to a physically meaningful one in which angular momentum is a good quantum number. Some properties of the Gel'fand basis, however, have direct applications to nuclear spectroscopy and are discussed in an Appendix. An elegant application of the Gel'fand basis has been made by Moshinsky (11) to show in the most direct way the reciprocity that exists between the unitary group $U(g)$ and the symmetric group $S(n)$ for a system of n particles, through his introduction of special Gel'fand states that give an explicit construction of the Young-Yamanouchi basis of $S(n)$ by unitary group techniques.

1b   Physically Relevant Subgroup Chains: Search for New Symmetries

As laid down by Racah (2, 3), the fundamental program for spectroscopy begins with a search for subgroups $G_1, G_2, \ldots$, which can be imbedded in the group $U(g)$: $U(g) \supset G_1 \supset G_2 \supset \ldots \supset R(3)$ where the lowest member of the chain is generated by the relevant angular momentum operator. The ideal case in which the groups $G_1, G_2, \ldots$ would be sufficient for a complete solution to the state labeling problem is hardly ever met. Yet in many cases the groups G that have been used in nuclear spectroscopy have not only led to useful classification schemes but have also been associated with physically relevant symmetries.

For an isolated single particle level j or a set of nearly degenerate single particle levels $j, j', \ldots$ of both neutrons ($m_t = +\frac{1}{2}$) and protons ($m_t = -\frac{1}{2}$) the chain starts with the group $U(g)$ with $g = 2(2\Omega) = 2[\Sigma(2j+1)]$. This group is generated by the infinitesimal operators $A_{\mu'\mu} = a^+_{j'm'm'_t} a_{jmm_t}$, whose commutator algebra is given by

$$[A_{\mu\nu}, A_{\lambda\sigma}] = \delta_{\nu\lambda} A_{\mu\sigma} - \delta_{\mu\sigma} A_{\lambda\nu}$$

The search for subgroups involves a search for subsets of these operators that are themselves closed under the commutation process. Since the groups generated by the physically relevant angular momenta must be imbedded in the group chain, it will be vital to organize the infinitesimal operators into sets of spherical tensors in both the space-spin and isospin parts of the full space and enumerate the infinitesimal operators in terms of unit tensor operators

$$U^{k\,k_t}_{q\,q_t}(j',j) = \sum_{mm_t} \langle jmkq \,|\, j'm' \rangle \langle \tfrac{1}{2}m_t k_t q_t \,|\, \tfrac{1}{2}m'_t \rangle a^+_{j'm'm'_t} a_{jmm_t}$$

The infinitesimal operators with $k = 0$, $k_t = 1$ form the three components of the isospin operator $\mathbf{T}$ that generate the isospin group $SU_T(2)$. These commute with the infinitesimal operators with $k_t = 0$, which in turn generate the group $U(2\Omega)$ $[= U(2j+1)$ for a single shell]. The direct product group $U(2\Omega) \times SU_T(2)$ is a subgroup of $U(4\Omega)$. The IRs $[f_1 f_2 \ldots f_{2\Omega}]$ of $U(2\Omega)$ and $[\tilde{f}_1 \tilde{f}_2]$ of $SU_T(2)$ must be related by a row $\leftrightarrow$ column interchange for the corresponding Young shapes since the IRs of $U(4\Omega)$ must belong to the totally antisymmetric representation $[111\ldots1] \equiv [1^n]$. The IRs of $U(2\Omega)$ are therefore restricted to those with $f_i \leq 2$; $[f] = [2^a 1^b]$ with $b = 2T$ and $a = \frac{1}{2}n - T$.

In a single j-shell, the infinitesimal operators with $k_t = 0$, $k = $ odd integer commute among themselves and thus generate a subgroup of $U(2j+1)$, the unitary symplectic subgroup $Sp(2j+1)$. The operators with $k_t = 0$ and $k = 1$ are, except for a multiplicative factor of $[j(j+1)]^{\frac{1}{2}}$, the components of the angular momentum vector $\mathbf{J}$ (again with a closed commutator algebra) leading to the subgroup chain $U(2j+1) \supset Sp(2j+1) \supset SU_J(2)$. The infinitesimal operators of $Sp(2j+1)$ commute with the $J = 0$ pair creation operators for nucleon pairs coupled to $J = 0$, $T = 1$

$$[a^+ \times a^+]^{J=0\;T=1}_{0\,M_T} = \sum_{mm_t} \langle jmj-m \,|\, 00 \rangle \langle \tfrac{1}{2}m'_t \tfrac{1}{2}m_t \,|\, TM_T \rangle a^+_{jmm_t} a^+_{j-mm_t}$$

This can be seen from the fact that a commutator of $U^{k0}_{q0}(j,j)$ with $[a^+ \times a^+]^{0\,1}_{0\,M_T}$

would yield a pair creation operator $[a^+ \times a^+]^{JT}_{MM_T}$ with $J = k$ and $T = 1$, but with $k =$ odd integer such an operator is identically zero since it is impossible to couple two nucleons *anti*symmetrically to odd J and $T = 1$. Zero-coupled pairs thus belong to the scalar IR of $Sp(2j+1)$, and the transformation properties of the n-particle functions under $Sp(2j+1)$ are determined by the v nucleons that remain after all zero-coupled pairs have been removed from the n-particle wave functions. The IRs of $Sp(2j+1)$ can thus be characterized by Young shapes with $\sigma_i \leq 2$ squares in the ith row $[\sigma_1 \sigma_2 \ldots \sigma_{j+\frac{1}{2}}] = [2^c 1^d]$ where $\sum \sigma_i = 2c+d = v$ (the seniority number), and $\frac{1}{2}d = t$ [the isospin of the v nucleons free of zero-coupled pairs, the reduced isospin introduced by Flowers (12)]. In the case of configurations of identical nucleons (neutrons only or protons only), the isospin quantum numbers m_t and vector couplings in isospin space can be omitted.

In the case of mixed configurations based on single particle orbits $j_1, j_2, \ldots$ with $\sum (2j+1) = 2\Omega$, two alternate subgroup chains can be used. In the more conventional j-j coupling description the group $U(2\Omega)$ is decomposed into the direct sum of unitary groups $U(2j_1+1) \dotplus U(2j_2+1) \dotplus \ldots$ with IRs characterized by quantum numbers $n_1, T_1; n_2, T_2; \ldots$. The classification scheme is then built on subgroup chains $U(2j_a+1) \supset Sp(2j_a+1) \supset SU_{J_a}(2)$ based on each of the unitary groups, where the IR labels of $Sp(2j_a+1)$ are characterized by the seniority quantum numbers v_a, t_a of each single particle orbit. Alternately the classification scheme can be based on the subgroup chain $U(2\Omega) \supset Sp(2\Omega) \supset SU_J(2)$, where the unitary symplectic group is generated by the infinitesimal operators

$$[2j_a+1]^{-\frac{1}{2}} U^{k\,0}_{q\,0}(j_a, j_b) + (-1)^{k+j_a+j_b+l_a+l_b}[2j_b+1]^{-\frac{1}{2}} U^{k\,0}_{q\,0}(j_b, j_a)$$

while the $Sp(2\Omega)$-invariant zero-coupled pair operators are given by

$$\sum_a [2j_a+1]^{\frac{1}{2}}(-1)^{l_a}[a^+_{j_a} \times a^+_{j_a}]^{J=0\,T=1}_{0\,M_T}$$

The IRs of $Sp(2\Omega)$ are characterized by generalized (or mixed configuration) seniority quantum numbers v, t, where the generalized seniority is related to the IR labels of $Sp(2\Omega)$ by $v = \sigma_1 + \sigma_2 + \ldots \sigma_\Omega$ and counts the number of nucleons free of the $Sp(2\Omega)$-invariant zero-coupled pairs.

In the lighter nuclei where space symmetry or the Wigner supermultiplet quantum numbers can be expected to correspond to a physically relevant symmetry, it is advantageous to base the group chain on the direct product $U(\Omega) \times SU(4)$ imbedded in $U(4\Omega)$; e.g. $U(24) \supset [U(6) \times SU(4)]$ for the $2s$-$1d$ shell. In terms of triple unit tensor operators $U^{k_l k_s k_t}_{q_l q_s q_t}(l_a l_b)$, with $l_a, l_b = 0$ or 2 in the $2s$-$1d$ shell, the group $U(\Omega)$ is then generated by the infinitesimal operators with $k_s = 0, k_t = 0$; while the Wigner supermultiplet group $SU(4)$ is generated by the infinitesimal operators with $k_l = 0, k_s k_t = 10, 01$, and 11, which, except for normalization factors of $3^{\frac{1}{2}}, 3^{\frac{1}{2}}$, and 3 are the operators $\sum \sigma_i, \sum \tau_i$, and $\sum \sigma_i \tau_i$. The space symmetry labels $[f_1 f_2 \ldots f_\Omega]$ [IRs of $U(\Omega)$] are related to the spin-isospin symmetry labels $[\tilde{f}_1 \tilde{f}_2 \tilde{f}_3 \tilde{f}_4]$ (supermultiplet quantum numbers) by the row $\leftrightarrow$ column interchange of the associated Young shapes. The space symmetry group could be further classified by the subgroup $U(2l_a+1) \dotplus U(2l_b+1) \dotplus \ldots$, a direct sum of unitary groups corresponding to the individual l subshells, or the subgroup $R(\Omega)$, the rotation group in Ω-dimensional space, whose IRs give the seniority classification of the space part of the n-particle wave function. Neither of these classifications lead to good nuclear symmetries (4). For the states of a major oscillator shell, l_a, $l_b = N$, $N-2,\ldots, 1$ (or 0), Elliott's SU(3) group, however, furnishes a physically relevant subgroup. This group is understood most simply in terms of Boson operators α_i^+, $i = x$, y, z, or $i = +1, 0, -1$, which create one oscillator quantum of the nuclear harmonic oscillator shell model. A single particle creation operator in the Nth oscillator shell is thus made up of N (symmetrically coupled) oscillator quanta operators. The oscillator quanta preserving operators $[\alpha^+ \times \alpha]_q^k$ of spherical tensor rank $k = 0$, 1, and 2 are respectively the quanta number operator which counts the total number of oscillator quanta, the orbital angular momentum operator $\mathbf{L}$, and the quadrupole moment operator Q_q^k (restricted to the subspace of a major oscillator shell). These generate the group U(3) and, if the quanta number operator is omitted, the group SU(3) whose IRs can be labeled by $[\lambda_1 \lambda_2 \lambda_3]$ or $(\lambda\mu) = (\lambda_1 - \lambda_2, \lambda_2 - \lambda_3)$, respectively (4). In the $2s$-$1d$ shell, therefore, the decomposition of IRs of U(6) into IRs of U(3) or SU(3) involves the question: how does an n-particle function of symmetry described by the Young shape $[f_1 f_2 \ldots f_6]$, which describes the symmetry under the permutation of the n-particle indices, induce the symmetries of the functions of $2n$ oscillator quanta where each particle contributes a symmetrically coupled pair of oscillator quanta, and the symmetries of the oscillator quanta functions are described by Young shapes $[\lambda_1 \lambda_2 \ldots]$ with $\Sigma \lambda_i = 2n$, with at most three rows since each of the $2n$ oscillator quanta has at most three quantum states. Extensive reviews have covered the SU(3) symmetry and its relation to the rotational character of nuclear spectra in the $2s$-$1d$ shell (Elliott, 13; Harvey, 14; and Kramer & Moshinsky, 15). In heavy deformed nuclei, however, the effect of the spin orbit interaction becomes so important that it alters the closing of major oscillator shells and the goodness of SU(3) symmetry breaks down. This failure of SU(3) symmetry in heavier nuclei has kept alive a continuing search for new symmetry groups.

It is remarkable that the single particle levels of the nuclear shell model that are nearly degenerate are very frequently pairs of levels such as $2d_{5/2} 1g_{7/2}$, $2f_{7/2} 1h_{9/2}$, $3s_{1/2} 2d_{3/2}, \ldots$, that is, pairs of levels of the type $l_j(l+2)_{j+1}$. Consider, for example, the doublet $2d_{5/2} 1g_{7/2}$ being filled by protons in nuclei with $Z > 50$ and magic neutron number $N = 82$. The conventional spectroscopic description for the configuration $(d_{5/2} g_{7/2})^n$ is based on either of the two subgroups of U(14): U(6)$\dotplus$U(8), a direct sum of the unitary groups for each single particle level, or Sp(14), the generalized seniority group. Recently a number of workers (Arima, Harvey & Shimizu, 16; Arvieu, 17; Hecht & Adler, 18, 19) have pointed out that an alternate symmetry classification, $U(2\tilde{l}+1) \times SU_{\tilde{s}}(2)$, may have physical relevance where this direct product group is attained by assigning to a single nucleon in the configuration $d_{5/2} g_{7/2}$ a pseudo orbital angular momentum $\tilde{l} = 3$ and a pseudo spin angular momentum $\tilde{s} = \frac{1}{2}$, or by considering the doublet $l_j(l+2)_{j+1}$ as a pseudo spin orbit doublet with $\tilde{l} = l+1$, $\tilde{s} = \frac{1}{2}$. The nucleon creation operators are then expressed by the combinations

$$a_{\tilde{l} \tilde{m}_l \frac{1}{2} \tilde{m}_s}^+ = \sum_{j = \tilde{l} \pm \frac{1}{2}} \langle \tilde{l} \tilde{m}_l \tfrac{1}{2} \tilde{m}_s | jm \rangle a_{jm}^+$$

and one-body operators can be expressed in terms of unit spherical tensor operators in pseudo orbital and pseudo spin space

$$U_{q\,q_s}^{\tilde{k}_l \tilde{k}_s} = \sum \langle \tilde{l}\,\tilde{m}_l\,\tilde{k}_l\,\tilde{q}_l \,|\, \tilde{l}\,\tilde{m}_l' \rangle \langle \tfrac{\tilde{1}}{2}\tilde{m}_s\,\tilde{k}_s\,\tilde{q}_s \,|\, \tfrac{\tilde{1}}{2}\tilde{m}_s' \rangle\, a_{\tilde{l}\,\tilde{m}_l'\,\frac{\tilde{1}}{2}\tilde{m}_s'}^{+}\, a_{\tilde{l}\,\tilde{m}_l\,\frac{\tilde{1}}{2}\tilde{m}_s}$$

where the operators with $\tilde{k}_s = 0$ generate the group $U(2\tilde{l}+1)$. The operators with $\tilde{k}_l = 0, \tilde{k}_s = 1$ (when multiplied with a factor $[3/4]^{\frac{1}{2}}$) are the three components of the pseudo spin operator $\tilde{\mathbf{S}}$, while the operators with $\tilde{k}_l = 1, \tilde{k}_s = 0$ (when multiplied with the factor $[\tilde{l}(\tilde{l}+1)]^{\frac{1}{2}}$) are the three components of the pseudo orbital angular momentum operator $\tilde{\mathbf{L}}$, where $\tilde{\mathbf{L}}+\tilde{\mathbf{S}} = \mathbf{J}$, or the vector coupling of $\tilde{L}$ and $\tilde{S}$ leads to the *real* total angular momentum J. The operators with $\tilde{k}_s = 0, \tilde{k}_l =$ odd integer generate the subgroup $R(2\tilde{l}+1)$ and commute with the pair operators for nucleon pairs coupled to $\tilde{L} = 0, \tilde{S} = 0$. Since this pair operator coincides with the $Sp(2\Omega)$-invariant $J = 0$-coupled pair operator, the seniority quantum number associated with the subgroup chain $U(2\tilde{l}+1) \supset R(2\tilde{l}+1) \supset R(3)$ coincides with the generalized seniority quantum number v associated with the group $Sp(2\Omega)$. The n-particle states of configurations $l_j(l+2)_{j+1}$ of identical nucleons can thus be classified by a pseudo $\tilde{L}\tilde{S}$ coupling scheme $|nv\tilde{L}\tilde{S}JM_J\rangle$. The single particle splitting of the levels $l_j, (l+2)_{j+1}$ can be described by a pseudo spin orbit coupling term in the one-body part of the Hamiltonian. Since the coefficients of this $\tilde{\mathbf{l}}\cdot\tilde{\mathbf{s}}$ term are in general much smaller than the coefficient of the real $\mathbf{l}\cdot\mathbf{s}$ term, the pseudo $\tilde{L}\tilde{S}$ scheme can be expected to be good if the effective two-body interaction is central in pseudo space ($\tilde{S} = 0$). The surface delta interaction, for example, satisfies this property exactly. Since it has proved to be a reasonably good effective interaction in many regions of the periodic table, we might expect the $\tilde{L}\tilde{S}$ scheme to be physically relevant. Shell model calculation for 82-neutron nuclei with even Z (19, 20) show that the lowest 0^+, 2^+, 4^+, 6^+ states are predominantly $\tilde{S} = 0$ (with $v = 0$ for 0^+, $v = 2$ for $J \neq 0$, and with $\tilde{S} = 0$ intensities between 69 and 97%). Moreover, the electric quadrupole operator (a true $L = 2, S = 0$ operator) is predominantly an $\tilde{L} = 2, \tilde{S} = 0$ operator. Calculations with Woods-Saxon $1g_{7/2}2d_{5/2}$ wave functions give $\tilde{S} = 1, \tilde{L} = 1$ and $\tilde{S} = 1$, $\tilde{L} = 3$ components with amplitudes of the order of 0.1 compared with the dominant $\tilde{S} = 0, \tilde{L} = 2$ component (19). (Note that the calculation of these amplitudes involves two angular momentum recoupling transformations, first from real LS to j-j, and then from j-j to $\tilde{L}\tilde{S}$ coupling.) Many of the observed $E2$ retardations can thus be understood in terms of the approximate selection rule $\Delta\tilde{S} = 0$. Analysis of recent inelastic α-scattering experiments in 82-neutron nuclei (21) can also be understood qualitatively in terms of a $\Delta\tilde{S} \simeq 0$ selection rule.

In configurations of both neutrons and protons, it may be advantageous to incorporate the pseudo spin $\tilde{S}$ and isospin T into a pseudo $SU(4)$ scheme. Since the magic number shell closures can all be made up of pseudo spin orbit doublets with $\tilde{l} = \tilde{l}_{max}, \tilde{l}_{max}-2, \ldots, 0$ (or 1), together with a single orbit of opposite parity and $j = \tilde{l}_{max}+5/2$, it is possible to assign to the natural parity part of a major shell configuration pseudo oscillator quantum numbers ($\tilde{N} = \tilde{l}_{max}$) and classify many-particle states by the IRs of a pseudo $SU(3)$ group which describe the symmetry of the $n \times \tilde{N}$ pseudo oscilator quanta associated with the n-particle state. Arima, Harvey & Shimizu (16), motivated by the similarities between the effective interactions and spectra for nuclei in the real $d_{5/2}, d_{3/2}, s_{1/2}$ shell and nuclei beyond ^{56}Ni, filling $1f_{5/2}, 2p_{3/2}, 2p_{1/2}$ shells, have pointed out the potential usefulness of a classification scheme in terms of a pseudo SU(3) group for this $\tilde{N} = 2$ or pseudo $\tilde{d}_{5/2}\tilde{d}_{3/2}\tilde{s}_{1/2}$ shell. The possible goodness of the pseudo SU(3) and SU(4) symmetries has been investigated by Strottman (22) through a detailed analysis of the 3- and 4-particle systems in this pseudo $\tilde{s}\text{-}\tilde{d}$ shell. The 4-particle $T = 0$ nuclei, ^{60}Zn for the pseudo $\tilde{s}\text{-}\tilde{d}$, and ^{20}Ne for the real $s\text{-}d$ shell are compared in Table 1. The table shows the percentages of the major IRs for the lowest $0^+, 2^+, 4^+$ states in these nuclei. It is well known that both SU(4) and SU(3) symmetries are very good for the ground state rotational band in ^{20}Ne. Although pseudo SU(3) symmetry is not nearly as good in ^{60}Zn, larger values of the pseudo SU(3) quantum numbers $(\tilde{\lambda}\tilde{\mu})$ are definitely favored. Strottman concludes that the pseudo SU(4) group is useless for a truncation scheme since states with [211] and lower symmetry may contribute close to 10% of the wave functions of the lowest states. On the other hand, a truncation in terms of pseudo SU(3) symmetry is feasible. A shell model calculation, using a truncation scheme in which only pseudo SU(3) IRs with Casimir invariants greater than that for $(\tilde{\lambda}\tilde{\mu}) = (40)$ have been retained, gives results in very good

Table 1   The 4-particle $T = 0$ system. Comparison of pseudo SU(4)-SU(3) with real SU(4)-SU(3) symmetry[a]

		^{60}Zn			^{20}Ne		
[f]	$(\lambda\mu)$	0_1^+	2_1^+	4_1^+	0_1^+	2_1^+	4_1^+
[4]	(80)	30.9	28.9	40.2	89.5 (78)	90.6 (82)	84.8 (76)
	(42)	0.8	0.7	1.0	1.6 (10)	1.8 (7)	6.1 (11)
	(04)	0.3	0.1		2.5	0.4	0.1
[31]	(61) $L = J, J \pm 1$	41.3	40.9	40.4	5.9 (7)	6.8 (5)	8.2 (11)
	(42) ^3L		2.6	1.3			0.2
	(23)	0.6	1.1	0.5	0.2	0.1	0.1
[22]	(42) $^1L + {}^5L$	15.4	15.5	10.8	0.1	0.1	0.1
	(04) $^1L + {}^5L$	0.9	0.1	0.1			
[211]	(50) ^3L	6.9	7.2	4.2			

[a] For ^{20}Ne [f], $(\lambda\mu)$, L refer to real space-symmetry, SU(3), and orbital angular momentum quantum numbers; for ^{60}Zn these labels refer to the corresponding pseudo quantum numbers. The entries give sums of squares of amplitudes times 100. The numbers for ^{60}Zn are taken from Strottman (22) and were calculated with Kuo-Brown matrix elements and single particle energies for the f-p shell. The numbers for ^{20}Ne are taken from Akiyama, Arima & Sebe (50) (modified central interaction); numbers in parentheses for ^{20}Ne are taken from McGrory (49) and give the three dominant components using Kuo-Brown matrix elements for the 2s-1d shell. Note that SU(3) representations (20) in [4]; (31), (12), and (20) in [31]; (31) and (20) in [22]; (23), (31), (12), and (01) in [211]; and (12) in [1111] have insignificant amplitudes.

agreement with a shell model calculation for the full $f_{5/2}p_{3/2}p_{1/2}$ basis. In addition, Arima, Harvey & Shimizu (16) point out that pseudo SU(3) symmetry may be expected to be better in the upper part of the $\tilde{s}$-$\tilde{d}$ shell, since the separation of the single particle energies ($p_{1/2}$ above the center of gravity of $f_{5/2}p_{3/2}$) is such that the symmetry-breaking terms in the single particle Hamiltonian can be expected to partially cancel the symmetry-breaking terms in the effective two-body interaction in the case of the spectra of a few holes rather than a few particles (just the opposite from the situation in the real s-d shell). Thus, the lowest 0^+, 2^+, 4^+ states in ^{66}Ni carry the leading pseudo SU(3) IR with intensities greater than 90%, while the corresponding intensities in ^{58}Ni may be as low as 76%.

Ratna Raju et al (23) exploit the equivalence between the $(1g_{7/2}2d_{5/2}2d_{3/2}3s_{1/2})$ part of the proton and the $(1h_{9/2}2f_{7/2}2f_{5/2}3p_{3/2}3p_{1/2})$ part of the neutron configurations in rare earth nuclei and pseudo oscillator shells $[(\tilde{f}_{7/2}\tilde{f}_{5/2}\tilde{p}_{3/2}\tilde{p}_{1/2})$ and $(\tilde{g}_{9/2}\tilde{g}_{7/2}\tilde{d}_{5/2}\tilde{d}_{3/2}\tilde{s}_{1/2})]$ to show that the low-lying natural parity rotational bands in strongly deformed nuclei can be described approximately by many-particle states coupled to the leading pseudo SU(3) IRs (maximum possible value of $2\tilde{\lambda}+\tilde{\mu}$) of these configurations. In particular, the model gives rotational spectra with the correct ordering of the K_J bands and remarkably good predictions for ground state magnetic moments.

Searches for even more exotic symmetry groups have been undertaken in the above spirit. No very good symmetry classifications are known for nuclei beyond ^{40}Ca with $T \ll T_{\text{max}}$. Although the $f_{5/2}$-$f_{7/2}$ spin orbit splitting of 6.4 MeV is not particularly large compared with the $d_{3/2}$-$d_{5/2}$ splitting of 5.08 MeV, it is known that the real SU(3) symmetry breaks down completely for nuclei near the beginning of the real f-p shell. The calculations of Bhatt & McGrory (24) for ^{44}Ti, the f-p shell analogue of ^{20}Ne, show that ^{44}Ti states contain significant components of many SU(3) IRs [the leading SU(3) IR $(\lambda\mu) = (12, 0)$ constitutes only 26% of the ground state wave function of ^{44}Ti, compared with an 80–90% content of $(\lambda\mu) = (80)$ in ^{20}Ne]. In the SU(3) scheme we can think of the single particle state of the f-p shell with $l = 1$ and 3 as built from three totally symmetrically coupled angular momentum-1 "objects" (the three oscillator quanta associated with the single particle state). It is also possible to think of the single particle state with $l = 1$ and 3 as built from two antisymmetrically coupled "spin-2 objects" (or "quarks"), leading to a classification scheme in terms of the subgroups SU(5) $\supset$ R(5) $\supset$ R(3) of U(10); or alternately as built from two symmetrically coupled "spin-3/2 objects," leading to a classification scheme in terms of the subgroups SU(4) $\supset$ Sp(4) $\supset$ R(3); where the two schemes are related since R(5) and Sp(4) have Lie algebras of the same structure. The potential usefulness of such classification schemes has been investigated in (25). Although there is some indication that the spectra of f-p shell nuclei are dominated by the simpler IRs of R(5) or Sp(4), these symmetries are far from good, and their usefulness is questionable. However, it might be hoped that a search for new symmetries in this spirit may prove fruitful for other mixed configurations of interest in nuclear spectra. In this connection an attempt by Vincent (26) to classify nuclei in the $d_{5/2}d_{3/2}s_{1/2}$ shell in terms of pseudo angular momenta of 1 and 3/2 should be mentioned.

1c   Noninvariance Groups, Quasispin, and Quasiparticle Groups

Some disadvantages are inherent in the conventional (Racah) spectroscopic classification of configurations such as j^n (fixed n) by the group chain $U(2\Omega) \supset \ldots \supset R(3)$. The highest symmetry group in the chain has IRs characterized by trivial quantum numbers such as n and T. Although these are strongly conserved quantities, they are related to mathematically simple symmetries that do not require the full complexities of the groups $U(2\Omega)$. [The group $U(2\Omega)$ has IRs characterized by 2Ω labels. The IRs that occur in nature, however, are characterized by one or two quantum numbers: n, or n, T. In addition, the highest symmetry group labeling is dependent on the j values, which is not true of the quantum numbers n and T.]

The symmetry groups in the conventional chains are generated by infinitesimal operators that conserve the particle number. An alternate possibility involves symmetry groups that include among their generators operators that do not conserve nucleon number. The simplest and perhaps most useful of these are the so-called quasispin groups, first introduced into nuclear physics by Kerman (27) and Helmers (28). For a single shell j of identical nucleons, the three operators

$$\mathscr{S}_+ = \sum_{m>0} (-1)^{j-m} a_{jm}^+ a_{j-m}^+, \quad \mathscr{S}_- = \sum_{m>0} (-1)^{j-m} a_{j-m} a_{jm},$$
$$\mathscr{S}_0 = \tfrac{1}{2}(N_{\text{op.}} - \Omega)$$

satisfy the usual angular momentum commutation relations and thus generate a group $SU_{\mathscr{S}}(2)$. The operator $\mathscr{S}_+$ that creates a pair of identical nucleons coupled to $J = 0$ commutes with the unit tensor operators $U_q^k(j,j)$ with *odd* k that generate the group $Sp(2j+1)$. The same holds for the operators $\mathscr{S}_-$, $\mathscr{S}_0$. The generators of $SU_{\mathscr{S}}(2)$ and $Sp(2j+1)$ thus commute with each other, making possible a classification under the direct product group $SU_{\mathscr{S}}(2) \times Sp(2j+1)$. In addition the family of all 2- and 1-particle creation and annihilation operators together with all number-preserving operators of rank 1 are closed under the commutation process and generate a group. In particular (for identical nucleons, leaving off subscripts j), the family of operators

$$a_{m_1}^+ a_{m_2}^+; \quad a_{m_2} a_{m_1}; \quad \tfrac{1}{2}(a_{m_1}^+ a_{m_2} - a_{m_2} a_{m_1}^+); \quad a_m^+; \quad a_m$$

a family of $(2j+1)(4j+3)$ operators in all, generate the rotation group in $(4j+3)$-dimensional space, $R(4j+3)$. If the operators a_m^+, a_m are omitted, the family of operators is still closed under the commutation process and generate the subgroup $R(4j+2)$. Judd (29) has shown that all states for the configuration j^n with $n = 0, \ldots, (2j+1)$ belong to the single IR $(\tfrac{1}{2}\tfrac{1}{2} \ldots \tfrac{1}{2}) \equiv (\tfrac{1}{2}^{2j+1})$ of $R(4j+3)$, which decomposes into the two IRs $(\tfrac{1}{2}\tfrac{1}{2} \ldots \tfrac{1}{2}(-1)^n)$ of $R(4j+2)$, one of which is for all states with n even, the other for all states with n odd. It would thus appear that the classification of the states j^n under the groups $R(4j+3) \supset R(4j+2)$ is almost spectroscopically empty. However, the fact that IRs of the group $SU_{\mathscr{S}}(2) \times Sp(2j+1)$ are imbedded in a single IR of a higher group has powerful consequences. For example, it is the basis for the fact (Helmers, 28) that IRs of $SU_{\mathscr{S}}(2)$ and $Sp(2j+1)$ are labeled by the same quantum number, the seniority v. The IRs of $SU_{\mathscr{S}}(2)$ can be labeled by the quasispin quantum number $\mathscr{S}$ [related to the eigenvalue $\mathcal{S}(\mathcal{S}+1)$ of $\mathcal{S}^2$], given by the highest weight (maximum possible value of $\mathcal{S}_0$), which for a state with seniority v is given by $\mathcal{S} = \frac{1}{2}(\Omega - v)$. States are thus labeled by $|\mathcal{S}\mathcal{M}_{\mathcal{S}}; \beta JM_J\rangle = |\frac{1}{2}(\Omega-v)\frac{1}{2}(n-\Omega); \beta JM_J\rangle$ where βJM_J can be regarded as subgroup labels for $\mathrm{Sp}(2j+1)$ [β distinguishes states in the case of multiple occurrences of J in an IR of $\mathrm{Sp}(2j+1)$]. Note that n, through the quantum number $\mathcal{M}_{\mathcal{S}}$, now plays the role of a trivial subgroup label in a simple group. The dependence on the quantum number n can thus be factored out of a full nuclear matrix element by a quasispin Wigner coefficient and a simple application of the Wigner Eckart theorem in quasispin space. For configurations of identical nucleons, the most detailed discussion of the n-dependence of nuclear matrix elements has been given from this point of view by Lawson and Macfarlane (30, 31). It is necessary to classify operators according to their spherical tensor character not only in ordinary space but in quasispin space as well. The single particle creation and annihilation operators for a fixed m form the basis for the 2-dimensional IR $\mathcal{S} = \frac{1}{2}$ of the quasispin group. In particular

$$a_{jm}^+ = T_{m\,\mathcal{M}_{\mathcal{S}}=+\frac{1}{2}}^{j\,\mathcal{S}=\frac{1}{2}}; \quad (-1)^{j-m}a_{j,-m} = T_{m\,\mathcal{M}_{\mathcal{S}}=-\frac{1}{2}}^{j\,\mathcal{S}=\frac{1}{2}}$$

where the first indices give the ordinary spherical tensor rank while the second indices give the rank in quasispin space. The n-dependence of one-particle fractional parentage coefficients can thus be obtained from the reduction formula

$$\langle j^n v'\beta'J'M'|a_{jm}^+|j^{n-1}v\beta JM\rangle$$
$$= \frac{C(n)}{C(\bar{v})}\langle JM\,jm|J'M'\rangle\langle j^{\bar{v}}v'\beta'J'\|a_j^+\|j^{\bar{v}-1}v\beta J\rangle$$
$$v' = v\pm 1,\ \bar{v} = \max(v', v+1)$$

where $C(n) = \langle\frac{1}{2}(\Omega-v)\frac{1}{2}(n-1-\Omega)\frac{1}{2}\,\frac{1}{2}|\frac{1}{2}(n-\Omega)\rangle$. The irreducible tensor rank of more complicated operators built from $\mathbf{a}^+$, $\mathbf{a}$ is obtained by straightforward vector coupling techniques. For example, one-body unit tensor operators $U_q^k(j,j)$ with odd k are quasispin scalars ($\mathcal{S} = 0$ operators); the basis for the well-known fact that matrix elements of such operators are n-independent in the seniority scheme. One-body operators $U_q^k(j,j)$ with even k, together with the pair creation and annihilation operators A_{kq}^+, $(-1)^q A_{k-q}$, form the three components of a quasispin vector ($\mathcal{S} = 1$ operator; with $\mathcal{S}_0 = 0, +1, -1$). In a single j-shell the two-body interaction can be written

$$V = \sum_{JM} V_J A_{JM}^+ A_{JM} = \frac{1}{2}\sum_{JM} V_J\{(A_{JM}^+ A_{JM} + A_{JM} A_{JM}^+)$$
$$+ (A_{JM}^+ A_{JM} - A_{JM} A_{JM}^+)\}$$

where the last term is an $\mathcal{S} = 1$ operator (antisymmetric coupling of two identical $\mathcal{S} = 1$ operators) but is reduced to the simple operator $\mathcal{S}_0$ via the commutation properties of $\mathbf{A}^+$ and $\mathbf{A}$; while the first term (built from two symmetrically coupled identical $\mathcal{S} = 1$ operators) gives a combination of quasispin tensor operators of rank $\mathcal{S} = 0$ and $\mathcal{S} = 2$. In particular

$$V = T_0^{\mathcal{S}=0} + T_0^{\mathcal{S}=2} + \frac{1}{\Omega}[\sum_J (2J+1)V_J]\frac{1}{2}(n-\Omega)$$

The seniority breaking part of the interaction is thus given solely by the $\mathscr{S} = 2$ tensor. This gives a simple explanation (30) of the goodness of the seniority quantum number v in the configuration $(7/2)^4$ of identical nucleons, for example. Of the two states with $J = 2$ (and $J = 4$) one has $v = 2$, hence $\mathscr{S} = 1$; the other $v = 4$, hence $\mathscr{S} = 0$; but an $\mathscr{S} = 2$-operator cannot connect states with $\mathscr{S} = 0$ and $\mathscr{S} = 1$.

In mixed configurations with $j = j_a, j_b, \ldots$, the total quasispin operator $\mathscr{S}$ can be constructed from the single-shell quasispin operators by $\mathscr{S} = \Sigma(-1)^{l_a}\mathscr{S}_a$ (Arvieu, 32; Ichimura and Arima 33, 34). Its components commute with the generators of $Sp(2\Omega)$, and its IRs are labeled by the generalized seniority quantum number v through $\mathscr{S} = \frac{1}{2}(\Omega - v)$. (Note, however, that $v \neq \Sigma v_a$; instead, $\mathscr{S}$ is related to the quantum numbers $\mathscr{S}_a = \frac{1}{2}(\Omega_a - v_a)$ by a vector coupling in quasispin space.) Note also that the $\mathscr{S} = 1$ operator in the general two-body interaction is now not related simply to the number operator, and the symmetry-breaking part of the two-body interaction includes $\mathscr{S} = 1$ and $\mathscr{S} = 2$ operators.

In configurations of both neutrons and protons the quasispin group is more complicated since the $J = 0$-coupled pair operator is a $T = 1$ operator with three components, $M_T = \pm 1, 0$. The quasispin group is generated by the ten operators

$$\mathbf{A}^+, \mathbf{A}, \mathbf{T}, (\tfrac{1}{2}N_{\text{op.}} - \Omega)$$

where $\mathbf{A}^+$, $\mathbf{A} = (\mathbf{A}^+)^+$, and $\mathbf{T}$ are vector operators in isospin space,

$$(A^+_{M_T} = \tfrac{1}{2}\Sigma_{m,m_t}(-1)^{j-m}\langle \tfrac{1}{2}m_t \tfrac{1}{2}m_{t'} \,|\, 1 M_T \rangle a^+_{jmm_t} a^+_{j-mm'_t})$$

These ten operators generate a rotation group in five dimensions, $R(5)$. This 5-dimensional quasispin group has been studied by a number of workers (Helmers, 28; Flowers & Szpikowski, 35; Ichimura, 36; Parikh, 37; Ginocchio, 38; Goshen & Lipkin, 39; Hecht, 40; and Hemenger, 41). Since the infinitesimal generators of $R(5)$ commute with the infinitesimal generators of $Sp(2j+1)$, the full classification scheme can again be based on a direct product group: $R(5) \times Sp(2j+1)$ where the IRs of $R(5)$ and $Sp(2j+1)$ are again labeled by the same quantum numbers; now v *and* t, since the quasispin group $R(5)$ is a group of rank 2 with IRs characterized by $(\omega_1\omega_2)$, the highest weights, which are given by the largest possible values of the two operators $H_1 = (\tfrac{1}{2}N_{\text{op.}} - \Omega)$ and $H_2 = T_0$: $(\omega_1, \omega_2) = (\Omega - \tfrac{1}{2}v, t)$. Since $R(5)$ is a 10-parameter group of rank 2, a set of $\frac{1}{2}(10 - 2) = 4$ subgroup labels are needed for a complete classification scheme. The mathematically canonical subgroup chain $R(5) \supset R(4) = [R_{\mathscr{S}^n}(3) \times R_{\mathscr{S}^p}(3)]$, corresponding to a decomposition into separate neutron and proton quasispin groups, furnishes such a set through the quantum number $\mathscr{S}^n\mathscr{M}_{\mathscr{S}^n}\mathscr{S}^p\mathscr{M}_{\mathscr{S}^p}$. However, $\mathscr{S}^n$ and $\mathscr{S}^p$ do not correspond to *good* quantum numbers. If the $R(5)$ quasispin group is to be used to extract the n-, T-dependent factors of nuclear matrix elements, the labeling scheme must include the physically relevant quantum numbers n, T, and M_T. Unfortunately, no simple fourth operator exists that commutes with the operators $N_{\text{op.}}$, $\mathbf{T}^2$, T_0, and the Casimir invariants of $R(5)$. Such a labeling problem (inner multiplicity problem) occurs in almost all of the physically relevant group chains of nuclear spectroscopy. For many particularly simple IRs, however, the physically relevant subgroup labels are often sufficient for a complete labeling of the states of the IR.

This is true for the IRs $(\omega_1 0)$, $(\omega_1 \frac{1}{2})$, and (tt) of R(5) in which there are no multiple occurrences of T for any n. (Note that these include the IRs with $v = 0$ and $v = 1$.) In more complicated IRs with multiple occurrences of T, it is always possible in any specific numerical case (ω_1, ω_2, n, T given by specific numbers) to construct the necessary number of states through some arbitrary choice of states and a numerical orthogonalization process. The states are then simply tagged with a fourth label α, with $\alpha = 1, 2, \dots, d$ in the case of a d-fold occurrence of n, T (Ginocchio, 38). This is the usual resolution of the inner multiplicity problem for physically relevant subgroup chains with missing quantum numbers. Often the physics of the problem gives a natural choice of the missing label, even though this label may not be related to the eigenvalues of some operator. States of a given v, t, n, T, for example, are built naturally from one set of v nucleons coupled to isospin t, where these v nucleons are entirely free of $J = 0$-coupled pairs, and another set of nucleons made up of $p = \frac{1}{2}(n - v)$ pairs of nucleons each coupled to $J = 0$, $T = 1$. These p pairs are coupled to isospin T_p, where $T_p = p$, $p-2$, $p-4, \dots$, and the total isospin T is the result of the vector coupling $\mathbf{T} = \mathbf{T}_p + \mathbf{t}$ [with $T = \omega_1 - m$ ($m = 0, 1, 2, \dots$), the number of occurrences of a given T is given by $\min(m+1, \omega_1 - t + 1)$]. Although the label T_p has physical significance, it is not related in any way to the eigenvalue of a Hermitean operator; hence two states with different values of T_p are in general not orthogonal to each other. Although there may be some merit in using a nonorthogonal basis (e.g. see Racah, 42), it is difficult to apply the formalism of an irreducible tensor calculus in such a basis. Alternate labeling schemes for the R(5) quasispin group have been suggested on physical grounds, where the fourth label is used to count the number of alpha-like groupings of four nucleons coupled to $J = 0$, $T = 0$ (Flowers & Szpikowski, 35; Parikh, 37) or the number of nucleons left after all such four-particle clusters have been removed from the n-particle state (Goshen & Lipkin, 39). All such schemes however would lead to R(5) Wigner coefficients of very complicated algebraic structure and suffer from the additional disadvantage that they have no definite symmetry properties under particle-hole conjugation. If the R(5) quasispin formalism is to be used as a tool for extracting the n-, T-dependence of nuclear matrix elements, it becomes important to find a useful orthogonal basis and a solution to the problem of the fourth operator. As in the analagous problem of the missing operator for the chain $SU(3) \supset R_L(3)$ (Racah, 42), the missing operator can be expressed in terms of two operators of degree 3 and 4, respectively, in the infinitesimal operators which generate the group. In the case of R(5), the two operators are (using vector notation for isospin space)

$$\mathbf{T} \cdot [\mathbf{A}^+ \times \mathbf{A}] \quad \text{and} \quad (\mathbf{A}^+ \cdot \mathbf{A}^+)(\mathbf{A} \cdot \mathbf{A})$$

Note that the operators are number-conserving and scalars in isospin space and hence commute with $N_{op.}$, $\mathbf{T}^2$, and T_0. As in the case of SU(3) (Racah, 42), it has not been possible to find a general function of these two operators that has simple rational eigenvalues. However, for specific classes of IRs Hemenger (41) has constructed combinations of the two operators with simple eigenvalues (labeled below by α), which have led to R(5)-Wigner coefficients of relatively simple algebraic structure, and with definite symmetry properties under particle-hole conjugation. Hemenger's solution is restricted to R(5) IRs with $(\omega_1, \omega_2) = (\Omega - \frac{1}{2}v, t) = (\omega_1, 1)$, $(\omega_1, 3/2)$, and $(t+1, t)$. However, together with the IRs $(\omega_1, 0)$, $(\omega_1, \frac{1}{2})$, and (t, t) (for which no fourth operator is needed), these cover all possible R(5) IRs for $\Omega \leq 5$, that is all R(5) IRs for simple shells j with $j \leq 9/2$. The R(5) quasispin formalism has therefore been refined into a usable tool for extracting the n-, T-dependence of nuclear matrix elements in the seniority scheme.

In configurations of both protons and neutrons the seniority quantum numbers are in general far from being good quantum numbers (for the $1f_{7/2}$ shell, see Ginocchio, 43). The seniority classification scheme $|(v, t); n\alpha T M_T; \beta J M_J\rangle$ nevertheless furnishes a very useful basis for calculations, particularly since the quantum numbers n, T, M_T have been made to play the role of simple subgroup labels through the introduction of the quasispin group, and a classification of the states by the direct product group R(5) × Sp(2j+1) has split the classification problem into two separate problems, one involving the structure of R(5) and the n-, T-dependence which is completely independent of j, the second involving the subgroup labels of Sp(2j+1) which carry the explicit j-dependent properties.

The symplectic group, G_1, and the quasispin groups, G_2, are examples of "complementary groups." This concept was introduced by Moshinsky & Quesne (44) for the direct product $G_1 \times G_2$, which is a subgroup of a higher group, such that there is a $1:1$ correspondence between all IRs of G_1 and G_2 contained in a single IR of the higher group. For the symplectic and quasispin groups, the higher group is the noninvariance group R(4j+2) for configurations of identical nucleons [or R(8j+4) for configurations of both protons and neutrons]. The IRs of the higher group are $(\frac{1}{2}\frac{1}{2}\ldots \pm\frac{1}{2})$ for even or odd numbers of nucleons, respectively. The generators of G_2 are formed by all the invariant operators with respect to G_1 that can be constructed from among the family of generators of the noninvariance group. Since the noninvarance groups R(4j+2) and R(8j+4) are themselves subgroups of higher noninvariance groups, additional possibilities of complementary groups suggest themselves. Moshinsky and Quesne have studied the case of identical nucleons in a single shell j in detail: In this case R(4j+2) is a subgroup not only of R(4j+3) but also of the unitary group in 2^{2j+1} dimensions, which is generated by the $(2^{2j+1})^2$ operators

$$a_{m_1}^+ a_{m_2}^+ \ldots a_{m_n}^+ a_{m_{n'}} \ldots a_{m_2'} a_{m_1'} \quad \text{with} \quad n, n' = 0, 1, \ldots, 2j+1$$

The family of *all* states

$$A_{n,\mu}^+ |0\rangle = a_{m_1}^+ a_{m_2}^+ \ldots a_{m_n}^+ |0\rangle$$

with particle numbers $n = 0, 1, \ldots, 2j+1$, span the single 2^{2j+1}-dimensional IR [1] of the noninvariance group $U(2^{2j+1})$; and the states of this IR can be further characterized by the IR labels of the subgroup chain

$$U(2^{2j+1}) \supset R(4j+3) \supset R(4j+2) \supset U(2j+1) \supset Sp(2j+1) \supset R(3)$$

Each of the subgroups in this chain can be considered as a group G_1 for a complementary pair $G_1 \times G_2$. The group R(3) whose IRs are labeled by J is of particular interest. Its complementary partner is generated by all possible R(3)-invariant operators that can be formed from the generators of $U(2^{2j+1})$. Moshinsky and Quesne show that these are the operators

$$C_{n\,v\,\beta\,J}^{n'v'\beta'J} = \sum_M C_{n\,v\,\beta\,JM}^{n'v'\beta'JM} \quad \text{with} \quad J = 0, j, \ldots$$

where

$$C_{n\,v\,\beta\,JM}^{n'v'\beta'J'M'} = P_{nv\beta JM}^{+} \prod_{r=1}^{2j+1-n'} \left(\frac{N_{\text{op.}} - r}{-r}\right) P^{n'v'\beta'J'M'}$$

and where $N_{\text{op.}} = \Sigma_m a_m^{+} a_m$, and $P_{nv\beta JM}^{+}$ is a linear combination of the operators $A_{n\mu}^{+}$ above, which create n-particle states of definite $v\beta JM$. Finally, $\mathbf{P} = (\mathbf{P}^{+})^{+}$. The operators $C_{n\,v\,\beta\,J}^{n'v'\beta'J}$ for fixed J form a set of $(d_J)^2$ operators which satisfy the commutation relations of a unitary group, where d_J equals the total number of occurrences of J, counting all possible n of the shell j. The group G_2 complementary to R(3) is thus the direct sum of the unitary groups

$$G_2 = U(d_{J_a}) \dotplus U(d_{J_b}) \dotplus \ldots$$

For example, for the $j = 5/2$ shell, with a 4-, 3-, 2-, 2-, 1-, and 1-fold occurrence of the J-values 0, 5/2, 2, 4, 3/2, and 9/2, the group G_2 is

$$G_2 = U(4) \dotplus U(3) \dotplus U(2) \dotplus U'(2) \dotplus U(1) \dotplus U'(1)$$

States belonging to a definite IR J of R(3) belong to the IR $[1]$ the group $U(d_J)$ with the same J and to the IR $[0]$ of all the other unitary groups in the direct sum.

So far the applications of these group classifications have been restricted to well-known results or results obtainable by more conventional techniques. For example, Quesne (45) has used properties of the generators of $U(d_J)$ above to determine the strengths of the effective p-body forces $(p = 0, \ldots, 2j+1)$ in the $1f_{7/2}$ shell. However, the fact that the quantum number J has been lifted from its conventional role as the IR label for the lowest subgroup of a group chain to the labeling of the highest group in the direct product $G_1 \times G_2$ may have important consequences and may perhaps lead to a canonical resolution of the inner multiplicity problem associated with multiple occurrences of a given J in the configuration j^n.

The analysis of the groups $U(2^{2\Omega})$ and its subgroups [in particular $Sp(2\Omega)$ and the quasispin group $SU_{\mathscr{S}}(2)$] is unchanged if the particle creation and annihilation operators $\mathbf{a}^{+}, \mathbf{a}$ are replaced by quasiparticle creation and annihilation operators $\mathbf{b}^{+}, \mathbf{b}$, where the $\mathbf{b}$s and $\mathbf{a}$s are related by the Bogoliubov Valatin transformation

$$b_{jm}^{+} = U a_{jm}^{+} - V(-1)^{j-m} a_{j-m}$$

provided that the Us and Vs are independent of j in the case of mixed configurations with $2\Omega = \Sigma(2j+1)$, a property satisfied exactly if the different single particle levels j are degenerate. In this case $\mathscr{S}$ or v serves as a good quantum number for either the particle or the quasiparticle basis. However, in this case the only nonspurious quasiparticle states (Arvieu, 32) are states with $v = n_{qp}$ ($n_{qp} = $ quasiparticle number). By setting $U = \cos\frac{1}{2}\theta$, a state of a definite quasiparticle number can be related to states of good particle number (or inversely) by a simple rotation in quasispin space (Macfarlane, 31),

$$|\mathscr{S}\mathscr{M}'_\mathscr{S} = \tfrac{1}{2}(n_{qp}-\Omega); \beta JM\rangle$$
$$= \sum_{\mathscr{M}_\mathscr{S}} |\mathscr{S}\mathscr{M}_\mathscr{S} = \tfrac{1}{2}(n-\Omega); \beta JM\rangle D^{\mathscr{S}}_{\mathscr{M}_\mathscr{S}\mathscr{M}'_\mathscr{S}}(0,\theta,0)$$

with $\mathscr{S} = \tfrac{1}{2}(\Omega-v) = \tfrac{1}{2}(\Omega-n_{qp})$ for nonspurious states.

A quite different quasiparticle group has recently been introduced by Elliott & Evans (46) for the single j-shell of both neutrons and protons, based on a quasiparticle formalism developed by Armstrong & Judd (47) for the atomic l-shell. Elliott and Evans introduce two types of quasiparticle operators

$$\frac{1}{\sqrt{2}}(a^+_{jmm_t}+(-1)^{j-m+\frac{1}{2}-m_t}a_{j-m-m_t}) \quad m = -j,\ldots,+j$$

$$\frac{1}{\sqrt{2}}(a^+_{jmm_t}-(-1)^{j-m+\frac{1}{2}-m_t}a_{j-m-m_t}) \quad m_t = \pm\tfrac{1}{2}$$

where $\mathbf{a}^+$ $(\mathbf{a})$ are real nucleon creation (annihilation) operators. In order to obtain a set of $4j+2$ *independent* quasiparticle creation operators, Elliott and Evans restrict the m quantum number to be positive. With a slight modification in point of view, Hecht & Szpikowski (48) restrict the isospin quantum number m_t instead and define the two types of quasiparticle operators

$$\lambda^+_m = \frac{1}{\sqrt{2}}(a^+_{jm+\frac{1}{2}}-(-1)^{j-m}a_{j-m-\frac{1}{2}})$$

$$\text{with } m = -j,\ldots,+j$$

$$\mu^+_m = \frac{1}{\sqrt{2}}(a^+_{jm-\frac{1}{2}}-(-1)^{j-m}a_{j-m+\frac{1}{2}})$$

where the $2j+1$ operators λ^+_m and $\lambda_m = (\lambda^+_m)^+$ satisfy the usual Fermion anticommutation relations; similarly for μ^+_m, (μ_m). In addition, any λ operator anticommutes with any μ operator; that is, these are distinguishable quasiparticles. The group theoretical classification scheme is then based on the group chain

$$R(8j+4) \supset [R_\lambda(4j+2) \times R_\mu(4j+2)]$$

where the group $R(8j+4)$ is generated by the full set of operators, $\mathbf{a}^+\mathbf{a}^+$, $\mathbf{aa}$, $\mathbf{a}^+\mathbf{a}$, and each factor in the direct product group is further classified by subgroup chains

$$R_\lambda(4j+2) \supset [Sp_\lambda(2j+1) \times SU_\lambda(2)] \supset [R_\lambda(3) \times SU_\lambda(2)]$$

similarly for $R_\mu(4j+2)$, where the generators and the IRs are shown in Table 2 for the λ-branch of the chain. The corresponding operators for the μ-branch of the chain are obtained by the replacement $\lambda^+_m \to \mu^+_m$, except that $(T_\lambda)_\pm \to (T_\mu)_\mp$ and $(T_\lambda)_0 \to -(T_\mu)_0$. Finally, the total angular momentum and isospin operators $\mathbf{J}$ and $\mathbf{T}$ are given by

Table 2 The λ-factor of the quasiparticle factorization

Group	Generators	IR Label
$R_\lambda(4j+2)$	$(\lambda^+\lambda^+)^{J_\lambda}, (\lambda\lambda)^{J_\lambda}, (\lambda^+\lambda)^{J_\lambda}$	$(\tfrac{1}{2}\tfrac{1}{2}\cdots \pm\tfrac{1}{2})$
$Sp_\lambda(2j+1)$	$(\lambda^+\lambda)^{J_\lambda=\text{odd}}$	$(1^{j+\frac{1}{2}-2T_\lambda}0^{2T_\lambda})$
$R_\lambda(3)$	$(\lambda^+\lambda)^{J_\lambda=1} = \mathbf{J}_\lambda,$ e.g. $(J_\lambda)_0 = \Sigma m\lambda_m^+\lambda_m$	J_λ
$SU_\lambda(2)$	$\{(\lambda^+\lambda^+)^0, \tfrac{1}{2}[(\lambda^+\lambda)^0+(\lambda\lambda^+)^0], (\lambda\lambda)^0\} = \mathbf{T}_\lambda$ e.g. $(T_\lambda)_+ = \displaystyle\sum_{m>0}(-1)^{j-m}\lambda_m^+\lambda_{-m}^+$	T_λ

$$\mathbf{J}_\lambda+\mathbf{J}_\mu = \mathbf{J} \quad \text{and} \quad \mathbf{T}_\lambda+\mathbf{T}_\mu = \mathbf{T}$$

The second equation explains the symbols T_λ, T_μ used for the IR labels of $SU_\lambda(2)$, $SU_\mu(2)$. Since the λ and μ operators are each *mathematically* equivalent to a set of identical nucleon operators, the group $SU_\lambda(2)$, for example, is a 3-dimensional quasispin group for the λ-particles, and we can make the identification $T_\lambda = \mathscr{S}_\lambda$ so that T_λ is related to an identical particle seniority number v_λ: $T_\lambda = \tfrac{1}{2}(j+\tfrac{1}{2}-v_\lambda)$. Mathematically, the relation between v_λ and J_λ is the usual one for the j-shell of identical particles and leads to the possible J_λ, T_λ values shown in Table 3. (It must be emphasized that v_λ is not to be confused with any real seniority, J_λ and T_λ do not correspond to any physical angular momentum and isospin, although J and T do refer to the physical total angular momentum and isospin.) For $j \leqq 7/2$ there is no multiple occurrence of J_λ values for any T_λ value, or of J_μ for any T_μ (Table 3). Hence the simple classification scheme

$$|(J_\lambda J_\mu)JM_J; (T_\lambda T_\mu)TM_T\rangle$$

is a complete classification scheme for shells $j \leqq 7/2$ of both protons and neutrons. Since the basis involves only ordinary angular momentum couplings, the only calculational tool required is ordinary Racah algebra. In calculating the matrix element of any operator, the technique is to express the operator in terms of λ^+, μ^+, λ, μ, rather than a^+, a. Only ordinary Racah algebra is needed to reduce the matrix element of any operators to a few reduced matrix elements of the type

$$\langle j^{v_\lambda+1}v_\lambda' J_\lambda' \| \lambda^+ \| j^{v_\lambda}v_\lambda J_\lambda\rangle$$

Table 3 The J_λ T_λ structure of each factor[a]

T_λ	0	1/2	1	3/2	2
$j = 1/2$	1/2	0	–	–	–
$j = 3/2$	2	3/2	0	–	–
$j = 5/2$	3/2, 9/2	2, 4	5/2	0	–
$j = 7/2$	2, 4, 5, 8	3/2, 5/2, 9/2, 11/2, 15/2	2, 4, 6	7/2	0

[a] Entries are the J_λ values for the T_λ values of the column headings (similarly for J_μ, T_μ).

which can be read from tables of identical nucleon cfps [the number of nontrivial reduced matrix elements is 4 for the $j = 5/2$ shell and 30 for the $j = 7/2$ shell; for details see (48)]. It is through this simplicity that the new quasiparticle technique gains advantages over more conventional spectroscopic techniques. The main disadvantage comes from the fact that nucleon number is not a good quantum number in the $|(J_\lambda J_\mu)JM_J;(T_\lambda T_\mu)TM_T\rangle$ scheme. However, techniques have been developed for effecting the transformation from the $|(J_\lambda J_\mu)JM_J;(T_\lambda T_\mu)TM_T\rangle$ scheme to states of good particle number [see (48) for the $j = 5/2$ shell]. Since the calculation of matrix elements is very simple in the $|(J_\lambda J_\mu)JM_J;(T_\lambda T_\mu)TM_T\rangle$ scheme, an alternate approach might involve the simultaneous diagonalization of the Hamiltonian and the number operator.

2 THE GOODNESS OF NUCLEAR SYMMETRIES

Since many of the symmetries associated with the vector spaces of the nuclear shell model are highly approximate, the question of the goodness of nuclear symmetries is important. Extensive shell model calculations have been carried out for nuclei in several regions of the periodic table, and the detailed IR structure of the eigenvectors for many of the low-lying states of such nuclei are now known. The greatest attention has been focused on the examination of the groups SU(3) and SU(4) in the 2s-1d shell and the seniority groups in heavier nuclei. As an example, Table 4 shows the major SU(3) components in the states of the ground state rotational band in ^{22}Na. This nucleus has been chosen since it is the $T = 0$ nucleus with the largest number of particles in the s-d shell for which a full shell model calculation has been carried out with realistic effective two-body matrix elements along with an SU(3)-SU(4) symmetry analysis (McGrory, 49; for $J = 3$ the dimension of the shell

Table 4   SU(3) content of ground state $K_J = 3$ band of ^{22}Na[a]

[f]	(λμ)		$J^\pi =$	3^+	4^+	5^+	6^+	7^+	8^+	9^+
[42]	(82)	$S = 1$ $L = J-1$		68(74)	62	53	53	44	50	59
		$S = 1$ $L = J$		8(0.6)	12	12	16	16	19	13
	(63)	$S = 1$ $L = J-1$		(7)	8	17	16	19	18	10
	(44)	$S = 1$ $L = J-1$		(5)				6		
[411]	(63)			6(1)						
	(90)			(3)						
[321]	(71)			(2)						10

[a] The numbers are taken from McGrory (49) (calculated with Kuo-Brown matrix elements, full s-d shell basis), and show the percentage probabilities for the states of space symmetry [f] and SU(3) labels (λμ) with greater than 5% probability. The numbers in parentheses for the 3^+ ground state are taken from Akiyama et al (50) (calculated with central interaction. SU(3)-truncated basis). For a pure $K_J = 3$ band of (λμ) = (82), the percentages of states with $S = 1$ $L = J$ relative to the percentages of $S = 1$ $L = J-1$ shown in line 1 would be: 18, 25, 25, 26, 20, 23, 19, with percentages of less than 4 for states with $L = J+1$.

model matrix is 366). However, the results are characteristic of all nuclei with $A \lesssim 25$. The leading SU(3) representation in the highest spatial symmetry $[f]$ carries about 70% of the total wave function for all members of the ground state rotational band. This result is fairly insensitive to the interactions used, although the calculations of Akiyama, Arima & Sebe (50), using a central effective interaction, tend to overestimate slightly the percentage of the leading IR. The remaining significant components are spread over IRs with $\lambda + 2\mu = 2n$ and large values of the SU(3) Casimir invariants, the exact percentages being somewhat interaction-dependent. The family relationship of the different members of the rotational band is quite apparent. In most of the nuclei of the $2s$-$1d$ shell with $A \lesssim 25$ (49–51), the first and second excited rotational bands are also dominated by a single IR of SU(3) with percentages from 50–80%. It is clear that a truncation of the full shell model space in terms of SU(3) amd SU(4) symmetry is physically meaningful. The successful shell model calculations that have been carried out in an SU(3) truncated basis (50–52) have involved approximately 10–20 different SU(3) IRs from among the 2–4 highest spatial symmetries. These include the leading SU(3) representations (maximum possible value of $2\lambda + \mu$, corresponding to intrinsic states in which Nilsson levels are being filled in order in the region of good asymptotic oscillator quantum numbers). In addition, it has been discovered (Harvey & Sebe, 53) that IRs with $\lambda + 2\mu = 2n$ generally play an important role. The remaining SU(3) IRs to be included in the basis are those that have strong connections to the above through the one-body spin orbit interaction with SU(3) irreducible tensor character $(\lambda \mu) = (11)$, and the dominant SU(3)-breaking piece of the two-body interaction which has SU(3) irreducible tensor character (22). Such detailed shell model calculations in an SU(3) truncated basis have been facilitated through a development of the needed SU(3) irreducible tensor calculus. In the case of many symmetries, however, it would be impossible to test the goodness of a symmetry through such detailed calculations, and it becomes important to develop simpler a priori tests for the goodness of nuclear symmetries.

2a Irreducible Tensor Character of Effective Interactions

One possibility involves the classification of the effective Hamiltonians according to their irreducible tensor character under the higher symmetry groups (Vincent, 54, 55) to compare the strengths of the symmetry-breaking and symmetry-preserving parts of the interaction. (Such a classification is of course also vital if calculations are to be carried out in an irreducible tensor formalism.)

It may be instructive to illustrate this with a very simple example. Nuclei in the upper half of the $2s$-$1d$ shell, $29 \leq A \leq 40$, have been analyzed in terms of the simple shell model space $2s_{1/2}1d_{3/2}$ by Glaudemans, Wiechers & Brussaard (56) who, assuming an inert ^{28}Si core...$(1d_{5/2})^{12}$, have calculated the needed 15 matrix elements for the effective two-body interaction for this configuration together with the two single particle energies from a fit to the experimental data. The levels $2s_{1/2}1d_{3/2}$ form the two members of a pseudo spin orbit doublet, a pseudo $\tilde{p}$-shell, so that it is also possible to classify the states and the interaction under a pseudo SU(3) × pseudo SU(4) symmetry group. The question arises whether the pseudo

SU(3) × SU(4) and pseudo $\tilde{L}\tilde{S}$ symmetries are physically meaningful for this doublet.

The symmetry classification of an operator begins with the symmetry classification of the single particle creation and annihilation operators $\mathbf{a}^+$ ($\mathbf{a}$), where $\mathbf{a}^+$ transforms according to the g-dimensional, fundamental IR [1] of U(g), whereas the annihilation operator $\mathbf{a}$ transforms according to the conjugate IR $[1^{g-1}]$. From the direct product

$$[1] \times [1^{g-1}] = [1^g] + [2 \ 1^{g-2}]$$

it can be seen that the one-body operator has an SU(g) scalar piece [0] (U(g) IR $[1^g]$), a multiple of the number operator, and an SU(g) tensor piece $[2 \ 1^{g-2}]$. For the pseudo $\tilde{p}$-shell it will be convenient to use a double tensor classification: $\mathbf{a}^+$ transforms according to the IR [1] of U(3) [we shall use SU(3) IR labels $(\lambda\mu) = (10)$]; and according to the IR [1] of SU(4); whereas $\mathbf{a}$ transforms according to $[1^2]$ of U(3), $(\lambda\mu) = (01)$, and according to $[1^3]$ of SU(4). More specifically, in terms of irreducible tensor operators $t^{(\lambda\mu); [f]}_{lm_l; \frac{1}{2}m_s \frac{1}{2}m_t}$ the tensorial character is given by

$$a^+_{lm_l m_s m_t} = t(a^+)^{(10); [1]}_{lm_l; \frac{1}{2}m_s \frac{1}{2}m_t}$$
$$a_{lm_l m_s m_t} = (-1)^{\lambda + l - m_l + \frac{1}{2} - m_s + \frac{1}{2} - m_t} t(a)^{(01); [1^3]}_{l - m_l; \frac{1}{2} - m_s \frac{1}{2} - m_t}$$

where the SU(3)-SU(4) dependence of the phase factor is (as always) a question of phase convections (57). We have also omitted tildes that refer to the fact that the above are pseudo SU(3), $\tilde{l}$, $\tilde{s}$ quantum numbers, since the mathematics is the same for the real and pseudo SU(3)... schemes. The classification of the effective interaction for the $2s_{1/2}1d_{3/2}$ shell in terms of pseudo SU(3) × SU(4) irreducible tensor operators begins with a transformation of the two-body matrix elements $\langle j_3 j_4 J T | V | j_1 j_2 J T \rangle$ to pseudo $\tilde{L}\tilde{S}$ coupling $\langle (\tilde{l}^2)\tilde{L}\tilde{S}JT | V | (\tilde{l}^2)\tilde{L}\tilde{S}JT \rangle$ via 9-j-j to $\tilde{l}$-$\tilde{s}$ recoupling transformations, so that the interaction is expressed in terms of $\tilde{l}$-$\tilde{s}$ coupled pair creation and annihilation operators to be classified in terms of $\tilde{p}$-shell symmetries. (Henceforth the coupling for the $\tilde{p}$-shell is mathematically equivalent to that for a real p-shell and the tildes will be omitted for brevity.) The pair creation operators $[a^+ \times a^+]$ coupled to $L = 0$ or 2 are symmetrically coupled in orbital (or pseudo orbital) space and hence antisymmetrically coupled in spin-isospin space and belong to $(\lambda\mu) = (20)$ and $[f] = [1^2]$; the operator $[a^+ \times a^+]$ coupled to $L = 1$ belongs to $(\lambda\mu) = (01)$ and $[f] = [2]$. The pair annihilation operators $[t(a) \times t(a)]$ belong to the corresponding conjugate representations (02), $[1^2]$ and (10), $[2^3]$, respectively. Fully coupled two-body irreducible tensor operators are then obtained by coupling the pair creation operators (IRs $(\lambda_2\mu_2)[f_2]$) to the pair annihilation operators (conjugate IRs $(\mu'_2\lambda'_2)[f'^*_2]$) to resultant irreducible tensor character $(\lambda_0\mu_0)L_0 M_0$; $[f_0]S_0 M_{S_0} T_0 M_{T_0}$, where L_0 and S_0 are further coupled to $J_0 = 0$ since the interaction is rotationally invariant. It is convenient to express the two-body interaction in terms of two-body unit tensor operators T_u.

$$H = \sum_{\substack{(\lambda_2\mu_2)(\mu'_2\lambda'_2) \\ (\lambda_0\mu_0)[f_0]L_0}} V([(\lambda_2\mu_2)(\mu'_2\lambda'_2)](\lambda_0\mu_0)L_0[f_0]) \, T_u[(\lambda_2\mu_2)(\mu'_2\lambda'_2)]^{(\lambda_0\mu_0); [f_0]}_{L_0; S_0 T_0; J_0}$$

where the unit tensor operators are defined through the coupling

$$T_u = \left[\left[[a^+ \times a^+]^{(\lambda_2\mu_2)[f_2]} \times [t(a) \times t(a)]^{(\mu_2'\lambda_2')[f_2'^*]} \right]^{(\lambda_0\mu_0);[f_0]}_{L_0; S_0 = L_0 T_0 = 0} \right]_{J_0 = 0}$$

$$= \sum_{\substack{L_2 L_2' S_2 S_2' T_2 T_2' \\ M_2 M_{S_2} M_{T_2} q}} \langle (\lambda_2\mu_2)L_2 ; (\mu_2'\lambda_2')L_2' \| (\lambda_0\mu_0)L_0 \rangle$$

$$\times \langle [f_2]S_2 T_2 ; [f_2'^*]S_2' T_2' \| [f_0]S_0 T_0 \rangle$$

$$\times \langle L_2 M_2 L_2' M_2' | L_0 q \rangle \langle S_2 M_{S_2} S_2' M_{S_2}' | S_0 - q \rangle$$

$$\times \langle T_2 M_{T_2} T_2' M_{T_2}' | 00 \rangle \langle L_0 q S_0 - q | J_0 = 00 \rangle$$

The coupling requires SU(3) and SU(4) Wigner coefficients where each of these is factored into a reduced (double-barred) Wigner coefficient and ordinary L-, S-, and T-space vector coupling coefficients. (For the simple IRs $(\lambda_2\mu_2)[f_2]$ needed for the p-shell, the quantum numbers L_2, $S_2 T_2$ are sufficient for a complete labeling of the states.) Reduced SU(3)/R(3) Wigner coefficients have now been calculated by a number of authors in very general form [Draayer & Akiyama (57) and references quoted there], while many of the reduced SU(4)/SU$_S$(2) × SU$_T$(2) Wigner coefficients needed for nuclear calculations can be found in (58). The possible irreducible tensor character $(\lambda_0\mu_0)L_0[f_0]$ of H follows from the decomposition of the direct products $(\lambda_2\mu_2) \times (\mu_2'\lambda_\mu')$ and $[f_2] \times [f_2'^*]$:

$$(20) \times (02) = (00) + (11) + (22) \rightarrow [1^2] \times [1^2] = [0] + [211] + [22]$$

$$(01) \times (10) = (00) + (11) \qquad\;\; \rightarrow [2] \times [2^3] = [0] + [211] + [422]$$

$$(20) \times (10) = (11) + (30) \qquad\;\; \rightarrow [1^2] \times [2^3] = [211] + [332]$$

$$(01) \times (02) = (11) + (03) \qquad\;\; \rightarrow [2] \times [1^2] = [211] + [31]$$

With $S_0 = L_0$ restricted to $S_0 = 0, 1, 2$, only IRs $(\lambda_0\mu_0) \rightarrow [f_0]$ with matching L_0, S_0 occur in H. The L and S, T structure of the IRs is given by

(00)	$L = 0$	[0]	$ST = 00$	Since $T_0 = 0$,
(11)	$L = 1, 2$	[211]	$ST = 10, \ldots$	only ST values
(22)	$L = 0, 2^2, 3, 4$	[22]	$ST = 00, 20, \ldots$	with $T = 0$ are
(30)(03)	$L = 1, 3$	[422]	$ST = 00, 20, \ldots$	shown in the
		[332][31]	$ST = 10, \ldots$	decomposition
				of SU(4) IRs

In all there are 15 Hermitean combinations $(\lambda_2\mu_2)(\mu_2'\lambda_2')(\lambda_0\mu_0) L_0 [f_0]$ corresponding to the 15 independent two-body matrix elements. The coefficients of these 15 pseudo [SU(3) $\supset$ R(3)] × SU(4) irreducible tensors are shown in Table 5 for the GWB interaction of the $2s_{1/2}1d_{3/2}$ or pseudo $\tilde{p}$-shell. [It should be pointed out that two different pseudo $\tilde{p}$-shell classification schemes are actually possible for an $s_{1/2}d_{3/2}$ shell. In the first (Table 5), the association between the $s_{1/2}d_{3/2}$ and $\tilde{p}$-shell is made as follows: $\mathbf{a}^+_{s_{\frac{1}{2}}} \rightarrow \mathbf{a}^+_{\tilde{p}_{\frac{1}{2}}}$, $\mathbf{a}^+_{d_{3/2}} \rightarrow \mathbf{a}^+_{\tilde{p}_{3/2}}$; while in the second: $\mathbf{a}^+_{s_{\frac{1}{2}}} \rightarrow \mathbf{a}^+_{\tilde{p}_{\frac{1}{2}}}$, $\mathbf{a}^+_{d_{3/2}} \rightarrow -\mathbf{a}^+_{\tilde{p}_{3/2}}$ instead. However, the pseudo SU(3) × SU(4) character of the two schemes turns out to be very similar.] From Table 5 the two-body interaction is immediately separated into symmetry-preserving and symmetry-breaking parts.

Table 5 Pseudo $[SU(3) \supset R(3)] \times SU(4)$—IR tensor decomposition of the effective two-body interaction for the $2s_{1/2} 1d_{3/2}$ shell[a]

	$(\lambda_2\mu_2)$	$(\mu_2'\lambda_2')$	$(\lambda_0\mu_0)$	L_0	$[f_0]$	V(MeV)
Scalar						
	(20)	(02)	(00)	0	[0]	4.932 (2.739)[d]
	(20)	(02)	(00)	0	[22]	−2.068
	(01)	(10)	(00)	0	[0]	−0.498 (−2.500)[d]
	(01)	(10)	(00)	0	[422]	−1.464
	(20)	(02)	(22)	0	[0]	−0.278
	(20)	(02)	(22)	0	[22]	0.806
Vector						
	(20)	(02)	(11)	1	[211]	−0.469
	(01)	(10)	(11)	1	[211]	+0.423
	(20)	(10)[b]	(11)	1	[211]	−0.371
	(20)	(10)[b]	(11)	1	[332]	−0.561
	(20)	(10)[b]	(30)	1	[211]	−0.710
	(20)	(10)[b]	(30)	1	[332]	−1.015
Tensor						
	(20)	(02)	(11)	2	[22]	−1.566
	(01)	(10)	(11)	2	[422]	−3.655
	(20)	(02)	(22)[c]	2	[22]	0.148

[a] The interaction is that of Glaudemans, Wiechers & Brussaard (56); the transformation to the pseudo $SU(3) \times SU(4)$ scheme has been carried out by Draayer.

[b] For entries that are not self-conjugate, the coefficient for the (missing) conjugate partner is $(-1)^{L_0}$ times that listed.

[c] Although there are two independent operators with $L_0 = 2$ in $(\lambda_0\mu_0) = (22)$, only the single Hermitean combination contributes to the Hamiltonian.

[d] Strength of the scalar terms after the two-body unit operator has been removed from the interaction.

Tensor operators with *either* $(\lambda_0\mu_0) = (00)$ *or* $[f_0] = [0]$ are symmetry-preserving since they cannot change the symmetry labels in either pseudo orbital or pseudo spin-isospin space. Thus the first five operators of the table are symmetry-preserving. This can be seen more directly since these five operators can be written as combinations of the two-body unit, T^2, $\tilde{S}^2$, $\tilde{L}^2$, and Majorana exchange operators [or alternately the two-body part of the SU(3) Casimir invariant]:

$$T[(20)(02)]^{(00)[0]} = -\frac{1}{6}\sum_{i<j}(1+P_{ij}^M) = -\frac{1}{9}n(n-1) - \frac{1}{18}(g_{\lambda\mu}-4n)$$

$$T[(01)(10)]^{(00)[0]} = -\frac{1}{\sqrt{30}}\sum_{i<j}(1-P_{ij}^M)$$

$$= -\frac{1}{3\sqrt{30}}[n(n-1)-(g_{\lambda\mu}-4n)]$$

$$T[(20)(02)]^{(00)[22]} = \frac{1}{12} \sum_{i<j} (\boldsymbol{\sigma}_i \cdot \boldsymbol{\sigma}_j - \boldsymbol{\tau}_i \cdot \boldsymbol{\tau}_j) = \frac{1}{6}[S(S+1) - T(T+1)]$$

$$T[(01)(10)]^{(00)[422]} = \frac{1}{2\sqrt{30}} \sum_{i<j} [\tfrac{2}{3}(\boldsymbol{\sigma}_i \cdot \boldsymbol{\sigma}_j)(\boldsymbol{\tau}_i \cdot \boldsymbol{\tau}_j) - (\boldsymbol{\sigma}_i \cdot \boldsymbol{\sigma}_j) - (\boldsymbol{\tau}_i \cdot \boldsymbol{\tau}_j)]$$

$$= -\frac{1}{18\sqrt{30}} \{7n(n-1) + 8(g_{\lambda\mu} - 4n)$$

$$+ 30[S(S+1) + T(T+1) - \tfrac{3}{2}n]\}$$

$$T[(20)(02)]^{(22)[0]} = -\frac{1}{6\sqrt{5}} \sum_{i<j} (4\mathbf{l}_i \cdot \mathbf{l}_j - 1 + 3P_{ij}^M)$$

$$= -\frac{1}{6\sqrt{5}} [2L(L+1) - g_{\lambda\mu} + 2n]$$

where the SU(3) Casimir invariant is $g_{\lambda\mu} = \lambda^2 + \lambda\mu + \mu^2 + 3(\lambda + \mu)$. Table 5 shows at once that some of the symmetry-breaking parts of the interaction, particularly the tensor terms ($\tilde{L}_0 = \tilde{S}_0 = 2$), are as important as the symmetry-preserving terms, although the question arises whether the strength coefficients V are weighted in the best possible way to give a good measure of the relative importance of the symmetry-breaking vs symmetry-preserving terms. Since it is the matrix elements in n-particle states of symmetry-breaking relative to symmetry-preserving terms which are important, Vincent (54) and French (59) suggest that H be expanded in terms of irreducible tensor operators T_a which form an orthonormal set under the scalar product which is defined by the full trace in the n-particle space

$$\{T_a \mid T_b\}_n = Tr\{T_a \mid T_b\}_n = \delta_{ab}$$

where

$$\{T_a \mid T_b\}_n = \sum_{\substack{(\lambda\mu)\kappa LM_L \\ \beta STM_SM_T}} \langle \tilde{p}^n(\lambda\mu)\kappa LM_L \beta STM_S M_T \mid T_a^+ T_b \mid \tilde{p}^n(\lambda\mu)$$

$$\times \kappa LM_L \beta STM_S M_T \rangle$$

in our example. The quantum numbers β and κ are needed only for $n > 2$ where there may be multiple occurrences of S, T in certain $[f]$ and multiple occurrences of L in $(\lambda\mu)$; note that $(\lambda\mu)$ determines $[f]$. Only SU(3), R(3), SU(4)-invariant operators survive under the trace operation so that operators T_a, T_b which transform according to different IRs $(\lambda_0\mu_0)$, L_0, $[f_0]$ are automatically orthogonal. For $n > 2$, operators with the same $(\lambda_0\mu_0)$, L_0 $[f_0]$ but different $(\lambda_2\mu_2)(\mu_2'\lambda_2')$ are not necessarily orthogonal. However, for $n = 2$ the two-body unit tensor operators defined above not only form an orthogonal set but are normalized according to $\{T_u \mid T_u\} = 1$. If $H = H_b + H_p$ is split into a symmetry-breaking and symmetry-preserving part, the scalar product $\{H_b \mid H_b\}$ for the two-particle space is given by the sums of squares of the coefficients of the symmetry-breaking terms in Table 5. Since the two-body unit operator only shifts the whole spectrum, its strength is irrelevant to the question of the goodness of a symmetry, and it is important to remove it from H, defining the remaining Hamiltonian as H' (see last column of Table 5). French (59) then suggests that a good measure of the goodness of a symmetry[3] is given by

$$\zeta(n) = \{H_b \mid H_b\}_n / \{H' \mid H'\}_n$$

The classification of any interaction according to unit irreducible tensor operators of the type T_u defined above will allow the immediate calculation of $\zeta(2)$. For the pseudo SU(3) symmetry of the $2s_{1/2} 1d_{3/2}$-shell, we obtain $\zeta(2) = 0.507$ for the two-body interaction of Table 5, and conclude that this may not be a good symmetry. The goodness of the $\tilde{L}\tilde{S}$ quantum numbers is expected to be no better; for pseudo R(3) symmetry $\zeta(2) = 0.492$. So far the analysis has ignored the one-body part of H. The $2s_{1/2} 1d_{3/2}$ splitting can be described by a one-body $\tilde{l}\cdot\tilde{s}$ term. (The numbers of GWB lead to a coefficient of $+0.812$ MeV.) The possibility exists that the symmetry-breaking parts of this contribution to H may partially cancel the symmetry-breaking vector terms in the two-body Hamiltonian. To split the one-body $\tilde{l}\cdot\tilde{s}$ term into a symmetry-preserving and -breaking part it is convenient to write

$$h = \sum_i \tilde{l}_i \cdot \tilde{s}_i = \tilde{L}\cdot\tilde{S} - \sum_{i<j} (\tilde{l}_i \cdot \tilde{s}_j + \tilde{l}_j \cdot \tilde{s}_i) = h_p + h_b$$

so that h can be expanded in terms of unit irreducible tensor operators of the above type. For $\tilde{l}\cdot\tilde{s}$ all terms have irreducible tensor character $(\lambda_0 \mu_0) = (11)$, $L_0 = 1$, $[f_0] = [211]$, and h_b, when expanded in terms of unit tensor operators, has coefficients $-[15/4]^{\frac{1}{2}}$, $-3/2$, and $-[9/2]^{\frac{1}{2}}$ for tensors with $(\lambda_2 \mu_2)(\mu_2'\lambda_2') = (20)(02)$, $(01)(10)$, and $(20)(10)$. Including these terms in H_b and H', we obtain $\zeta(2) = 0.625$; that is, the pseudo SU(3) symmetry still appears to be badly broken in the $2s_{1/2} 1d_{3/2}$ shell.

Several analyses of the irreducible tensor character of the effective interactions used for the $2s$-$1d$-shell have recently been carried out. Vincent (54) and Vergados (60) give the SU(3) reduction for a few simple central interactions. Pluhar (61) gives the major components for the full $[SU(6) \supset SU(3) \supset R(3)] \times SU(4)$-reduction for the Kuo-Brown interaction, and Draayer (51) gives an $[SU(3) \supset R(3)] \times SU(4)$-reduction for the Kuo-Brown interaction and several of its modifications. These analyses indicate that the strongest SU(3) symmetry-breaking tensors have IR character $(\lambda\mu) = (22)$, although pairing effects may make themselves felt through SU(3) (44) tensors (Vincent), and $(S_0 = L_0 = 2)$ "tensor" terms make important contributions. Table 6 shows the $\zeta(2)$ values calculated by Draayer for the SU(3) symmetry. ζ_V gives the symmetry-breaking measure for the fully renormalized Kuo-Brown interaction. This criterion indicates that SU(3) symmetry may be expected to be reasonably good as far as the two-body part of the interaction is concerned. The

---

[3] Ginocchio (43), in his analysis of the goodness of symplectic symmetry in the $1f_{7/2}$ shell, defines a somewhat similar criterion for measuring the goodness of a symmetry, but removes not only the two-body unit operator but two-body $\mathbf{T}^2$ and $\mathbf{J}^2$ operators as well in arriving at his equivalent of $\{H' \mid H'\}$ in order to remove from the symmetry analysis the influence of the strongly conserved quantities n, T, and J. With this modification Ginocchio finds $\zeta(2)$ values of ~ 0.6 to 0.8 for symplectic symmetry in the $1f_{7/2}$ shell.

Table 6 $\zeta(2)$—Values for SU(3) symmetry in the $2s$-$1d$ shell

$\zeta(2)_V$	0.188
$\zeta(2)_{V+l^2}$	0.132
$\zeta(2)_{V+l\cdot s}$	0.709
$\zeta(2)_{V+l^2+l\cdot s}$	0.680

one-body contributions to the effective Hamiltonian, Σl_i^2 and $\Sigma l_i \cdot s_i$, have been split into symmetry-breaking and symmetry-preserving parts as indicated above. Their separate contributions to $\zeta(2)$ are shown in Table 6. If only Σl_i^2 is included ζ_{V+l^2} is decreased; that is, the symmetry-breaking parts of the one-body l^2-term partially cancel the symmetry-breaking parts of the two-body interaction V. It is known that the sign of the one-body term is important for this cancellation (center of gravity of $1d$ levels above the $2s_{\frac{1}{2}}$ level). In a fictitious H, in which the sign of the one-body l^2 term is changed arbitrarily, ζ_{V-l^2} is increased to a value of 0.49. The calculations of Bhatt & McGrory (24) for ^{44}Ti indicate that it may be the opposite sign of the l^2 term in the $1f$-$2p$ shell (center of gravity of $2p$ above $1f$) that is largely responsible for the complete failure of SU(3) symmetry in this shell. Finally, it can be seen that the one-body $l \cdot s$ term makes drastic contributions to the SU(3) symmetry-breaking terms in the $2s$-$1d$ shell.[4]

An analysis of the goodness of generalized seniority in mixed configurations of identical nucleons has been carried out by Bohigas, Quesne & Arvieu (62) who use a somewhat modified $\zeta(2)$ as a measure of the goodness of this symmetry for many different interactions in many shells, through a quasispin analysis (comparison of the strengths of quasispin breaking $\mathscr{S} = 1$ and $\mathscr{S} = 2$ tensor terms with the strengths of the quasispin preserving terms). The results show $\zeta(2)$ values that are very small for many of the effective interactions used in light nuclei, $\zeta(2)$s of the order of 0.02, e.g., for the Cohen-Kurath $1p$-shell interaction and the GWB $2s_{1/2}1d_{3/2}$ inter-action, and of the order of 0.1 for the Kuo-Brown $2s$-$1d$ shell interaction (for identical nucleon configurations, $T = T_{\max}$). Generalized seniority can thus be expected to be a good quantum number in nuclei such as ^{22}O. In the richer configurations of heavier nuclei, however, the $\zeta(2)$ values for generalized seniority are much larger.

2b   The Important Role Played by Infinitesimal Operators

The irreducible tensor analysis presented above shows that there are two criteria for isolating the symmetry-preserving parts of an operator. There are two types of symmetry-preserving operators: 1. Invariant operators, which transform according to the scalar IR of the symmetry group [e.g. $(\lambda_0\mu_0) = (00)$-tensor operators for SU(3)], and 2. Operators built from the infinitesimal operators that generate the group $[(\lambda_0\mu_0) = (11)$-operators for SU(3)], since the matrix elements of the infinitesimal operators are diagonal in the IRs of the symmetry group. The

---

[4] There are, however, indications that the contributions of the one-body $l \cdot s$ term to $\zeta(n)$, with $n > 2$, are relatively less important, so that $\zeta(2)$ tends to overemphasize the symmetry breaking of the one-body part of H.

operator $\mathbf{L} \cdot \mathbf{L}$ is therefore an SU(3) preserving operator even though it contains both $(\lambda_0 \mu_0) = (00)$ and (22) components. Note, however, that an operator built from products of two infinitesimal operators is in general a mixed two- and one-body operator. In the p-shell, however, even the two-body part of $\mathbf{L} \cdot \mathbf{L}$, $(\mathbf{L} \cdot \mathbf{L})_{2b} = \sum_{i<j} \mathbf{l}_i \cdot \mathbf{l}_j$, leaves SU(3) invariant since the one-body part of $\mathbf{L} \cdot \mathbf{L}$ is effectively a multiple of the number operator, namely $nl(l+1) = 2n$. In the $2s$-$1d$ shell, on the other hand, the two-body operator $(\mathbf{L} \cdot \mathbf{L})_{2b}$ can be written as a combination of $(\mathbf{L} \cdot \mathbf{L})$, an SU(3) symmetry-preserving operator, and a one-body operator, $-\sum_i \mathbf{l}_i^2$, with $(\lambda_0 \mu_0) = (00)$ and (22) components, where the latter can now break SU(3) symmetry. (Even here the possibility exists that these induced one-body symmetry-breaking terms of the two-body operator can be partially canceled by the corresponding pieces in the true one-body part of an interaction. Table 6 shows that this is indeed the case for the $2s$-$1d$ shell.)

French (63) has used the above two criteria to write the general symplectic symmetry-preserving two-body interaction for the single j-shell in terms of arbitrary coefficients α_0, α, $\alpha_{k=\text{odd}}$:

$$\alpha_0 (\mathbf{U}^0 \cdot \mathbf{U}^0)_{2b} + \alpha \sum_k (2k+1)(\mathbf{U}^k \cdot \mathbf{U}^k)_{2b} + \sum_{k=\text{odd}} \alpha_k (\mathbf{U}^k \cdot \mathbf{U}^k)_{2b}$$

where, in terms of the unit tensor operators introduced in section 1, $(\mathbf{U}^k \cdot \mathbf{U}^k) = \Sigma_q (-1)^q U_{q0}^{k0}(j, j) U_{q0}^{k0}(j, j)$. The first two terms are Casimir invariants for $U(1) \times SU(2j+1)$ and can be written in terms of two-body unit and $\mathbf{T}^2$ operators. They are symplectic invariants [transform according to IR $(0 \ldots 0)$]. The last terms with $k = \text{odd}$ can be related to the products of two infinitesimal generators for $\mathrm{Sp}(2j+1)$, $(\mathbf{U}^k \cdot \mathbf{U}^k)$, since the additional induced one-body parts of these operators are simple multiples of the number operator (the only $k = 0$ unit tensor for the single j-shell). Since the most general symplectic symmetry-preserving two-body interaction has $(j + 5/2)$ arbitrary coefficients, there are in general $(j - 3/2)$ conditions on the two-particle energies that must be satisfied by a symmetry-preserving interaction. French shows that these collapse to a single condition for shells of identical nucleons with $j = 9/2$, $11/2$. Lanford (64) has recently made an analysis of the $1g_{9/2}$, $1h_{9/2}$, and $2g_{9/2}$ shells and shows that this one condition is very nearly satisfied in these shells. The root mean square deviation between the energy levels given by a seniority-preserving interaction and the experimental two-particle energies in these shells is of the order of 7–12 keV, so that seniority can be expected to be a very good quantum number in these $j = 9/2$ shells.

In their analysis of generalized seniority in mixed configurations of identical nucleons, Bohigas, Quesne & Arvieu (62) show that the most general quasispin symmetry-preserving interaction can be written as

$$H = T_0^{\mathscr{S}=0} + \lambda(N_{\text{op.}} - \Omega) + \alpha[\mathscr{S} \times \mathscr{S}]_0^2$$

that is, an interaction made up of a quasispin scalar ($\mathscr{S} = 0$) term, a term proportional to $\mathscr{S}_0$, and a tensor ($\mathscr{S} = 2$) term built from two quasispins (infinitesimal operators) coupled to spherical tensor rank $\mathscr{S} = 2$ in quasispin space. Note that this $\mathscr{S} = 2$ term is a pure two-body term since one-body operators can have a quasispin tensor rank of at most 1. Bohigas et al give the coefficients λ and α in terms of general two-particle matrix elements and enumerate the conditions that must be satisfied by the two-particle matrix elements of a general interaction if it is to be a symmetry-preserving interaction of the above form.

2c   Spectral Distribution Methods

A criterion for the goodness of a symmetry can be given by the condition $\zeta(2) \ll 1$. The detailed calculations for SU(3) symmetry, however, show that even with $\zeta(2)$ values of ~ 0.7 we may be dealing with a good symmetry that can be used successfully as the basis for a truncation scheme for shell model calculations. It may therefore be important to look for more reliable measures of symmetry breaking. The most promising new approach is due to J. B. French and collaborators who have applied spectral distribution methods (65) to the study of the goodness of nuclear symmetries. The most detailed application to the higher symmetry groups have involved the group SU(4) and space symmetry in the $2s$-$1d$ shell (French & Parikh, 66; and Parikh, 67). In the distribution method one considers the energy distribution of the intensities of the various IRs of a group. These are given by the moments of the Hamiltonian averaged over the states of the IR $[f]$. The pth moment of H is given by

$$\langle H^p \rangle^{[f]} = \frac{1}{d[f]} \sum_{\alpha \varepsilon [f]} \langle [f] \alpha | H^p | [f] \alpha \rangle$$

where $d([f])$ is the dimension of the IR. (For a unitary group the IR labels f_i designate both the particle number n and the IR of the group.) If H is of low particle rank such as the $0+1+2$-body interactions of nuclear physics, the distributions (over fixed IR subspaces as well as over all states of a given n) are approximately normal (Gaussian) (65), and are in this approximation determined by the first two moments ($p = 1, 2$). The first moment gives the centroid energy of the IR

$$E_c([f]) = \langle H \rangle^{[f]}$$

while the width is given by the variance

$$\sigma^2([f]) = \langle H^2 \rangle^{[f]} - (\langle H \rangle^{[f]})^2$$

Only the scalar parts of H and H^2 survive in the averaging over states of a specific IR. For SU(4) symmetry and a $0+1+2$-body H (with a maximum particle rank of 4 for H^2), $\langle H \rangle$ and $\langle H^2 \rangle$ can be expressed solely in terms of the Casimir invariants of the group

$$\langle H \rangle^{[f]} = P_2(n) + P_0(n)G_2 = a + bn + cn^2 + dG_2$$
$$\langle H^2 \rangle^{[f]} = P_4(n) + P_2(n)G_2 + P_1(n)G_3 + P_0(n)G_4 + P_0(n)G_2^2$$

where $P_k(n)$ are polynomials of degree k in n, and G_k are the Casimir invariants of degree k in the infinitesimal operators. [Note that the IRs of U(4) are specified by n and the three Casimir invariants G_2, G_3, G_4 which are simple functions of the f_i (7, 67).] The 0- to 4-particle space serves as a defining space, and the above formulae can serve to propagate information from the defining space to the larger

n-particle spaces. For SU(4) the propagation is simple. The number of IRs in the defining space is equal to the number of coefficients in the expansions of $\langle H \rangle$ and $\langle H^2 \rangle$ in terms of the G_ks so that only the U(4) Casimir invariants are needed for the propagation process. Detailed expressions for $E_c([f])$ and $\sigma^2([f])$ in terms of the E_cs and σ^2s for the 12 IRs $[f]$ for $n \leq 4$ have been given by Parikh (67). (Actually, to obtain forms for the traces which are less sensitive to inaccuracies in the input values, French and Parikh have replaced some of the $n \leq 4$ input IRs in favor of IRs with n near 4Ω.) However, once the coefficients in the $P_k(n)$ have been determined from the input net of 12 IRs for a particular H, it is a trivial matter to calculate the centroids E_c and widths σ for any IR $[f]$ in the shell. A symmetry can be considered very good if the widths σ are much smaller than the separations D between the centroids E_c of different $[f]$. Some of the results of French and Parikh are summarized in Table 7. For the 2s-1d shell and a modified Kuo interaction the widths σ are from 7–10 MeV. For the two lowest-lying centroids the values of D are typically from 4–6 MeV, so that the ratios do not satisfy the criterion $(\sigma/D) \ll 1$.

The distribution method can also be used to give a rough measure of the mixing of IRs in the low energy domain of the spectrum. From the dimensionalities, centroids, and widths the Gaussian approximations to the intensity distributions of the individual $[f]$s are constructed. The relative magnitudes of these distributions in the ground state domain are then taken as a measure of the relative intensities of the various IRs, where these relative intensities can also be used as a rough

Table 7 SU(4) width to spacing ratio (σ/D) and approximate intensities in the ground state domain as estimated by spectral distribution techniques for nuclei in the first half of the 2s-1d shell.[a]

n	(σ/D)			Estimated intensity in percent for the four highest space symmetries		
	R	R spe $= 0$	$K + 12fp$	R	R spe $= 0$	$K + 12fp$
4	0.7	0.4	1.1	92-8	100	70-27-3
5	1.0	0.6	1.5	90-9-1	100	75-20-4-1
6	1.7	0.9	2.6	87-9-4-1	99	73-17-5-5
7	1.2	0.7	1.9	80-19-1	100	50-45-3-1
8	1.0	0.6	1.5	79-19-1	100	36-49-11-1
9	1.3	0.8	2.0	82-16-1	100	52-35-9-0
10	2.0	1.2	3.2	80-10-8-2	98-1	57-15-10-16
11	1.4	0.8	2.1	79-19-1	99-1	43-44-7-4
12	1.1	0.7	1.6	78-20-1	100	28-52-14-2

[a] Taken from French & Parikh (66). n = particle number. R = central Rosenfeld interaction; for spe = 0, single particle energies have been put to zero. $K + 12fp$ = Kuo interaction as modified by Halbert et al. σ = average width of the two lowest-lying $[f]$s; D is their centroid separation.

measure of the admixture of different IRs in the ground state. The estimates of French and Parikh for the four highest space symmetries are also shown in Table 7, although this method seems to overestimate the mixing of representations, particularly for distributions with $(\sigma/D) > 1$ as for the Kuo interaction. For example, for $n = 4$ the spectral distribution estimates predict an admixture of 27% [31]-space symmetry to the dominant 70% [4]-space symmetry. The detailed calculations of McGrory (also with the Kuo interaction) for the 0_1, 2_1, 4_1 states of ^{20}Ne indicate that these numbers should be closer to 10% and 90%. For $n = 8$, the spectral distribution estimates for the [44]- and [431]-space symmetries are 36 and 49. The diagonalization of the Kuo interaction for the 0^+ states of ^{24}Mg in the full $2s$-$1d$ shell model space indicates that the exact numbers are 80% and 20%. The calculations of McGrory (49) show that the 0^+ ground state of ^{24}Mg contains 73% of SU(3) IR(84), the leading $(\lambda\mu)$ in the [44]-space symmetry. For $n = 9$, the spectral distribution estimates give percentages of 52 and 35 for the two highest space symmetries [441] and [432], whereas (with a similar H) the shell model calculations of Draayer (51) give percentages of 90 and 10 for the ground state of ^{25}Mg, although in this case the percentages in the lower space symmetries, [4311],..., have been arbitrarily set equal to zero by a truncation of the shell model space. It may, however, be possible to use spectral distribution methods to make more reliable estimates of the mixing percentages if the SU(4) distributions are extended to those with fixed isospin.

Parikh & Wong (68) have suggested that the widths be decomposed into partial widths and separated into external widths (responsible for symmetry admixing) and an internal (diagonal) contribution

$$\sigma^2([f_\alpha]) = \sum_{\beta \neq \alpha} \sigma^2([f_\alpha]; [f_\beta]) + \sigma^2([f_\alpha]; [f_\alpha])$$

The ratios $\chi^2(\alpha, \beta) = \sigma^2([f_\alpha]; [f_\beta])/(E_c([f_\alpha]) - E_c([f_\beta]))^2$ provide a measure for symmetry breaking. Using a central Rosenfeld interaction for the $2s$-$1d$ shell, Parikh and Wong find $\chi^2([4], [31]) = 0.35$, for example. Since the space symmetry [4] should be very good with this interaction, the criterion $\chi^2 \ll 1$ may again be too stringent.

In summary, it may be concluded that more work needs to be done to find a reliable, *simple* criterion for the goodness of a symmetry. Although the simple tests $\zeta(2) \ll 1$, $(\sigma/D) \ll 1$, or $\chi^2(\alpha, \beta) \ll 1$ certainly furnish sufficient conditions for the goodness of a symmetry, we have seen several examples of reasonably good symmetries for which these numbers are quite large, so that they may not be necessary conditions in the case of many useful symmetries.

3   THE IRREDUCIBLE TENSOR FORMALISM AS A CALCULATIONAL TOOL

Even in cases where a symmetry has only approximate validity, the irreducible tensor formalism for the higher symmetry groups may be very useful as a calculational tool.

The ordinary angular momentum calculus through Racah's development of the algebra of (spherical) tensor operators has provided us with simple techniques for the calculation of the matrix elements of such operators. Consider, for example, the matrix element of a one-body operator of definite spherical tensor rank, $F_q^k = \sum_i f_q^k(i)$, connecting two n-particle states in a simple j-shell. Using the Wigner-Eckart theorem such a matrix element can be written

$$\langle j^n \alpha' J'M' | F_q^k | j^n \alpha JM \rangle = \sum_{\alpha''J''} \sum_{M''mm'} \langle J''M''jm | JM \rangle \langle J''M''jm' | J'M' \rangle$$
$$\times \langle jmkq | jm' \rangle \langle \alpha J \| a_j^+ \| \alpha''J'' \rangle \langle \alpha'J' \| a_j^+ \| \alpha''J'' \rangle \langle j \| f^k \| j \rangle$$

where the double-barred matrix elements of a_j^+ are (except for trivial factors such as $[n]^{\frac{1}{2}}$) coefficients of fractional parentage. The dependence on the subgroup labels M sits only in the product of the three Wigner coefficients. The sum over subgroup labels M'', m, m' can be carried out and expressed in terms of a single Wigner coefficient and a Racah coefficient, so that the above matrix element takes the simple form

$$\sum_{\alpha''J''} (-1)^k \frac{\langle JMkq | J'M' \rangle}{[2k+1]^{\frac{1}{2}}} \frac{U(JjJ'j; J''k)}{U(JjJj; J''0)} \langle \alpha J \| a_j^+ \| \alpha''J'' \rangle$$
$$\times \langle \alpha'J' \| a_j^+ \| \alpha''J'' \rangle \langle j \| f^k \| j \rangle$$

where the Racah or U-coefficient is given in unitary form. (The Racah coefficient in the denominator with a k of zero is a convenient way of giving simple dimensionality and phase factors.)

For the higher symmetries, where the n-particle state vectors are classified according to some group chain such as $U(g) \supset G_1 \supset G_2 \supset \ldots \supset R(3)$, it will be possible to extend this technique to the higher symmetry groups if the operators are classified as irreducible tensor operators under $U(g)$, G_1, G_2, ... as well as $R(3)$. Matrix elements of operators such as $\mathbf{a}^+$ can then be expressed in terms of products of reduced $U(g)/G_1$, G_1/G_2, ...-Wigner coefficients and ordinary $R(3)$-Wigner coefficients. This follows Racah's factoring of the cfp (2) into reduced coefficients, each factor corresponding to one step in the group chain. For matrix elements of 1- (or 2-) body operators the sums over $n-1$ ($n-2$)-particle subgroup labels of $U(g)$ can then be carried out in analogy with the sum over M'', (m, m') subgroup labels above; and the resultant matrix element connecting two n-particle states can be expressed in terms of a single product of reduced Wigner coefficients and recoupling (Racah) coefficients for the higher symmetry groups. Details can be found in references 23, 51, 58, and 60. As a specific example, consider the matrix element of the two-body unit operators of $[SU(3) \supset R(3)] \times [SU(4) \supset SU_S(2) \times SU_T(2)]$ irreducible tensor character introduced in the earlier discussion of the p- (or $\bar{p}$-) shell. Using the technique outlined above, the matrix element of such an operator connecting two n-particle states can be expressed as

$$\langle p^n(\lambda'\mu')\kappa'L'; [f']\beta'S'T; JM_J M_T | T[(\lambda_2\mu_2)(\mu_2'\lambda_2')]_{L_0; S_0T_0=0; J_0=0}^{(\lambda_0\mu_0); [f_0]}$$
$$\times |p^n(\lambda\mu)\kappa L; [f]\beta ST; JM_J M_T \rangle$$

$$= -n(n-1)(-1)^{\phi}U(JLS'L_0; SL')\left[\frac{(2S'+1)}{(2S_0+1)(2S+1)}\right]^{\frac{1}{2}}$$

$$\times \sum_{(\lambda''\mu'')[f'']}\left[\frac{D^2[f'']}{d(\lambda_2\mu_2)d[f_2]D[f]D[f']}\right]^{\frac{1}{2}}$$

$$\times \sum_{\rho}\langle(\lambda\mu)\kappa L; (\lambda_0\mu_0)L_0\|(\lambda'\mu')\kappa'L'\rangle_{\rho}$$

$$\times \frac{U\big((\lambda\mu)(\mu'_2\lambda'_2)(\lambda'\mu')(\lambda_2\mu_2); (\lambda''\mu'')(\lambda_0\mu_0)\rho\big)}{U\big((\lambda\mu)(\mu_2\lambda_2)(\lambda\mu)(\lambda_2\mu_2); (\lambda''\mu'')(00)\big)}$$

$$\times \sum_{\bar{\rho}}\langle[f]\beta ST; [f_0]S_0T_0\|[f']\beta'S'T'\rangle_{\bar{\rho}}$$

$$\times \frac{U\big([f][f_2'^*][f'][f_2]; [f''][f_0]\bar{\rho}\big)}{U\big([f][f_2^*][f][f_2]; [f''][0]\big)}$$

Here the $d[f]$ and $D[f]$ are dimensionality factors for the unitary and permutation groups, respectively. The phase factor ϕ is partly dependent on phase conventions (57). The double-barred coefficients are reduced SU(3)/R(3) and SU(4)/SU(2) × SU(2) Wigner coefficients. The $(\lambda\mu)$- and $[f]$-dependent U-coefficients are SU(3) and SU(4) Racah coefficients, both given in unitary form in a notation that is a straightforward generalization of that for the ordinary angular momentum U-coefficients. The expressions are complicated by the sums over multiplicity labels ρ and $\bar{\rho}$ which arise whenever the Kronecker products $(\lambda\mu)\times(\lambda_0\mu_0)\rightarrow(\lambda'\mu')$ and $[f]\times[f_0]\rightarrow[f']$ are not simply reducible but are such that $(\lambda'\mu')$ and $[f']$ occur with a d $(\bar{d})$-fold multiplicity, with $\rho = 1,\ldots,d$; and $\bar{\rho} = 1,\ldots,\bar{d}$. (Note that in the above example the other products in the recoupling process, such as $(\lambda\mu)\times(\mu'_2\lambda'_2)$, $(\lambda''\mu'')\times(\lambda_2\mu_2)$, and $(\mu'_2\lambda'_2)\times(\lambda_2\mu_2)$, with $(\lambda_2\mu_2) = (20)$ or (01), are all simply reducible and thus all free of multiplicity labels.) For the groups SU(3) and SU(4) the needed Wigner and Racah coefficients are now available (57, 58). Even for the higher unitary groups, however, the above general form of the matrix elements may be useful since the unitary group Racah coefficients are independent of subgroup labels and can thus be evaluated in the canonical U(g) ⊃ U(g−1)...subgroup chain. Very general expressions in terms of Casimir invariants are now being derived for such higher unitary group Racah coefficients (see Louck & Biedenharn, 7). Most of the difficulties associated with the above form of the matrix elements arise from the sums over multiplicity labels, particularly for the higher unitary groups where there may be an intertwining of reduced Wigner coefficients through multiplicity labels (see Appendix). For shell model calculations in the 2s-1d shell, however, where SU(3) and SU(4) symmetries furnish a meaningful basis for a truncation scheme, the detailed application of the SU(3) and SU(4) irreducible tensor formalism has been very useful as a practical calculational tool. It is to be expected that new useful applications will be made of the unitary group irreducible tensor formalism, e.g. in the calculation of the partial widths needed for a detailed application of spectral distribution methods to the higher unitary groups.

ACKNOWLEDGEMENT
It is a pleasure to acknowledge many stimulating discussions with J. P. Draayer during the writing of this article and to thank J. P. Draayer and J. B. McGrory for permission to quote from unpublished work.

APPENDIX: THE GEL'FAND BASIS

A complete solution to the state labeling problem for the group $U(g)$ is given through the labeling by the IRs of the canonical subgroup chain $U(g) \supset U(g-1) \supset U(g-2) \supset \ldots \supset U(1)$ where each of the subgroups $U(g-s)$ is labeled by $g-s$ integers $f_{i(g-s)}$, $i = 1, \ldots, (g-s)$. It is common to label state vectors for the IRs $[f_{1g}f_{2g}\ldots f_{gg}]$ of $U(g)$ with a triangular Gel'fand pattern

$$
\left|
\begin{array}{ccccccc}
f_{1g} & f_{2g} & \cdots & & \cdots & f_{g-1g} & f_{gg} \\
 & f_{1g-1} & f_{2g-1} & \cdots & f_{g-2g-1} & & f_{g-1g-1} \\
 & & f_{1g-2} & \cdots & & f_{g-2g-2} & \\
 & & & \cdots & & & \\
 & & & f_{12} & f_{22} & & \\
 & & & & f_{11} & &
\end{array}
\right\rangle = \left| \begin{array}{c} [f_{ig}] \\ (f_{ig-1}) \end{array} \right\rangle
$$

The possible values of the subgroup labels are given by the "betweenness condition" (Weyl branching rule): $f_{ij} \geq f_{ij-1} \geq f_{i+1,j}$. Although the Gel'fand basis has few *direct* applications to problems in nuclear physics, some of the recent progress in the detailed study of this basis (7–9) is useful for such problems. The generalization by Biedenharn & Louck (7) of the concept of unit tensor operators (Wigner operators) to the higher unitary groups should be mentioned. It will be useful to organize the infinitesimal generators of $U(g)$ into operators of definite irreducible tensor rank under all of the subgroups of the chain. For the physically relevant subgroup chains containing $R(3)$ the notion of unit spherical tensor operators, e.g. $T_q^k = \sum_m a_{j'm'}^+ a_{jm} \langle jmkq | j'm' \rangle$, is familiar. Unit tensor operators of this type are to be labeled in addition with the shift index Δ (besides IR label k and subgroup label q). The matrix element of such an operator is given by

$$
\langle j'm' | T\left\langle \begin{array}{c} \Delta \\ k \\ q \end{array} \right\rangle | jm \rangle = \langle jmkq | j'm' \rangle \delta_{j',j+\Delta}
$$

that is, the reduced matrix element is given by $\delta_{j',j+\Delta}$. The shift index Δ runs over the same set of values as the subgroup label q ($\Delta = k, \ldots, -k$), but is subject to modification rules if it acts on nongeneric states (e.g. when acting on a state with $j = 0$, Wigner operators with negative values of Δ have zero matrix elements). The generalization of the concept of unit tensor operators to the group $U(g)$ leads to an operator labeled by 1. the IR labels $[F_{ig}]$, 2. a set of subgroup labels given by a lower Gel'fand pattern, and 3. a set of shift labels that can be given by an upper Gel'fand pattern, where the labels in the upper pattern satisfy the same "betweenness condition" as the F_{ij} of the lower pattern, so that the shift indices of the upper pattern run over the same set of numbers as the subgroup labels of the lower pattern (an example of Weyl's addition theorem). A Wigner operator for the group U(3), for example, would then be labeled by

$$
\mathbf{T}\left\langle
\begin{matrix}
 & & \Gamma_{11} & & \\
 & \Gamma_{12} & & \Gamma_{22} & \\
F_{13} & & F_{23} & & F_{33} \\
 & F_{12} & & F_{22} & \\
 & & F_{11} & &
\end{matrix}
\right\rangle
$$

where the Γ_{ij}, together with the IR labels $F_{i3}(\equiv\Gamma_{i3})$, give the shift effected by the operator when acting on a state vector with IR $\lfloor f_{ig}\rfloor$; where the tensor operator will connect IRs $[f_{ig}]$ only to IRs $[f'_{ig}]$ with

$$f'_{1g} = f_{1g}+\Gamma_{11}, \quad f'_{2g}+\Gamma_{12}+\Gamma_{22}-\Gamma_{11},$$
$$f'_{3g} = f_{3g}+\Gamma_{13}+\Gamma_{23}+\Gamma_{33}-\Gamma_{12}-\Gamma_{22}$$

Since the direct (Kronecker) product of $[f_{ig}]$ with $[F_{ig}]$ is in general not simply reducible, a specific IR $[f'_{ig}]$ can be reached by several shift operators Γ_{ij} with the same value for the sums $\Sigma_i \Gamma_{ij}$, and the number of different possible shift operators with fixed $\Sigma_i \Gamma_{ij}$ corresponds to the multiplicity of the Kronecker product. It will be convenient to express all U(g) irreducible tensor operators in terms of unit tensor operators of the above type. In a series of papers (7–9) Biedenharn & Louck substantiate the claim that the decomposition of U(g) irreducible tensor operators according to this prescription will put the generalization of the Wigner-Eckart theorem to the higher unitary groups on a sound group theoretical foundation. The upper pattern labels will therefore have applications to problems in nuclear spectroscopy. They can also be used to make direct contact with Littlewood's rules for the outer multiplication of two IRs of the permutation group [coupling of n-particle functions with symmetry labels $[f_{ig}]$ with p-particle functions with symmetry labels $[F_{ig}]$; (see e.g. (71))]. By labeling the shift indices Γ_{1i} by a, Γ_{2i} by b, Γ_{3i} by $c,\ldots$, the number of squares added to the ith row of the Young tableau $[f_{ig}]$ labeled a in the first step of the Littlewood process is given by $\Gamma_{1i}-\Gamma_{1(i-1)}$. The number of squares labeled b in the second step is then given by $\Gamma_{2i}-\Gamma_{2(i-1)}$, etc. The process is illustrated by the direct product $[420]\times[210]$ for U(3) below, where the upper Gel'fand patterns for $[210]$ are shown alongside the Young shapes for the product symmetries:

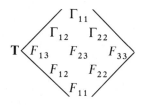

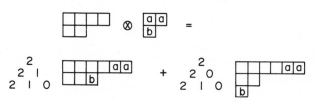

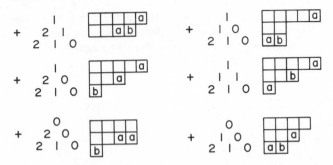

The betweenness conditions for the Γs automatically take care of all additional restrictions embodied in the Littlewood rules for operators acting on generic states. If certain $f_{ig} - f_{i+1g}$ are not sufficiently large, however, the product IRs are subject to modification rules that can best be expressed by the Littlewood rules: no two as can sit in the same column of a final tableau, as are added before bs,....

A U(g) unit tensor operator can be decomposed into a reduced Wigner operator and a U($g-1$) Wigner operator, where the matrix elements of the reduced Wigner operator depend only on the IR labels of U(g) and U($g-1$), and this process can be iterated to decompose a U(g) unit tensor operator into reduced U(g), U($g-1$), U($g-2$),... operators and finally an ordinary U(2) Wigner operator. For a U(4) unit tensor operator, e.g.

$$
\mathbf{T}\left\langle \begin{matrix} & & \Gamma_{11} & & \\ & \Gamma_{12} & & \Gamma_{22} & \\ \Gamma_{13} & & \Gamma_{23} & & \Gamma_{33} \\ F_{14} & F_{24} & & F_{34} & F_{44} \\ & F_{13} & F_{23} & F_{33} & \\ & F_{12} & & F_{22} & \\ & & F_{11} & & \end{matrix} \right\rangle = \sum_{\gamma'_{ij}} \mathbf{T}\left[\begin{matrix} & & \Gamma_{11} & & \\ & \Gamma_{12} & & \Gamma_{22} & \\ \Gamma_{13} & & \Gamma_{23} & & \Gamma_{33} \\ F_{14} & F_{24} & & F_{34} & F_{44} \\ & F_{13} & F_{23} & F_{33} & \\ & \gamma'_{12} & & \gamma'_{22} & \\ & & \gamma'_{11} & & \end{matrix} \right]
$$

$$
\times \sum_{\gamma''_{i1}} \mathbf{T}\left[\begin{matrix} & & \gamma'_{11} & & \\ & \gamma'_{12} & & \gamma'_{22} & \\ F_{13} & & F_{23} & & F_{33} \\ & F_{12} & & F_{22} & \\ & & \gamma''_{11} & & \end{matrix} \right] \times \mathbf{T}\left\langle \begin{matrix} & \gamma''_{11} & \\ \gamma''_{12} & & F_{22} \\ & F_{11} & \end{matrix} \right\rangle
$$

where the tensor operators with square brackets are reduced Wigner operators, and the sums over the γ_{ij}s run over all the integers allowed by the betweenness conditions. The matrix elements of such a unit tensor operator connecting the full Gel'fand state vector f_{ij} to a similar Gel'fand state vector f'_{ij} can thus be factored into a product of reduced Wigner coefficients and an ordinary U(2) Wigner coefficient. The matrix element of the above unit tensor operator then has the form:

$$\sum_{\substack{\gamma'_{12}, \gamma'_{22} \\ \text{fixed } (\gamma'_{12}+\gamma'_{22})}} \left\langle \begin{matrix} [f_{14}\; f_{24}\; f_{34}\; f_{44}] \\ (f_{13}\; f_{23}\; f_{33}) \end{matrix} \quad \begin{matrix} [F_{14}\; F_{24}\; F_{34}\; F_{44}] \\ (F_{13}\; F_{23}\; F_{33}) \end{matrix} \right.$$

$$\gamma' \left\| \begin{matrix} [f'_{14}\; f'_{24}\; f'_{34}\; f'_{44}] \\ (f'_{13}\; f'_{23}\; f'_{33}) \end{matrix} \right\rangle_{\Gamma}$$

$$\times \left\langle \begin{matrix} [f_{13}\; f_{23}\; f_{33}] \\ (f_{12}\; f_{22}) \end{matrix} \quad \begin{matrix} [F_{13}\; F_{23}\; F_{33}] \\ (F_{12}\; F_{22}) \end{matrix} \; \right\| \left. \begin{matrix} [f'_{13}\; f'_{23}\; f'_{33}] \\ (f'_{12}\; f'_{22}) \end{matrix} \right\rangle_{\gamma'}$$

$$\times \left\langle \begin{matrix} [f_{12}\; f_{22}] \\ f_{11} \end{matrix} \quad \begin{matrix} [F_{12}\; F_{22}] \\ F_{11} \end{matrix} \; \right| \left. \begin{matrix} [f'_{12}\; f'_{22}] \\ f'_{11} \end{matrix} \right\rangle$$

where the double-barred Wigner coefficients are reduced U(4)/U(3) and U(3)/U(2) coefficients respectively, and the single-barred U(2) coefficient can be identified with the ordinary SU(2) Wigner coefficient $\langle lm\, LM \,|\, l'm' \rangle$ with $l = \frac{1}{2}(f_{12}-f_{22})$ and $m = f_{11} - \frac{1}{2}(f_{12}+f_{22})$. The reduced U(g)/U(g−1) Wigner coefficients are functions only of the quantum numbers of U(g) and U(g−1). They can be evaluated by taking the matrix elements of the reduced U(g)-(square bracket) Wigner operators between Gel'fand states of U(g) which are of highest weight in U(g−1),.... The f'_{ij} are related to the f_{ij} by the upper Gel'fand pattern shift indices

$$f'_{j4} = f_{j4} + \Sigma_i\, \Gamma_{ij} - \Sigma_i\, \Gamma_{i(j-1)} \quad (\text{with } \Gamma_{i4} \equiv F_{i4} \text{ and } \Gamma_{i,-1} \equiv 0)$$

$$f'_{13} = f_{13} + \gamma'_{11};\; f'_{23} = f_{23} + \gamma'_{12} + \gamma'_{22} - \gamma'_{11};\; f'_{33} = f_{33} + \Sigma_i\, F_{i3} - \gamma'_{12} - \gamma'_{22}$$

$$f'_{12} = f_{12} + \gamma''_{11};\; f'_{22} = f_{22} + F_{12} + F_{22} - \gamma''_{11}$$

The values of the labels Γ_{ij} are fixed by the nature of the operator. With *fixed* values of the quantum numbers f_{ij} *and* f'_{ij} for the ket and the bra of the matrix element, the labels γ_{ij} are also fixed with the exception of γ'_{12}, γ'_{22} if the product $[f_{i3}] \times [F_{i3}] \to [f'_{i3}]$ has a multiplicity greater than 1, that is, if the betweenness conditions allow more than one set of values for γ'_{12}, γ'_{22} for fixed $\gamma'_{12} + \gamma'_{22}$. Note that the matrix element involves a sum over these multiplicity labels with resultant intertwining of the reduced U(4)/U(3) and U(3)/U(2) Wigner coefficients. Note that the U(3)/U(2) and U(2) Wigner coefficients are free of the label γ'' since γ''_{11} is determined uniquely by f'_{12} and f_{12}. Although the above considerations apply to the canonical Gel'fand subgroup chain, a similar reduction process can be used for the physically relevant subgroup chains of nuclear physics. Irreducible unit tensor operators can again be defined whose matrix elements are products of reduced Wigner coefficients, one for each step in the group chain. Whenever the group chain includes unitary groups, however, it will be advantageous to use upper Gel'fand pattern labels in the classification of irreducible tensor operators and gain a theoretically sound group characterization of the multiplicity labels. This will again result in an intertwining of the reduced Wigner coefficients for the group chain, since there will be sums over labels such as γ', which are needed as part of both the ket and the bra, respectively, of successive Wigner coefficients of the chain.

Literature Cited

1. Wigner, E. P. 1937. *Phys. Rev.* 51:106
2. Racah, G. 1949. *Phys. Rev.* 76:1352
3. Racah, G. 1965. *Group Theory and Spectroscopy,* 37:28–84. Berlin: Springer
4. Elliott, J. P. 1958. *Proc. Roy. Soc. A* 245:128, 562; Elliott, J. P., Harvey, M. 1963. *Proc. Roy. Soc. A* 272:557
5. Mottelson, B. 1960. *International School of Physics "Enrico Fermi,"* Course XV, ed. G. Racah, 54–72. New York: Academic
6. Gel'fand, I. M., Tseitlin, M. L. 1950. *Dokl. Akad. Nauk. SSSR* 71:1071
7. Louck, J. D., Biedenharn, L. C. 1970. *J. Math. Phys.* 11:2368
8. Biedenharn, L. C., Louck, J. D. 1968. *Commun. Math. Phys.* 8:89
9. Holman, W. J., Biedenharn, L. C. 1971. In *Group Theory and its Applications,* ed. E. M. Loebl, Vol. II, 1–73. New York: Academic
10. Moshinsky, M., Devi, V. S. 1969. *J. Math. Phys.* 10:455
11. Moshinsky, M. 1966. *J. Math. Phys.* 7:691
12. Flowers, B. H. 1952. *Proc. Roy. Soc. A* 212:248
13. Elliott, J. P. 1963. In *Selected Topics in Nuclear Theory,* ed. F. Janouch, 157–208. Vienna: IAEA
14. Harvey, M. 1968. In *Advances in Nuclear Physics,* ed. M. Baranger, E. Vogt, 1:67–182. New York: Plenum
15. Kramer, P., Moshinsky, M. 1968. In *Group Theory and its Applications,* ed. E. M. Loebl, 340–468. New York: Academic
16. Arima, A., Harvey, M., Shimizu, K. 1969. *Phys. Lett. B* 30:517
17. Arvieu, R. 1969. In *International School of Physics "Enrico Fermi,"* Course XL, ed. M. Jean, R. A. Ricci, 629–42. New York: Academic
18. Hecht, K. T., Adler, A. 1969. *Nucl. Phys. A* 137:129
19. Hecht, K. T. 1973. In *Proceedings of 15th Solvay Conference in Physics,* ed. I. Prigogine. New York: Gordon and Breach
20. Jones, W. P., Borgman, L. W., Hecht, K. T., Bardwick, J., Parkinson, W. C. 1971. *Phys. Rev. C* 4:580
21. Baker, F. T., Tickle, R. 1972. *Phys. Rev. C* 5:182
22. Strottman, D. 1972. *Nucl. Phys. A* 188:488
23. Ratna Raju, R. D., Draayer, J. P., Hecht, K. T. 1973. *Nucl. Phys. A* 202:433
24. Bhatt, K. H., McGrory, J. B. 1971. *Phys. Rev. C* 3:2293
25. Hecht, K. T. 1971. *Particles Nucl. 2:* 117
26. Vincent, C. M. 1967. *Phys. Rev.* 159:869
27. Kerman, A. K. 1961. *Ann. Phys. (New York)* 12:300
28. Helmers, K. 1961. *Nucl. Phys.* 23:594
29. Judd, B. R. 1968. In *Group Theory and its Applications,* ed. E. M. Loebl, 183–220. New York: Academic
30. Lawson, R. D., Macfarlane, M. H. 1965. *Nucl. Phys.* 66:80
31. Macfarlane, M. H. 1966. In *Lectures in Theoretical Physics,* ed. P. D. Kunz et al, 8C:583–677. Boulder: Univ. Colorado
32. Arvieu, R. 1969. In *Cargese Lectures in Physics,* ed. M. Jean, 3:183–250. New York: Gordon and Breach
33. Arima, A., Ichimura, M. 1966. *Prog. Theor. Phys.* 36:296
34. Ichimura, M. 1969. In *Progress in Nuclear Physics,* ed. D. M. Brink, J. H. Mulvey, 10:307–53. Oxford: Pergamon
35. Flowers, B. H., Szpikowski, S. 1964. *Proc. Phys. Soc. (London)* 84:193, 86:672
36. Ichimura, M. 1964. *Progr. Theor. Phys.* 32:757, 33:215
37. Parikh, J. C. 1965. *Nucl. Phys.* 63:214
38. Ginocchio, J. N. 1965. *Nucl. Phys.* 74:794
39. Goshen, S., Lipkin, H. J. 1968. In *Racah Memorial Volume. Spectroscopic and Group Theoretical Methods in Physics,* ed. F. Bloch, S. G. Cohen, A. deShalit, S. Sambursky, I. Talmi, 245–73. Amsterdam: North Holland
40. Hecht, K. T. 1965. *Phys. Rev. B* 139:794; Ibid 1965. *Nucl. Phys.* 63:177; Ibid 1967. *Nucl. Phys. A* 102:11
41. Hemenger, R. P., Hecht, K. T. 1970. *Nucl. Phys. A* 145:468
42. Racah, G. 1964. In *Lectures of the Istanbul Summer School in Theoretical Physics,* ed. F. Gursey, 31–36. New York: Gordon and Breach
43. Ginocchio, J. N. 1965. *Nucl. Phys.* 63:449
44. Moshinsky, M., Quesne, C. 1970. *J. Math. Phys.* 11:1631
45. Quesne, C. 1970. *Phys. Lett. B* 31:7
46. Elliott, J. P., Evans, J. A. 1970. *Phys. Lett. B* 31:157
47. Armstrong, L., Judd, B. R. 1970. *Proc. Roy. Soc. A* 315:27, 39
48. Hecht, K. T., Szpikowski, S. 1970. *Nucl. Phys. A* 158:449
49. McGrory, J. B. Unpublished

50. Akiyama, Y., Arima, A., Sebe, T. 1969. *Nucl. Phys. A* 138:273
51. Draayer, J. P. *Nucl. Phys.* In press
52. Flores, J., Perez, R. 1967. *Phys. Lett. B* 26:55
53. Harvey, M., Sebe, T. 1969. *Nucl. Phys. A* 136:459
54. Vincent, C. M. 1968. *Nucl. Phys. A* 106: 35, 104, 199
55. Vincent, C. M. 1967. *Phys. Rev.* 163: 1044
56. Glaudemans, P. W. M., Wiechers, G., Brussaard, P. J. 1964. *Nucl. Phys.* 56:529
57. Draayer, J. P., Akiyama, Y. *J. Math. Phys.* In press
58. Hecht, K. T., Pang, S. C. 1969. *J. Math. Phys.* 10:1571
59. French, J. B. 1967. *Phys. Lett. B* 26:75
60. Vergados, J. D. 1968. PhD thesis. Univ. Michigan (*Tech. Rep. 07591-2-T*)
61. Pluhar, Z. 1971. *Nucl. Phys. A* 167:33
62. Bohigas, O., Quesne, C., Arvieu, R. 1968. *Phys. Lett. B* 26:562
63. French, J. B. 1960. *Nucl. Phys.* 15:393
64. Lanford, W. A. 1969. *Phys. Lett. B* 30: 213
65. French, J. B. 1967. In *Nuclear Structure*, eds. A. Hossain, Harum-ar-Raschid, M. Islam, 85–123. Amsterdam: North Holland; French, J. B., Ratcliff, K. F. 1971. *Phys. Rev. C* 3:94, 117; Chang, F. S., French, J. B., Thio, T. H. 1971. *Ann. Phys. (New York)* 66:137
66. French, J. B., Parikh, J. C. 1971. *Phys. Lett. B* 35:1
67. Parikh, J. C. 1971. *Lectures on Group Symmetries in Nuclear Structure (Univ. Rochester Rep. UR-874-350*); Parikh, J. C. 1973. *Ann. Phys. (New York)* 76:202
68. Parikh, J. C., Wong, S. S. M. 1972. *Nucl. Phys. A* 182:593
69. Weaver, L., Biedenharn, L. C. 1972. *Nucl. Phys. A* 185:1
70. Ogura, H. 1973. *Nucl. Phys. A* 207:161
71. Hamermesh, M. 1962. *Group Theory and its Application to Physical Problems*, 249–54. Reading, Mass: Addison-Wesley

✂ 5537

LINEAR RELATIONS AMONG NUCLEAR ENERGY LEVELS

Daniel S. Koltun[1]

Department of Physics, University of Rochester, Rochester, New York

CONTENTS

INTRODUCTION

One of the features of the shell model of nuclear structure is that different nuclei "in the same shell" are expected to have closely related properties: e.g. regularities of the ground-state spins and moments of odd-A nuclei. The energy spectra of

[1] This work was supported in part by the US Atomic Energy Commission.

163

different nuclei "in the same shell" are expected to depend on the same *effective interaction* in the shell, and therefore be related to one another. If one considers simple shell model configurations (e.g. in *j-j* coupling), one finds that different spectra may be connected by certain linear relations that depend very little on the details of the effective interaction, but only, for example, on whether it is a two-body interaction.

These *linear spectroscopic relations* for nuclear spectra associated with simple configurations are the main subject of this article. We shall investigate them from two points of view, thus making contact with the two ways in which effective interactions are treated in shell model theory.

First, we shall consider the interactions as theoretically unknown, and use experimental spectra to determine features of the interaction. This empirical study may be considered an extension of earlier studies of the *j-j* coupling shell model by Talmi (1) and others. Rather than attempting to find the best two-body effective interaction from the spectra, we shall look for evidence of effective many-body interactions by testing the linear spectroscopic relations. The number of simple shell model spectra experimentally known has increased in the last few years, and makes such a review possible.

There is also a many-body theory of the effective interaction that connects the shell model of nuclear spectra back to the Schrodinger equation for nuclei with interacting nucleons. The application of this theory has made considerable advances in recent years (e.g. 2–4). One is now able to produce effective interactions, and from them calculate shell model spectra similar to the experimental spectra to provide some confidence in the approach. One test of this theory is whether it reproduces relationships for spectra of different nuclei. We shall see that this question can also be formulated in terms of effective many-body interactions, and the linear spectroscopic relations.

1   SHELL MODEL CONFIGURATIONS

1.1   Configurations

In the shell model of nuclei, as of atoms, we classify states according to *configurations* (5). The configuration is defined by giving a number of single-particle orbital states, and assigning a fixed number of particles to each orbit. The specification of the orbital state for the spherical shell model includes the radial (principal) quantum number n and the orbital angular momentum l. Since spin-orbit splitting is important in nuclei, one also includes the total angular momentum $j = l \pm \frac{1}{2}$, as well as the particle type (proton or neutron) in the labeling of orbital states. This defines a j-shell; if the configuration contains v particles in the j-shell we designate that by (j^v), with the nl understood. The maximum value of the integer v is $N = 2j+1$ for particles of the same type; for $v = N$ we have a closed shell.

A configuration is specified by $(j_1^{v_1}, j_2^{v_2} \ldots j_n^{v_n})$ where $\sum v_i = A$, the total number of particles. A single configuration in general contains many distinct states, which may be distinguished by the total angular momentum of the nucleus, J, and x, which stands for all other labels (we may ignore the angular momentum projection M_J).

Since closed shell states are nondegenerate, configurations with the largest number of closed shells have the fewest distinct states. These normally correspond to nuclear states of low excitation.

The assignment of specific configurations to spectral levels in nuclei is clearly an approximation to the physical situation. A state of simple configuration could be a solution of a one-body hamiltonian equation, such as appears in the Hartree-Fock approximation. In this approximation, all the states belonging to a given configuration are degenerate. Improving the approximation by including the interaction among nucleons (in perturbation theory) will lead to a breaking of the degeneracy, and a separation of the energy levels $E_{J_x}(j_1^{x_1} \ldots j_n^{x_n})$ for each configuration. Of course, the same interaction will also usually mix different configurations, so that eigenstates no longer have pure configurations.

There is probably considerable configuration mixing in most nuclear states. There is evidence from nuclear direct reactions involving particle transfer that configurations lying nearby in unperturbed energy mix into low energy states. The strong, short-range repulsion and the tensor force are expected to admix into these low levels, configurations quite high in unperturbed energy. One could deal with this problem by enlarging the shell model space to include the strongly interacting configurations. This may be practical for a few nearby configurations, but the dimension of the degenerate perturbation problem grows rapidly with addition of configurations.

1.2   Effective Interaction

In spite of the large amount of configuration mixing, the nuclear shell model based on simple configurations often works surprisingly well. We shall see particular examples of this later. This leads one to reformulate the shell model problem so that the features of the pure configurations remain, by introducing an *effective interaction*. This effective interaction is a hamiltonian operator defined to act only in the *model space* of states of a single configuration (or a small number of configurations), but whose eigenvalues correspond to certain physical energy levels of the nucleus. The eigenstates in the model space are much simpler than the physical states to which they correspond, since the former are chosen to belong to one configuration. The effective interaction must be quite complicated, since it carries all the information about configuration mixing. It will usually be a many-body operator, involving two, three, four, or more particles, even though the interaction in the original Schrodinger problem may have been a two-particle potential.

Calculation of the effective interaction from the true interaction is a formidable problem; we shall discuss some of its aspects below, in section 4. However, it may be possible to invert the spectroscopic problem by treating the effective interaction as an *unknown* to be determined from the experimental energy levels supposedly corresponding to the model configuration. This approach was first introduced to atomic spectroscopy by Bacher & Goudsmit (6). It has often been used in the nuclear shell model (1, 7) not only to avoid the problem of configuration mixing, but also because of ignorance of the details of the nucleon-nucleon interaction.

This approach is generally more interesting for those cases in which the effective interaction turns out to be simple, for example those involving only few-body operators (i.e. two- or three-body), since these have the fewest undetermined parameters and therefore lend themselves to the possibility of predicting new levels. In contrast, if for the spectra of nuclei corresponding to the configurations (j^v) (plus closed shells) we always require v-body forces, one can make no predictions for other nuclei.

For the most part, analysis of shell model spectra has been made in terms of two-body effective interactions. We shall see that this is sometimes a good approximation, but that there is evidence for three- and more-body interactions from shell model spectra.

1.3   Simple Configurations

We shall consider configurations with one or two open j-shells, plus a core of closed shells, and denote them by (j^v) or (j^v, k^μ), ignoring the cores. The j-shell is filled for $v = N$ and the k-shell for $\mu = M$. We shall treat proton and neutron shells as distinct, rather than introducing isospin notation. Isospin is a redundant label for the states we shall consider, and is not convenient for some particle-hole cases.

The model states (labeled by Jx) are eigenstates of the effective interaction $\mathcal{H}$ in the model space of the configuration, with energies E_{Jx}:

$$E_{Jx}(j^v) = \langle j^v; Jx | \mathcal{H}(j) | j^v; Jx \rangle \qquad \text{1.1a.}$$
$$E_{Jx}(j^v, k^\mu) = \langle j^v k^\mu; Jx | \mathcal{H}(jk) | j^v k^\mu; Jx \rangle \qquad \text{1.1b.}$$

As discussed above, $\mathcal{H}$ is a many-body operator, and depends on the choice of model space. We write $\mathcal{H}$ in terms of operators of definite particle rank α

$$\mathcal{H}(j) = \sum_{\alpha=0}^{N} \mathcal{V}(\alpha) \qquad \text{1.2a.}$$

or particle rank in each orbit

$$\mathcal{H}(jk) = \sum_{\alpha=0}^{N} \sum_{\beta=0}^{M} \mathcal{V}(\alpha, \beta) \qquad \text{1.2b.}$$

where, for example, $\mathcal{V}(\alpha, \beta)$ is an interaction among α particles in the j-shell and β particles in the k-shell. When substituted into equations 1, only the terms of $\mathcal{H}$ with $\alpha \leq v$, $\beta \leq \mu$ can contribute. For example, the core energy ($v = \mu = 0$) is given by

$$E(0) \equiv E_0(j^0) = \langle \text{core} | \mathcal{V}(0) | \text{core} \rangle$$
$$E(0,0) \equiv E_0(j^0, k^0) = \langle \text{core} | \mathcal{V}(0,0) | \text{core} \rangle$$

and the one-particle energy by

$$E(j) \equiv E_j(j^1) = \langle j; J = j | \mathcal{V}(0) + \mathcal{V}(1) | j; J = j \rangle$$
$$= E(0) + \langle j; j | \mathcal{V}(1) | j; j \rangle$$

and similarly for $E(j,0) \equiv E_j(j^1, k^0)$ and $E(0,k) \equiv E_k(j^0, k^1)$. Note that we have simplified the notation by dropping the Jx labels when the state is unique.

A configuration j^{N-v} may also be treated as a *hole* configuration with v-holes in the j-shell (written: j^{-v}) with respect to a core which now includes the filled j-shell: j^N. The hole energy spectrum is given by

$$E_{Jx}(j^{-v}) = \left\langle j^{N-v}; Jx \left| \sum_{\alpha=0}^{N-v} \mathcal{V}(\alpha) \right| j^{N-v}; Jx \right\rangle \qquad 1.3.$$

and similarly for $E_{Jx}(j^{-v}, k^{\mu})$.

There are a number of simple spectra we shall need later. They are:

(a) the closed shell energies:

$$E(0),\ E(N),\ E(0,0),\ E(0,M),\ E(N,M) \qquad 1.4.$$

where we now use N for j^N, M for k^M.

(b) the one-particle-like spectra:

$$E(j),\ E(j^{-1}),\ E(j,0),\ E(0,k),\ E(j,N),\ E(0,k^{-1}),\ E(j^{-1},N) \qquad 1.5.$$

where we omit some possibilities of interchanging $j \leftrightarrow k$, $N \leftrightarrow M$.

(c) the two-particle-like spectra:

$$E_J(j^2),\ E_J(j^{-2}),\ E_J(j^2, N),\ E_J(j^{-2}, N),\ \text{etc with } J = 0, 2, 4, \ldots, 2j-1;$$
$$E_J(j,k),\ E_J(j, k^{-1}),\ E_J(j^{-1}, k^{-1}) \text{ with all integral } |j-k| \leqq J \leqq j+k \qquad 1.6.$$

2 LINEAR SPECTROSCOPIC RELATIONS

For a given effective interaction $\mathcal{H}$ in a single configuration, say j^v, all the energy levels $E_{Jx}(j^v)$ are determined for all numbers $v = 0, 1, \ldots, N$ of nucleons in the shell. If interactions of all particle ranks up to N are present in $\mathcal{H}$, then the $E_{Jx}(j^v)$ are all independent, in the sense that any one energy level can be varied independently of the others by some variation of the interaction. If, however, $\mathcal{H}$ has a maximum particle rank $\alpha < N$, then this is no longer true, and relationships may exist that determine the dependence of some E_{Jx} on others. These relationships are in general highly nonlinear, since the E_{Jx} are roots obtained by diagonalizing matrices, whose elements are linear in $\mathcal{H}$.

It is possible to find some linear spectroscopic relations among the E_{Jx}, under different assumptions about $\mathcal{H}$. Most of the relations that are known hold if the maximum particle rank is two; this is the usual assumption made in shell model spectroscopy. There are also linear spectroscopic relations that hold for interactions of maximum particle rank greater than two. (We refer to these as many-body effective interactions.)

In the following subsections we review a number of linear spectroscopic relations that hold for energy spectra of nuclei with simple configurations: (j^v) or (j^v, k^{μ}). In section 3 we shall exhibit cases of nuclear spectra to which these spectroscopic relations may be applied and tested. If we find that relations dependent on a two-body interaction do not hold exactly, we will have a quantitative measure of the importance of many-body effective interactions in these spectra.

2.1   Two-Body Effective Interaction

In this subsection we assume that there are no terms in the effective interaction of particle rank greater than two, that is, $\alpha \leq 2$ in equation 1.2a and $\alpha + \beta \leq 2$ in equation 1.2b.

It is convenient to separate the effects of the interactions of different rank (here 0, 1, 2) by introducing *interaction energies*:

$$\varepsilon(j) = E(j) - E(0) = \langle j; j | \mathscr{V}(1) | j; j \rangle \quad [\text{or} = E(j,0) - E(0,0), \text{etc}]$$
$$\text{2.1a.}$$

$$W_J(j^2) = E_J(j^2) + E(0) - 2E(j) = \langle j^2; J | \mathscr{V}(2) | j^2; J \rangle \qquad \text{2.1b.}$$

$$W_J(j,k) = E_J(j,k) + E(j,0) - E(0,k)$$
$$= \langle j,k; J | \mathscr{V}(1,1) | j,k; J \rangle \qquad \text{2.1c.}$$

Since there is only one state of (j^2) or (j,k) with a given value of $J(M_J)$, the matrix elements $\varepsilon(j)$ and W_J [and the closed shell energy $E(0)$ or $E(0,0)$] completely specify the effective interaction (of rank ≤ 2). Then, all the energy levels of the shell $E_{Jx}(j^v)$ or $E_{Jx}(j^v, k^\mu)$ are determined by these quantities. However, as mentioned above, if there is more than one state x for a given J, E_{Jx} will not be linear in ε and W, since matrix diagonalization in the different x-states is involved.

SPECTROSCOPIC AVERAGE   However, one finds a simple linear relation for the *average* of an energy spectrum, which is defined by

$$\bar{E} \equiv \sum_{Jx} [J] E_{Jx} / \sum_{Jx} [J] \qquad \text{2.2.}$$

where the sum is over all states, with the weighting $[J] \equiv 2J + 1$ to count each magnetic substate M_J once. The relations for the averages are

$$\bar{E}(j^v) = \tfrac{1}{2} v(v-1) \bar{W}(j^2) + v\varepsilon(j) + E(0), \ (v \geq 2) \qquad \text{2.3a.}$$
$$\bar{E}(j^v, k^\mu) = v\mu \bar{W}(j,k) + \bar{E}(j^v, 0) + \bar{E}(0, k^\mu) - E(0,0), \ (v, \mu \geq 1) \qquad \text{2.3b.}$$

where $\bar{E}(j^v, 0)$ and $\bar{E}(0, k^\mu)$ can be further reduced by using equation 2.3a. The forms of equations 2.3a and 2.3b are intuitive: the one-particle energy $\varepsilon(j)$ [or $\varepsilon(k)$] is multiplied by the number of j- (or k-) particles; the average two-particle interaction energy $\bar{W}$ is multiplied by the number of particle pairs. The derivation of these relations is discussed by deShalit & Talmi (8, pp. 337–38) and by French (9). The averages reduce to single energy values for the simple spectra of equations 1.4 and 1.5; e.g. for $v = N$ or $N-1$, $\mu = M$.

PARTICLE-HOLE RELATIONS   Linear relations obtain between the spectra of v-holes and v-particles in a single j-shell [see (8), p. 229–30 and (5), Ch. 12].

$$E_{Jx_c}(j^{-v}) - v\varepsilon(j^{-1}) - E(N) = E_{Jx}(j^v) - v\varepsilon(j) - E(0) \qquad \text{2.4.}$$
$$\varepsilon(j^{-1}) = E(j^{-1}) - E(N) \qquad \text{2.1d.}$$

where Jx_c defines the (j^{-v}) state *conjugate* (under particle-hole transformation) to the state Jx of (j'). Every state x has its conjugate x_c; it is the state (of v-particles) that completes the closed shell j^N when it is coupled to the state $(j^{N-v})Jx$ (equation 1.3). [See (8), and French (10) where the particle-hole transformation in second-quantized form is discussed.] Equation 2.4 says that the v-hole excitation spectrum (i.e. measured from the ground state) is identical with the v-particle excitation spectrum.

A special case of equation 2.4 occurs when v equals 2:

$$W_J(j^{-2}) = W_J(j^2) \qquad\qquad 2.5.$$

where we define the two-hole interaction energy

$$W_J(j^{-2}) = E_J(j^{-2}) + E(N) - 2E(j^{-1}) \qquad\qquad 2.1e.$$

For two shells, one has the Pandya relation (11) for particle-hole nuclei:

$$W_J(j,k^{-1}) = W_J(j^{-1},k) = - \sum_I [I] W(jkkj; JI) W_I(j,k) \qquad\qquad 2.6.$$

where we use the Racah coefficient $W(jkkj; JI)$, and define the particle-hole interaction energy

$$W_J(j,k^{-1}) = E_J(j,k^{-1}) + E(0,M) - E(j,M) - E(0,k^{-1}) \qquad\qquad 2.1f.$$

The hole-hole case (j^{-1},k^{-1}) can be related either to the particle-particle case or the particle-hole case by

$$W_J(j^{-1},k^{-1}) = W_J(j,k) \qquad\qquad 2.7a.$$
$$W_J(j^{-1},k^{-1}) = T_J\{W_I(j,k^{-1})\} \qquad\qquad 2.7b.$$

where we have introduced the *Pandya* transform of a spectrum:

$$T_J\{W_I(j,k)\} \equiv - \sum_I [I] W(jkkj; JI) W_I(j,k) \qquad\qquad 2.8.$$

and where

$$W_J(j^{-1},k^{-1}) = E_J(j^{-1},k^{-1}) + E(N,M) - E(N,k^{-1}) - E(j^{-1},M) \qquad 2.1g.$$

We note that the same transformation (equation 2.8) takes $(j,k) \rightleftarrows (j^{-1},k)$.

MULTIPOLES The particle-hole relations of equations 2.6 and 2.7 can be put in more symmetrical form in terms of the coefficients of the *multipole expansion* of the interaction energy. We define the *multipole coefficients*

$$\alpha^\lambda(j,k) = ([\lambda]/[j][k])^{1/2} \sum_J (-1)^{j+k-\lambda} [J] W(jkjk; J\lambda) W_J(j,k) \qquad 2.9.$$

and similarly for $\alpha^\lambda(j,k^{-1})$, $\alpha^\lambda(j^{-1},k^{-1})$. These coefficients enter naturally in the expansion of the effective interaction $\mathscr{V}(j,k)$ in terms of multipole tensors in the separate shells, and are discussed in detail by French (10). Comparison of the multipole coefficients for a number of two-body spectra was initiated by Moinester, Schiffer & Alford (12) and has been extended by Schiffer (13).

When rewritten in multipole form, equations 2.6 and 2.7 become

$$\alpha^{\lambda}(j,k) = (-1)^{\lambda+1}\alpha^{\lambda}(j,k^{-1}) = (-1)^{\lambda+1}\alpha^{\lambda}(j^{-1},k) = \alpha^{\lambda}(j^{-1},k^{-1}) \qquad 2.10.$$

In the Pandya relation, one needs the complete spectrum for (j,k) to get each level of (j,k^{-1}), while equation 2.10 shows that each $\alpha^{\lambda}(j,k)$ goes into one $\alpha^{\lambda}(j,k^{-1})$. Of course one requires a complete spectrum to obtain the α^{λ}.

The coefficient with $\lambda = 0$ (monopole) plays a special role, since

$$\alpha^{0}(j,k) = ([j][k])^{-1}\sum_{J}[J]W_{J}(j,k) = \bar{W}(j,k) \qquad 2.11.$$

We shall make use of this below when we compare particle-hole spectra with spectral averages.

SENIORITY   Under further assumptions about the effective two-body interaction $\mathscr{V}(2)$ in the (j^{n}) configuration (or if $j \leq 7/2$), seniority (v) may be a good quantum number $(1,9)$. In this case there will be a linear relation for the average of the energy spectrum over states of fixed seniority v for v-particles

$$\bar{E}_{v}(v) \equiv \sum_{Jx}[J]E_{vJx}/\sum_{Jx}[J] \qquad 2.12.$$

For two particles $v = 0$ or 2 we define

$$\begin{aligned}\bar{V}_{2} &\equiv \sum_{J>0}[J]W_{J}(j^{2})/\sum_{J>0}[J]\,; \;(J\text{ even}) \\ \bar{V}_{0} &\equiv W_{0}(j^{2}) \qquad (J=0\text{ only})\end{aligned} \qquad 2.13.$$

The linear relation for the seniority average may be written (8, p. 334)

$$\begin{aligned}\bar{E}_{v}(v) &= \tfrac{1}{2}v(v-1)\bar{V}_{2} + \tfrac{1}{2}(v-v)(2j+3-v-v)(2j+1)^{-1}(\bar{V}_{0}-\bar{V}_{2}) \\ &\quad + v\varepsilon(j) + E(0)\end{aligned} \qquad 2.14.$$

For ground states we have unique seniority (and J), and equation 2.14 can be simplified to give

$$E_{g.s.}(j^{2}) = E(0) + v\varepsilon(j) + \tfrac{1}{2}\alpha v(v-1) + \beta[v/2] \qquad 2.15.$$

where $[v/2]$ is the largest integer $\leq \tfrac{1}{2}v$, and $\alpha = (2j\bar{V}_{2}-\bar{V}_{0})(2j+1)^{-1}$, $\beta = (2j+2)(\bar{V}_{0}-\bar{V}_{2})(2j+1)^{-1}$.

EXTENSIONS AND OTHER RELATIONS   There are other possible linear spectroscopic relations that go beyond the scope of this article. For example, one can introduce averaging for other symmetries (than seniority), providing the interaction is appropriately invariant. One can also consider averaging over several orbits, and obtain a generalization of equation 2.3. However, one does not usually have sufficient experimental information to make use of such relations.

2.2   Many-Body Effective Interactions

We now consider many-body effective interactions, interactions whose highest particle rank (see equation 1.2) is greater than two. These have received very little attention in shell model theory, but some relations for energy averages and particle-hole related spectra are known.

SPECTROSCOPIC AVERAGE   The generalization of equation 2.3a for the (j^v) configuration can be written

$$\bar{E}(j^v) = \sum_{m=1}^{v} \binom{v}{m}\bar{W}(j^m) \qquad 2.16.$$

where we define the average m-body interaction energy inductively

$$\bar{W}(j^m) = \bar{E}(j^m) - \sum_{l=1}^{m} \binom{m}{l}\bar{W}(j^{m-l}) \qquad 2.17.$$

with $W(j^0) \equiv E(0)$ and $W(j) \equiv \varepsilon(j)$. Clearly $\bar{W}(j^m)$ is the average m-body matrix element of the m-body rank interaction

$$\bar{W}(j^m) = \sum_{Jx} [J]\langle j^m; Jx | \mathscr{V}(m) | j^m; Jx \rangle / \sum_{Jx} [J] \qquad 2.18.$$

The generalization of equation 2.3b may be similarly obtained. Equation 2.16 is an example of scalar averaging of many-body operators that has been introduced and developed by French (9). This equation holds for any rank interaction, but cannot be tested against experimental spectra unless the highest particle rank is less than the particle number.

PARTICLE-HOLE RELATION   A generalized particle-hole relation, which relates the four nuclei of equation 2.6 (Pandya relation) and 2.7, has been derived by Koltun & West (14, 15).

$$W_J(j,k) + W_J(j^{-1},k^{-1}) = -\sum_{I} [I] W(jkkj; JI)\{W_I(j,k^{-1}) + W_I(j^{-1},k)\} \qquad 2.19.$$

This equation holds for an interaction $\mathscr{V}(\alpha, \beta)$ (equation 1.2b) for which either $\alpha \leq 1$ or $\beta \leq 1$, which includes all particle ranks through *three*, and selected higher rank interactions. Relation 2.19 is not practical as it stands, since at least one of the nuclei has a large proton excess, and will not be accessible. However, if j and k are proton and neutron shells with the same (nlj), then (j,k^{-1}) and (j^{-1},k) are mirror nuclei, and $W_I(j,k^{-1}) = W_I(j^{-1},k)$, up to Coulomb effects. Equation 2.19 then takes the form

$$W_J(j,k) + W_J(j^{-1},k^{-1}) = -2\sum_{I} [I] W(jkkj; JI) W_I(j,k^{-1}) \qquad 2.20.$$

In multipole form, this becomes

$$\alpha^\lambda(j,k) + \alpha^\lambda(j^{-1},k^{-1}) = 2(-1)^{\lambda+1}\alpha^\lambda(j,k^{-1}) \qquad 2.21.$$

SENIORITY   An expression for the average energy $\bar{E}_v(j^v)$ for states of fixed seniority, in terms of an effective interaction of given rank α, has been given by Quesne (16). This generalization of equation 2.14 is given not as an algebraic expression in the quantum numbers, but in terms of the coefficients of fractional percentage.

3   EXPERIMENTAL SPECTRA

3.1   Spectra for Simple Configurations

We now consider the interpretation of some experimental energy spectra in terms of the relations for states of simple configurations, which we discussed in section 2. Unless otherwise specified, we use the most recent energy spectral information in the *Nuclear Data Sheets* (17), and binding energies in the tables of Wapstra & Gove (18). For the (j^ν) configurations, we discuss examples from the $1d_{3/2}$-, $1f_{7/2}$-, and $2d_{5/2}$-shells (principal quantum numbers start from 1). We use (p) or (n) to denote a proton or neutron shell. For the (j^ν, k^μ) configurations, interesting examples occur in the $1d_{3/2}(\mathrm{p})$-$1d_{3/2}(\mathrm{n})$, $1d_{3/2}(\mathrm{p})$-$1f_{7/2}(\mathrm{n})$, $1d_{3/2}(\mathrm{p})$-$2p_{3/2}(\mathrm{n})$, $1f_{7/2}(\mathrm{p})$-$1f_{7/2}(\mathrm{n})$, $1f_{7/2}(\mathrm{p})$-$2p_{3/2}(\mathrm{n})$, and $1g_{9/2}(\mathrm{p})$-$2d_{5/2}(\mathrm{n})$ shells.

NUCLEI WITH CONFIGURATION (j^ν) Consider the spectra of Figure 1 as an example of the assignment of simple configurations to energy levels. The simple configuration assignment would be $(j^{\pm 2})$ for these cases, with $j = 1f_{7/2}(\mathrm{n})$ for Ca and $j = 2d_{5/2}(\mathrm{n})$ for Zr: ^{42}Ca as two $(1f_{7/2})$ neutrons and a ^{40}Ca core, etc. The assigned configuration spectra are shown with heavy lines: $J = 0, 2, 4, 6$ for $(1f_{7/2}^2)$ and $J = 0, 2, 4$ for $(2d_{5/2}^2)$. We know from particle transfer reactions that the configurations are not pure; in particular, there is mixing in the $J = 0$ and 2 states shown in Figure 1. However, the model states are those in which the chosen configuration predominates, and, as here, are generally the lowest levels of each possible spin (J) and appropriate parity.

To test the particle-hole relation (equation 2.5), we compare the interaction

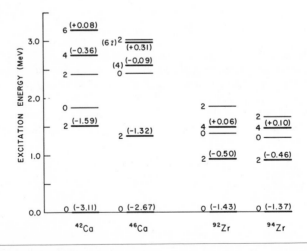

Figure 1   Comparison of the $(j^{\pm 2})$ energy spectra (heavy lines) of ^{42}Ca with ^{46}Ca, and of ^{92}Zr with ^{96}Zr. Interaction energies are indicated in parentheses.

energies $W_J(j^{\pm 2})$ (equations 2.1b and 2.1e) that are indicated in parenthesis in Figure 1, for the two pairs shown. We see that the $W_J(j^{-2})$ are consistently higher than the $W_J(j^2)$, by about 200 keV for the Ca isotopes, and by about 50 keV for the Zr cases. The nuclei ^{50}Ti and ^{54}Fe (not illustrated), which are $(j^{\pm 2})$ for $1f_{7/2}$(p), have spectra quite similar to the $1f_{7/2}$(n)-nuclei ^{42,46}Ca, and again $W_J(j^{-2})-W_J(j^2) \sim$ 200–300 keV. We have no test in the $1d_{3/2}$-shell, since the (j^2) and (j^{-2}) configurations occur in the same nuclei. The deviations from equation 2.5 indicate many-body effective interactions.

We can also compare the averages of spectral energies $\bar{E}(j^v)$ for nuclei with (j^v) configurations. We have equation 2.3a, which holds for two-body interactions, or equation 2.16, which holds for all ranks. The latter, with equations 2.17 and 2.18, would provide a complete decomposition into interactions of different particle rank if we had the complete (j^v) for all $v = 1.2,\ldots,N$. It is convenient to invert equation 2.3a in the form

$$B_j(v) = 2(v-1)^{-1}\{v^{-1}[\bar{E}(j^v) - E(0)] - \varepsilon(j)\} \qquad 3.1.$$

with $v \geq 2$, where $B_j(v)$ may be computed from the average of the experimental (j^v) spectrum, and the ground-state energies for $v = 0, 1$. For equation 2.3a to hold, the experimentally determined values of $B_j(v)$ must be independent of v and equal to $\bar{W}(j^2)$. Variation in the value of $B_j(v)$ with v is an indication of the presence of many-body interactions. In this case, only $B_j(2) \equiv \bar{W}(j^2)$, and $B_j(v)$ is a polynomial in v, as may be seeen from equations 2.16 and 2.17: for example, a three-body interaction leads to a linear v-dependence:

$$B_j(v) = \tfrac{1}{3}(v-2)\bar{W}(j^3) + \bar{W}(j^2), \; v \geq 3 \qquad 3.2.$$

The practical difficulty with equation 3.1 is the requirement of complete (j^v) spectra to define the average. However, this requires only the ground-state energies for the filled shell $v = N$, and $N-1$, which have unique states. (We must also have the ground-state energies for $v = 0, 1$.) Two spectra, which are available in several shells, are $v = 2$ and $N-2$, such as those in Figure 1.

In Table 1 we display the values of $B_j(v)$ obtained for a number of j-shells in which the $\bar{E}(j^v)$ can be obtained for $v = 0, 1, 2, N-2, N-1,$ and N. The most striking feature is the change in $B_j(v)$ with v, generally to more positive values

Table 1 Average interaction energies for (j^v) configuration nuclei

v	$j = 1d_{3/2}$(n) Nucleus	$B_j(v)^a$	$j = 1d_{3/2}$(p) Nucleus	$B_j(v)^a$	$j = 1f_{7/2}$(n) Nucleus	$B_j(v)^a$	$j = 1f_{7/2}$(p) Nucleus	$B_j(v)^a$	$j = 2d_{5/2}$(n) Nucleus	$B_j(v)^a$
2	^{34}S	−0.998	^{38}Ar	−0.051	^{42}Ca	−0.470	^{50}Ti	0.068	^{92}Zr	−0.222
$N-2$	^{34}S	−0.998	^{38}Ar	−0.051	^{46}Ca	−0.320	^{54}Fe	0.292	^{94}Zr	−0.131
$N-1$	^{35}S	−0.372	^{39}K	0.050	^{47}Ca	−0.260	^{55}Co	0.310	^{95}Zr	−0.125
N	^{36}S	−0.394	^{40}Ca	0.034	^{48}Ca	−0.252	^{56}Ni	0.320	^{96}Zr	−0.125

^{a} $B_j(v)$ is defined in equation 3.1, and is given in MeV. Data were obtained from references 17 and 18.

with increasing v. The approximate linearity (in v) of the $1f_{7/2}$-cases means that the $B_j(v)$ can be fit rather well with a three-body (repulsive) interaction, as in equation 3.2. The other cases are less linear (see Table 4).

NUCLEI WITH CONFIGURATION (j^v, k^μ) We now consider spectral averages for (j^v, k^μ) configurations. The relation that holds for two-body interactions was given in equation 2.3b. For purposes of comparing spectra, we introduce a number of quantities derived from experimental energies:

$$
\begin{aligned}
A(j,k) &= \bar{E}(j,k) + E(0,0) - E(j,0) - E(0,k) \\
A(j,k^{-1}) &= -\bar{E}(j,k^{-1}) - E(0,M) + E(j,M) + E(0,k^{-1}) \\
A(j,M) &= M^{-1}[E(j,M) + E(0,0) - E(0,M) - E(j,0)] \\
A(j^{-1},k^{-1}) &= \bar{E}(j^{-1},k^{-1}) + E(N,M) - E(j^{-1},M) - E(N,k^{-1}) \\
A(j^{-1},M) &= -M^{-1}[E(j^{-1},M) - E(N,M) - E(j^{-1},0) + E(N,0)] \\
A(N,k^{-1}) &= -N^{-1}[E(N,k^{-1}) - E(N,M) - E(0,k^{-1}) + E(0,M)]
\end{aligned}
\qquad 3.3.
$$

Each quantity A is a second difference of energy averages, referring to four different nuclei. Note that different closed shell energies appear in the different A: $E(0,0)$, $E(N,0)$, $E(0,M)$ and $E(N,M)$ [as compared to the $B_j(v)$ which all refer to the same $E(0)$]. The A are defined so that they must all be equal for a given set of (j^v, k^μ) if equation 2.3b is to hold, and all A equal $\bar{W}(j,k)$. Deviation from equality implies many-body interactions.

The values of the various A for several shells are displayed in Table 2. The variation of the different A for the $1d_{3/2}(p)$-$1d_{3/2}(n)$ and $1f_{7/2}(p)$-$1f_{7/2}(n)$ shells is quite noticeable; the differences are as big as a few hundred keV. For the other cases,

Table 2   Average interaction energies for (j^v, k^μ) configuration nuclei

Orbits	$1f_{7/2}$-$1d_{3/2}$		$1f_{7/2}$-$1f_{7/2}$		$1f_{7/2}$-$2p_{3/2}$		$1g_{9/2}$-$2d_{5/2}$	
	Nucleus[a]	A[b]	Nucleus[a]	A[b]	Nucleus[a]	A[b]	Nucleus[a]	A[b]
$A(j,k)$	^{38}Cl	-1.031	^{42}Sc	-1.400	^{50}Sc	-0.597	^{92}Nb	-0.401 $(-0.412)^c$
$A(j,k^{-1})$	^{40}K	-0.991	^{48}Sc	-0.872			^{96}Nb	-0.399 $(-0.337)^c$
$A(j,M)$	^{41}Ca	-1.012	^{49}Sc	-0.942			^{97}Nb	-0.385 $(-0.386)^c$
$A(j^{-1},k^{-1})$			^{54}Co	-0.783				
$A(N,k^{-1})$	^{47}K	-0.980	^{55}Ni	-0.772				
$A(j^{-1},M)$					^{59}Co	-0.587		

[a] Nucleus corresponding to (j^v, k^μ).

[b] The energies A are defined in equation 3.3, and are given in MeV. Data were obtained from references 17 and 18.

[c] Data from reference 22.

the differences are only a few tens of keV. (Note that the average two-body interaction $A(j,k) = \bar{W}(jk)$ is also smaller in these cases.) For the most part, the averages A get more positive (repulsive) as the particle number is increased, as was true for the $B_j(v)$.

Now we look at the particle-hole relations for $(j^{\pm 1}, k^{\pm 1})$, given in equations 2.6, 2.7, and 2.19. Comparisons are illustrated for three cases in Figures 2–4. The (j, k^{-1}) spectra have been transformed to $T_l[W_J(j, k^{-1})]$ using the Pandya transformation (equation 2.8), so that a direct comparison of $W_l(j,k)$ with $T_l[W_J(j,k^{-1})]$ is a test of the particle-hole relation (equation 2.6). Figure 2 shows the spectra of ^{38}C and ^{40}K. The lower multiplets of levels, with $J = 2, 3, 4, 5$ are assigned to the $[1d_{3/2}(p)^{\pm 1}, 1f_{7/2}(n)]$ configurations. This is the oldest known case of a pair of particle-hole related spectra. Goldstein & Talmi (19) and Pandya (11) independently derived linear relations for these energy levels, based on a two-body effective interaction, and pointed out that the deviation of the multiplets from these relations was small (10–50 keV). Pandya's relation was that given in equation 2.6, and is applicable to any $(j, k^{\pm 1})$ pair, while the derivation of (19) was particular to a $j = 3/2$ shell.

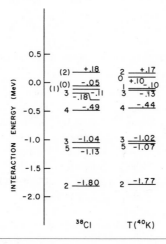

Figure 2 Comparison of energy levels of ^{38}Cl with Pandya transform of energy levels of ^{40}K.

Figure 2 also shows the recently assigned levels (20, 21) $J = 0, 1, 2, 3$ pertaining to the $[1d_{3/2}(p)^{\pm 1}, 2p_{3/2}(n)]$ configurations. For the latter case, one does not have the interaction energy, but the excitation energy above the lower multiplet. The $d_{3/2}^{\pm 1}$-$p_{3/2}$ multiplet shows somewhat larger deviations (50–100 keV) than the $d_{3/2}^{\pm 1}$-$f_{7/2}$ multiplet.

For ^{92}Nb-^{96}Nb (Figure 3) we have used the interaction energies of Comfort et al (22). These include a measured mass difference for ^{96}Nb-^{96}Zr, which differs from that given in *Nuclear Data Sheets*; had we used the latter, the transformed spectrum

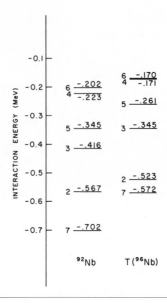

Figure 3   Comparison of energy levels of ^{92}Nb with Pandya transform of energy levels of ^{96}Nb.

$T(^{92}$Nb) would be shifted down by ~ 60 keV from that illustrated, bringing the levels into somewhat closer average agreement. This is also reflected in Table 2, in which one sees closer agreement for the A for the Nuclear Data masses, than for those of Comfort et al. However, on theoretical grounds (see section 5) one might expect an average shift of 20–40 keV between ^{92}Nb and ^{96}Nb. The relative level spacings differ by up to 100 keV. It is interesting that if one instead compares $T(^{92}$Nb) with ^{96}Nb, as shown by Comfort et al, the main disagreement with the particle-hole relation (now equation 2.6) is for the $J = 2$ levels, which now lie highest.

For the $1f_{7/2}{}^{\pm}$(p)-$1f_{7/2}{}^{\pm}$(n) configurations, one has the three spectra shown in Figure 4. The absolute energy for ^{54}Co is uncertain, since the mass of ^{55}Ni is unknown, but it has been estimated from Coulomb energy differences (15), with an uncertainty of ± 50 keV. Here one can test not only the particle-hole relations (equations 2.6 and 2.7) between pairs of spectra, but also the relation (equation 2.19) for three spectra. Here we find that the deviations from the relations are of the order of several hundreds of keV, which is consistent with the behavior of the spectral averages in Table 2.

It is also useful to compare the spectra in multipole form, using the coefficients α^{λ} defined in equations 2.9. This is done in Table 3, where for the particle-hole nuclei we list the values of $(-1)^{\lambda+1}\alpha^{\lambda}(j, k^{-1})$, using the particle-hole relation of equation 2.10. Now we have separated the average energies ($\lambda = 0$), which were also given in Table 2, from the multipole terms ($\lambda > 0$), which describe the splitting.

Table 3 Multipole coefficients for particle-hole nuclei (in keV)[a]

Nuclei	³⁸Cl	T(⁴⁰K)	³⁸Cl	T(⁴⁰K)	⁴²Sc	T(⁴⁸Sc)	⁵⁴Co	⁹²Nb	⁹⁶Nb
Orbits	$1f_{7/2}$-$1d_{3/2}$		$2p_{3/2}$-$1d_{3/2}$		$1f_{7/2}$-$1f_{7/2}$			$1g_{9/2}$-$2d_{5/2}$	
λ									
0	−1031	−991	−	−	−1400	−872	−772	−401(−412)[b]	−399(−337)[b]
1	153	167	−71	−145	−93	−249	−309	−77	−56
2	−391	−389	−128	−95	−738	−763	−879	−162	−146
3	−41	−48	−59	−5	−256	−224	−228	−35	−16
4					−454	−360	−284	−59	−48
5					−330	−224	−162	−43	−31
6					−300	−222	−154		
7					−101	−161	+9		

[a] Coefficients $\alpha^{\lambda}(j,k)$, $(-1)^{\lambda+1}\alpha^{\lambda}(j,k^{-1})$, and $\alpha^{\lambda}(j^{-1},k^{-1})$ (for ⁵⁴Co only), of equations 2.9–10, using data of Reference 17.

[b] Using masses of Reference 22.

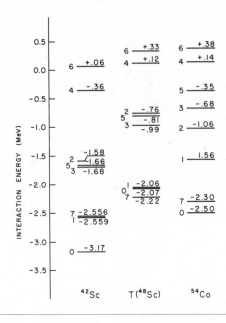

Figure 4 Comparison of energy levels of ^{42}Sc and ^{54}Co with Pandya transform of energy levels of ^{48}Sc.

The deviations from equation 2.10 for $\lambda > 0$ run from ~ 5 keV in ^{38}Cl-^{40}K to ~ 100 keV for the $f_{7/2}$-shell.

3.2   Interpretation

It is clear from the previous section that the spectra for many single-configuration nuclei require many-body effective interactions. The most systematic interpretation would be to assume that the interaction has all particle ranks, and determine the interaction as much as possible from the spectra. This approach has the problem that one does not generally have sufficient spectroscopic information for all the necessary nuclei. It also has the problem that it may try to explain too much. For example, suppose among the spectral levels of the nuclei with configurations (j^v) there are a few levels with large configuration mixing, but very little for the other levels. An attempt to fit this set uniformly would lead to a very "unstable" effective interaction, with components of high particle rank that cancel each other for most levels. Such a "nonconvergent" description of the shell would not be particularly useful.

An alternative approach is to assume that the interaction has a definite maximum rank, and see how consistently one can determine the interaction. Here we may test the more general linear spectroscopic relations of section 2.2 For example, we can interpret the differences between the $B_j(v)$ of Table 1, for given j, by introducing $\bar{W}(j^m)$ with $m = 3$ or 4, in equations 2.16–18. This generalizes the tests of the two-body interaction relations of section 3.1 to many-body effective interactions of lower rank.

THREE-BODY EFFECTIVE INTERACTIONS   Let us assume only three-body interactions. Then we can obtain values for $\bar{W}(j^3)$ from the $B_j(v)$ of Table 1 by inverting equation 3.2:

$$\bar{W}^{(v)}(j^3) = 2(v-2)^{-1}[B(v) - B(2)] \qquad 3.4.$$

If the assumption of only three-body interactions were correct, the $\bar{W}^{(v)}(j^3)$ would be independent of v, and equal to $\bar{W}(j^3)$. We show some values of $\bar{W}^{(v)}(j^3)$ in Table 4.

Table 4   Average three-body interaction energies $\bar{W}^{(v)}(j^3)$ for (j^v) configurations[a]

Orbit v	$1d_{3/2}(n)$	$1d_{3/2}(p)$	$1f_{7/2}(n)$	$1f_{7/2}(p)$	$2d_{5/2}(n)$
$N-2$	1.878	0.297	0.113	0.168	0.136
$N-1$	1.878	0.297	0.126	0.145	0.097
N	0.996	0.125	0.110	0.126	0.073

[a] $\bar{W}^{(v)}(j^3)$ is defined in equation 3.4. Values (in MeV) are calculated from data of Table 1.

The results show the three-body effective interaction to be repulsive. For the $1f_{7/2}$-shells the average strength is rather stable at about 120 keV per triplet of particles, leaving little room for a four-body interaction. The $2d_{5/2}$-shell is similar, but more v-dependent. The $1d_{3/2}$ neutron shell looks rather poorly behaved, since the data call for important higher rank interactions.

We can similarly try fitting three-body interactions to the (j^v, k^μ) average spectra of Table 2. Here we obtain possible values of $\bar{W}(j, k^2)$ and $\bar{W}(j^2, k)$ from the A of equation 3.3:

$$\begin{aligned}
\bar{W}^{(1)}(j, k^2) &= 2(M-1)^{-1}[A(j, M) - A(j, k)] \\
\bar{W}^{(2)}(j, k^2) &= (M-1)^{-1}[A(j, k^{-1}) - A(j, k)] \\
\bar{W}^{(1)}(j^2, k) &= (N-1)^{-1}[A(j^{-1}k^{-1}) - A(j, k^{-1})] \\
\bar{W}^{(2)}(j^2, k) &= (N-1)^{-1}[A(j^{-1}, N) - A(j, N)]
\end{aligned} \qquad 3.5.$$

Some values of these quantities are shown in Table 5, for the nuclei of Table 2.

Table 5   Average three-body interaction energies $\bar{W}^{(n)}$ for (j^v, k^μ) configurations[a]

Orbits	$1d_{3/2}\text{-}1d_{3/2}$	$1f_{7/2}\text{-}1d_{3/2}$	$1f_{7/2}\text{-}1f_{7/2}$	$1g_{9/2}\text{-}2d_{5/2}$
$\bar{W}^{(1)}(j, k^2)$	0.284	0.013	0.131	0.006 (0.010)
$\bar{W}^{(2)}(j, k^2)$	0.209	0.014	0.075	0.0004 (0.015)
$\bar{W}^{(1)}(j^2, k)$	−0.158		0.014	
$\bar{W}^{(2)}(j^2, k)$	0.303	0.004	0.023	

[a] $\bar{W}^{(n)}$ are defined in equation 3.5. Values in MeV are calculated from data of Table 2.

Again, the average three-body interaction is generally repulsive. For the $1f_{7/2}$-$1d_{3/2}$ and $1g_{9/2}$-$2d_{5/2}$ cases the strength is quite small ($\lesssim 10$–15 keV per triplet). For the $1f$-$1d$ case, $\bar{W}^{(1)} \cong \bar{W}^{(2)}$, while for $1g$-$2d$ the disagreement is the same order as the values, for either choice of masses. The $1d_{3/2}$-$1d_{3/2}$ cases look less regular than the others. For the $1f_{7/2}$-$f_{7/2}$ nuclei, $W(j, k^2)$ is similar to $W(j^3)$ of Table 4, i.e. about 120 keV. However, the values of $W(j^2, k)$ that come from the end of the shell are smaller.

A three-body interaction $\mathcal{V}(j, k^2)$ can be found to remove the disagreement between $W_J(j, k)$ and the transformed particle-hole spectrum $T\{W_J(j, k^{-1})\}$ for any pair of nuclei (j, k), (j, k^{-1}). The need for higher rank interactions can be found only by including the (j^{-1}, k^{-1}) spectrum, and testing equations 2.20 and 2.21. This can be done for the $1f_{7/2}$-$1f_{7/2}$ shells using the multipole coefficients listed in Table 3. The second difference

$$\Delta\alpha_4^\lambda = \tfrac{1}{2}[\alpha^\lambda(j, k) + \alpha^\lambda(j^{-1}, k^{-1})] - (-1)^{\lambda+1}\alpha^\lambda(j, k^{-1}) \qquad 3.6.$$

would be zero if there were only two- and three-body interactions, as we have seen in section 2.2. The values of $\Delta\alpha_4^\lambda$ [which may be found in Table II of (15)] are not zero, but are indeed smaller on the average than the particle-hole differences, like $\Delta\alpha_1^\lambda = (-1)^{\lambda+1}\alpha^\lambda(j, k^{-1}) - \alpha^\lambda(j, k)$. Therefore, the higher rank interactions are apparently less important than three-body interactions.

SENIORITY   Quesne (16) has attempted to decompose the (j^ν) spectra for the $1f_{7/2}$-nuclei (neutron and proton, separately), in terms of a many-body interaction of good seniority. She was able to extract the explicit contributions from $\mathcal{V}(\alpha)$ with $\alpha = 1$, 2, 3 (from $\nu = 1$, 2, 3 nuclei) and find the residual effect of $\Sigma\mathcal{V}(\alpha)$ for $\alpha = 4$–8 (not separately) from the other nuclei ($\nu = 4$–8). The matrix elements of $\mathcal{V}(3)$ (for $\nu = 3$) are $\pm(200$–500$)$ keV, and the $\mathcal{V}(4)$ matrix elements (for $\nu = 4$) are even larger. There may be some "instability" of the sort mentioned at the beginning of this subsection. However, the results for the average three-body interaction $\bar{W}(j^3)$ are similar to those in Table 4 for $1f_{7/2}$-nuclei; Quesne obtained $\bar{W}(j^3) = 145$ keV for neutrons and 330 keV for protons. [The results in (16) are tabulated in such a way that the many-body effects appear very large; this is because of factors of $\binom{\nu}{\alpha}$ for α-body interactions.]

"REAL" MANY-BODY INTERACTIONS   The discussion of the effective interaction so far has been from the point of view of interpreting sets of nuclear energy levels in terms of simple configurations. The introduction of many-body interactions simply insures sufficient degrees of freedom (in the sense of free parameters) to fit the spectra exactly. We remarked earlier that a two-body potential between nucleons can generate a many-body effective interaction, through configuration mixing. But there is also the possibility that there are "real" many-body interactions between nucleons, independent of configurations. These would show up in the linear spectroscopic relations in just the same way as the many-body effects of configuration mixing.

Therefore, one can consider the results of this section as providing an upper estimate of the strength of "real" many-body interactions. For example, one could conclude from Tables 4 and 5 that a three-body interaction can have three-body matrix elements no bigger than ~ 100 keV in magnitude. This assumes no "accidental" cancellation between "real" and configuration mixing effects, an assumption in which presumably one may have more confidence, the more cases on which one has to base the estimates.

4   THEORY OF THE EFFECTIVE INTERACTION

Up to now we have treated the effective interaction as a set of quantities of unknown origin that serve to correlate the spectroscopic information for energy levels assigned to simple configurations. Now we consider the theory by which the effective interaction can be calculated, at least in principle, given the hamiltonian for the nucleus (e.g. in terms of a two-nucleon potential).

We shall sketch the background theory briefly, following the model-space approach of Bloch & Horowitz (23) and Feshbach (24). This theory has been reviewed and further developed by Brandow (25). We shall be particularly interested in the separation of interactions of different particle rank, and in the application of the theory to the linear spectroscopic relations.

4.1   Model Space

Suppose the nuclear hamiltonian is H and the nuclear states and energies are given by $H\psi_\lambda = E_\lambda \psi_\lambda$. We require of an effective interaction that it reproduce a finite part of the spectrum E_λ on a (finite) set of model states ϕ_λ:

$$\mathcal{H} \phi_\lambda = E_\lambda \phi_\lambda \qquad\qquad 4.1.$$

For our case, the ϕ_λ are single configuration states $\phi_{Jx}(j^\nu,\ldots)$: the set of states ϕ_{Jx} define (and span) the model space.

The model states are usually chosen to be degenerate solutions of some hamiltonian H_0 (e.g. simple particle shell model) $H_0\phi_\lambda = E_0\phi_\lambda$. In most theories (with perturbation theory in mind), the model state ϕ_λ for a given E_λ is assumed to be given by the part of the state vector ψ_λ which lies within the model space. This may be accomplished formally with projection operators P, Q [following Feshbach (24)] defined by

$$P+Q = 1, \quad P^2 = P, \quad Q^2 = Q, \quad PQ = QP = 0, \quad P\phi_\lambda = \phi_\lambda, \quad Q\phi_\lambda = 0$$
$$4.2.$$

The assumption is that $P\psi_\lambda = C_\lambda \phi_\lambda$ where C_λ is a normalizing constant. Writing $\psi_\lambda = (P+Q)\psi_\lambda$, equation 4.1 can be written in coupled form

$$PHP\psi_\lambda + PHQ\psi_\lambda = E_\lambda P\psi_\lambda$$
$$QHP\psi_\lambda + QHQ\psi_\lambda = E_\lambda Q\psi_\lambda$$
$$4.3.$$

This can be solved for $P\psi_\lambda$ to give equation 4.1, with the effective interaction $\mathcal{H}$ given by

$$\mathcal{H} = P\big(H + HQ(E_\lambda - QHQ)^{-1}QH\big)P \qquad\qquad 4.4.$$

This operator in the model space has a complicated dependence on H, and also on the choice of the model space through P and Q, and apparently on E_λ. It is also known that the solutions of equation 4.1 are not orthogonal, so that $\mathscr{H}$ is not hermitian in the model space; this is not an essential difficulty (25).

To study $\mathscr{H}$ as a many-body operator, say in the (j^v) configuration (see equation 1.2a), one would have to define P_v, Q_v, (and $\mathscr{H}_v$) for each case v. Then $\mathscr{H}_0$ defines the zero-particle interaction, which is just the ground-state energy $E(0)$ of the closed shell nucleus $v = 0$. Then $\mathscr{H}_1 = E(0) + \mathscr{V}(1)$, and the matrix element of $\mathscr{V}(1)$ $[=\varepsilon(j)]$ defines the one-body interaction $\mathscr{V}(1)$. Similarly, one would define $\mathscr{V}(n)$ from the matrix elements (in the ϕ_λ) for all $\alpha = 0, 1, \ldots, n$.

The effective interaction of equation 4.4 could actually be calculated in part, if one were able to solve, for example, a multiconfiguration shell model hamiltonian problem. Then H would be the hamiltonian (matrix) in the (larger) multiconfiguration space, and P would be the projector into a single configuration. One could calculate $\mathscr{H}$ from the ψ_λ, E_λ using equation 4.4 or an equivalent one. In order to separate the different particle rank operators $\mathscr{V}(\alpha)$, one would have to perform the larger space matrix calculation for each v. This often becomes an impractically large problem for $v > 3$ or 4 if the multiconfiguration space is, say, a major oscillator shell (e.g. $2p$-$1f$ shell).

4.2   Linked Cluster Perturbation Theory

An alternative to the direct calculation of equation 4.4 is a perturbation expansion in $V = H - H_0$:

$$\mathscr{H} = H_0 + P \sum_{n=0}^{\infty} V \left(\frac{Q}{E_\lambda - H_0} V \right)^n P \qquad \qquad 4.5.$$

where we have used the fact that P and Q commute with H_0. The series form may be modified by use of the linked cluster theorem (26) to give an expansion (linked cluster) in which one may explicitly separate the contributions to $\mathscr{H}$ of definite particle rank. Clearly, this is a particularly useful technique for treating the spectroscopic relations of section 2. It is easy to keep track both of orders of V and of particle rank in the linked cluster expansion with the help of diagrams.

We illustrate the linked cluster expansion in diagrams somewhat schematically in Figure 5 for the configuration (j^v). The solid directed line segments (*valence lines*) stand for particles in the j-orbit; for the (j^v) case each diagram begins at the bottom and ends at the top with v such lines. The dashed lines represent particles in excited orbits (upward) or holes in the core (downward). Matrix elements of V (here assumed to be two-body only) are represented by the vertices (or nodes). Each diagram stands for a specific contribution to the matrix elements $\langle (j^v)\lambda | \mathscr{H} | (j^v)\lambda' \rangle$ where λ, λ' give the coupling of the (j^v) lines at the top or bottom, respectively, of the diagram.

To obtain the value of the contribution of a specific diagram, the orbit of each line segment must be specified. The rules for calculating a given diagram are derived from Wick's rules (23, 25, 26) for matrix elements of equation 4.5 in a v-particle state. The theory also specifies the number and type of diagram in each order. The

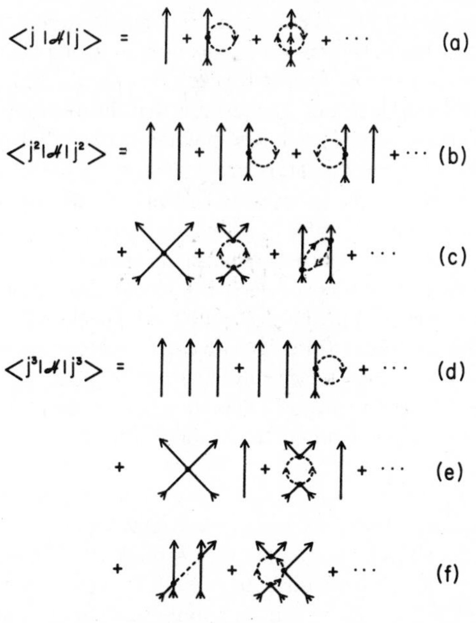

Figure 5 Some of the diagrams representing the linked cluster expansion of the matrix elements of $\mathscr{H}$.

effective interaction given by the linked cluster expansion can be put in a form that is explicitly independent of the energy eigenvalues (E_λ) by the inclusion of "folded diagrams," introduced by Brandow and others (25, 27, 28). This development also allows a clean separation of particle ranks of the effective interaction.

Returning to the examples of Figure 5, the first row (a) represents the first few terms in $\langle j|\mathscr{H}|j\rangle$, namely, from the left: the unperturbed one j-particle energy (no interaction), the interaction with holes involving no excitations (Hartree-Fock potential), and a second-order term with a two-particle–one-hole excitation. The sum of all such terms gives the one-body energy $\varepsilon(j)$, relative to the closed shell $(v = 0)$ ground state, for which there is no diagram. The closed shell energy $E(0)$ is added to the sum of diagrams.

For j^2 we show some diagrams in rows (b) and (c); the valence lines could be coupled: $\lambda = \lambda' = J$. In row (b) we have grouped the diagrams in which the two valence lines have no interactions (vertices) in common. It can be shown (25) that all the interactions fall on one valence line or the other, so that the sum of terms for row (b) gives twice that of row (a), that is $2\varepsilon(j)$. In row (c) we have diagrams with two valence lines interacting; the sum gives the interaction energy $W_J(j^2)$ as defined in equation 2.1b.

Similarly for j^3 it is possible to collect the diagrams into groups with only one valence particle interacting [row (d)], two valence particles interacting [row (e)], or all three interacting [row (f)]. There are no interference terms: these have been cancelled by "folded" diagrams (25). The terms of row (d) add up to $3\varepsilon(j)$. Further, the terms of row (e) are simply related to terms in row (c), which differ only by removal of the inactive valence line. The result is that the diagrams of Figure 5 can be interpreted as giving matrix elements of interactions $\mathscr{V}(\alpha)$ of particle ranks $\alpha = 1, 2, 3$. Diagrams for the operators are shown in Figure 6. The one-particle interaction has one valence line entering and leaving the diagram, and so on. The generalization to more than one type of valence particle $(j^{\nu}, k^{\mu}, \ldots)$ is direct.

The grouping of diagrams by particle rank is equivalent to selective configuration mixing. For example, in the independent pair approximation one allows a pair of valence nucleons to be excited (by interaction V) to higher configurations, in which the two particles may continue to interact with each other, but with no third particle. The sum of such terms (which include the first two diagrams of Figure 6b) gives the Brueckner reaction matrix, which is therefore part of the two-particle effective interaction $\mathscr{V}(2)$. The second order "core polarization" term (third diagram of Figure 6b) is also part of $\mathscr{V}(2)$.

THREE-BODY DIAGRAMS   Two low order contributions to $\mathscr{V}(3)$ are shown in Figure 6c. The first involves the excitation of a single particle out of the valence shell; such an excitation can also contribute to $\mathscr{V}(2)$. That excitation of single valence particles to second order (in V) can be represented by a three-body effective interaction was recognized by Bacher & Goudsmit for atoms in 1934 (6), and more recently by Osnes (29) and Bertsch (30). This term is the lowest order perturbation contribution to the many-body effective interaction, and can be expected to have important effects on spectroscopic relations, as we shall see. The second diagram in Figure 6c is a simple correction to the independent pair approximation, in which one of the excited particles interacts with a third valence particle.

If only a few low orders of perturbation are important, it is possible to calculate explicitly the effective many-body interaction. This is considerably easier than calculating configuration mixing to the same perturbation order, which would also include corrections to $\mathscr{V}(1)$ and $\mathscr{V}(2)$. The point is that spectra calculated with

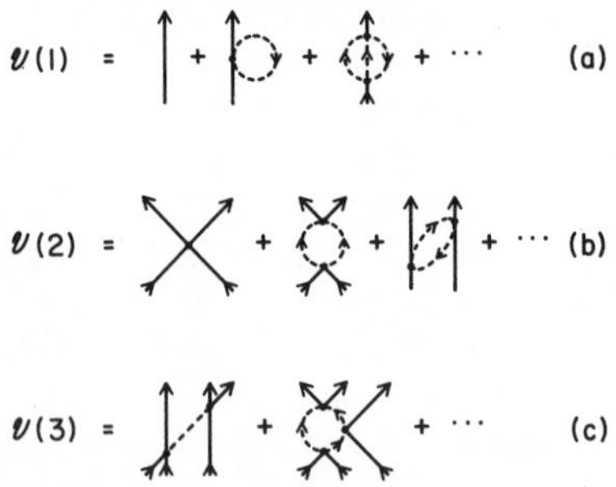

Figure 6   Diagrams for interactions of definite particle rank.

any approximate $\mathscr{H} = \mathscr{V}(1) + \mathscr{V}(2)$ must obey all the spectroscopic relations of section 2.1, which hold for two-body effective interactions. If we are not interested directly in the spectra, but in the deviations from the spectroscopic relations, we do not need $\mathscr{V}(1) + \mathscr{V}(2)$, but only $\mathscr{V}(\alpha)$, $\alpha > 2$.

If the perturbation expansion is not rapidly convergent, the diagrams of Figure 6c may still provide a guide to the relative importance of mixing different configurations. For example, we might expect one-particle excited configurations to be more important than two- (or three-) particle excited configurations, with respect to contributions to $\mathscr{V}(3)$.

AVERAGING DIAGRAMS   It is also possible to skip the effective interaction and calculate directly from the diagrams quantities of interest in the spectroscopic relations. For example, consider the energy of the filled shell configuration, (j^N). In Figure 7 we illustrate the linked cluster expansion in nondegenerate [Goldstone (26)] form. One notices that the open diagrams of Figure 6 have been converted into closed diagrams in Figure 7 by connecting the open ends of the valence lines

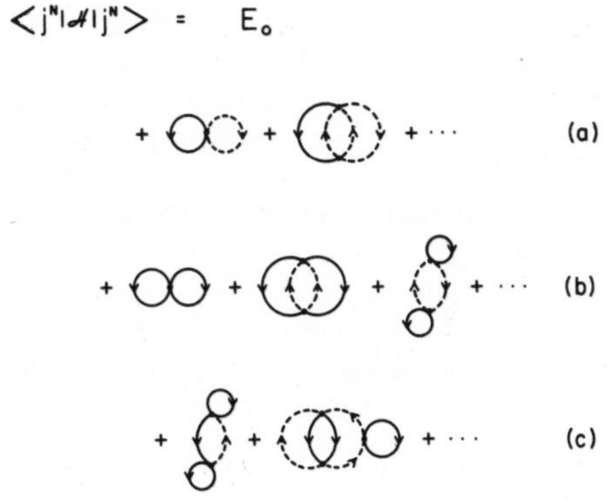

Figure 7   Diagrams for filled shell energy $E(N)$ in terms of averages of interaction diagrams of Figure 6.

to make *valance hole* lines. Diagrams 7a, b, and c give contributions to the average interaction energies (equation 2.18) $\bar{W}(j)$, $\bar{W}(j^2)$, and $\bar{W}(j^3)$ respectively, and the entire expansion corresponds to the general relation of average energies given in equation 2.16. This technique for direct calculation of averages has been introduced by Goode & Koltun (31).

PARTICLE-HOLE TRANSFORMATION OF DIAGRAMS   Similarly, there is a method for direct diagrammatic calculation of the quantities needed for the particle-hole relations (32). We consider (j,k) and (j,k^{-1}). In Figure 8a and c we illustrate two kinds of diagrams for the interaction energy $W_J(j,k^{-1})$, using boxes to represent excited lines and all vertices. The downgoing solid lines represent the valence hole

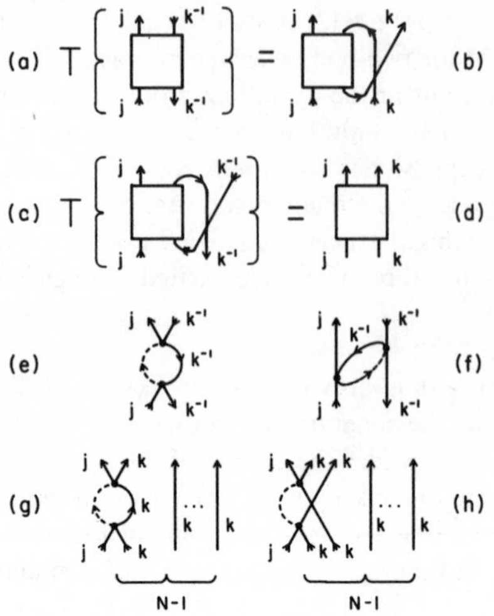

Figure 8 Particle-hole and particle-particle diagrams which are related by the Pandya transform: (*a*) and (*b*), (*c*) and (*d*). Diagrams (*e*) and (*f*) cannot be transformed: both contain two- and three-particle interactions, shown in (*g*) and (*h*).

orbit (k^{-1}). The symbol T applied to a diagram means taking the Pandya transformation (equation 2.8) of the valence diagram $W_J(j, k^{-1})$. Then the Figure expresses the statement that the transform T of diagram (*a*) equals diagram (*b*), and T of (*c*) equals (*d*), for any diagrams of the form (*a*) or (*c*), which contain no hole-line segments (k^{-1}), connecting vertices within the box. Thus, for these diagrams

$$T\{W_J(j, k^{-1})\} = W_J(j, k) \qquad\qquad 4.6.$$

Two second-order diagrams that violate this condition are shown in Figure 8*e* and *f*: such terms will violate the Pandya relation 4.6. This implies that (*e*) and (*f*) have contributions from many-body effective interactions, which can be seen by redrawing the diagrams as for the equivalent configuration (j, k^{N-1}) using only *particle* lines. Both (*e*) and (*f*) then become combinations of (*g*), which we recognize as a two-body interaction diagram, and (*h*), a three-body interaction of the form $\mathscr{V}(j, k^2)$ (see Figure 6).

The point is that one can calculate corrections to equation 4.6 directly from diagrams in the form 8 (*e*) or (*f*), rather than calculating the three-body effective interaction $\mathscr{V}(j, k^2)$ and then transforming its matrix elements in the (j, k^{-1}) configuration. One must, however, subtract the two-body contribution (*g*). One can then ignore all the *transforming* diagrams [(*a*) and (*c*)] since these satisfy equation 4.6; these contain no many-body interactions. It may then be possible to obtain the difference between two interaction spectra, $W_J(j, k)$ and $T_J\{W_I(j, k^{-1})\}$, without having to calculate the spectra to the same degree of accuracy.

5 CALCULATIONS OF MANY-BODY EFFECTIVE
INTERACTIONS

There have been several calculations of many-body effects to explain the disagreement between experimental spectra for simple configurations, and the predictions of the spectroscopic relations of section 2.1. Several of these are similar in starting point, in that they begin with configuration mixing in second order of perturbation theory, and notice that a three-body interaction is generated by allowing one excited particle (or hole) outside the valence orbit. As we have discussed above, this is the kind of term given by the first diagram of Figure 6c. This term may be expected to be significant if the excited orbits lie quite close in single-particle energy to the valence orbit. This is often the case for the other orbits in a major (oscillator) shell; for example, the $1f_{5/2}$-, $2p_{3/2}$-, and $2p_{1/2}$-orbits lie within about 6 MeV of the $1f_{7/2}$-orbit for $1f_{7/2}$-shell nuclei.

The first calculation of this type appeared in a paper of Pandya & French (33) shortly after the $1f_{7/2}$-$1d_{3/2}$ spectra of ^{38}Cl and ^{40}K were first known and discussed (11, 19). The idea was to use the close similarity of the (transformed) spectra (Figure 2) to set limits on the effective three-body interaction. First they calculated the second-order effect of exciting the $1f_{7/2}$-neutron to the nearby $2p_{3/2}$- and $1f_{5/2}$-orbits. As we mentioned, this is equivalent to calculating the lowest order three-body diagram of Figure 6c (or of Figure 8e). They found these terms to be of the order of 100 keV, which is considerably larger than the differences of 20–40 keV in the experimental spectra. They then introduced a simple zero-range three-body interaction to cancel the calculated three-body terms. Presumably, this extra interaction could represent higher order configuration mixing, or some "real" three-body force. The apparent cancelation of effects is puzzling; we return to this case below.

Osnes (29) included the same diagrams in a calculation of the three-body effective interaction to the binding energy of the Calcium isotopes. He allowed the excited single-particle (or hole) to go up (or down) $2\hbar\omega$ in oscillator shells. It is not clear that he also included the nearby orbits in the same shell, particularly since his calculated results came out much smaller than those of other very similar calculations (discussed below). It is possible to extract a value of the average $\bar{W}(j^3)$ from his quoted value of the three-body shift of 81 keV in $E(0)$ for ^{48}Ca: $\bar{W}(j^3) = 81$ keV$/8 \cdot 7 = 1.6$ keV $[j = 1f_{7/2}(n)]$. This is about 10^{-2} smaller than the needed $\bar{W}(j^3) = 110$ keV (see Table 4).

Bertsch (30) calculated a three-body interaction to improve the agreement of the closed shell energy $E(j^N)$ with the two-body interaction energies $W_J(j^2)$ obtained from the empirical spectrum (see equations 2.3a, 3.1, and 3.2). He also based his calculation on the second-order diagrams of Figure 6c, with the excited particle only in orbits of the same major shell. His method involved calculating a modified two-body interaction by matrix diagonalization within the major (oscillator) shell for two particles. It can be shown that calculation of the closed shell energy $E(j^N)$ from equation 2.3a with this interaction is approximately equivalent to calculation of the second-order three-body term of Figure 6c with modified vertices; the new vertices are effective two-body interactions of higher order, which include pair excitation within the major shell.

Bertsch's results can be put in a form easily compared with Tables 4 and 5:

$$\bar{W}(j^3) = 0.225 \text{ MeV for } 1d_{3/2}(\text{p})$$
$$= 0.060 \text{ MeV for } 1f_{7/2}(\text{n})$$
$$\bar{W}(j^2, k) + \bar{W}(j, k^2) = 0.191 \text{ MeV for } 1d_{3/2}\text{-}1d_{3/2}$$
$$= 0.046 \text{ MeV for } 1f_{7/2}\text{-}1f_{7/2}$$

$$5.1.$$

The numbers are about a factor $\frac{1}{2}$ smaller than the empirical results. The positive sign (repulsive interaction) is fixed by second-order perturbation theory.

West & Koltun (32) and Goode, Koltun & West (15) applied the perturbation theory of many-body effective interactions to explain the differences among the (transformed) spectra of ^{42}Sc, ^{48}Sc, and ^{54}Co (Figure 4 and Table 3). The calculations were based on the second-order diagrams of Figure 8(e) and (f), with single excitations from the $1f_{7/2}$- to the $1f_{5/2}$-, $2p_{3/2}$-, and $2p_{1/2}$-orbits. However, the mixing of the excited configurations is sufficiently strong, particularly in the case of ^{42}Sc, that they were forced to include higher order terms. This was done by transforming the calculation to diagonalization of shell model matrices, which contain only the basic configuration and those with one excited particle. Specifically, these include the configurations

$$(j, k), (j, \alpha), (\alpha, k) \text{ for } {}^{42}\text{Sc}$$
$$(j, k^{-1}), (\alpha, k^{-1}), (j, \alpha, k^{-2}) \text{ for } {}^{48}\text{Sc}$$
$$(j^{-1}, k^{-1}), (j^{-1}, \alpha, k^{-2}), (\alpha, j^{-2}, k^{-1}) \text{ for } {}^{54}\text{Co}$$

with α ranging over $1f_{5/2}$, $2p_{3/2}$, and $2p_{1/2}$, while $j = 1f_{7/2}(\text{p})$, $k = 1f_{7/2}(\text{n})$. This matrix method introduces small many-body effects of rank higher than three. Similar effects also come from the use of different excitation energies (from experiment) for the single-particle levels α, for the three different nuclei.

The results of this calculation are given in multipole form in Table 6. We compare the differences of the experimental α^λ of Table 3:

$$\Delta\alpha_1^\lambda = (-1)^{\lambda+1}\alpha^\lambda(j, k^{-1}) - \alpha^\lambda(j, k)$$
$$\Delta\alpha_2^\lambda = \alpha^\lambda(j^{-1}, k^{-1}) - (-1)^{\lambda+1}\alpha^\lambda(j^{-1}, k)$$

$$5.2.$$

The calculated $\Delta\alpha^\lambda$ generally agree with the experimental values in sign and roughly in magnitude (i.e. to a factor of 2). The calculation of the monopole differences $\Delta\alpha_1^0$ and $\Delta\alpha_2^0$ is equivalent to Bertsch's calculation of the closed shell binding (equation 5.1 for ^{56}N and ^{48}Ca), and also underestimates the experimental effect (by about 1/3). The results of these two calculations do depend on the choice of the effective two-body interaction for the perturbing interaction. Both calculations made use of the Kuo-Brown renormalized interaction (34) within the $1f$-$2p$ shell.

It would seem that for the $1f_{7/2}$-shell, the many-body correction to the spectro-

Table 6   Calculated multipole differences for $(1f_{7/2}{}^{\pm 1}, 1f_{7/2}{}^{\pm 1})$ nuclei (in MeV)

| λ | $\Delta\alpha_1^\lambda = T(^{48}\text{Sc}) - {}^{42}\text{Sc}$ | | | $\Delta\alpha_2^\lambda = {}^{54}\text{Co} - T(^{48}\text{Sc})$ | |
	Expt.[a]	Theor.[b]	Theor.[c]	Expt.[a]	Theor.[b]
0	0.529	0.346	0.32	0.100	0.072
1	−0.156	−0.065	−0.14	−0.061	−0.022
2	−0.024	−0.019	−0.06	−0.116	−0.042
3	0.032	0.051	0.06	−0.004	−0.017
4	0.094	0.071	0.02	0.076	−0.007
5	0.106	0.098	0.11	0.061	−0.010
6	0.078	0.068	0.03	0.068	0.017
7	−0.060	0.053	−0.04	0.170	0.054

[a] Experimental values of $\Delta\alpha^\lambda$, defined in equation 5.2, obtained from data of Table 3.
[b] Results of Reference 15 with excitations in $1f$-$2p$ shells only.
[c] Results of Reference 35 with excitations from $1d_{3/2}$ to $1f_{7/2}$ as well as within $1f$-$2p$ shells.

scopic relations is to a good extent accounted for by an effective three-body inter-action generated in second order by one-particle excitations to nearby orbits. A complete calculation has not been made for the $1g_{9/2}$-$2d_{5/2}$ case (^{92}Nb-^{96}Nb); one difficulty is the lack of information in the single-particle energies for the $1g$-$2d$-$3s$ shells. A preliminary estimate (by the author and T. Mizutani) gives the monopole shift $\Delta\alpha_1^0$ for this case as 20–40 keV, assuming an average excitation energy of ~ 2 MeV, which is a reasonable order of magnitude (see Table 3).

It is perhaps curious that the same three-body interaction doesn't seem to work for the $1d_{3/2}$-$1f_{7/2}$ case, ^{38}Cl-^{40}K. We mentioned that Pandya and French calculated the effect of such an interaction, and found terms of the order 100 keV, which is several times the experimental result. Goode (35) has recently found similar size terms for this interaction, calculated with the Kuo-Brown matrix elements; his numbers are listed in Table 7, in multipole form ($\Delta\alpha_1^\lambda$) and may be compared to the experiment $\Delta\alpha^\lambda$ from Table 3.

Table 7   Calculated multipole differences for ^{38}Cl $- {}^{40}$K (in MeV)[a]

λ	Expt.	Theor.[b]	Theor.[c]	Theor.[d]
0	0.03	0.13	−0.08	0.01
1	0.02	0.05	−0.03	0.01
2	0.00	0.01	−0.01	0.02
3	−0.01	−0.05	−0.01	−0.03

[a] Values of $\Delta\alpha_1^\lambda$ (equation 5.2), from Reference 35.
[b] Excitation within major shells only.
[c] Excitation of $1d_{3/2}$ to $1f_{7/2}$ only.
[d] "Hybrid model": both excitations included.

Goode has also considered an alternative configuration mixing which could apply to this case. It is quite possible to excite pairs of $1d_{3/2}$-particles into the $1f_{7/2}$-shell, both because the single-particle energy differences is small (about 3 MeV in ^{33}S) and because the relevant two-body interaction is reasonably strong. Goode has therefore considered a two-shell (fd) model with particles in the $1d_{3/2}$- and $1f_{7/2}$-shells. This is a model with considerable configuration mixing. From the single $1d_{3/2}^{\pm 1}$(p)-$f_{7/2}$(n) configuration point of view, there may be high rank many-body interactions. In perturbation theory, the leading term would be given by the third-order diagram of Figure 6c: two d-neutrons are excited to the f-shell; one interacts with the third (f-proton) and then the pair de-excite. This term should be attractive, as opposed to the second-order term, which is repulsive [on the average: $\bar{W}(jk^2) > 0$]. The results for this model are shown in the third column of Table 7. The agreement with experiment is no better than for the second-order term, however, and now the monopole shift ($\lambda = 0$) is negative.

Goode has therefore combined both effects into a "hybrid model" in which the fd model is calculated with a new effective interaction, in which the second-order three-body interaction is included. This leads to the results in the fourth column of Table 7. The agreement with experiment is now improved, apparently from the partial cancellation of the two kinds of configuration mixing.

This "hybrid" model also has some effects on the nuclei ^{42}Sc-^{48}Sc, since it can produce excitations of the ^{40}Ca-^{48}Ca cores. Here, the f-d effects are much smaller than the second-order effects calculated by Goode, West, and Koltun: they are shown in Table 6. The most notable improvement is in the change of sign of $\Delta\alpha_1^\lambda$ for $\lambda = 7$.

In summary, there are now a few calculations of many-body corrections to spectroscopic relations, based on very limited configuration mixing into neighboring shells. Although based on the perturbation ideas of section 4, the calculations have included higher orders by diagonalization of limited shell model matrices (on the limited configurations). However, these calculations do not correspond completely to multishell matrix calculations of energy spectra: only those configurations are included which, in perturbation theory, generate many-body interactions. The result is that these calculations are generally more successful in explaining the spectroscopic differences than in predicting the spectra themselves.

It should be pointed out that although the agreement of the calculated differences with the experimental ones is good to $\pm 50\%$, the absolute accuracy is 20–50 keV. This is considerably better than the absolute accuracy of, say, the two-body spectra calculated from the same effective two-body interactions which are used [e.g. Kuo-Brown (34)] in the many-body calculation. In this sense, one has a successful theory of differences of spectra, which is more accurate than the theory of the spectra themselves.

It should also be noted that the only calculations of this sort are for nuclei in the mass neighborhood of ^{40}Ca, involving the $1d$-$2s$ and $1f$-$2p$ shells. This is also the region in which most of the spectroscopic relation information is presently available.

Lastly, we should note some recent work of Barrett, Halbert, & McGrory (36, and private communication) on constructing many-body effective interactions without perturbation theory. They are able to solve certain multishell matrix diagonalizations with a two-body interaction, for, say, two or three particles in several j-shells (spanning a major oscillator shell). From the solutions they can select the matrix elements within one (or two) configurations corresponding to the effective interaction. This is essentially the projection method of equations 4.1–4.4. By subtraction they can obtain separately the two- and three-body effective interactions. Some preliminary results have been reported for the $1d$-$2s$ shells.

It would be possible to use the interactions so obtained to calculate, for example, the particle-hole spectra, closed shell energies, and so on. It would be interesting to see how this compares to the extended perturbation methods, which include less complete three-body, but some higher-body interactions.

6 CONCLUSIONS

The linear spectroscopic relations provide a rather sensitive tool for the study of effective interactions in the nuclear shell model. It is not just that the effects are small: energy differences are typically 20–100 keV, as we have seen. The interesting point is that it seems to be possible to understand these small numbers from the theory of the effective interaction. Only Coulomb effects in shell model spectra are treated with similar accuracy.

One could certainly use more experimental spectra, but these may not be easily obtained. There are new data for ^{56}Co and ^{58}Co (37) which were not included in Table 1. However, many other interesting nuclei seem inaccessible—for example, the particle-hole partners in the $1g_{9/2}$-$1g_{9/2}$ shell of ^{90}Nb, or of the (j, k) or (j, k^{-1}) spectra for nuclei in the region around ^{208}Pb. Perhaps more spectra of excited configurations, like $2p_{3/2}$-$1d_{3/2}$ in ^{38}Cl and ^{40}K, can be found.

The theoretical work has really just begun; only a few cases have been studied, and in limited approximations. It should be remarked that there is a question about the rapid convergence of perturbation theory for the effective interaction (31, 38). One would like to know how much of this possible difficulty would appear in the differences of effective interactions discussed in sections 4 and 5.

We have used the model-state formulation of the theory of the effective interaction. The result was that a given spectrum exactly corresponds to some many-body effective interaction, but the model state function is not the wave function of the physical state. To study other properties than energies of the states, one must introduce effective operators for transitions, moments, and so on. These may also be many-body operators. The method of linear spectral relations may be extended to these operators; this has been initiated by Goode & West (39).

ACKNOWLEDGMENTS
The author gratefully acknowledges helpful communication on this subject over several years with W. P. Alford, J. B. French, P. Goode, J. P. Schiffer, and B. J. West. T. Mizutani helped in preparing the data tables.

Literature Cited

1. Talmi, I., Unna, I. 1960. *Ann. Rev. Nucl. Sci.* 10:353
2. Brown, G. E. 1971. *Unified Theory of Nuclear Models and Forces.* Amsterdam: North-Holland. 3rd ed.
3. Bertsch, G. F. 1972. *The Practitioner's Shell Model.* New York: Elsevier
4. Baranger, M. 1969. *Proc. Int. Sch. Phys. Enrico Fermi,* Course 40, ed. M. Jean, R. A. Ricci, 511–614. New York: Academic
5. Condon, E. U., Shortley, G. H. 1935. *The Theory of Atomic Spectra.* London: Cambridge
6. Bacher, R. F., Goudsmit, S. 1934. *Phys. Rev.* 46:948
7. Talmi, I. 1970. *Theory of Nuclear Structure: Trieste Lectures 1969,* 455. Vienna: IAEA
8. de Shalit, A., Talmi, I. 1963. *Nuclear Shell Theory.* New York: Academic
9. French, J. B. 1967. *Nuclear Structure,* ed. A. Hossain et al, 85–124. Amsterdam: North-Holland
10. French, J. B. 1966. *Proc. Int. Sch. Phys. Enrico Fermi,* Course 36, ed. C. Bloch, 278–375. New York: Academic
11. Pandya, S. P. 1956. *Phys. Rev.* 103:956
12. Moinester, M., Schiffer, J. P., Alford, W. P. 1969. *Phys. Rev.* 179:984
13. Schiffer, J. P. 1972. *The Two-Body Force in Nuclei,* ed. S. M. Austin, G. M. Crawley, p. 205. New York: Plenum
14. Koltun, D. S., West, J. B. 1969. *Contributions to the International Conference on the Properties of Nuclear States,* p. 218. Montreal: Univ. Montreal
15. Goode, P., Koltun, D. S., West, J. B. 1971. *Phys. Rev. C* 3:1527
16. Quesne, C. 1970. *Phys. Lett. B* 31:7
17. *Nucl. Data Sheets* (through 1972, Vol. 7)
18. Wapstra, A. H., Gove, N. B. 1971. *Nucl. Data Tables* 9:267
19. Goldstein, S., Talmi, I. 1956. *Phys. Rev.* 102:589
20. Freeman, R. M., Gallman, A. 1970. *Nucl. Phys. A* 156:305
21. Engelbertink, G. A. P., Olness, J. W. 1972. *Phys. Rev. C* 5:431
22. Comfort, J. R., Maher, J. V., Morrison, G. C., Schiffer, J. P. 1970. *Phys. Rev. Lett.* 25:383
23. Bloch, C., Horowitz, J. 1958. *Nucl. Phys.* 8:91
24. Feshbach, H. 1962. *Ann. Phys. (New York)* 19:287
25. Brandow, B. H. 1967. *Rev. Mod. Phys.* 39:771
26. Goldstone, J. 1956. *Proc. Roy. Soc. London A* 239:267
27. Johnson, M. B., Baranger, M. 1971. *Ann. Phys. (New York)* 62:172
28. Kuo, T. T. S., Lee, S. Y., Ratcliff, K. F. 1971. *Nucl. Phys. A* 176:65
29. Osnes, E. 1968. *Phys. Lett. B* 26:274
30. Bertsch, G. F. 1968. *Phys. Rev. Lett.* 21:1694
31. Goode, P., Koltun, D. S. 1972. *Phys. Lett. B* 39:159
32. West, B. J., Koltun, D. S. 1969. *Phys. Rev.* 187:1315
33. Pandya, S. P., French, J. B. 1957. *Ann. Phys. (New York)* 2:166
34. Kuo, T. T. S., Brown G. E. 1968. *Nucl. Phys. A* 114:241
35. Goode, P. *Nucl. Phys.* In press
36. Barrett, B. R., Halbert, E. C., McGrory, J. B. 1972. *Bull. Am. Phys. Soc.* 17:553
37. Schneider, M. J., Daehnick, W. W. 1972. *Phys. Rev. C* 5:1330
38. Barrett, B. R., Kirson, M. W. 1970. *Nucl. Phys. A* 148:145
39. Goode, P., West, B. J. 1972. *Particles Nucl.* 4:26

INTERMEDIATE STRUCTURE IN NUCLEAR REACTIONS

× 5538

Claude Mahaux
University of Liège, Liège, Belgium

CONTENTS

I INTRODUCTION

Right above the lowest particle threshold, nuclear reactions show a large density of very narrow resonances. This narrowness indicates that the resonances are due to

193

the formation of a compound system with a long lifetime: according to Bohr (1), the incident particle shares its energy with many nucleons of the target; a long time elapses before one nucleon gets enough energy to escape from the nucleus. Hence, the nuclear states have, in this "compound nucleus model," a very complicated structure, which cannot be understood theoretically in full detail. The complexity of the compound states suggests that statistical assumptions may be used in the compound nucleus model. The validity of these assumptions can in principle be checked in the region of isolated resonances. At higher energy the resonances cannot be resolved and often overlap. The statistical assumptions can nevertheless be tentatively conserved, giving rise to the statistical model of nuclear reactions (2). This model appears to be valid for most reactions involving intermediate and heavy nuclei (A > 40). Deviations appear when the lifetime of the compound nucleus is less than the time required to achieve random sharing of the excitation energy among all degrees of freedom of the nucleus. This can occur in two ways (2): (a) The emission of the particle takes place within times that do not greatly exceed the nuclear transit time. This is the characteristic feature of direct reactions. (b) Certain simple states are preferentially excited and the subsequent randomization is slow. Then deviations from the statistical model may occur "locally," in the vicinity of the energy of these simple states. The expression *intermediate structure* (IS) denotes a deviation of the latter type, i.e. a departure from the statistical model that takes place in a limited domain of excitation energy. The size ΔE of this domain is related to the lifetime τ of the associated simple state by the relation $\tau = \hbar/\Delta E$. Since IS is related to the excitation of a simple state, its experimental investigation yields most valuable information on the existence of simple modes of motion for the nucleus at high excitation energy. Moreover, the decay of these simple states may provide interesting results concerning the states of the residual nucleus. This fully justifies the detailed experimental and theoretical investigation of IS.

Historically, the first identified IS is the giant dipole resonance (3), which consists of the concentration of most of the dipole radiative strength in a limited domain of excitation energy. This IS is due to the existence of a simple mode of motion where the groups of neutrons and of protons oscillate out of phase about the center-of-mass of the nucleus (4–6). The wave function of this "giant dipole state" is obtained by acting with the dipole operator $E1$ on the ground-state wave function ψ_0:

$$|\Omega_{gd}\rangle = E1\,|\psi_0\rangle \qquad\qquad 1.1.$$

We implicitly assume that Ω_{gd} is normalized. This giant dipole state contains the full dipole strength. The configuration Ω_{gd} is not an eigenstate of the full Hamiltonian; it is mixed by the residual interaction with many other configurations. However, this mixing is sizable only in the vicinity of the energy of the giant dipole state. Hence, the nuclear resonances lying in this energy domain have an enhanced radiative width. Experimentally, the giant dipole resonance appears as a single broad resonance whose width reaches a few MeV since the individual compound resonances are not resolved.

The observation and theoretical understanding of the giant dipole resonance thus result from two features: (a) A configuration can be found which exhausts the dipole strength. (b) The residual interaction spreads this configuration only locally. A similar situation may exist in other nuclear reactions, which would not necessarily involve photons. For instance, the sum of the spectroscopic factors as measured in stripping reactions reaches unity [feature (a)] and the full single-particle strength is usually concentrated at low energy (this reflects the low density of states and the weakness of the residual interaction). Thus, local enhancements of the spectroscopic factors may be expected for low-lying states in medium and heavy nuclei (7, 8). This phenomenon is not always considered as IS proper for two reasons. Firstly, it pertains to bound states, and the expression IS is often reserved for compound nuclear resonances. Secondly and mainly, the number of bound states involved in the phenomenon is usually too small to imply a statistically meaningful departure from the statistical model, whose validity at low excitation energy is moreover doubtful. Nevertheless, the description of the spreading of a simple bound configuration shares all the essential features of IS. Hence, this problem will be studied in section II.

Before proceeding, let us give a few definitions and results. In most theoretical descriptions a resonance results from the coupling, by a "residual" interaction (V), of a bound state (ψ_λ) of some model Hamiltonian (H_0), to the scattering eigenstates (χ_E^c) of H_0. In first approximation, the probability amplitude for decay of ψ_λ into channel c is proportional to the matrix element (section III)

$$V_\lambda^c(E) = \langle \chi_E^c | V | \psi_\lambda \rangle \qquad\qquad 1.2.$$

By analogy with the giant dipole state, this suggests that giant resonances may appear in the particle channel c because of the existence of the "giant channel state"

$$\Omega_c = \sum_\lambda |\psi_\lambda\rangle \langle \psi_\lambda| V |\chi_E^c\rangle \qquad\qquad 1.3.$$

which exhausts the coupling strength to χ_E^c. We recall that the giant dipole state is given by equation 1.1, which may be written in the form

$$\Omega_{gd} = \sum_\lambda |\psi_\lambda\rangle \langle \psi_\lambda| E1 |\psi_0\rangle \qquad\qquad 1.4.$$

There exists no example, however, where Ω_c produces a giant resonance in channel c. This is because the residual interaction V usually mixes Ω_c with other bound states lying far away in energy; hence, the spreading of Ω_c is not local. In other words, the lifetime of Ω_c is too short to justify a physical interpretation of this configuration.

Around 1963 it was proposed that local enhancements of partial widths in particle channels may, nevertheless, take place (9, 10). As in the case of photon channels this phenomenon would be due to simple configurations ("doorway states"), locally spread. The work of Kerman, Rodberg & Young (10, 11) is intimately related to the dynamical description of a resonance process proposed by Feshbach, Weisskopf and collaborators (12–15). A similar model was already used in the mid-fifties (16) to describe the giant resonances discovered by Barschall (17) in the energy dependence of the average total neutron cross section: it was proposed then (16) that these giant resonances correspond to the spreading of a single-particle state.

Let us describe the basic idea of the doorway state model with a simple example. We assume that the target in channel c is a doubly closed shell nucleus, described by the independent particle model. The configuration χ_E^c is sketched in Figure 1(a); E is the energy of the incident nucleon and c denotes the other quantum numbers of the channel (orbital and total angular momenta). Let H_0 be the Hartree-Fock potential. If the residual interaction is a sum of two-body potentials, only the two particle-one hole (2p-1h) states, among all the bound shell-model configurations, are coupled to χ_E^c. In a more picturesque language, the first collision of the incident nucleon with a target nucleon gives rise only to 2p-1h configurations, shown in Figure 1(b). Further collisions lead to more complicated states [Figure 1(c)]. A 2p-1h configuration is called a *doorway state* (15). If every doorway configuration is only locally spread and if their spreading domains do not overlap, a given doorway configuration will exhaust the full coupling strength to channel c in its own spreading domain, and may produce a local enhancement of the partial widths. As in the case of the giant channel states Ω_c, it is found, however, that the 2p-1h states are usually so strongly spread that they overlap (18). Nevertheless, the basic ingredients of this model may remain valid: special configurations (usually not pure 2p-1h configurations) can exist, which exhaust most of the coupling strength to a channel in a given range of excitation energy, and which are not too strongly spread. They may then give rise to local enhancements of the partial widths, i.e. to IS. In practice, it appears that particularly favorable conditions must be fulfilled for the spreading to remain small and the enhancement to be pronounced. A small spreading may, for instance, result from selection rules which prevent the mixing with other configurations. This is the case for the isobaric analogue resonances (19) and the IS seen in neutron-induced fission on some elements (20, 21). We note that a sizable enhancement requires a large coupling strength of the special configuration to channel c: the special configuration must have a "simple" structure when compared to the channel wave function χ_E^c. This confirms that IS provides evidence for the existence of simple modes of motion of nuclei at high excitation energy.

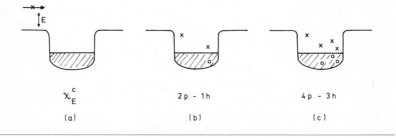

Figure 1   Scattering (*a*), doorway (*b*), and complicated (*c*) model configurations, for a doubly closed shell target.

Many data presented as experimental evidence for IS were received with disbelief, and this field remained a controversial one for a long time: (a) Firstly, different meanings are attached to the expression "intermediate structure." For instance, some people associate IS with the possibility of accounting for the total coupling strength of a group of resonances to a channel (i.e. of reproducing the sum of the partial widths) with a single configuration. In this interpretation, the "group" of resonances may for instance contain only 2 or 3 resonances, or may not be characterized by any enhancement. Without denying the interest of this type of analysis, we believe that it is not always appropriate to associate it with IS, since the data need not display any particular feature. Rather, this analysis reflects the success of a theoretical calculation of the sum of the resonance widths. (b) Secondly, and mainly, local enhancements of the partial widths can always occur with a finite probability within the framework of the statistical model itself: the partial widths are randomly distributed around their average, and "accidental" bunchings of large partial widths may take place. Hence, the apparatus of statistical analysis must usually be used to calculate with what probability the data imply a deviation from the statistical model.

The fact that doorway states exist or, equivalently, that the compound nucleus may decay before complete randomization, implies for instance that the spectrum of emitted particles presents an excess of high-energy particles. The theory of this "pre-equilibrium emission" was developed by Griffin (22).

An elegant discussion of IS in the frame of Feshbach's projection operator formalism (13, 14) has been given by Feshbach, Kerman & Lemmer (23).

Section II is devoted to IS and simple configurations seen in *bound states*. Section III deals with the theoretical description of IS in compound nuclear reactions. Several useful methods of statistical analysis are listed in section IV. Examples of IS in compound nuclear reactions are given in section V, and section VI contains a discussion of the usefulness of the concept of doorway states in the absence of IS proper.

II   INTERMEDIATE STRUCTURE IN BOUND STATES

The theoretical analysis of IS in the case of bound states is somewhat simpler than in the case of resonances. However, both the validity of the statistical model and the accuracy of the experimental data are then usually worse than in the case of resonance reactions. Accordingly, the purpose of the present section is mainly to introduce the basic theoretical concepts in a simple and familiar case.

II. 1   Configuration Mixing

In the present section we recall a few relations concerning the mixing of a configuration ϕ_0 with other configurations ϕ_m $(m = 1, \ldots, M)$. For simplicity, we assume that the set $\{\phi_j\}$ $(j = 0, \ldots, M)$ is orthonormal. The configurations $\{\phi_j\}$ span a *model* Hilbert space in which we can define a model Hamiltonian H_0 by associating an energy E_j with each ϕ_j:

$$H_0 = \sum_{j=0}^{M} |\phi_j\rangle E_j \langle\phi_j| \qquad\qquad 2.1.$$

We write the full Hamiltonian H in the form

$$H = H_0 + V \qquad\qquad 2.2.$$

thereby defining the residual interaction V.

Let $\mathbf{O}$ denote the orthogonal matrix which diagonalizes the Hamiltonian matrix

$$\mathbf{H} = H_{jk} = \langle\phi_j| H_0 + V |\phi_k\rangle \qquad\qquad 2.3.$$

The eigenstates of $\mathbf{H}$ are given by

$$\psi_\lambda = O_{\lambda 0}\phi_0 + \sum_{m=1}^{M} O_{\lambda m}\phi_m \qquad\qquad 2.4.$$

We note that ψ_λ is only an approximation to the true physical state, since it is entirely contained in the model space $\{\phi_j\}$. The observable quantities, $F_{\lambda c}^2$, are usually given in the model by the square of the projection of ψ_λ on some state θ_c, whose nature depends upon each particular experiment:

$$F_{\lambda c} = \langle\theta_c|\psi_\lambda\rangle = O_{\lambda 0}\langle\theta_c|\phi_0\rangle + \sum_{m=1}^{M} O_{\lambda m}\langle\theta_c|\phi_m\rangle \qquad\qquad 2.5.$$

II. 2   Simple Configurations

The word "simple" covers a rather subjective notion. A configuration can be simple with respect to some reference state, and appear complicated when compared to another reference state. Let us take one example. If a stripping reaction (d, p) is performed on a doubly closed shell target (for instance ^{208}Pb), the state θ_c in equation 2.5 is a single-particle state, and the observable $F_{\lambda c}^2$ is the spectroscopic factor of the state ψ_λ in the corresponding single-particle state. In this case, the simple configurations are single-particle states. If the target is a one-hole state, for instance ^{207}Pb, the simple configurations for stripping are rather one particle-one hole (1p-1h) states. In the case of (^{3}He, p) on a doubly closed shell target, the simple configurations are two particle states, etc. Expressing this in a more formal way, the configuration ϕ_0 is "simple" with respect to θ_c if

$$\omega_{0c} = \langle\theta_c|\phi_0\rangle \neq 0 \qquad\qquad 2.6.$$

while the "complicated" configurations ϕ_m are characterized by

$$\omega_{mc} = \langle\theta_c|\phi_m\rangle = 0 \qquad\qquad 2.7.$$

Equations 2.6 and 2.7 describe an extreme situation, and in practice, ϕ_0 is said to be simpler than ϕ_m if

$$\omega_{0c}^2 \gg \omega_{mc}^2 \qquad\qquad 2.8.$$

For simplicity let us take the idealized case described by equations 2.6 and 2.7. Then we have

$$F_{\lambda c}^2 = O_{\lambda 0}^2 \langle \theta_c | \phi_0 \rangle^2 \qquad\qquad 2.9.$$

$$\sum_{\lambda=0}^{M} F_{\lambda c}^2 = \langle \theta_c | \phi_0 \rangle^2 \qquad\qquad 2.10.$$

where we used the orthogonality of **O**. By analogy with the terminology used in reaction theory, ϕ_0 may be called a "doorway state" (15). Choosing a configuration ϕ_0, one computes $\langle \theta_c | \phi_0 \rangle$ and substitutes its value in equation 2.10. If this relation is fulfilled, the following assumptions are substantiated:

(a) The configuration ϕ_0 is the only one that has an overlap with θ_c.
(b) The configuration ϕ_0 is entirely contained within the states ψ_λ.
(c) The experiment yields the projection of λ on θ_c.

We note that since $F_{\lambda c}^2$ is obtained from experiment, these assumptions actually apply to the true nuclear state λ, rather than to the model wave function ψ_λ. In the simple example of a stripping reaction on a doubly closed shell target, equation 2.10 translates to the familiar statement that the sum of spectroscopic factors of all the levels containing a fraction of the single-particle state of interest is equal to unity. The single relation 2.10 only provides a weak test for the above set of assumptions. This test is thus meaningful only if one or more of the assumptions can be taken as granted from our general understanding of nuclear physics.

II. 3   Spreading of a Simple Configuration

Equation 2.4 shows that the configuration ϕ_0 is "fragmented" among the levels ψ_λ. Let us admit that assumptions (a)–(c) of section II.2 are justified. From the measurement of the quantities $F_{\lambda c}^2$ and from the knowledge of $\langle \theta_c | \phi_0 \rangle^2$, one can deduce the value of $O_{\lambda 0}^2$ that gives the amount of admixture of ϕ_0 in ψ_λ. A meaningful theoretical calculation of the fragmentation is quite difficult since it involves the "complicated states" ϕ_m. Instead of computing the individual mixing coefficients $O_{\lambda 0}^2$, one can attempt to evaluate their average distribution, i.e. the dependence of their mean upon energy. This is possible only if the number of states ψ_λ is large in an energy interval in which the mean of $O_{\lambda 0}^2$ does not vary appreciably. In practice, this is rarely the case for bound states, since the density of levels with given angular momentum and parity is fairly low. The problem is nevertheless interesting because it involves the main concepts and tools of the description of IS. In the present section we follow the derivation given in (24). Let us diagonalize **H** in the basis $\{\phi_m\}$ ($m = 1, \ldots, M$). We call the resulting eigenvalues ε_m and the corresponding configurations γ_m ($m = 1, \ldots, M$). The states γ_m are, a fortiori, "complicated." Since the configurations γ_m are "complicated," it appears legitimate to treat them in a statistical way. We next diagonalize **H** in the space $\{\phi_0, \gamma_m\}$. We call the resulting eigenvalues ζ_λ and, in agreement with our previous notation, the resulting eigenstates ψ_λ ($\lambda = 0, \ldots, M$). We have

$$\psi_\lambda = \hat{O}_{\lambda 0}\phi_0 + \sum_{m=1}^{M} \hat{O}_{\lambda m}\gamma_m \qquad\qquad 2.11.$$

with

$$\hat{O}_{\lambda 0} = O_{\lambda 0} \qquad\qquad 2.12.$$

We introduce the quantities

$$\varepsilon_0 = E_0 + \langle \phi_0 | V | \phi_0 \rangle \tag{2.13.}$$

$$v_m = \langle \gamma_m | H | \phi_0 \rangle \tag{2.14.}$$

In the basis $\{\phi_0, \gamma_m\}$, the matrix $\mathbf{H}$ has the form

$$\mathbf{H} = \begin{bmatrix} \varepsilon_0 & v_m \\ v_n & \varepsilon_m \delta_{mn} \end{bmatrix} \tag{2.15.}$$

It is easy to check that

$$O_{\lambda 0}^2 = \left\{ 1 + \sum_{m=1}^{M} v_m^2 / (\zeta_\lambda - \varepsilon_m)^2 \right\}^{-1} \tag{2.16.}$$

which implies the identity

$$\left[z - \varepsilon_0 - \sum_{m=1}^{M} v_m^2 / (z - \varepsilon_m) \right]^{-1} = \sum_{j=0}^{M} O_{j0}^2 / (z - \lambda_j) \tag{2.17.}$$

Let us take the identity 2.17 at the complex value $z = E + iI$. If the energy interval I is larger than the average distance d between two consecutive energies $\varepsilon_m, \varepsilon_{m+1}$, we have

$$\sum_{m=1}^{M} v_m^2 / (E + iI - \varepsilon_m) \simeq P \int [v^2(E') / (E - E')] \, dE' - i\pi v^2(E) \tag{2.18.}$$

Here, $v^2(E)$ is the quantity

$$v^2(E) = [v_m^2]_E \, \rho(E) \tag{2.19.}$$

where $[v_m^2]_E$ denotes the average of v_m^2 in an energy interval of length I centered on E, and $\rho(E) = d^{-1}$ is the density of energies ε_m in that interval. In the same way, we find

$$I \sum_{\lambda=0}^{M} O_{\lambda 0}^2 / [(E - \zeta_\lambda)^2 + I^2] \simeq \pi [O_{\lambda 0}^2]_E \, \rho(E) \tag{2.20.}$$

assuming that the densities of λ_j and ε_j are equal. Equations 2.17–2.20 yield

$$[O_{\lambda 0}^2]_E = \frac{1}{2\pi\rho(E)} \frac{\Gamma^\downarrow(E) + 2I}{(E - \tilde{\varepsilon}_0)^2 + \frac{1}{4}[\Gamma^\downarrow(E) + 2I]^2} \tag{2.21.}$$

where

$$\tilde{\varepsilon}_0 = \varepsilon_0 + P \int [v^2(E') / (E - E')] \, dE' \tag{2.22.}$$

while the "spreading width" $\Gamma^\downarrow$ is given by

$$\Gamma^\downarrow(E) = 2\pi v^2(E)\rho(E) \tag{2.23.}$$

II. 4   Intermediate Structure

The derivation of equation 2.21 is fairly general for the extreme situation described by equations 2.6 and 2.7. We only assumed that many levels are contained in the averaging interval and that the quantity v^2 (E) is a smooth function of energy. Equation 2.21 implies that the observable quantities $F^2_{\lambda c}$ are then distributed along a Lorentzian. Their distribution can, however, be different in the more realistic case described by inequality 2.8. Then we have

$$F^2_{\lambda c} = O^2_{\lambda 0} \left\{ \omega_{0c} + \sum_{m=1}^{M} \omega_{mc} [v_m/(\zeta_\lambda - \varepsilon_m)] \right\}^2 \qquad 2.24.$$

Proceeding in the same way as in section II.4, one finds

$$[F^2_{\lambda c}]_E = [(E - \varepsilon_0 + \Delta)^2 + \phi^2]/[(E - \tilde{\varepsilon}_0) + \tfrac{1}{4}(\Gamma^\downarrow + 2I)] \, [\omega^2_{mc}]_E \qquad 2.25.$$

where we have

$$\Delta = [(v_m \, \omega_{mc})_E/(\omega^2_{mc})_E] \omega_{0c} \qquad 2.26.$$

$$\phi^2 = \left\{ \tfrac{1}{4}(\Gamma^\downarrow + 2I) + \frac{d}{2\pi} \frac{\omega^2_{0c}}{[\omega^2_{mc}]_E} \right\} \left\{ \Gamma^\downarrow + 2I - \frac{2\pi}{d} \frac{[v_m \, \omega_{mc}]^2_E}{[\omega^2_{mc}]_E} \right\} \qquad 2.27.$$

We now consider two special cases. (a) If $\Gamma^\downarrow \gg I$, and if the relation

$$[v_m \, \omega_{mc}]^2_E = [v^2_m]_E [\omega^2_{mc}]_E \qquad 2.28.$$

holds, equation 2.27 shows that one has then $\phi^2 = 0$. Therefore, the average $[F^2_{\lambda c}]_E$ then vanishes at $\varepsilon_0 - \Delta$. It may appear quite unlikely that relation 2.28 is fulfilled, since one would usually expect that no correlation exists between the matrix elements v_m and ω_{mc}. A correlation may, however, occur if the state vectors $|\theta_c\rangle$ and $V|\phi_0\rangle$ are similar. This can happen, as we shall see in section V, for the isobaric analogue states. In that case, the asymmetry and its dynamical interpretation were discovered by Robson (25). (b) In the limit $\omega^2_{0c} \simeq [\omega^2_{mc}]_E$, we have naturally $[F_{\lambda c}]^2 = [\omega^2_{mc}]_E$: no enhancement exists if the configuration ϕ_0 is not simpler than γ_m with respect to channel c.

II. 5   Statistical Model

In section I we defined IS as a local deviation from the assumptions of the statistical model. In the case of bound states, the statistical model is based on the assumption that the matrix $\mathbf{O}$ in equation 2.4 is a random matrix, and that many configurations $\{\phi_j\}$ contribute to the observable quantity $F_{\lambda c}$. The consequence of these "microscopic" assumptions is that the following properties hold, where $-\!\!-\!\!-\!^\lambda$ denotes an ensemble average over many levels in an energy interval centered on E:

$$\overline{F_{\lambda c} F_{\lambda c'}}^{\,\lambda} = f_c(E) \delta_{cc'} \qquad 2.29.$$

$f_c(E)$ is a monotonic function of E \qquad 2.30.

Equations 2.29 and 2.30 can be taken as the basic assumptions of the statistical model for bound states. We have seen that equation 2.30 can be violated if simple modes of motion exist. If θ_c and $\theta_{c'}$ share the same isolated doorway state, equation 2.5. gives

$$F_{\lambda c}/F_{\lambda c'} = \langle \theta_c | \phi_0 \rangle / \langle \theta_{c'} | \phi_0 \rangle \qquad\qquad 2.31.$$

thus leading to a violation of assumption 2.29. We show in section III that a similar situation holds for compound nuclear resonances.

II. 6   Examples

II. 6.1 SIMPLE CONFIGURATIONS WITHOUT FRAGMENTATION   Bloch, Cindro & Harar (8) give some experimental evidence for the excitation of simple bound configurations by direct reactions, in the vicinity of closed shells. One particle-one hole (1p-1h) configurations are preferentially excited in (d, p) or (d, n) reactions on one-hole states targets, or in (p, p'γ_0) on closed shell targets. Boch et al (8) reviewed some data for nuclei with $Z = 20$; $Z, N = 28$; $Z, N = 50$; $Z, N = 82$. In these cases, the 1p-1h configurations do not appear to be fragmented. This is because of experimental limitations, which hinder the measurement of spectroscopic factors smaller than about 0.2. Usually, however, some fragmentation is expected. This is observed in ^{59}Ni, for instance, where it has been interpreted by Lande & Brown (7) as corresponding to the coupling of a single-particle state (ϕ_0) to the complicated states (ϕ_m) obtained by coupling single-particle states to 2^+ and 3^- core vibrations. Bloch et al (8) also review some data showing that a group of unresolved states around 4-MeV excitation energy is often excited in the inelastic scattering of α particles on the isotones $N = 20$ and on the isotopes $Z = 28$ (Ni) and $Z = 50$ (Sn). They suggest that these groups correspond to the spreading of collective 3^- excitations of the core. The detailed nature of this collective state is difficult to find, since a microscopic description of the (α, α') process is not available and because it lies at rather high excitation energy. Theoretical calculations by Veje (26) predict several 3^- levels near 5-MeV excitation energy in ^{58}Ni (for instance).

II. 6.2 α-PARTICLE TRANSFER   The theoretical proposal that modes of motion with quartet configurations may exist (i.e. be selectively excited and narrowly spread) in medium-weight and heavy nuclei (27, 28) is an interesting and controversial subject. It has been related to striking experimental data obtained in α-particle transfer reactions (29–32). Let us take one example: Strong peaks are found between 4- and 10-MeV excitation energy in ^{58}Ni in (^{16}O, ^{12}C) α-particle transfer reactions on ^{54}Fe. It was argued by Robson (33) that this specific reaction proceeds via compound nucleus formation rather than via a direct reaction mechanism. Then, the observed peaks could not be interpreted as corresponding to the existence of quartet states, i.e. of low-lying deformed excited states having, very roughly speaking, the structure of an α particle added to a spherical core. However, it appears that the same groups of levels are excited with alpha transfer from various projectiles, namely ^{18}O, ^{16}O, ^{14}N, ^{12}C, and ^{6}Li (34, 35). This, added to the narrowness of the peaks, strongly suggests that they are not due to a Q-dependence effect, i.e. to the favorable matching of the incoming and capture trajectories in a semiclassical picture. Similar results have recently been obtained in ^{94}Mo (29, 36). Thus, an interpretation in terms of a direct transfer is at least plausible. This is supported by (^{16}O, ^{12}C) reactions in the 2s-1d shell (37–40), where the residual nuclei are better known and the measurements more accurate. Even there, however, the interpretation of the data is not straightforward and the statistical model may be valid (41). The main argument in favor of associating this tentative IS with "quartet states" is that the latter are predicted to lie in the appropriate region of excitation energy (28, 42). Similar energies are found in a more phenomenological model where these peaks are associated to bound or resonant single-particle levels in the optical-model potential for α particles (43).

II. 6.3 PYGMY RESONANCES   The γ-ray spectra in electric dipole radiative neutron capture shows an anomalous shape near $E_\gamma \simeq 6$ MeV, for $120 < A < 210$ (44). The precise nature of the deviation from the statistical model is not clear. It had often been interpreted as a bump above the prediction of the statistical model but recent measurements (45) indicate that the anomaly rather consists in a dip at $E_\gamma \simeq 3$ MeV, followed by a steep rise to the normal value. The anomaly persists at about the same photon energy when the neutron energy varies. A presumably related anomaly is seen in (p, γ), (d, pγ), (γ, γ), and (γ, γ') reactions (46, 47). The anomaly does not appear in γ-ray spectra following other reactions as in (p, p'γ), (n, n'γ) (48), and μ-capture (49). Several theoretical interpretations of these features have been propounded, all of which involve the existence at the energy of the bump of one or several modes of motion which are "simple" with regard to the photon channel (46, 50–53). It does not appear, however, that these models are fully satisfactory and the proper theoretical interpretation of the data probably remains to be found, or at least confirmed.

III   RESONANCE REACTIONS

III. 1   Statistical Assumptions

We write the scattering matrix in the following form in the energy range of interest (ΔE).

$$S_{cc'}(E) = S_{cc'}^{BG} - i\sum_\lambda \gamma_{\lambda c}\gamma_{\lambda c'}/(E - \zeta_\lambda + \tfrac{1}{2}i\Gamma_\lambda) \qquad 3.1.$$

where $S_{cc'}^{BG}$ is a smooth function of energy and the sum over λ runs over all resonances contained in ΔE. Using the notation of section II.5, we can express the main assumptions of the statistical model in the following form:

(a) the partial widths in different channels are not correlated

$$\overline{\gamma_{\lambda c}\gamma_{\lambda c'}}^\lambda = s_c(E)\delta_{cc'} \qquad 3.2.$$

(b) $s_c(E)$ is a monotonic function of energy   3.3.

(c) $S_{cc'}^{BG}$ is diagonal in the channel indices   3.4.

Since $S_{cc'}^{BG}$ is a smooth function of E, assumption 3.4 cannot be violated locally, and IS can only consist in a deviation from assumptions 3.2 or 3.3.

If we normalize χ_E^c in equation 1.2 according to

$$\langle \chi_E^c | \chi_{E'}^{c'} \rangle = 2\pi\delta(E - E')\delta_{cc'} \qquad 3.5.$$

we shall see below that

$$\gamma_{\lambda c} \simeq V_\lambda^c \qquad 3.6.$$

in first, but good approximation. Comparing equations 1.2 and 2.5, the similarity between assumptions 2.29 and 2.30 on the one hand, and 3.2 and 3.3 on the other hand, is striking, with the correspondence

$$|\theta_c\rangle \leftrightarrow V |\chi_E^c\rangle \qquad 3.7.$$

The physical picture underlying relation 3.6 is the following: If scattering would not be taken into account, the nuclear states ψ_λ would be obtained by diagonalizing H in the space $\{\phi_j\}$, as described in section II. The coupling of ψ_λ to the scattering states χ_E^c transforms them into metastable states, i.e. into resonances, whose probability of decay into channel c is proportional to $|\gamma_{\lambda c}|^2$. Actually, equation 3.6 is only an approximation, valid for isolated resonances and obtained by neglecting the coupling between two bound states via the scattering states, i.e. quantities of the type

$$\int dE' \langle \phi_j | V | \chi_{E'}^c \rangle (E^+ - E')^{-1} \langle \chi_{E'}^c | V | \phi_n \rangle \qquad 3.8.$$

Taking relation 3.8 into account complicates the final expressions, but adds nothing essential from the physical point of view. We shall thus neglect it, except in sections III.4 and V.3. It is also necessary, for equation 3.6 to hold, to neglect the possibility of a direct transition from channel c to channel c', via

$$V_{cc'}^{EE'} = \langle \chi_E^c | V | \chi_{E'}^{c'} \rangle \qquad 3.9.$$

We shall indicate in section III.4 how this direct channel-channel coupling modifies the expressions given in sections III.2 and III.3.

III. 2   Doorway States

By analogy with the terminology used in section II.2, we call a "doorway configuration for channel c" a model state ϕ_0 which is such that the quantity

$$\langle \chi_E^c | V | \phi_0 \rangle = \Gamma_{0c}^{\frac{1}{2}} \qquad 3.10.$$

is different from zero. We gave a simple example in section I. If ϕ_0 is the only doorway state for channel c which has to be taken into account in the energy region of interest, the relations given in section II.3 can be used with the correspondence 3.7. Equation 2.10 becomes

$$\sum_{\lambda = 0}^{M} \gamma_{\lambda c}^2 = \Gamma_{0c} \qquad 3.11.$$

From equations 1.2, 2.21, and 3.6 we find

$$[\gamma_{\lambda c}^2]_E = \frac{d}{2\pi} \frac{(\Gamma^\downarrow + 2I)\Gamma_{0c}}{(E - \tilde{\epsilon}_0)^2 + \frac{1}{4}(\Gamma^\downarrow + 2I)^2} \qquad 3.12.$$

Let us assume that the spreading width

$$\Gamma^\downarrow(E) = (2\pi/d)v^2(E) = (2\pi/d)\overline{\langle\gamma_m| H |\phi_0\rangle^2}^m \qquad 3.13.$$

is a smooth function of energy, and is smaller than the energy domain ΔE where other doorway configurations can be neglected. Equation 3.12 shows that the partial widths are then enhanced in the vicinity of $\tilde{\epsilon}_0$, thus implying a deviation from statistical assumption 3.3.

If channels c and c' share the same isolated doorway state, we have

$$\gamma_{\lambda c}^2/\gamma_{\lambda c'}^2 = \Gamma_{0c}/\Gamma_{0c'} \qquad 3.14.$$

and assumption 3.2 is violated, even if $\Gamma^\downarrow$ is not a smooth function of energy.

In summary, we found that the conjunction of the following two circumstances leads to IS:

(a) Only one doorway configuration must be taken into account in some domain ΔE.

(b) The spreading width $\Gamma^\downarrow$ is smooth.

Assumption (a) implies that

$$\Gamma^\downarrow \ll \Delta E \qquad 3.15.$$

It was first thought that IS would be a fairly common phenomenon. This belief was essentially based on the simple example of 2p-1h doorway states given in section I, and on the fact that the density of 2p-1h configurations is much smaller than that of more complicated states. It turned out, however, that favorable circumstances must be met for inequality 3.15 to hold (18). We return to this point in section V. We emphasize, moreover, that the nature of the doorway configurations is essentially model-dependent since the choice of the basis states $\{\phi_j\}$, or equivalently of the model Hamiltonian H_0, is largely arbitrary. In particular, it is always possible to perform a unitary transformation in the space $\{\phi_j\}$ in such a way that only one of the resulting configurations, namely the configuration Ω_c in equation 1.3, is a doorway for channel c. Thus, assumption (a) above does not by itself imply any particular behavior for the data, and the addition of assumption (b) is essential. This will be confirmed below where we study in more detail a model where only one channel and one doorway state ϕ_0 exist, along with M complicated states γ_m.

III. 3   One Doorway State Model

We consider a one-channel case (elastic scattering) with only one doorway state ϕ_0 and M complicated states γ_m (equations 2.11–2.14). It is then easy to derive the following expression for the scattering function (54):

$$S_{cc} = \exp(2i\delta_c) \frac{\text{c.c.}}{E - e_0 + \frac{1}{2}i\Gamma_{0c} - \sum\limits_{m=1}^{M} v_m^2/(E - \varepsilon_m)} \qquad 3.16.$$

where δ_c is the potential scattering phase shift associated with χ_E^c and where

$$e_0 = \varepsilon_0 + (P/2\pi)\int dE'(E - E')^{-1}\Gamma_{0c}(E') \qquad 3.17.$$

Here, and below, c.c. denotes the complex conjugate of the denominator. Equation 3.16 can be written in the forms

$$S_{cc} = \exp(2i\delta_c) \frac{\text{c.c.}}{1 + i\sum\limits_{\lambda=1}^{M+1} \Gamma_{\lambda c}(E - \alpha_\lambda)^{-1}} \qquad 3.18.$$

$$= \exp(2i\delta_c)\left[1 - i\sum\limits_{\lambda=1}^{M} \frac{\gamma_{\lambda c}^2}{E - \beta_\lambda + (i/2)\delta_\lambda}\right] \qquad 3.19.$$

Hence, the cross section exhibits $M + 1$ resonances, as expected from the existence of $M + 1$ bound configurations. The sum rules

$$\sum_\lambda \gamma_{\lambda c}^2 = \sum_\lambda \delta_\lambda = \frac{1}{2}\sum_\lambda \Gamma_{\lambda c} = \Gamma_{0c} \qquad 3.20.$$

can easily be checked from equations 3.16–3.19.

It is well known that any elastic scattering data involving at most $M + 1$ resonances can always be fitted with expression 3.18, and hence also with 3.16. In other words, it is always possible to find a model that reproduces the data and where only one doorway configuration ϕ_0 exists. This is not surprising in view of the remark made at the end of section III.2. It is thus clear that the existence of an isolated doorway state does not imply the existence of IS. Nevertheless, the identification of the doorway configuration ϕ_0 is often of great interest.

In order to obtain IS from the one doorway state model, it is necessary to add one feature to the model, namely assumption (b) of section III.2. Then, the average value of S_{cc} reads

$$S_{cc}(E + iI) = \exp(2i\delta_c)\{1 - i\Gamma_{0c}/[E - e_0 + \frac{1}{2}i(\Gamma_{0c} + \Gamma^\downarrow + 2I)]\} \qquad 3.21.$$

Hence, the average total cross section displays a bump of width $\Gamma_{0c} + \Gamma^\downarrow + 2I$. From the identity of expressions 3.16 and 3.18, one easily finds that

$$[\Gamma_{\lambda c}]_E = \frac{d}{4\pi} \frac{\Gamma_{0c}(\Gamma^\downarrow + 2I)}{(E - e_0)^2 + \frac{1}{4}(\Gamma^\downarrow + 2I)^2} \qquad 3.22.$$

In the limit of isolated resonances, $\Gamma_{\lambda c} \simeq \frac{1}{2}\gamma_{\lambda c}^2$ and equation 3.22 yields 3.12. We emphasize that the assumptions that $\Gamma^\downarrow$ is independent of energy and that equations 3.21 and 3.22 hold are three equivalent statements: if one of them is true, it implies the other two.

III. 4   Correlation between Channel Widths

We showed in section III.2 that a common isolated doorway state gives rise to a deviation from assumption 3.2, although 3.3 is not necessarily violated. In the present section we show that direct reactions, i.e. the fact that $V_{cc'}^{EE'}$ (equation 3.9) does not vanish, may lead to a correlation between partial widths. The direct transition 3.9 modifies the expression 3.6 for the level width as follows (55):

$$\gamma_{\lambda c} = V_\lambda^c(E) + (2\pi)^{-1} \sum_{c'} \int dE'(E^+ - E')^{-1} V_{EE'}^{cc'} V_\lambda^{c'}(E') \qquad 3.23.$$

The origin of the second term on the right-hand side of equation 3.23 is that the state ψ_λ is coupled to channel c indirectly, via channel c'. The imaginary part of this term reads

$$-\tfrac{1}{2}i \sum_{c'} V_{EE}^{cc'} V_\lambda^{c'}(E) \qquad 3.24.$$

where the sum runs over open channels. It is clear that this contribution produces a correlation between $\gamma_{\lambda c}$ and $\gamma_{\lambda c'}$ (56). Similar expressions hold for photonuclear reactions, where direct capture gives rise to a correlation between particle and photon widths (57–59). This recently received experimental confirmation (52).

We note that direct reactions are smoothly dependent upon energy. Hence, they usually do not produce IS. However, if IS is present in channel c, the conjunction of 3.2. and 3.3 may imply that $s_{c'}$ is not a monotonic function of energy.

III. 5   Asymmetry

In the preceding sections we made the extreme assumption 2.7. In practice, the states γ_m are weakly coupled to channel c because of distant doorway states. The situation is then similar to the one described by equations 2.24–2.27. Moreover, the matrix element 3.8 implies that the complicated states γ_m are then coupled to ϕ_0 not only directly, via v_m (equation 2.14), but also indirectly, via channel c. The imaginary part of 3.8 reads

$$-i\pi \sum_{c^+} \langle \phi_0| V |\chi_E^c\rangle \langle \chi_E^c| V |\gamma_m\rangle \qquad 3.25.$$

In the case of isobaric analogue resonances, v_m may be much smaller than 3.25, since ϕ_0 and γ_m have different isospins. Then, the quantities Δ and ϕ in equation 2.25 vanish if only one channel is open, and the distribution of partial widths displays a characteristic asymmetry, as first predicted by Robson (25).

III. 6   Conclusions

In the one-channel case, IS consists of a violation of assumption 3.3. If the quantity $s_c(E)$ has a Lorentzian shape, or, equivalently, if the total cross section in channel c has a Breit-Wigner shape, one can interpret the data in terms of an isolated doorway configuration, mixed with complicated states that can be treated statistically. The identification of IS thus consists in exhibiting that the mean of the partial widths is energy-dependent in a nonmonotonic way, or, equivalently, that the mean of the matrix elements v_m^2 in equation 3.13, i.e. essentially the spreading width $\Gamma^\downarrow$, is (almost) independent of energy. Thus, this model for IS retains the standard statistical assumptions, after having singled out, and treated on a separate footing, the doorway configuration. We emphasized that one can always assume that a single doorway state exists, for one given channel, regardless of the detailed nature of the data. IS arises only if the additional assumption is made that $\Gamma^\downarrow$ is constant.

IV   STATISTICAL ANALYSIS

IV. 1   Introduction

The observable quantities (spectroscopic factors, partial widths, total cross sections, etc) undergo statistical fluctuations around their average. Hence, it is not easy to know whether an apparent enhancement of partial widths (for instance) is due to an accidental bunching of large widths, or whether it implies the existence of IS. We mentioned in section I that this ambiguity is at the basis of the controversy associated with the interpretation of some data in terms of IS. Hence, it is useful to develop statistical methods for evaluating the significance level of a tentative deviation from the statistical assumptions. In the present chapter we present a very brief survey of most of the available methods (60).

IV. 2   Enhancement of Partial Widths

The tests mentioned in the present section are applicable to the identification of deviations from assumptions 2.30 and 3.3. Let $S(\Gamma) = \{\Gamma_1, \ldots, \Gamma_M\}$ denote the sequence of observed partial widths in a given channel, ordered according to increasing resonance energies. The problem is finding the probability of occurrence of $S(\Gamma)$, when the quantities Γ are assumed to be random variables with some continuous probability distribution. The tests involve the following qualities, respectively.

(a) *Number of runs about the median.* This test, developed by James (61), involves the number of "runs" about the median. A run is defined as an unbroken sequence of values above or below the median. This method is independent of the probability distribution of the variable Γ_λ.

(b) *Longest run above the median.* Moore (62) suggested that the length of the longest run above the median may imply a deviation from randomness. The feasibility of this test is demonstrated in (60), on the basis of previous work by Olmstead (63).

(c) *Longest run above the line of optimal run length.* This line is such as to maximize the shortest of the longest run above and below it. This test was proposed and used in (60).

(d) *Number and length of runs up and down.* A run up is an unbroken sequence of values that form an increasing set. This test appears well adapted to the identification of IS (64).

(e) *Mean square successive difference.* The ratio

$$\eta = \sum_{j=1}^{M-1} (x_{j+1} - x_j)^2 \bigg/ \left\{ \sum_{j=1}^{M} (x_j - \bar{x})^2 \right\}$$ 4.1.

may indicate the existence of a nonmonotonic energy dependence of the average of a *normal* population. Since the distribution of $\{\Gamma_{\lambda c}\}$ is not normal (65), this test can only be applied to other quantities, like the ratio of two sums of many partial widths (60).

(*f*) *Serial correlation coefficient for lag 1.* This coefficient is given by

$$R_1 = \left[\sum_{\lambda=1}^{M} \Gamma_\lambda \Gamma_{\lambda+1} - M^{-1} \left(\sum_{\lambda=1}^{M} \Gamma_\lambda \right)^2 \right] \bigg/ \left[\sum_{\lambda=1}^{M} \Gamma_\lambda^2 - M^{-1} \left(\sum_{\lambda=1}^{M} \Gamma_\lambda \right)^2 \right]$$ 4.2.

The value of this quantity can imply a deviation from randomness, but appears to be quite sensitive to the possible omission of only one value, even in a fairly large sample (60).

(*g*) *Large adjacent values.* The occurrence of a run of large values may imply a deviation from the statistical assumptions, as shown in (60).

IV. 3   Correlation between Widths

Here we mention two tests that can establish the existence of deviations from assumptions 2.29 or 3.2.

(*a*) *The rank-correlation coefficient* yields information concerning the existence (but not the degree) of a correlation between $\gamma_{\lambda c}^2$ and $\gamma_{\lambda c'}^2$ ($\lambda = 1, \ldots, M$). Each $\gamma_{\lambda c}^2$ is assigned a number corresponding to its ranking in size. If d_λ denotes the difference between the ranks of $\gamma_{\lambda c}^2$ and $\gamma_{\lambda c'}^2$, the rank correlation coefficient is given by

$$\rho_{cc'} = 1 - 6(M^3 - M)^{-1} \sum_{\lambda=1}^{M} d_\lambda^2$$ 4.3.

This test is used in (66).

(*b*) *The product-moment correlation function* is given by

$$\alpha_{cc'} = \frac{M \sum_\lambda \gamma_{\lambda c}^2 \gamma_{\lambda c'}^2 - \sum_\lambda \gamma_{\lambda c}^2 \sum_\mu \gamma_{\mu c'}^2}{\left[M \sum_\lambda \gamma_{\lambda c}^4 - \left(\sum_\lambda \gamma_{\lambda c}^2 \right)^2 \right]^{\frac{1}{2}} \left[M \sum_\lambda \gamma_{\lambda c'}^4 - \left(\sum_\lambda \gamma_{\lambda c'}^2 \right)^2 \right]^{\frac{1}{2}}}$$ 4.4.

Its value indicates the degree of correlation (66).

V   EXPERIMENTAL DATA

V. 1   Introduction

We mentioned in section I that there exist three well-established examples of IS, namely the giant dipole resonances, the isobaric analogue resonances, and IS in sub-threshold neutron-induced fission. Excellent specialized reviews are available for these cases; we shall therefore only briefly describe their nature. We then turn to a discussion of more disputed examples of IS.

From the preceding chapters, we conclude that the analysis of an IS phenomenon normally involves the following steps:

(a) Do the data imply a significant deviation from the assumptions of the statistical model?

(b) Is the IS characterized by definite values of the angular momentum and the parity?

(c) Can one identify the structure of the doorway configuration?

(d) Is the IS seen in several channels?

The first two questions concern the identification of IS, the last two its interpretation.

V. 2 Giant Dipole Resonance

The main features of this IS have been reviewed (67–69). The microscopic structure of the doorway state Ω_{gd}, which gives rise to IS in the photon channel, is given by equation 1.4. In medium-weight and heavy nuclei, the average cross section shows some substructure, around a Lorentzian shape. This is ascribed to the coupling of Ω_{gd} with surface vibrational states (70, 71). The value of the spreading width $\Gamma^\downarrow$ has recently been computed for ^{208}Pb (72, 73), with moderate success. The dependence of $\Gamma^\downarrow$ upon mass number is fairly characteristic (74) and ought to be understood theoretically.

V. 3 Isobaric Analogue Resonance

There are several recent reviews of the isobaric analogue resonances (IAR) (19, 54, 75, 76). The presentation adopted in (54, 76) is closest to ours. Let $\psi_P = |_Z A_N\rangle$ denote the wave function of a low-lying state of a medium-weight or heavy nucleus and T_- the isospin lowering operator. The configuration $\psi_> = T_-|\psi_P\rangle$ would be an eigenstate of H for nucleus $_{Z+1}A_{N-1}$ if the interactions were charge independent. This is not the case, mainly because of the Coulomb force, whose main effect is twofold: (a) It shifts the energy of $\psi_>$ with respect to that of ψ_P by the Coulomb energy, which is roughly equal to 1.5 $ZA^{-\frac{1}{3}}$ MeV. Hence, the energy of $\psi_>$ corresponds to a high excitation energy in the nucleus $_{Z+1}A_{N-1}$. Because of its simple structure, $\psi_>$ can play the role of a doorway configuration for the proton channel. This doorway state produces the IAR. (b) The Coulomb interaction mixes $\psi_>$ with the complicated configurations γ_m, which have an isospin one unit smaller than $\psi_>$. The direct mixing $\langle\psi_>|H|\gamma_m\rangle$ is small, since γ_m and $\psi_>$ have different isospins. Hence, the indirect coupling via the proton channel (equation 3.25) may be larger than the direct coupling. This leads to the characteristic asymmetric enhancement predicted by Robson (25), as discussed in section III.5. The theory reproduces fairly well the energy of the IAR and its partial widths in the proton channels. Until very recently, the calculated values of the spreading width $\Gamma^\downarrow$ were much too large. The inclusion of the mixing of $\psi_>$ with the giant-monopole doorway apparently greatly improves the agreement between theory and experiment (76, 77). From the partial widths of the IAR, one can determine the spectroscopic

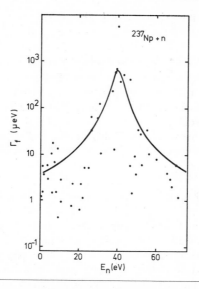

Figure 2   Fission widths of ^{237}Np (n, f). The resonances with no detectable fission width have been omitted [adapted from (79)].

factors of the "parent state" ψ_p with an accuracy comparable to that achieved in stripping experiments.

V. 4   Subthreshold Neutron-Induced Fission

Striking IS has been discovered (20, 21, 78, 79) in neutron-induced fission on some elements, with a very strong (factor 10^3) local enhancement of the partial widths in the fission channel f. Figure 2 shows the measurable fission widths for ^{237}Np (n, f), near an IS at 40 eV (79). The smallest widths presumably correspond to resonances with an angular momentum different from that of the IS. The doorway states responsible for this IS are shape isomers, i.e. excited states whose deformation is very different from that of the ground state (80, 81). These shape isomers also produce IS below neutron threshold. This can be seen in (d, pf) stripping reactions; in ^{240}Pu, for instance, the 0^+, 2^+, 4^+, 6^+, and 8^+ members of a rotational band based on a shape isomer have been observed; the associated moment of inertia is twice smaller than that of the ground-state rotational band (82). The discovery of IS in fission has greatly improved our understanding of the dynamics of the fission process, and splendidly confirmed the importance of Strutinski's deformation-dependent "shell corrections" (83).

V. 5   Discussion

IS can be observed in a channel c only if the enhancement of the partial widths or of the average total cross section is sufficiently pronounced. The conditions for this are: (a) The escape width Γ_{0c} is larger than or comparable to the spreading width

$\Gamma^{\downarrow}$ (equation 3.22). (*b*) The doorway state is isolated, i.e.

$$\Gamma^{\downarrow} \ll D_d \qquad\qquad 5.1.$$

where D_d is the distance to the closest doorway state. It was estimated by Payne (18) that condition 5.1 is not likely to be fulfilled for doorway states of the 2p-1h type (section I). In practice, inequality 5.1 implies that ϕ_0 and γ_m have very different structures, i.e. that a selection rule renders $\Gamma^{\downarrow}$ exceptionally small. This is true for the three examples given above. In the following section we describe IS seen in heavy ion reactions, whose identification and origin are not yet clear. Then we turn to a brief review of a number of other tentative examples of IS.

V. 6   Heavy Ion Reactions

Almqvist, Bromley & Kuehner (84) observed three pronounced bumps in $^{12}C + {}^{12}C$, at $E_{cm} \simeq 6$ MeV, for several outgoing channels (n, α, γ, and p). The first two bumps were shown to have definite angular momentum and parity (2^+ and 4^+), respectively. Surprisingly, the total width of these peaks is nearly equal to the sum of the partial widths (85), which indicates that the spreading width is very small. This, in turn, implies that the structure of the doorway states responsible for the IS is very different from that of the surrounding "complicated states." It was proposed (86–88) that the doorway configurations are molecular states composed of two ^{12}C ions. This interpretation was abandoned after the discovery of the existence of several IS at lower energy (89). This led Michaud & Vogt (90, 91) to propose that one of the ^{12}C ions is disassociated into three α particles. These authors can reproduce the average $^{12}C + {}^{12}C$ total cross section at low energy, and explain the large value of the $^{20}Ne + \alpha$ branching ratio. This interpretation also appears to account qualitatively for some aspects (92) of the IS observed in $^{12}C + {}^{16}O$ below the Coulomb barrier (93). It also appears possible that one IS exists in ^{14}N (^{14}N, γ) ^{28}Si (94), but no satisfactory interpretation has been propounded. It is remarkable that none of nine other reactions studied at present with ions of comparable masses exhibits IS. This striking fact has been emphasized by Stokstad in a recent review of the experimental data of IS in heavy ion reactions (95). It could be due to the fact that, in these other reactions, any IS is damped by the large number of open channels (95), although the possible existence of an IS in $^{14}N + {}^{14}N$ is then hard to explain.

Bumps have also been observed in heavy ion reactions above the Coulomb barrier. A recent review of the corresponding experimental data has been given by Stokstad (95). For instance, strong oscillations are seen in $^{12}C + {}^{12}C$ just above the Coulomb barrier ($E_{cm} = 7$–14 MeV). They appear compatible with the statistical model (96). At higher energies ($E_{cm} = 16$–40 MeV), broad bumps ($\simeq 2$-MeV wide) exist in $^{12}C + {}^{12}C$, $^{16}O + {}^{12}C$, $^{16}O + {}^{16}O$. They are usually interpreted as a potential scattering phenomenon. On these broad bumps are superimposed narrower peaks, whose identification as IS (97) is hazardous, since they may be compatible with the statistical model (98).

There have been claims that IS has been observed in the compound nucleus in transfer reactions initiated with heavy ions. Many of these appear to still need corroboration (95). A possible exception is the case ^{16}O (^{12}C, α) ^{24}Mg, where an IS probably exists at 36.5-MeV excitation energy in ^{28}Si (99). Evidence for nonstatistical effects, in the form of channel correlations, was also found in ^{16}O (^{16}O, ^{12}C) ^{20}Ne. This was interpreted as IS due to high spin doorway states in ^{32}S (100), but it could also be due to the fact that the channels corresponding to the $0^+, 2^+$, and 4^+ levels of ^{20}Ne are very similar (101).

Strong backward enhancements were observed in the elastic scattering of α particles from α-particle nuclei of mass $12 \leq A \leq 48$. They were interpreted (102, 103) as corresponding to the formation of a molecular state where one of the ions would be an α particle. However, an interpretation in terms of a direct process cannot be excluded (104). A recent review of these phenomena and of their interpretation can be found in Ref. 105.

In conclusion, there exists converging evidence towards the existence of quasi-molecular doorway states below the Coulomb barrier, but the nature of the constituents is debatable. It seems that α-particle entities play a particular role. The field of heavy ion reactions appears to be very promising for the discovery of simple modes of motion through the systematic investigation of IS.

V. 7   Other Examples

Besides the cases reviewed above, few convincing examples of IS have been discovered. The reasons have been discussed in section V.5. Here we give a list of these cases; their statistical significance is discussed in (60).

(a) ^{19}F + n. Six bumps exist in the average total cross section, for $0.3 \leq E_n \leq 2.0$ MeV (106). A definite angular momentum can be attached to most of the bumps, but the statistical significance of the IS is low. A proposal for the structure of the doorway configurations (107) was only moderately successful.

(b) ^{56}Fe + n. The neutron widths are enhanced between 200 and 600 keV (108), but the spreading width $\Gamma^{\downarrow}$ depends upon energy (24).

(c) ^{206}Pb + n. The neutron widths are enhanced near $E_n = 500$ keV, with definite angular momentum and parity (109). Correlation has been claimed between neutron and photon widths (66) but this is disputed (110). A theoretical interpretation of the IS has been given in terms of a vibrational state of ^{206}Pb coupled to a single-particle neutron state (111, 112).

(d) ^{28}Si + n. A correlation exists between neutron and photon widths, between 0.2 and 1.2 MeV, with an apparent enhancement (113, 114).

(e) ^{115}In (n, γ). A local enhancement (in an energy interval of only 100 eV) of radiative widths for high-energy photons has been reported in (115). The angular momenta of the resonances have not been measured.

(f) ^{187}Re (n, γ). An enhancement similar to the preceding case in an interval only 30-eV wide has been claimed (116). It was recently disproved (117).

(g) ^{70}Ge (p, p). Narrow substructures have been detected within an IAR (118). The significance level of these IS is hard to evaluate (60).

(h) ^{40}Ca (p, p′). Twelve $(5/2^+)$ resonances lie between 6 and 8 MeV, among which five, grouped between 6.95 and 7.30 MeV, have large widths (119). The test (g) of section IV.2 shows that this IS is statistically significant (60).

VI   DOORWAY STATES WITHOUT INTERMEDIATE STRUCTURE

VI. 1   Introduction

We mentioned earlier that the concept of doorway states may be useful even in the absence of IS proper. This is illustrated in the present chapter. In section VI.2 we deal with the identification of a doorway configuration that can account for the sum of the partial widths of a group of resonances. Section VI.3 is devoted to the calculation of the neutron strength function and of the imaginary part of the optical-model potential. Finally, we discuss pre-equilibrium emission (see section I) in section VI.4.

VI. 2   Identification of Doorway States

We saw in section III.3 that it is always possible to interpret any sample of partial widths in channel c in terms of a model in which only one doorway configuration exists for channel c. The identification of this doorway configuration may be of interest, even in the absence of IS. In light nuclei, for instance, the giant dipole state and the isobaric analogue states are mixed with only a few complicated states, and no IS proper exists. Nevertheless, the interpretation of the cross section in terms of a single doorway state is clearly very natural. This is also true for some other cases: the tentative IS observed in ^{206}Pb + n (section V.7) may not be meaningful from a statistical point of view, but the construction of a doorway configuration by Beres & Divadeenam (111, 112) is instructive. This remark also applies to ^{40}Ca (p, p') (119). In ^{56}Fe (n, γ), three (3/2$^-$) resonances are found near $E_n = 230$ keV; a very tentative proposal has been given for the doorway configuration (120).

VI. 3   Strength Function and Optical-Model Potential

The strength function is defined by the ratio

$$S_c = s_c/d \qquad\qquad 6.1.$$

of the average partial width (equation 3.2) to the average level spacing. The following sum rule can be proved, in the one-channel case (13, 54)

$$\sum_\lambda \gamma_{\lambda c}^2 = \sum_j \langle \phi_j | V | \chi_E^c \rangle^2 \qquad\qquad 6.2.$$

Equation 3.11 is a particular case of the latter relation. Only the doorway states contribute to the right-hand side of equation 6.2. We have thus

$$S_c = \langle \Gamma_{0c} \rangle / D \qquad\qquad 6.3.$$

where $\langle \Gamma_{0c} \rangle$ represents the average escape width (see equation 3.10) and where D is the average spacing of the doorway states in the energy interval ΔE where the average is performed. Equation 6.3 is valid only if the influence of the doorway states lying outside ΔE can be neglected. If this is true, equation 6.3 shows that the strength function can be calculated from these doorway states alone. This was pointed out by Block & Feshbach (15) and Shakin (121), who calculated the strength function in the tin isotopes and in ^{209}Pb. Recently, the strength functions of ^{41}Ca (122) and ^{89}Sr (123) were successfully analyzed in the same way.

The strength function is intimately related to the imaginary part W of the optical-model potential at low energy. Brueckner, Eden & Francis (124) and Shaw (9) computed W from equation 6.2, in the case of nuclear matter. In the latter model, expression 6.2 becomes the sum of the matrix elements between 1p and 2p-1h states, taking energy and momentum conservation and the Pauli principle into account. This approach to the calculation of W has recently received renewed interest (125–129).

VI. 4 Pre-equilibrium Decay

Around 1963, Sidorov (130) and Barschall and collaborators (131, 132) found that the high-energy part of the neutron spectrum in some (p, n) reactions disagrees with the predictions of the statistical model. Griffin (22) explained this discrepancy by taking into account the possibility that nucleons are emitted by the compound system before it reaches complete randomization. A recent review of these phenomena and of their interpretation has been given by Blann (133). In the doorway state model of IS, two widths are attached to a doorway configuration. The first one, Γ_{0c} (equation 3.10), is called the "escape width," and is related to the probability that the doorway configuration decays back into channel c. The other one, the spreading width $\Gamma^{\downarrow}$ (equation 3.13), is connected with the probability of reaching more complicated configurations. In the pre-equilibrium model, each configuration ϕ_j (equation 2.4) is assumed to have a finite escape width in addition to its spreading width for decay into a more complicated configuration. A configuration Φ_j with a low complexity mainly emits high-energy particles which leave the residual nucleus in a low excited state. This model accounts remarkably well for the experimental data, but involves a few parameters whose physical interpretation is not straightforward. Some authors (134) associate one of these parameters with the average spreading width of the doorway states, which they find equal to about 350 keV near A = 90 and to 1 MeV for A $\simeq$ 180.

Literature Cited

1. Bohr, N. 1936. *Nature* 137:344
2. Bodansky, D. 1962. *Ann. Rev. Nucl. Sci.* 12:79
3. Baldwin, G. C., Klaiber, G. S. 1948. *Phys. Rev.* 73:1156
4. Wilkinson, D. H. 1959. *Ann. Rev. Nucl. Sci.* 9:1
5. Goldhaber, M., Teller, E. 1948. *Phys. Rev.* 74:1046
6. Brink, D. 1957. *Nucl. Phys.* 4:215
7. Lande, A., Brown, G. E. 1966. *Nucl. Phys.* 75:344
8. Bloch, C., Cindro, N., Harar, S. 1969. *Progr. Nucl. Phys.*, ed. D. M. Brink, J. H. Mulvey 10:77
9. Shaw, G. L. 1959. *Ann. Phys.* 8:509
10. Kerman, A. K., Rodberg, L. S., Young, J. E. 1963. *Phys. Rev. Lett.* 11:422
11. Rodberg, L. S. 1961. *Phys. Rev.* 124:210
12. Weisskopf, V. F. 1961. *Phys. Today* 14(7):18
13. Feshbach, H. 1958. *Ann. Phys.* 5:357
14. Feshbach, H. 1962. *Ann. Phys.* 19:287
15. Block, B., Feshbach, H. 1963. *Ann. Phys.* 23:47
16. Lane, A. M., Thomas, R. G., Wigner, E. P. 1955. *Phys. Rev.* 98:693
17. Barschall, H. H. 1952. *Phys. Rev.* 86:431

18. Payne, G. L. 1968. *Phys. Rev.* 174:1227
19. Robson, D. 1966. *Ann. Rev. Nucl. Sci.* 16:119
20. Paya, D., Derrien, H., Fubini, A., Michaudon, A., Ribon, P. 1967. *Nuclear Data for Reactors.* Vienna: IAEA II:128
21. Migneco, E., Theobald, J. P. 1968. *Nucl. Phys. A* 122:603
22. Griffin, J. J. 1966. *Phys. Rev. Lett.* 17:478
23. Feshbach, H., Kerman, A. K., Lemmer, R. H. 1967. *Ann. Phys.* 41:230
24. Jeukenne, J. P., Mahaux, C. 1969. *Nucl. Phys. A* 136:49
25. Robson, D. 1965. *Phys. Rev. B* 137:535
26. Veje, C. I. 1966. *Kgl. Dan. Vidensk. Selsk, Mat. Fys. Medd.* 35:1
27. Danos, M., Gillet, V. 1967. *Phys. Rev.* 161:1034
28. Jaffrin, A. 1970. *Phys. Lett. B* 32:448
29. Faivre, J. C. et al 1970. *Phys. Rev. Lett.* 24:1188
30. Faraggi, H. et al 1971. *Ann. Phys.* 66:905
31. Faraggi, H., Lemaire, M. C., Loiseaux, J. M., Mermaz, M. C., Papineau, A. 1971. *Phys. Rev. C* 4:1375
32. Faraggi, H. 1972. *Comments Nucl. Particle Phys.* 5:79
33. Robson, D. 1972. *Comments Nucl. Particle Phys.* 5:16
34. Cassagnou, Y., Faraggi, H., Morrison, G., Papineau, A. Quoted in Ref. 32
35. Mermaz, M. C., Lemaire, M. C. Quoted in Ref. 32
36. Bohn, H. et al 1972. *Phys. Rev. Lett.* 29:1337
37. Middleton, R., Garrett, J. D., Fortune, H. T. 1970. *Phys. Rev. Lett.* 24:1436
38. Middleton, R., Garrett, J. D., Fortune, H. T. 1971. *Phys. Rev. Lett.* 27:950
39. Gobbi, A. et al 1971. *Phys. Rev. Lett.* 26:396
40. Maher, J. V., Erb, K. A., Wedberg, G. H., Ricci, J. L., Miller, R. W. 1972. *Phys. Rev. Lett.* 29:291
41. Greenwood, L. R. et al 1972. *Phys. Rev. C* 6:2112
42. Jaffrin, A. 1972. *Nucl. Phys. A* 196:577
43. Dudek, A., Hodgson, P. E. 1971. *J. Phys. C* 32:6-185
44. Starfelt, N. 1964. *Nucl. Phys.* 53:397
45. Bollinger, L. M., Loper, G. D., Thomas, G. E. 1972. *Bull. Am. Phys. Soc.* 17:580
46. Bartholomew, G. A. 1969. *Proc. Int. Symp. Neutron-Capture Gamma-Ray Spectrosc.,* Studsvik, Sweden, p. 553. Vienna: IAEA
47. Chrien, R. E. 1972. *Int. Conf. Nucl. Structure Study with Neutrons, Budapest.* In press
48. Bartholomew, G. A., Earle, E. D., Lone, M. A., Ferguson, A. J. 1972. *Int. Conf. Nucl. Structure Study with Neutrons, Budapest.* In press
49. Earle, E. D., Bartholomew, G. A. 1971. *Nucl. Phys. A* 176:363
50. Kuo, T. T. S., Blomqvist, J., Brown, G. E. 1970. *Phys. Lett. B* 31:93
51. Perez, S. M. 1970. *Phys. Lett. B* 33:317
52. Lane, A. M. 1971. *Ann. Phys.* 63:171
53. Brzosko, J. S., Piotrowski, J., Soltan, A. Jr., Szeflinski, Z. 1972. *Nucl. Phys. A* 189:545
54. Mahaux, C., Weidenmüller, H. A. 1969. *Shell-Model Approach to Nuclear Reactions.* Amsterdam: North-Holland
55. Mahaux, C. 1972. *Int. Conf. Nucl. Structure Study with Neutrons, Budapest.* In press
56. Hüfner, J., Mahaux, C., Weidenmüller, H. A. 1967. *Nucl. Phys. A* 105:489
57. Lane, A. M., Lynn, J. E. 1960. *Nucl. Phys.* 17:563
58. Estrada, L., Feshbach, H. 1963. *Ann. Phys.* 23:123
59. Boridy, E., Mahaux, C. 1973. *Nucl. Phys.* In press
60. Baudinet-Robinet, Y., Mahaux, C. 1973. *Phys. Rev.* In press
61. James, G. D. 1971. *Nucl. Phys. A* 170:309
62. Moore, M. S. 1972. *Statistical Properties of Nuclei,* ed. J. B. Garg, p. 155. New York: Plenum
63. Olmstead, P. S. 1958. *Monogr. 2937.* New York: Bell Telephone Syst. Tech. Publ.
64. Baudinet-Robinet, Y., Mahaux, C. 1972. *Phys. Lett. B* 42:392
65. Porter, C. E., Thomas, R. G. 1956. *Phys. Rev.* 104:483
66. Baglan, R. J., Bowman, C. D., Berman, B. L. 1971. *Phys. Rev. C* 3:2475
67. Brenig, W. 1965. *Advan. Theoret. Phys.* 1:59
68. Danos, M., Fuller, E. G. 1965. *Ann. Rev. Nucl. Sci.* 15:29
69. Spicer, B. 1969. *Advan. Nucl. Phys.,* ed. M. Baranger, E. Vogt. 2:1. New York: Plenum
70. Le Tourneux, J. 1965. *Kgl. Dan. Vidensk. Selsk. Mat. Fys. Medd.* 34:11
71. Huber, M. G., Danos, M., Weber, H. J., Greiner, W. 1967. *Phys. Rev.* 155:1073
72. Davidson, A. M. 1972. *Nucl. Phys. A* 180:208
73. Dover, C. B., Lemmer, R. H., Hahne, F. J. W. 1972. *Ann. Phys.* 70:458
74. Hayward, E. 1970. Photonuclear Reactions. *NBS Monogr. 118.* Wash, DC: US Dept. Commerce
75. Wilkinson, D. H., Ed. 1969. *Isospin in*

Nuclear Physics. Amsterdam: North-Holland
76. Auerbach, N., Hüfner, J., Kerman, A. K., Shakin, C. M. 1972. *Rev. Mod. Phys.* 44:48
77. Auerbach, N., Bertsch, G. 1973. *Phys. Lett. B* 43:175
78. Fubini, A., Blons, J., Michaudon, A., Paya, D. 1968. *Phys. Rev. Lett.* 20:1373
79. Michaudon, A. 1972. *Statistical Properties of Nuclei*, ed. J. B. Garg, p. 149. New York: Plenum
80. Lynn, J. E. 1968. In *Nuclear Structure*, p. 463. Vienna: IAEA
81. Weigmann, H. 1968. *Z. Phys.* 214:7
82. Specht, H. J., Weber, J., Konecny, E., Heunemann, D. 1972. *Phys. Lett. B* 41:43
83. Strutinsky, V. M. 1968. *Nucl. Phys. A* 122:1
84. Almqvist, E., Bromley, D. A., Kuehner, J. A. 1960. *Phys. Rev. Lett.* 4:515
85. Almqvist, E., Bromley, D. A., Kuehner, J. A., Wholen, B. 1963. *Phys. Rev.* 130:1140
86. Vogt, E. W., McManus, H. 1960. *Phys. Rev. Lett.* 4:518
87. Davis, R. H. 1960. *Phys. Rev. Lett.* 4:521
88. Imanishi, B. 1968. *Nucl. Phys. A* 125:33
89. Patterson, J. R., Winkler, H., Zaidins, C. S. 1969. *Astrophys. J.* 157:367
90. Michaud, G., Vogt, E. W. 1969. *Phys. Lett. B* 30:85
91. Michaud, G., Vogt, E. W. 1972. *Phys. Rev. C* 5:350
92. Vogt, E. W., Hartmann, G., Helb, H. D., Ischenko, G., Silber, F. Unpublished
93. Patterson, J. R., Nagorcka, B. N., Symons, G. D., Zuk, W. M. 1971. *Nucl. Phys. A* 175:545
94. Almqvist, E., Bromley, D. A., Kuehner, J. A. 1960. *Int. Conf. Nucl. Structure, Kingston, Canada*, p. 258. Amsterdam: North-Holland
95. Stokstad, R. G. 1973. *Conf. Intermed. Processes Nucl. Reactions, Yugoslavia*. Heidelberg: Springer Verlag
96. Bondorf, J. 1973. *Nucl. Phys. A* 202:30
97. Fink, H. J., Scheid, W., Greiner, W. 1972. *Nucl. Phys. A* 188:259
98. Low, K. S., Tamura, T. 1972. *Phys. Lett. B* 40:32
99. Stokstad, R. et al 1972. *Phys. Rev. Lett.* 28:1523
100. Singh, P. P., Sink, D. A., Schwandt, P., Malmin, R. E., Siemssen, R. H. 1972. *Phys. Rev. Lett.* 28:1714
101. Shaw, R. W., Norman, J. C., Vandenbosch, R. 1969. *Phys. Rev.* 184:1040
102. Oeschler, H. et al 1972. *Phys. Rev. Lett.* 28:694

103. Rinat, A. S. 1972. *Phys. Lett. B* 38:281
104. Agassi, D., Wall, N. S. 1973. Unpublished
105. Budzanowski, A. 1973. *Conf. Intermed. Processes Nucl. Reactions, Yugoslavia*. Heidelberg: Springer Verlag
106. Monahan, J. E., Elwyn, A. J. 1967. *Phys. Rev.* 153:1148
107. Afnan, I. R. 1967. *Phys. Rev.* 163:1016
108. Elwyn, A. J., Monahan, J. E. 1969. *Nucl. Phys. A* 123:33
109. Farrell, J. A., Kyker, G. C. Jr., Bilpuch, E. G., Newson, H. W. 1965. *Phys. Lett.* 17:286
110. Allen, B. J., Macklin, R. L., Fu, C. Y., Winters, R. R. 1973. *Phys. Rev. C* 7:2598
111. Beres, W. P., Divadeenam, M. 1970. *Phys. Rev. Lett.* 25:596
112. Beres, W. P., Divadeenam, M. 1973. *Phys. Rev. C* 7:862
113. Schwartz, R. B., Schrack, R. A., Heaton, H. T. 1971. *Bull. Am. Phys. Soc.* 16:495
114. Jackson, H. E., Toohey, R. E. 1972. *Phys. Rev. Lett.* 29:379
115. Coceva, C., Corvi, F., Giacobbe, P., Stefanon, M. 1970. *Phys. Rev. Lett.* 25:1047
116. Stolovy, A., Namenson, A. I., Godlove, T. F. 1971. *Phys. Rev. C* 4:1466
117. Stolovy, A., Namenson, A. I., Harvey, J. A. 1973. *Bull. Am. Phys. Soc. II* 18:592
118. Temmer, G. M., Maruyama, M., Mingay, D. W., Petrascu, M., Van Bree, R. 1971. *Phys. Rev. Lett.* 26:1341
119. Mittig, W. 1973. *Conf. Intermed. Processes Nucl. Reactions, Yugoslavia*. Heidelberg: Springer Verlag
120. Jackson, H. E., Strait, E. N. 1971. *Phys. Rev. C* 4:1314
121. Shakin, C. M. 1963. *Ann. Phys.* 22:373
122. Hay, W. D. 1969. PhD thesis. Oxford Univ. Unpublished. Quoted in Hodgson, P. E. 1971. *Nuclear Reactions and Nuclear Structure*, p. 558. Oxford: Clarendon
123. Divadeenam, M., Beres, W. P. Quoted by Newson, H. 1972. *Statistical Properties of Nuclei*, ed. J. B. Garg, p. 309. New York: Plenum
124. Brueckner, K. A., Eden, R. J., Francis, N. C. 1955. *Phys. Rev.* 98:1445
125. Reiner, A. S. 1964. *Phys. Rev.* 133:1105
126. Azziz, N. 1970. *Nucl. Phys. A* 147:401
127. Kidwai, H. R., Rook, J. R. 1971. *Nucl. Phys. A* 169:417
128. Hüfner, J., Mahaux, C. 1972. *Ann. Phys.* 73:525
129. Jeukenne, J. P., Lejeune, A., Mahaux, C. 1972. *Conf. Present Status Novel Developments in Nucl. Many-Body*

Problem, Rome. In press
130. Sidorov, V. A. 1962. *Nucl. Phys.* 35:253
131. Holbrow, C. H., Barschall, H. H. 1963. *Nucl. Phys.* 42:264
132. Wood, R. M., Borchers, R. R., Barschall, H. H. 1965. *Nucl. Phys.* 71:529

133. Blann, M. 1973. *Conf. Intermed. Processes Nucl. Reactions, Yugoslavia.* Heidelberg: Springer Verlag
134. Grimes, S. M., Anderson, J. D., Pohl, B. A., McClure, J. W., Wong, C. 1971. *Phys. Rev. C* 4:607

✖ 5539

PRODUCTION MECHANISMS OF TWO-TO-TWO SCATTERING PROCESSES AT INTERMEDIATE ENERGIES

Geoffrey C. Fox[1]
C. C. Lauritsen Laboratory of High Energy Physics, California Institute of Technology, Pasadena, California

C. Quigg[2]
Institute for Theoretical Physics, State University of New York, Stony Brook, Long Island, New York

Dedicated to Bruno Renner

CONTENTS

[1] Work supported by the US Atomic Energy Commission under Contract AT(11-1)-68.
[2] Research supported in part by the National Science Foundation under Grant No. GP-32998X.

219

1   INTRODUCTION

Together with hadron spectroscopy, the study of two-body reactions has until very recent times been the goal of most experiments on the high energy collisions of elementary particles, and the source of most of our understanding of strong interaction dynamics. However, we must confess to some embarrassment at finding ourselves charged with the task of summarizing briefly the status of a vast subject which is experiencing a prolonged and painful adolescence. In a few pages, it is difficult to convey accurately the sense of exasperation now felt by many workers in the field without giving short shrift to the concepts which form, we feel sure, a sturdy foundation (1) for a future theory of two-body collisions, and, indeed, of hadronic interactions in general. It is all too easy, when bewailing the lack of dramatic progress in the theory of high energy processes, to overlook the steady accumulation of insights that make up our present incomplete understanding of two-body reactions. We therefore feel fortunate to be able to refer to the article by Chiu (2) in the preceding volume of *Ann. Rev. Nucl. Sci.* in which the principal successes of two-body phenomenology have been recounted. Although we shall mention the underlying ideas in the next section, we will for the most part assume familiarity with the contents of Chiu's chapter. In addition, we are able to direct the reader to numerous reviews (3–13) that explore in depth the successful applications of these ideas. Likewise, the failures of the present-day Regge phenomenology (which occur in quantitative detail and not, we stress, in gross structure) have been the subject of other reviews (14–22), and we shall not dwell on them at length here. Instead we will concentrate on the experimental systematics and the gems of theoretical insight that organize them into a pattern which will be the basis of future deeper understanding.

For the rest of the introduction, we first continue on a spiritual level and explain the obstacles facing the development of the subject. Then we define the problem and finally present the details of our article.

1A Impasse

There are three basic points to be made. Firstly, in some respects the failures of our Regge-based phenomenology are indeed due to weakness of the theory. Here we have in mind the lack of any technique for computing Regge pole residues or the contributions of Regge cuts to scattering amplitudes.

Secondly, it is likely that in many instances the absence of experimental systematics is due to the existence of small effects which are often irrelevant and simply serve to obscure the simplicity of the principal two-body reaction mechanisms. We need not go far afield to find a dramatic (and well-understood) example: In the case of elastic πd scattering, a deep dip in the dominant contribution to the differential cross section is completely camouflaged (23) by a contribution from transitions between the s-wave and d-wave components of the deuteron wave function although the deuteron is only 7% d-wave.

Thirdly, for a large number of former aficionados of two-body reactions, the appeal has gone out of model building because there are not enough data of detailed enough nature to select among many "plausible" but unconvincing alternative theoretical schemes. Thus the belief is widespread that it is about as pointless at this time to concoct a model for two-body collisions as it is to build another theory of CP violation. There is the difference in degree that models for high energy collisions can be refuted by experiment within a time short on the scale of a physicist's productive years, whereas models for CP violation apparently cannot! In both cases, some new clue is needed to point the way to a "convincing" model.

It is the demise of explicit models that has led to the current vogue for "amplitude analysis," by which is meant the experimental determination of individual scattering amplitudes, rather than the mere measurement of incoherent sums of absolute squares of amplitudes as they occur in cross sections. It is disconcerting that we have come to this, and Sonderegger (24) has spoken poignantly of amplitude analysis "which must be the ultimate aspiration of the experimenter, but which has in fact become the last resort of theorists."

Lest the reader despair, we remember that the intuition gleaned from two-body collisions is the basis (25–27) for virtually all our expectations for the multiparticle production experiments that become more common as the eager physicist gains access to higher energy accelerators. For instance, the rise of a Regge phenomenology of inclusive reactions (28, 29) following Mueller's discovery of the Generalized Optical Theorem (30) serves as a reminder of the utility of our basic theoretical structure for organizing vast amounts of experimental information. The same ideas employed (with semiquantitative success) in two-body studies are providing for the first time a framework for the understanding and appreciation of multiparticle production. Even here, we must expect eventually to encounter the same difficulties that now confound the analysis of two-body reactions.

1B The Subject

Let us now say a few words about the scope of our problem. The partition of the total cross section into elastic scattering, inelastic two-body and quasi two-body channels, and "true" multiparticle production is summarized in Table 1 for three energy regimes. We will use the notation

$$a+b \rightarrow a+b \qquad\qquad 1.1.$$

to denote elastic scattering;

$$a+b \rightarrow c+d$$
$$\qquad\quad \big\lfloor \quad \rightarrow (\gamma\delta\ldots)$$
$$\qquad \big\lfloor \rightarrow (\alpha\beta\ldots) \qquad\qquad 1.2.$$

to indicate production of the (possibly unstable) particles c and d (where $\alpha\beta\gamma\delta\ldots$ are the products of subsequent decays of c and d);

$$a+b \rightarrow 1+2+3+\ldots+n \qquad\qquad 1.3.$$

to specify n-particle production; and

$$a+b \rightarrow c+\text{anything} \qquad\qquad 1.4.$$

to represent a single-particle inclusive reaction. Our attention will be confined

Table 1   Composition of the total cross section

Scattering Process	Fraction of Total Cross Section			Observables
	Low Energy	~2 GeV/c	High Energy	
Total Cross Section	1	1	1	σ_{tot} (essentially constant at high energy).
Elastic Scattering	>0.5	0.15–0.20	0.15–0.20	$d\sigma/dt$ (essentially constant at high energy); polarization parameters P, R, A.
Two (Stable) Particle Inelastic Channels	<0.5	0.10–0.15	→0	$d\sigma/dt$ (falls like 1/beam momentum at high energy); polarization parameters P, R, A.
Two (Unstable) Particle Inelastic Channels: Single or Double Resonance Production	0	0.70–0.75	<0.20	$d\sigma/dt$ (falls with incident momentum for most channels); decay angular distributions.
Genuine Multiparticle Channels	0	0	0.60–0.80	Production cross section; differential cross section as a function of $3n-4$ kinematic variables.

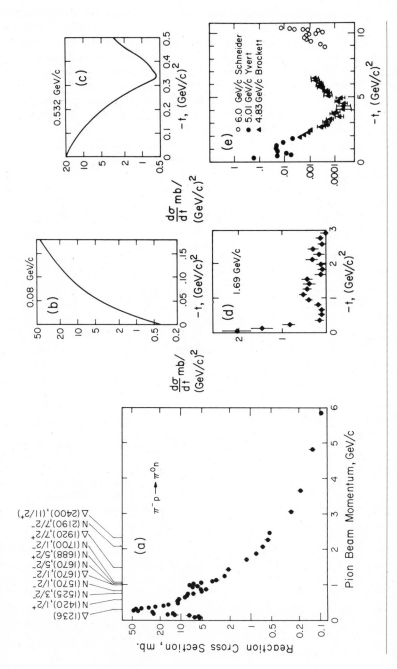

Figure 1 Integrated reaction cross section (297) and differential cross section (118, 206, 298–300) for $\pi^- p \to \pi^0 n$ at various laboratory momenta.

mainly to reactions 1.1 and 1.2, which are classified further (according to the properties of the final-state particles) in section 5.

Figure 1 summarizes the evolution of the differential cross section for the charge-exchange reaction $\pi^- p \to \pi^0 n$ as the beam momentum is increased from threshold through the resonance region to the intermediate-energy regime. The qualitative features are common to nearly all two-body reactions. (Exceptions to this generalization are explained in section 2.) From the obligatory s-wave angular distribution at threshold, the cross section turns into a nearly background-free p-wave distribution in the neighborhood of the isolated $P_{33}[\Delta(1236)]$ resonance, becomes more complicated at momenta of a few GeV/c where several resonances overlap, and finally assumes a stable shape at momenta above 3 GeV/c where the reaction takes place through many partial waves. The production angular distribution is typified by a forward "peripheral" peak and a backward peak much smaller in magnitude. It is the theoretical description of these peripheral peaks for incident momenta in the range from approximately 3 GeV/c to 30 GeV/c that forms the subject of this article. We shall refer to this range as the "intermediate-energy regime," in anticipation of different systematics at still higher energies.

1C   The Article

In the next section, we continue on a descriptive note and summarize the qualitative features of peripheral and Regge ideas. We shine a tarnished sun on the many dusty and discarded Regge models; dawn breaks and points the way to future sections; reflection adds perspective to past articles (2). The third section details dull formalism: We attack the sloppy and inconsistent conventions that make the study of resonance reactions unnecessarily hard. This brief polemic is supplemented by a separate preprint that lists and relates all experimental observables (polarizations, density matrix elements). Section 4 is a critical review of the many amplitude analyses of πN and hypercharge exchange reactions. The former are qualitatively similar while the latter are all found wanting. We examine the evidence for "peripheral" imaginary parts and find the evidence much less certain than often assumed. In particular, the so-called dual absorption model (31) is in some difficulties (32, 33).

Section 5 lists all known experimental systematics of forward scattering while section 6 outlines the pearls of theoretical wisdom that organize these results. We also suggest some fruitful areas for experimental investigation and finally conclude in section 7 with dreams for the future. The nonspecialist should probably omit the rather technical sections 3–5 on a first perusal.

2   SELF-EVIDENT TRUTHS AND THEIR EXTRAPOLATIONS

2A   Validity of the Peripheral Exchange Picture

In our introductory remarks, we have touched upon the outstanding characteristic of differential cross sections for two-body reactions: They are *peripheral*, i.e. sharply peaked about the forward or backward directions. This feature may be appreciated in terms of a geometrical picture in which the relative angular momentum l of the

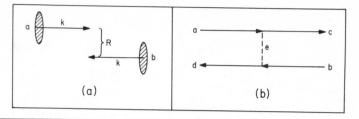

Figure 2 Complementary views of two-body reactions: (*a*) geometrical picture; (*b*) peripheral exchange picture.

incoming particles dictates the partial waves through which the reaction takes place (Figure 2*a*) and, in turn, the shape of the production angular distribution as $d\sigma/d\Omega_{cm} \sim [P_l(\cos\theta_{cm})]^2$. From the shape of peripheral peaks, one may roughly "measure" l and, therefore, knowing the cm momentum k of the colliding particles, deduce a semiclassical interaction radius R from the connection $l \approx kR$. This sort of reasoning typically implies an interaction radius of about 1 fm. This is large or comparable to hadron Compton wavelengths (1.4 fm for pions, 0.2 fm for nucleons), which may be expected to define a scale, and fully consistent with the range of nuclear forces (35).

While the geometrical picture provides a fount of intuition not only for two-body reactions (in which context we shall revisit it), but as well for more complicated hadron-hadron collisions (36), the keystone of the theory of two-body reactions is the peripheral exchange picture. The long apparent range of interaction makes it natural to associate scattering with the exchange (Figure 2*b*) of the least massive particle carrying the requisite quantum numbers. This natural association is also correct: Without important exception, all occurrences or nonoccurrences of peripheral peaks in two-body reactions can be understood in terms of the quantum numbers of the known [SU(3) multiplets of] mesons and baryons. The detailed verification of this hypothesis ranks as the outstanding achievement in the development of two-body phenomenology (37, 38).

To make concrete the implications of the peripheral exchange picture, let us introduce two-body kinematics and consider some specific cases. The scattering of spinless particles $1+2 \to 3+4$, represented in Figure 3, is described by a quantum mechanical amplitude $A(s, t)$, which is an analytic function of Mandelstam's energy variables

$$s = (p_1 + p_2)^2 \qquad\qquad 2.1.$$

$$t = (p_1 - p_3)^2 \qquad\qquad 2.2.$$

$$u = (p_1 - p_4)^2 \qquad\qquad 2.3.$$

where p_i is the four-momentum of particle i. In the usual experimental situation with particle 1 incident on particle 2 at rest,

$$s = m_1^2 + m_2^2 + 2m_2(p_{lab}^2 + m_1^2)^{\frac{1}{2}} \to 2m_2 p_{lab} \qquad\qquad 2.4.$$

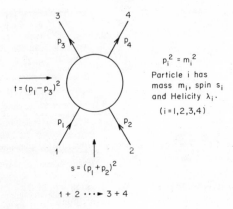

Figure 3 Labeling of particles for a two-body to two-body collision. The momenta are labeled p_i, masses m_i, particle spins s_i, and helicities λ_i.

at high energies. The squared momentum transfer is related to the cm scattering angle through

$$\cos\theta_s = \left[2st + s^2 - s\sum_{i=1}^{4} m_i^2 + (m_1^2 - m_2^2)(m_3^2 - m_4^2) \right]\bigg/ S_{12}S_{34} \to 1 + 2t/s$$

$$2.5.$$

at high energies. Here we use

$$S_{ij} = [s - (m_i + m_j)^2]^{\frac{1}{2}}[s - (m_i - m_j)^2]^{\frac{1}{2}} \qquad 2.6.$$

The principle of crossing (e.g. 3, 5) relates the six reactions

$$1 + 2 \leftrightarrow 3 + 4 \quad [s\text{-channel}] \qquad 2.7.$$
$$1 + \bar{3} \leftrightarrow \bar{2} + 4 \quad [t\text{-channel}] \qquad 2.8.$$
$$1 + \bar{4} \leftrightarrow 3 + \bar{2} \quad [u\text{-channel}] \qquad 2.9.$$

to the same function $A(s, t)$ (in the spinless case). Because $s + t + u = m_1^2 + m_2^2 + m_3^2 + m_4^2$, where m_i is the mass of particle i, it is convenient to display the Mandelstam variables in triangular coordinates, as we have done in Figure 4 for the reactions

$$s: \; K^- p \to \pi^+ \Sigma^- \qquad 2.10.$$
$$t: \; \bar{p}\Sigma^- \to \pi^- K^- \qquad 2.11.$$
$$u: \; \pi^- p \to K^+ \Sigma^- \qquad 2.12.$$

In the s- and u-channels, resonance formation is prominent at low energies; the resonances masses are indicated as lines of constant s or u. In contrast, no known resonances have the quantum numbers of the t-channel. Beyond the resonance region energies, forces for the scattering processes are provided by the available exchange particles. The absence of t-channel resonances thus implies the absence of forward peaks in the s- and u-channel reactions. On the other hand, the s-channel resonances are available as peripheral exchanges for backward scattering in the u-channel, and vice versa. The unmeasurable t-channel reaction has the u-channel

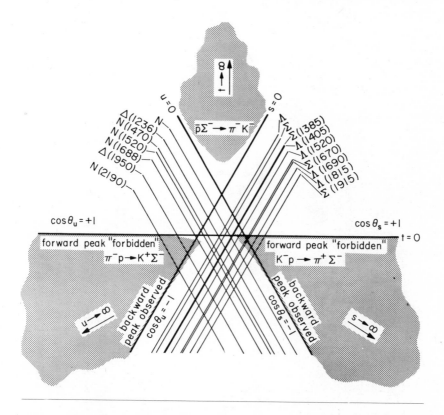

Figure 4 Mandelstam plot for the reactions $K^-p \to \pi^+\Sigma^-$, $\pi^-p \to K^+\Sigma^-$, and $\bar{p}\Sigma^- \to \pi^-K^-$, showing the correspondence between crossed-channel resonances and peripheral peaks in differential cross sections. Physical regions for the s-, t-, and u-channels are shaded.

resonances to drive forward scattering and the s-channel resonances to drive backward scattering.

More generally, it is a spectroscopic fact of life that all mesons fit into SU(3) singlets and octets, and all baryons fit into singlets, octets, and decimets, as prescribed by the elementary quark model (34, 39). The implied restriction on resonance quantum numbers allows one to predict occurrences of peripheral peaks. Some illustrative cases are collected in Table 2. [Other similar tabulations

Table 2 Correspondence between peripheral peaks and nearby poles

Reaction	$|Q_t|$	I_t	Y_t	nearby poles[a]	peak?[b]	$|Q_u|$	I_u	Y_u	nearby poles	peak?
$\bar{p}p \to \bar{p}p$	0	0,1	0	$[\mathcal{P}], f, \rho, \omega, A_2$	✓	2	1	2	–	0
$pp \to pp$	0	0,1	0	$[\mathcal{P}], f, \rho, \omega, A_2$	✓	forward-backward symmetric				✓
$\pi^-p \to \pi^-p$	0	0,1	0	$[\mathcal{P}], f, \rho$	✓	2	3/2	1	Δ	✓
$\pi^+p \to \pi^+p$	0	0,1	0	$[\mathcal{P}], f, \rho$	✓	0	1/2, 3/2	1	N, Δ	✓
$K^-p \to K^-p$	0	0,1	0	$[\mathcal{P}], f, \rho, \omega, A_2$	✓	2	1	2	–	0
$K^+p \to K^+p$	0	0,1	0	$[\mathcal{P}], f, \rho, \omega, A_2$	✓	0	0,1	0	Λ, Σ	✓
$\pi^-p \to \pi^0n$	1	1	0	ρ	✓	1	1/2, 3/2	1	N, Δ	✓
$\pi^-p \to K^+\Sigma^-$	2	3/2	-1		0	0	0,1	0	Λ, Σ	✓
$\pi^-p \to \eta^0n$	1	1	0	A_2	✓	1	1/2	1	N	✓
$\pi^+p \to \omega^0\Delta^{++}$	1	1	0	B, ρ	✓	1	1/2	1	N	✓
$\pi^+p \to \rho^0\Delta^{++}$	1	1	0	π, A_2	✓	1	1/2, 3/2	1	N, Δ	✓
$\bar{p}p \to \bar{\Sigma}^-\Sigma^+$	0	1/2, 3/2	1	K, K^*, K^{**}	✓	2	3/2	1	–	0
$\bar{p}p \to \bar{\Sigma}^+\Sigma^-$	2	3/2	1		0	0	1/2, 3/2	1	–	0
$K^-p \to \pi^-\Sigma^+$	0	1/2, 3/2	-1	K^*, K^{**}	✓	2	3/2	1	Δ	✓
$K^-p \to \phi\Lambda$	1	1/2	-1	K^*, K^{**}	✓	1	1/2	1	N	✓[c]

[a] $\mathcal{P}$ denotes Pomeron Exchange (sections 5A–C).

[b] The "exotic" reactions which have no particle to exchange, do, in fact, have a *small* peripheral peak. Section 5M considers this more carefully.

[c] Suppressed by the quark-model selection rule $N\bar{N} \not\to \phi$. See e.g. (34) and section 5M.

are given in (3) and (37).] The remainder of our review is devoted to attempts to extend this framework to a quantitative theory of peripheral cross sections.[3]

2B   The Regge Pole Framework

To the known SU(3) multiplets of resonances correspond families of Regge trajectories which obey a linear relation between particle spins (α) and masses squared:

$$\alpha(m^2) = \alpha_0 + \alpha'm^2 \qquad\qquad 2.13.$$

The slope parameter α' is universal (for all Regge trajectories) and equal to (0.85–1) $(\text{GeV}/\text{c})^{-2}$. The interpretation of this universality is one of the appealing results of dual resonance models (41). For dynamical reasons which either imply or are implied by the absence of exotic[4] states, the forces which build the hadron spectrum are exchange degenerate (42), so a particle occurs for each integer value of spin along a meson trajectory. A familiar example of this feature of the spectrum is the well-established set of $I = 1$ natural parity mesons ρ, A_2, g which occur at spins 1, 2, 3 on the trajectory $\alpha(m^2) \approx \frac{1}{2} + m^2$.

Let us now discuss in the Regge pole language the features we have already noticed in Figure 1. At low energies for the reaction $\pi^- p \to \pi^0 n$, the reaction cross section is characterized by the excitation of states on (for example) the Δ trajectory (Figure 5a). In the t-channel, there occur the states $\rho, g \ldots$ appearing on the ρ trajectory, which mediate forward scattering at high energies (Figure 5b) and give rise to the characteristic $s^{\alpha(t)}$ behavior of the near-forward scattering amplitude. Similarly, backward scattering proceeds at high energies (Figure 5c) by exchange of the u-channel trajectories, such as the nucleon trajectory, which generate the $s^{\alpha(u)}$ behavior of the near-backward scattering amplitude. For charge exchange, which is quite typical in this respect, the backward peak is an order of magnitude smaller than the forward peak. The study of the transition region, exemplified by Figure 1d, has led in recent years to a detailed understanding of the interplay of direct-channel resonances and crossed channel Regge trajectories. The beautiful work of Dolen, Horn & Schmid (43) showed conclusively that the direct-channel (resonance) and crossed-channel (Regge pole) descriptions are complementary. Thus the resonances cooperate to build up the high energy Regge behavior, whereas the Regge pole description is valid in an average sense, even in the resonance region

[3] An analogy may help the reader who is not a high energy physicist to understand the present state of our art. If one had invented nonrelativistic quantum mechanics, quantized the radiation field, and discovered that the fine structure constant is small, he could correctly enumerate allowed and forbidden transitions between atomic states but would be unable to compute every rate precisely. In two-body scattering, the peripheral exchange picture and the quark model for the resonance spectrum allow one to catalog allowed and forbidden peaks, but do not confer the ability to compute cross sections precisely. We do not know why the quark model works, but neither does quantum electrodynamics explain the size of α. Our goal, however, is to explain scattering and spectrum (i.e. everything) at once (40).

[4] By exotic states, we mean mesons with quantum numbers not given by quark-antiquark pairs or baryons with quantum numbers not given by three quark states.

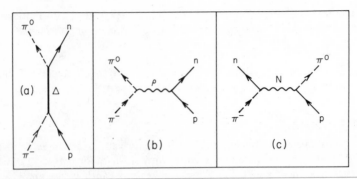

Figure 5 (*a*) Low energy pion-nucleon charge exchange is characterized by the production of direct-channel nucleon resonances among which are the $\Delta(1236)$, $\Delta(1950)$, and $\Delta(2420)$ lying on the Δ trajectory. (Compare Figure 1*a*.) (*b*) The high energy forward peak shown in Figure 1*e*, is driven by *t*-channel exchange of the ρ trajectory. (*c*) The high energy backward peak, also shown in Figure 1*e*, is driven by *u*-channel exchange of several baryon trajectories, among them the nucleon trajectory.

below 2 GeV/c. The equivalence of the two modes of description is known as duality and is further discussed in sections 4C and 4D. The remarkable cooperation between resonances, which leads to familiar high energy features, is illustrated in Figure 6 for the πN scattering amplitudes corresponding to isovector exchange in the *t*-channel. The first zeros of the Legendre functions associated with direct channel resonance contributions to the spinflip amplitude (dominant in πN charge exchange) all occur near $t = -0.6$ (GeV/c)^2, where the pronounced dip appears (see Figure 1*e*) in the high energy cross section, while the first zeros associated with the nonflip amplitude are gathered near $t = -0.2$ (GeV/c)^2, where the crossover between $\pi^+ p$ and $\pi^- p$ cross sections takes place (see Figure 7).

The mutual constraints among *s*-channel, *t*-channel, and *u*-channel trajectories are embodied in the dual resonance model given by Veneziano's inspired guess (44), wherein zero-width resonances appear as poles on the real axis. The Veneziano model therefore gives a caricature of the dual behavior of hadrons which cannot be quantitatively reliable. Odorico (45) has advocated the viewpoint that even when the transition is made from the dual resonance model idealization to real world situations, artifacts of the simple structure of the Veneziano amplitude remain. As one example, Figure 8 shows that the differential cross section for $K^- p \to \bar{K}^0 n$ exhibits structure at fixed u in the low energy ($\lesssim 2\text{-}GeV/c$) regime of the kind suggested by the Veneziano formula. Successes of this sort have led others (8, 46) to support the contention that *qualitative* lessons are indeed to be learned from dual resonance model amplitudes. Another implication of the dual resonance model is that $1/\alpha'$ sets the scale for hadronic phenomena both for the mass spectrum and for scattering dynamics. This suggests that for $s \gtrsim 10$ GeV^2 we are in the (asymptotic) Regge pole regime. It now seems that there are complications (Regge cuts?) which become dominant only at much higher energies, perhaps for

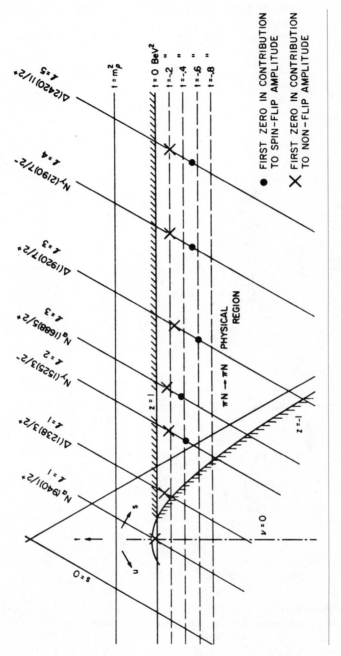

Figure 6 First zeros of the prominent resonances on the Mandelstam plot for the πN problem [from (43)].

$s \gtrsim 100 \, \text{GeV}^2$. If this be so, the study of two-body reactions at the National Accelerator Laboratory may provide many surprises.

In the intermediate energy regime, we may confront the distinctive predictions

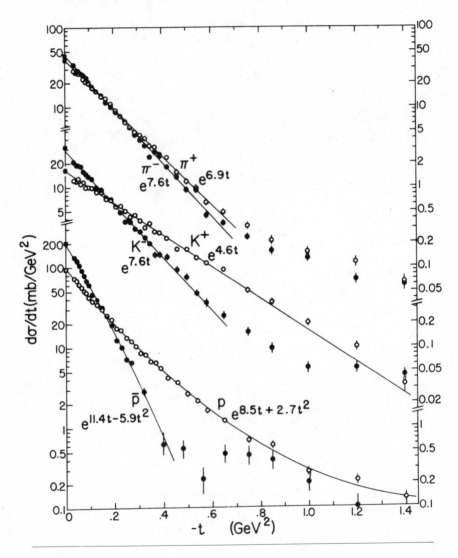

Figure 7 Differential cross sections for the scattering of six different particles from protons at a laboratory momentum of 5 GeV/c [from (103)]. The crossover which occurs in $\pi^{\pm} p$ scattering near $t = -0.2 \, (\text{GeV/c})^2$ is ascribed to the vanishing of the imaginary part of the nonflip ρ-exchange amplitude.

$$K^-p \rightarrow \bar{K}^0 n$$

u scale

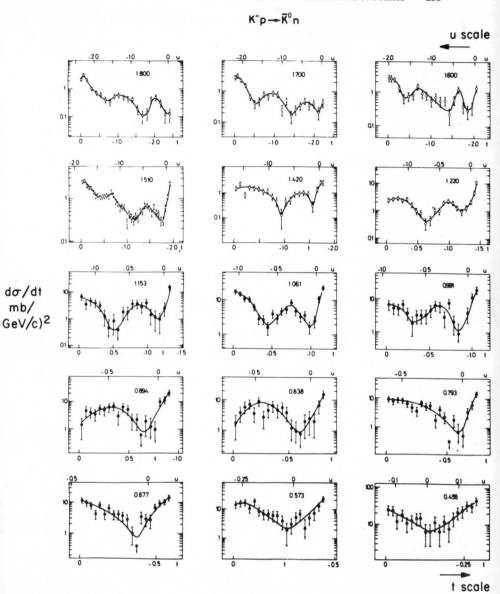

$d\sigma/dt$
mb/
$(GeV/c)^2$

t scale

Figure 8 $K^-p \rightarrow \bar{K}^0 n$ angular distributions between 0.45 and 1.8 GeV/c plotted vs t and u (the beam momentum is indicated in each frame). The solid lines are hand-drawn to guide the eye. The figure shows three dips at constant positions of $u \approx -0.1$, -0.7, and -1.7 (GeV/c)^2 [from (45)].

of Regge theory with experiment. As a typical Regge pole amplitude, let us consider that for πN charge exchange:

$$A(s,t) = g_{\pi\pi}(t)g_{N\bar{N}}(t)\Gamma\left(1 - \alpha_\rho(t)\right)\left(e^{-i\pi\alpha_\rho(t)} - 1\right)(\alpha's)^{\alpha_\rho(t)} \qquad 2.14.$$

which displays the basic properties of interest. The most characteristic of these is "shrinkage." Recalling that $d\sigma/dt \propto s^{-2}|A|^2$, and using the linear trajectory $\alpha(t) = \alpha_0 + \alpha't$ (2.13), we have the prediction

$$\frac{d\sigma}{dt}(t)\bigg/\frac{d\sigma}{dt}(0) = (\text{Function of } t) \times \exp(2\alpha't\log s) \qquad 2.15.$$

of a forward peak which becomes more sharply peaked ("shrinks") as the primary energy is increased. One way of exhibiting shrinkage is by fitting the experimental cross sections to the form

$$\frac{d\sigma}{dt}(s,t) = (\text{Function of } t)\cdot(\alpha's)^{2\alpha_{\text{effective}}(t) - 2} \qquad 2.16.$$

If the reaction is dominated by the exchange of a single Regge pole, the effective trajectory inferred from the scattering data will match the trajectory determined by the resonance spectrum. For πN charge exchange, this is indeed the case (2).

The second prediction of exchange-degenerate Regge theory is of the occurrence of zeros in the scattering amplitude (hence in the cross section) at wrong signature nonsense points (47), e.g. at $\alpha_\rho = 0$ in the case of πN charge exchange. Again this provides a successful explanation for the dip in $d\sigma/dt$ at $t \approx -0.6$ (GeV/c)^2 [cf. Figure 1e]. Another consequence of the exchange degeneracy of meson trajectories is the prediction that vector and tensor exchange amplitudes should be 90° out of phase so that pairs of reactions such as

$$K^-p \to \bar{K}^0 n \qquad 2.17a.$$
$$K^+n \to K^0 p \qquad 2.17b.$$

for which the corresponding amplitudes are

$$\text{Vector} + \text{Tensor} \qquad 2.18a.$$
$$-\text{Vector} + \text{Tensor} \qquad 2.18b.$$

should have equal cross sections. Such line-reversal relations are well enough satisfied that systematics of the *breaking* of exchange degeneracy have been cataloged.

A final prediction of Regge theory is that Regge pole residues should factorize. For example, the ρ-exchange contributions to the reactions $\pi^-\pi^0 \to \pi^0\pi^-$, $\pi^-p \to \pi^0 n$, and $np \to pn$ will be in the ratios

$$[g_{\pi\pi}(t)]^2 : g_{\pi\pi}(t)g_{\bar{N}N}(t) : [g_{\bar{N}N}(t)]^2 \qquad 2.19.$$

(where we have suppressed helicity indices). Factorization has been tested extensively, especially for the Pomeranchuk singularity (48, 49) in both two-body and inclusive reactions, but never (for lack of precise data) in situations which provide very stringent tests. Since factorization is derivable from S-matrix axioms in the case of isolated Regge poles, any test probes the nature of the J-plane singularities.

2C   Some Contemporary Models

Although we shall avoid in this paper detailed quantitative confrontations of particular models with experimental results, it will be helpful to have in mind some of the divergent viewpoints represented by contending theories.

THE DUAL ABSORPTION MODEL   That the examination of scattering amplitudes in the impact parameter plane leads to important insights into reaction mechanisms has been stressed in modern times by Ross, Henyey & Kane (50). Their original discussions of the systematics of s-channel helicity amplitudes and the b-plane interpretation of those systematics did not, however, make contact with the concurrent developments in the field of duality. A synthesis of the Michigan school's s-channel-impact parameter view with the predictive power of duality was proposed by Harari (31) who named his scheme the dual absorption model. The idea is a throwback—with additional sophistication—to the old "ring model" (51). Harari guessed that the imaginary parts of scattering amplitudes (in nonexotic channels) were built up, through duality, by resonances concentrated on a peripheral ring of radius $R \approx 1$ fm, and should therefore be given by the amplitudes for Fraunhofer diffraction from an illuminated ring, which are proportional to $J_n[R(-t)^{\frac{1}{2}}]$, where n is the net s-channel helicity flip. We examine the evidence for this idea in section 4F.

COMPLEX REGGE POLES[5]   A general result of multiperipheral models is that the collision of J-plane singularities gives rise to complex conjugate Regge poles on unphysical sheets which may be approximated by a pair of first-sheet poles over a limited energy range. It has been hoped that complex poles might provide a compact parameterization for "input" Regge poles and the associated Pomeron-Reggeon branch cuts.

EFFECTIVE ABSORPTION MODELS[6]   In an effort to retain the intuitive appeal of the absorption recipe for Regge cuts while circumventing the embarrassments of line-reversal inequalities and πN charge exchange polarization, a number of authors (33, 54–60) have replaced the conventional elastic rescattering amplitude with an effective rescattering amplitude constructed to reproduce the observable quantities and without any a priori motivation. Apparently there is little predictive power in such an approach, although it might be hoped that one could understand how amplitudes get to be as they are. It is too early to judge (84) even the interpolative value of such models but it has already been pointed out that they all violate finite energy sum rules (61) so even their pedagogical value is to be doubted. It may be expected that the effective absorption models take into account Reggeon-Reggeon cuts as well as Reggeon-Pomeron cuts, but here again duality constraints

[5] A compact review is given by Roy (52).
[6] For a brief review, see Tran Thanh Van (53).

raise doubts. Finkelstein (62) and Worden (63) have obtained selection rules for Reggeon-Reggeon cuts which suppress the contributions to KN charge exchange and πN charge exchange, where their effects would have been particularly welcome. Although these arguments can always be breached or ignored, we are coming to believe that a pertubative approach to Regge cuts may be futile. The confrontation between experiments now in progress[7] on double charge exchange reactions and the Reggeon-Reggeon cut predictions of Michael (64) and Quigg (65) may provide a final verdict.

THE STRONG CENTRAL ABSORPTION PRESCRIPTION   One modified absorption scheme that has a possibility of being derivable from physical arguments, and not merely adjusted to agree with data, is the strong central absorption prescription (SCAP) advocated by Chiu (2), which is similar in effect (though not in motivation) to the $\pi N \rightarrow \rho N$ model of Williams (66), whose generalization to the arbitrary π-exchange process has come to be called the Poor Man's Absorption Model (67). Arguments drawing on the statistical model for particle production support the ancient absorptive peripheral model lore (68) that small impact parameter collisions do not contribute appreciably to production of two-body final states. The original proposal (2) disagrees with experiment. However, the so-called weak SCAP hypothesis that complete central absorption occurs in and only in reactions described by planar duality diagrams (69, 70), e.g. $K^- p \rightarrow \bar{K}^0 n$, $\pi^+ p \rightarrow K^+ \Sigma^+$, does appear consistent [(71) and section 4F] but of curtailed predictive power. To sharpen the distinction between models, we remember that Harari (31) postulates peripheral behavior for all imaginary parts and all spin amplitudes, whereas Chiu (2, 71) postulates peripheral behavior for both real and imaginary parts in the nonflip amplitude of a subset of reactions.

3 DEFINITIONS AND CONVENTIONS

In this section we introduce the constructs needed to extract and present in organized fashion the experimental data on two-body scattering. First, we introduce the scattering amplitudes with which the processes are described theoretically. Then we express the experimental observables in terms of amplitudes for the simplest (realizable) case of $0^- \frac{1}{2}^+ \rightarrow 0^- \frac{1}{2}^+$ scattering. The two final subsections, which are devoted to definitions and observables for resonance production reactions, may be skipped by the casual reader. The observables are uncomplicated in principle but have been made to appear arcane by a host of different conventional axes and phases used in the literature. To the problem of extracting a resonance cross section from the observed bump on background in the mass spectrum of its decay products, however, there is no absolutely correct solution. The practical necessity of comparing different experiments in any case calls for a uniform procedure (72). To this end we describe in the final subsection some of the methods used previously

---

[7] J. Guillaud, Orsay, Private communication. D. Mayer, Ann Arbor, Private communication.

and support a reasonable standard technique, the universal adoption of which would bring about consistently defined and hence more useful data.

3A   Amplitudes for $0^- \frac{1}{2}^+ \to 0^- \frac{1}{2}^+$ Scattering

The scattering of spinless particles, to which we alluded in section 2A, though theoretically trouble free, is not of practical importance for the simplest laboratory reactions (e.g. $\pi N \to \pi N$) involve two spin-$\frac{1}{2}$ particles. To describe reactions of interest, we must add spin labels to our amplitudes; for every choice of a spin basis, we obtain a distinct set of amplitudes. We need only mention here s-channel and t-channel helicity amplitudes ($H_s^{\lambda_1 \lambda_2 \to \lambda_3 \lambda_4}$ and $H_t^{\lambda_1 \lambda_3 \to \lambda_2 \lambda_4}$, respectively), which are the usual Jacob-Wick (73) amplitudes for reactions 2.7 and 2.8. The spin axes appropriate to particle 3 are shown in Figure 9. These sets of amplitudes are related

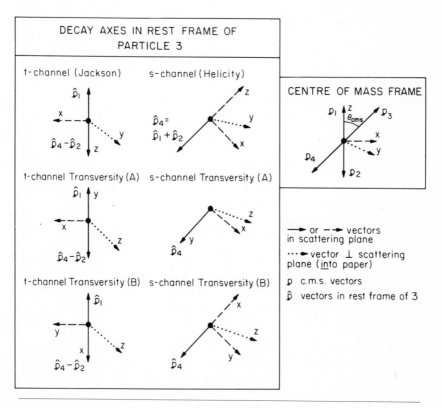

Figure 9 Some coordinate systems used to describe the spin-states and momenta for $1+2 \to 3+4$. Indicated are the overall center mass frame and six of the more common frames used to define spin-states and decay angles of particle 3. There are analogous sets of six axes defined in the rest frames for the other three particles (1, 2, and 4) in the reaction.

by a rotation (of the spin axes)—called the Trueman-Wick crossing relation (74)—whose exact form need not concern us here.

In the simple but important special case of meson-baryon elastic and (hyper)-charge-exchange scattering, particles 1 and 3, the mesons, have spin-parity 0^- and particles 2 and 4, the baryons, have spin-parity $\frac{1}{2}^+$. There are consequently two independent amplitudes which we choose as the s-channel helicity nonflip amplitude N and helicity flip amplitude F:[8]

$$
\begin{aligned}
N &= H_s^{+\to+} = H_s^{-\to-} \\
F &= H_s^{+\to-} = -H_s^{-\to+}
\end{aligned}
\qquad 3.1.
$$

The experimental observables are

$$d\sigma/dt = K\{|N|^2 + |F|^2\} \qquad 3.2.$$

$$\sigma_{\text{total}} = K'\,\text{Im}\,N(s, t = 0) \qquad 3.3.$$

$$P\,d\sigma/dt = 2K\,\text{Im}\,(NF^*) \qquad 3.4.$$

$$\hat{R}\,d\sigma/dt = 2K\,\text{Re}\,(NF^*) \qquad 3.5.$$

$$\hat{A}\,d\sigma/dt = K\{|N|^2 - |F|^2\} \qquad 3.6.$$

where we have defined the kinematic factors

$$K = 0.3893\,[\text{mb}/(\text{GeV})^2]/64\pi m_2^2 P_{\text{lab}}^2 \qquad 3.7.$$

and

$$K' = 0.3893\,[\text{mb}/(\text{GeV})^2]/2m_2 P_{\text{lab}} \qquad 3.8.$$

The differential cross section $d\sigma/dt$ and total cross section σ_{tot} off an unpolarized target require no discussion. The polarization P is related to the asymmetry in the scattering off a polarized target; $\hat{R}$ and $\hat{A}$ are in general linearly related to, and asymptotically become equal to, the Wolfenstein (75) parameters R and A which are the nonzero components of polarization of the final baryon scattered off a polarized target. The observables 3.2–3.4 and 3.5, 3.6 are then listed in ascending order of measurement difficulty. In particular, $\hat{R}$ and $\hat{A}$ have only been measured (85) in elastic scattering, for which the cross section is large. In the immediate future, we can anticipate $\hat{R}$ and $\hat{A}$ measurements in associated production reactions leading to self-analyzing (76) final-state hyperons (Λ^0, Σ^+).

3B   Density Matrix Elements and Decay Angular Distributions

If the produced particles are unstable, the spin structure of the scattering amplitudes may be deduced from the angular distributions of their decay products (77–82). For instance, for particle 3 a ρ^0 meson decaying into $\pi^+\pi^-$, the angular distribution of π^+ momenta in the ρ^0 rest frame, is given by

$$
\begin{aligned}
W(\theta, \phi) = (1/4\pi)\{ & 1 + \tfrac{1}{2}(1 - 3\rho^{00})(1 - 3\cos^2\theta) \\
& - 3[\rho^{1,-1}\sin^2\theta\cos 2\phi + \sqrt{2}\,\text{Re}\,\rho^{10}\sin 2\theta\cos\phi]\}
\end{aligned}
\qquad 3.9.
$$

---

[8] We designate $\pm\frac{1}{2}$ by $\pm$.

The spherical polar angles (θ, ϕ) specify the direction of the π^+ momentum $\vec{q}$ in the ρ^0 rest frame with respect to a specified coordinate system. The form 3.9 is valid for any orientation of the coordinate axes; each choice leads to a particular set of density matrix elements which are expressible in terms of corresponding scattering amplitudes. Referring $\vec{q}$ to the t-channel ("Jackson") frame, we obtain[9]

$$\rho_t^{mm'} \, d\sigma/dt = [(-1)^{m-m'} K/(2s_1+1)(2s_2+1)] \sum_{\lambda_1 \lambda_2 \lambda_4} H_t^{\lambda_1 m \to \lambda_2 \lambda_4} [H_t^{\lambda_1 m' \to \lambda_2 \lambda_4}]* \qquad 3.10.$$

in terms of t-channel helicity amplitudes. Alternatively, referring $\vec{q}$ to the s-channel ("helicity") frame, we have

$$\rho_s^{mm'} \, d\sigma/dt = [K/(2s_1+1)(2s_2+1)] \sum_{\lambda_1 \lambda_2 \lambda_4} H_s^{\lambda_1 \lambda_2 \to m \lambda_4} [H_s^{\lambda_1 \lambda_2 \to m' \lambda_4}]* \qquad 3.11.$$

In either case,

$$d\sigma/dt = [K/(2s_1+1)(2s_2+1)] \sum_{\lambda_1 \lambda_2 \lambda_3 \lambda_4} |H^{\{\lambda\}}|^2 \qquad 3.12.$$

The formulae pertaining to the decay of particle 4 are entirely analogous. Observe, however, that in equation 3.9 the element ρ^{10} changes sign if the conventional y-direction is reversed. Although there is universal agreement to choose the normal to the production plane ($\hat{y}$) along $\vec{p}_1 \times \vec{p}_3$ for analysis of particle 3, no such consensus exists in the case of particle 4. We endorse the same definition in both situations: $\hat{y} = \vec{p}_1 \times \vec{p}_3 = \vec{p}_2 \times \vec{p}_4$. Many experimental reports now leave the choice of the normal to the production plane as a conundrum for the reader, and occasionally incorrect expressions for the connection between density matrix elements and decay angular distributions in complicated cases occur in the literature. With the low-statistics data now extant such inconsistency is merely annoying, but in the coming era of multiparticle spectrometers continued casualness will lead to significant wasted effort. To ameliorate the situation, we have prepared a compendium of guaranteed formulae for relations analogous to equations 3.9–3.12 which we will supply on request.

Figure 9 also depicts the so-called transversity frame (83) for which the quantization axis is normal to the production plane. There are even two variants (marked A and B in Figure 9) of the transversity axes! Uniformity being desirable, we recommend the adoption of variant A which is related by a rotation through $\pi/2$ about the x-axis to the longitudinal frame pictured above it. Finally, for completeness, let us mention that the angular distributions are frequently described in terms of statistical tensors $T_{m_3}^{s_3}$, $T_{m_3 m_4}^{s_3 s_4}$, which are related to the density matrix elements in (80).

To illustrate the importance of density matrix element information, we note from equation 3.10 that $\rho^{00} \, d\sigma/dt$ only receives contributions from amplitudes with $\lambda_3 = 0$. As Gottfried & Jackson (78) proved, this fact allows the separation of $d\sigma/dt$ into the contributions from natural- and unnatural-parity exchange mechanisms. Taking for definiteness the reaction $\pi N \to \rho N$, we recall that $\rho^{00} \, d\sigma/dt$

---

[9] The phase $(-1)^{m-m'}$ is a consequence of the "particle 2" convention of (73).

is given in terms of unnatural parity (π, A_1) exchanges and that to leading order in s the combinations $(\rho^{11} \pm \rho^{1,-1}) d\sigma/dt$ correspond respectively to natural parity (A_2) and unnatural parity (π, A_1) exchanges.

3C   Extraction of Resonance Cross Sections

It is very well to discuss a resonance reaction such as $\pi^- p \to \rho^0 n$, but in fact, all we can measure is the three-body final state, $\pi^- p \to \pi^+ \pi^- n$ wherein, as shown in Figure 10, the presence of the ρ^0 resonance is evidenced by a bump in the $\pi^+ \pi^-$ mass spectrum. The observed spectrum consists of both resonance and background contributions which must be disentangled. Many ingenious techniques have been devised for making the separation. The common feature of the methods is the parameterization of the cross section as

$$\sigma = \alpha^2 |T_R(m_{\pi\pi})|^2 + \beta^2 |T_B(m_{\pi\pi})|^2 \qquad\qquad 3.13.$$

where the resonant amplitude T_R is a (modified) Breit-Wigner form (79, 86) and the incoherent background amplitude T_B is a smooth (usually polynomial) function of $m_{\pi\pi}$. The parameter α and hence the resonance production cross section is determined by fitting equation 3.13 to the data.

The basic method may be refined in many ways, for example, by the introduction of t-cuts or exponential t-dependences, by mass conjugation (87, 88), by the inclusion of many resonances in different subchannels of the final state (89), by the imposition

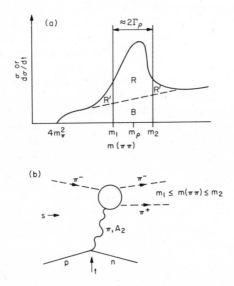

Figure 10   (a) Various ways (72) of defining a resonance cross section using a mass-cut $m_1 \leqq m(\pi\pi) \leqq m_2$ with possible background (B) and resonances tail correction (R'). (b) A Regge-pole exchange diagram for mass-cut data: m_1, m_2 are held fixed as $s \to \infty$.

of longitudinal phase space cuts, and ultimately by examination of prism plot displays (90) and of detailed partial-wave analyses (91, 92). Unfortunately, none of these techniques is free from ambiguity because the problem is inevitably "ill-posed." Therefore the fact that data from virtually every experiment are treated in a distinct way presents a decided obstacle to the comparison of resonance data from different sources. This is a more serious problem than the different choice of axes for density matrix elements; the latter may be corrected, a posteriori, by the user as the different axes are related by well-defined kinematic rotations.[10]

There are yet more ways (72) of defining cross sections as when an experiment contains only a few events, the fit methods are no longer applicable and it is usual to make a simple mass-cut, taking all data with $m_1 \lesssim m(\pi\pi) \leqq m_2$; there are no universally adopted values of m_1, m_2—commonly used are $m_1 \approx m_R - \Gamma, m_2 \approx m_R + \Gamma$ where the resonance has mass m_R and width Γ. This choice is the regions $R + B$ on Figure 10a; also used are "background subtraction" (R only) or "background subtraction + correction for resonance tails" ($R + R'$ only). The latter two resonance definitions are subject to the same arbitrariness and lack of uniformity as the fit methods. If universal values of m_1 and m_2 were agreed on, the simple mass-cut ($R + B$) would allow a *uniform* definition of a "resonance" cross section. It would not be a faithful representation of a theoretical single resonance cross section; however, this is of no great import because any theory of, say $\pi N \to \rho N$, should also be able to describe $\pi N \to (\pi\pi)N$ where the $\pi\pi$ system in a fixed mass-cut has both a $J^P = 1^-$ resonant ρ and a big $J^P = 0^+$ background amplitude. In particular, Regge pole exchanges with their characteristic energy dependence should describe such mass-cut data (Figure 10b). We therefore urge that in the future all resonance results be quoted for a simple mass-cut. The more complicated fit methods should *also* be used and are indeed vital for many applications [e.g. extraction of pure $I = 0$ exchange by forming $\sigma(\pi^+ p \to \rho^+ p) + \sigma(\pi^- p \to \rho^- p) - \sigma(\pi^- p \to \rho^0 n)$].

We should also warn the reader against the confusing practice of finding the overall production cross section by a fit method. Then, as each t bin contains but a few events, it is common to define $d\sigma/dt$ by a mass-cut but renormalize by a t-independent factor λ so that

$$\lambda \int_{-\infty}^{0} dt\, d\sigma/dt \bigg|_{\text{mass-cut}} = \sigma_{\text{prod}} \bigg|_{\text{fit}} \qquad 3.14.$$

Again this leads to lack of uniformity and it would be helpful if the unrenormalized $d\sigma/dt_{\text{mass-cut}}$ were always given as well.

There are also many techniques (87–92) developed for finding density matrix elements in the presence of background. We can mention the use of control regions on either side of the resonance, mass conjugation, or more generally maximum likelihood fits to the data with competing resonances removed. It appears that there is less difference in the various methods than for the overall cross section. However,

---

[10] All that is lost is an accurate evaluation of the errors as the full error matrix is rarely given for the density matrix elements—only the usual diagonal elements giving the separate error in each observable.

a different approach will be necessary with the high-statistics data now becoming available. For instance, the quark model predicts the vanishing of a number of statistical tensors. The predictions are roughly satisfied by existing data, within quite large errors. It is not known whether deviations indicate failures of the model for "pure" resonance production or confusion due to background. In future analyses the complete ($\geqq 3$-body) final state should be parameterized in a fashion consistent with all the common theoretical constraints upon resonance and background production, and the parameterization confronted with the original data sample of four-momenta. The popular procedure of "model-independent" extraction of the resonance signal carries an uncontrolled systematic error of about 10%, which cannot significantly be reduced.

4   PHENOMENOLOGICAL WEAPONS

In this section, we discuss methods of extracting from experimental data information on the s, t, spin, and partial-wave dependence of amplitudes underlying the observables. Such an approach has the considerable advantage that amplitude structure can be compared directly with theoretical expectations, and provides precise tests of (for instance) the distinct zero patterns for different amplitudes suggested by the absorption model (10, 11, 16, 93). In contrast, in making brute force fits of models to experimental observables (which are bilinear in the amplitudes), one may find comparable fits with many different sets of amplitudes, or may have a successful parameterization for one amplitude obscured by the poor description of another.

This powerful amplitude approach is unfortunately rarely applicable; it requires a complete set of observables and at present these are only available in πN elastic scattering. In other cases, we can try to use it but we must (partially) commit the sin of the model fit and introduce extra and probably wrong assumptions.

4A   Amplitude Analysis of πN Elastic Scattering

Pion nucleon elastic and charge exchange scattering can be described in terms of four complex amplitudes. These are the nonflip N and spinflip F amplitudes corresponding to two possible isospin states which we take to be the usual $I = 0$ and 1 t-channel exchanges. (These are normalized to be conventional $+$ and $-\pi N$ isospin states respectively.)

The amplitudes of the three measurable reactions are:

$$H(\pi^- p \to \pi^- p) = H(I = 0) + H(I = 1)$$
$$H(\pi^+ p \to \pi^+ p) = H(I = 0) - H(I = 1)$$
$$H(\pi^- p \to \pi^0 n) = -\sqrt{2} H(I = 1)$$

where the above holds separately for H as N or F—the two spin amplitudes.

At 6 GeV/c, a complete set of measurements exists for $-t \lesssim 0.5$ (GeV/c)^2 and, using the notation of section 2 for polarization observables, these are P, R, and $d\sigma/dt$ for $\pi^\pm p \to \pi^\pm p$, and P plus $d\sigma/dt$ for $\pi^- p \to \pi^0 n$. A is also known for

$\pi^- p \to \pi^- p$ but, in fact, this only gives sign information as its magnitude is known to greater accuracy from the relation $P^2 + R^2 + A^2 = 1$. The eight experimental observables are (one) more than enough to determine the four complex amplitudes $N(I = 0, 1)$, $F(I = 0, 1)$ within the usual unobservable overall phase.

Four amplitude analyses of this data have been performed recently[11] (94–97) and we summarize their results in Figures 11 and 12. We have recast all analyses in the same units [i.e. so that $d\sigma/dt \equiv \{|N|^2 + |F|^2\}$ mb/(GeV/c)^2] and have chosen the overall phase in the same way for each analysis. Thus $N(I = 0)$ is prescribed to have the same phase as given by the Barger-Phillips (98) Regge pole fit to all high energy πN data. This fit was constrained by finite energy (FESR) and continuous moment (CMSR) sum rules and so its amplitude structure ought to be reliable. In fact, its value for small $-t (\lesssim 0.2$ (GeV/c)^2) for the $I = 0$ phase is confirmed by dispersion theory calculations (99, 100). However, at large $-t$ there is no such supportive information and the basic FESR/CMSR input is less reliable; correspondingly the curves in Figures 11 and 12 may require rotation by the common (t-dependent) unobservable phase for $-t \gtrsim 0.2$ (GeV/c)^2.

The three early analyses [marked HM (94), Kelly (95), and Saclay (96) in the figures] were based on essentially the same data set. The ANL group (97) substituted their own new measurements (101) of $\pi^- p \to \pi^0 n$ polarization, which gave values considerably smaller than those of the CERN data (102) used by the other groups (see Figure 13). They also made use of the beautiful new data on $\pi^\pm p$ elastic scattering (103) that we have already displayed in Figure 7, which, because of the precise determination of relative normalization, specify rather well the small difference between the $\pi^\pm p$ cross sections near the forward direction.

HM and Saclay use similar methods: the data points are interpolated to the same t values and the amplitudes are determined separately for each t; Saclay uses many more t values than HM. ANL and Kelly both parameterize the amplitudes as a function of t and fit the data at the measured positions. Further, Kelly parameterizes all eight quantities [Re, Im N, $F(I = 0, 1)$] while, in recognition of the unknown overall phase, ANL fixes Re $N(I = 0) = 0$. The former seems best in principle as one should parameterize the true amplitudes and not those subject to an arbitrary rotation. However, the overall phase should still be poorly determined and the implementation in (95) must be wrong as this uncertainty was not reflected in the quoted error analysis. The difference between ANL and Kelly techniques is illustrated in Figure 11d for the small Re $F(I = 0)$ amplitude. The zero at $t \approx -0.15$ (GeV/c)^2 in the ANL solution was not present in their original "Re $F(I = 0)$" but is produced by the rotation to the Barger-Phillips overall phase. Clearly the smoothness assumption implicit in the ANL/Kelly method is not reliable for small amplitudes.

While we are discussing the general technique, it is worth noting that the qualitative similarity in the four analyses plotted in Figures 11 and 12 suggests that there is little point in repeating them without attempting to estimate the

[11] See also B. Wicklund, Private communication and Report at XVI International Conference on High-Energy Physics (Batavia, 1972).

possible overall phase rotation for $-t \gtrsim 0.2$ $(\text{GeV}/c)^2$. The latter can only be constrained through dispersion relations (or their relative FESR/CSMR), which can transmit to high energies the phase information contained in low energy unitarity. Some attempt in this direction has been made by Pietarinen (104, 105) but although his analysis technique may be impeccable, the results are not presented in a useful form. Similar sentiments have been expressed by Lovelace (106) who also has a useful review of parameterization techniques.

Concurrently with theoretical refinements, further experimental data are needed. A Monte Carlo simulation study to determine the most useful new measurements is described by Fox (107).

4B   Lessons from the πN Amplitude Analyses

First we define the partial wave amplitudes

$$f_{N,F} = 1/(32\pi\sqrt{s}k) \int_0^\infty N, F(s,t)J_n(b\sqrt{-t})d(-t) \qquad 4.1.$$

$$n = (0, 1) \text{ for (Nonflip, spinFlip) amplitude}$$

Here we have put k equal to the cms momentum and defined $b = l/k$ as the impact parameter discussed qualitatively at the beginning of section 2. N, F are normalized as in equation 3.2 and f is normalized so that it would be $e^{i\delta}\sin\delta$ for a

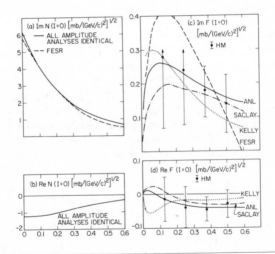

Figure 11   Results of πN amplitude analyses at 6 GeV/c for the $I = 0$ t-channel exchange. N and F refer to s-channel nonflip and spinflip amplitudes respectively (section 3A). The four analyses come from References 97 (ANL), 94 (HM), 95 (Kelly), and 96 (Saclay). The crude finite energy sum rule estimate of the imaginary parts (marked FESR) are described in section 4C. For clarity, we omit the error estimates on all except the HM analysis.

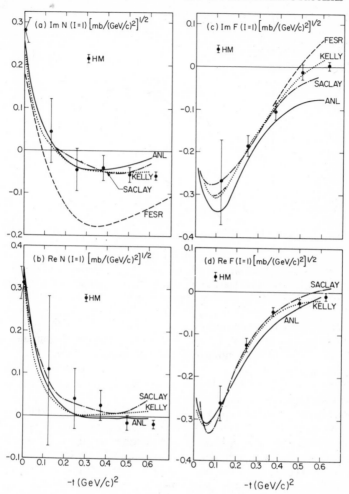

Figure 12 Results of πN amplitude analyses at 6 GeV/c for $I = 1$ t-channel exchange. Designations are the same as in Figure 11.

channel of definite s-channel isospin and spin. For instance, $f_N(b = 0) = i/2$ corresponds to complete s-wave absorption.

In Figure 14, we present the results of Kelly (95) who used equation 4.1 to partial wave decompose his amplitudes. This work has a nice treatment of errors but suffers from the disadvantage of an incorrect overall phase, which differs by $\sim 20°$ from the Barger-Phillips value shown in Figures 11 and 12; so the reader should make allowance for this extra error in interpreting Figure 14. Similar partial wave analyses are presented in (108—HM solution) and (109—Barger-Phillips fit).

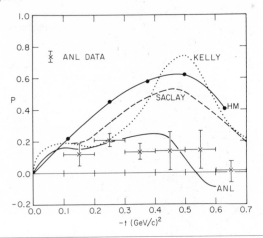

Figure 13 Comparison of the four 6-GeV/c πN amplitude analyses with the new 5-GeV/c $\pi^- p \to \pi^0 n$ polarization data from (101). The analyses are designated as in Figure 11. We do not show the CERN $\pi^- p \to \pi^0 n$ polarization data (102) which are similar to three analyses (HM, Kelly, Saclay) based upon them.

We can make the following (by now trite) observations from our figures:

(*a*) Im $N(I = 0)$ dominates—its approximate e^{2t} dependence translates into a Gaussian in Figure 14*a* concentrated at small impact parameter *b*.

(*b*) The reduced ratio $|F/N\sqrt{-t}|$ for $I = 0$ is 0.17 ± 0.02 at 6 GeV/c and 0.14 ± 0.03 at 16 GeV/c (96), both for the *t* range $0.2 \lesssim -t \lesssim 0.5$ (GeV/c)^2. The corresponding ratio for *t*-channel amplitudes is much larger at 0.8. Correspondingly *s*-channel helicity conservation ($F_s = 0$) is reasonably satisfied[12] while *t*-channel helicity conservation is ruled out (110, 111).

(*c*) The similar phase of N and $F(I = 0)$ indicates that Pomeron and P' have similar spin structure; in particular, the latter must also satisfy *s*-channel helicity conservation. The similarity between P and P' couplings was already indicated by the near-mirror symmetry of the polarization in $\pi^\pm p$ elastic scattering.

(*d*) Turning to $I = 1$, we find spinflip dominance. The current analyses do not extend far enough in *t* to confirm it, but the spinflip amplitudes are very near the Regge predictions: Re F proportional to $\alpha_\rho^2(t)$ with a double zero at $t \approx -0.6$ (GeV/c)^2 and Im F proportional to α_ρ (compare equation 2.14).

(*e*) On the other hand, Im N—quite contrary to the Regge zero at -0.6—has a zero at -0.15 (GeV/c)^2, to achieve the crossover (again, see Figure 7). Re N is also intriguing—if it vanishes at all for $-t \lesssim 0.5$ (GeV/c)^2, its zero lies at larger $-t$ than the zero of Im N. In any case, its very small size for $-t \gtrsim 0.2$ (GeV/c)^2 is suggestive as an artifact of the Regge double zero at $\alpha_\rho = 0$. In the *b*-plane, Im f_N and Im f_F both peak at $b \approx 1$ F while the real parts are more central. This observation was part of the phenomenological basis of the dual absorption model (section 2) and we will return to it in section 4F.

[12] However, this does contradict (119), which suggested that $F_s \to 0$ as $s \to \infty$.

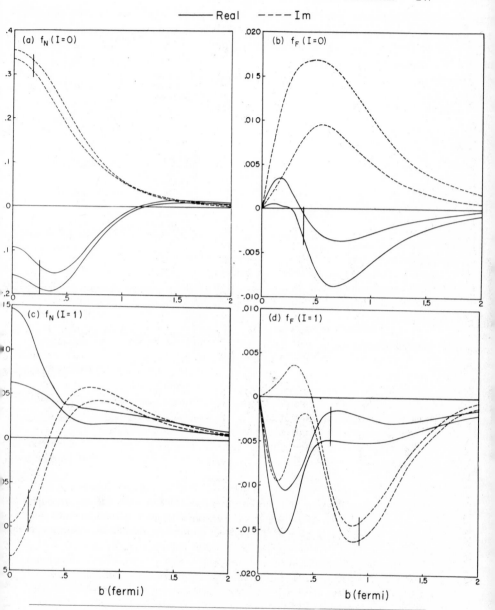

Figure 14   Partial wave (impact parameter) analysis of the πN amplitudes of Figures 11, 12 from the Kelly analysis (95). The curves represent statistical error bounds. However, for impact parameters smaller than the values indicated by vertical bars, systematic uncertainty from the unknown large $-t$ behavior of the amplitudes is larger than the indicated errors.

(*f*) Both Re and Im $N (I = 1)$ have a rapid variation away from $t = 0$ which is confirmed by dispersion relations (112, 113). In the *b*-plane, this corresponds (10, 11, 108) to a tail at large impact parameter which can be calculated quite rigorously from 2π exchange (114, 115).

4C   Finite Energy and Continuous Moment Sum Rules

If we assume Regge asymptotic behavior at fixed t

$$\text{Im } T \sim \sum_{i=1}^{m} r_i v^{\alpha_i(t)}$$ 4.2.

where $v = (s-u)/2$, then analyticity implies the finite energy sum rule (FESR) (43)

$$\int_0^{v_c} dv \, v^n \, \text{Im } T = \sum_{i=1}^{m} r_i(v_c^{\alpha_i+n+1}/\alpha_i+n+1)$$ 4.3.

In equation 4.3, n is any even (odd) integer for an invariant amplitude T that is odd (even) under crossing $v \to -v$. Choosing v_c so that the left-hand side of equation 4.3 can be evaluated from low energy phase shifts, the right-hand side gives useful constraints on the residues r_i of the Regge poles assumed to dominate the high energy amplitude. Typical of the qualitative utility of finite energy sum rules are the curves marked FESR in Figures 11 and 12. These come from evaluating equation 4.3 using the old CERN phase shift solution[13] (116) up to $s = 4.8$ (GeV/c)^2 and $\alpha(I = 0) = 0.8$, $\alpha(I = 1) = 0.58 + t$. This gives one an effective r_i for each amplitude which can be used in equation 4.2 to estimate Im T at 6 GeV/c. It is clear from the figures that the FESR curves, and hence low energy data plus analyticity, have predicted all the qualitative features of the 6-GeV/c amplitudes. FESRs have the great advantage that they directly predict individual amplitudes but the disadvantage that they can never be exact, as the Regge asymptotic expansion will always be an approximation at 2 GeV/c.

Similar constraints on the real part, and indeed on arbitrary mixtures of real and imaginary parts, may be found by replacing Im T by Im $\{(v^2 - v_0^2)^\varepsilon T\}$ in equations 4.2 and 4.3 (any real v_0, ε). The resultant relations are called continuous moment sum rules (CMSRs).

As we have already mentioned, Barger & Phillips (98) fitted all available πN high energy data in a Regge pole model constrained by FESRs and CMSRs. Although some of their poles (P'', ρ') are frankly phenomenological and surely no true asymptotic property of the πN amplitudes, the success of their fit is very striking. (It essentially *predicted* the amplitudes shown in Figures 11 and 12.) Their model has enjoyed much greater longevity and utility than many fits which, at the time, seemed to have greater theoretical motivation. Still the fact that their success was a triumph more for analyticity than for Regge poles does mean that it is only useful for interpolating and interpreting medium energy ($P_{lab} \lesssim 10$ GeV/c) data; deviations are inevitable and present at higher energies (120).

[13] The errors—especially in the small amplitude with many cancellations Im $N(I = 1)$—are quite large. FESR calculations using more recent analyses (117, 118) were not available.

Finite energy sum rules have also proved useful in photoproduction (121–123) and KN scattering (124), but the low energy phase shifts (and hence data) must be improved before the results are uniformly convincing.

4D   Duality

In deriving equation 4.3, one only assumes the Regge approximation 4.2 as an average from $v = v_c$ to ∞. There are some more speculative uses of low energy data to deduce results on the high energy amplitude, which essentially assume that equation 4.2 is valid at $v = v_c$, and so one is allowed to equate high and low energy amplitudes. As v_c is lowered, one must average the low energy data over a few hundred MeV (a typical resonance width) to avoid local resonance oscillations. This is usually justified by the totally abused word "duality" or the fine sounding "semi-local duality." It should be stressed that these deductions are even less quantitative than FESRs; and, in particular, they are often wrong (125) in the Veneziano model which is the basis of the current theoretical formulation of duality. Nevertheless, such duality analyses are useful because they allow much more explicit deductions than FESRs. To believe them, one must at least ask that a given deduction be both independent of v_c and insensitive to the trivial ambiguities of the extrapolation of high energy asymptotic expansions to low values of s.

Typical of a reasonable use of duality is the identification (126, 127) of the resonance contribution to low energy scattering with the Regge $(P', \omega, \rho, A_2, \ldots)$ "pole" contribution and the remaining background with the Pomeron term. This was, for instance, used by Harari & Zarmi (128) to discuss helicity conservation and deduce results in qualitative agreement with those found in section 4C from the high energy amplitude analyses. Fukugita & Inami (129) have tried to use this idea to discuss the partial wave structure of $\mathrm{Im}\,N(P')$. We know (sections 2 and 4B) that there is solid evidence that $\mathrm{Im}\,N(\omega, \rho)$ has a peak at impact parameter $b \simeq 1$ fm. EXD would predict a similar feature for $\mathrm{Im}\,N(P', A_2)$, but it turns out there is no direct evidence for this (see section 4F). However, plots of the partial wave amplitudes (Figure 15a) for $\bar{K}N$ and Figure 16 for πN) do show a peak at $b \approx 1$ fm for the direct channel resonance component: "duality" then implies such a peak for $\mathrm{Im}\,N(P')$. This is not conclusive, for at small impact parameter it is difficult to tell (broad) resonances from background and the Pomeron part (identified with the background) may be concealing a relatively large P' contribution at low b. We can test whether P' and ρ have the same impact parameter structure by asking if

$$\left[\mathrm{Im}\,f_N(P': b = 1\ \mathrm{fm})/\mathrm{Im}\,f_N(\rho: b = 1\ \mathrm{fm})\right]$$
$$= \left[\mathrm{Im}\,N(P': t = 0)/\mathrm{Im}\,N(\rho: t = 0)\right] \qquad 4.4.$$

The right-hand side of this is well determined by total cross-section data; the left-hand side is consistent with equation 4.4, but data fluctuations make it impossible to be quantitative.[14] Better is to use equation 4.4 with ρ replaced by ω so the right-hand side is 1. Then if $\mathrm{Im}\,f_N(P') < \mathrm{Im}\,f_N(\omega)$ at $b = 1$ fm (as we shall later

[14] M. L. Griss, Private communication (1973).

suggest), the decomposition $\bar{K}N(I_t = 0) = P + P' + \omega$, $KN(I_t = 0) = P + P' - \omega$ would suggest a depression of the KN partial wave amplitude at $b \approx 1$ fm. The relevant data are shown in Figures 15b–d. There is no real evidence for the suggested suppression and the small size of both the KN and $\bar{K}N$ background

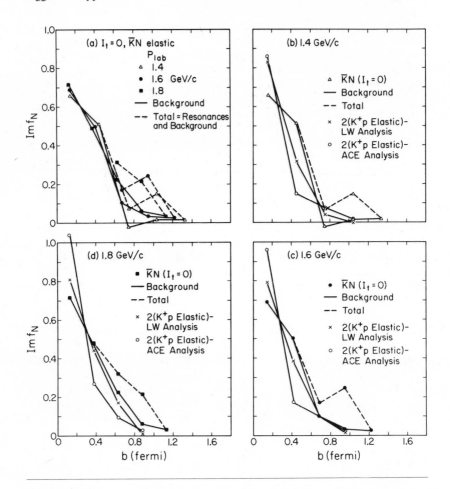

Figure 15 Partial wave structure of KN and $\bar{K}N$ scattering. (a) Background and resonance contributions to $\bar{K}N(I_t = 0)$ partial wave amplitude Im f_N—normalized as in equation 4.1—at various lab momenta. The figure is adapted from (129) and the data are from (301). The $I_t = 0$ isospin amplitude is normalized so that halving it gives its contribution to say $K^- p \to K^- p$. (b)–(d) Comparison of the same $\bar{K}N$ amplitude as in (a) with twice $K^+ p$ amplitude from (302) (marked LW, plotted is solution IA) and (303) (marked ACE). $K^+ n$ data are not available so we were forced to use $2(K^+ p)$ to estimate $I_t = 0$ for KN scattering.

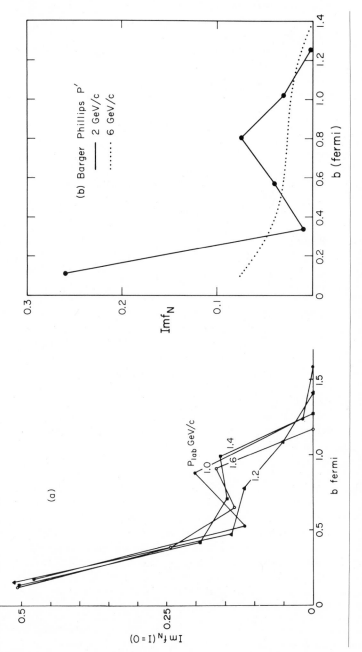

Figure 16 (*a*) Plot [adapted from (129)] of the partial wave decomposition—normalized as in equation 4.1—for Im $N(I = 0)$ in πN scattering at various lab momenta. The data come from (304). The curves—unlike those in Figure 15—are not separated into resonance and background terms. Had they been, the peak at $b \approx 0.8$ fm would be found to correspond to resonance contributions. (*b*) Partial wave analysis at 2 and 6 GeV/c (P. R. Stevens, Private communication) for the P' contribution to Barger and Phillips Regge pole fit to πN data (98).

amplitudes at $b \approx 1$ fm give equality there of $\operatorname{Im} f_N(P')$ and $\operatorname{Im} f_N(\omega)$ to within 20%.

One must use duality to make such arguments because the small size of the P' and ω at higher energies washes out the pronounced 1-fm peak. Furthermore, there is the difficulty that we are not seeing a true asymptotic effect at $1 \to 2$ GeV/c and the partial wave structure changes with energy. This is illustrated by the Barger-Phillips πN fit where the P' has a peak at $b \approx 1$ fm at 2 GeV/c but this disappears at 6 GeV/c (Figure 16b).

4E   Amplitude Analyses of Hypercharge Exchange Reactions

Consider the reactions $\pi N \to K(\Lambda, \Sigma)$, $\bar{K}N \to \pi(\Lambda, \Sigma)$ where for definiteness we shall usually choose the hyperon to be Λ.[15] $\pi^- p \to K^0 \Lambda$ and $K^- n \to \pi^- \Lambda$ can be described in terms of amplitudes of definite symmetry under crossing as

$$H(\pi^- p \to K^0 \Lambda) = H_T + H_V$$
$$H(K^- n \to \pi^- \Lambda) = H_T - H_V$$

4.5.

where in the simple Regge pole approximation H_V corresponds to K^* and H_T to K^{**} exchange. Invoking SU(3) and EXD, one can easily prove (130–135) that $\operatorname{Im} H_T = \operatorname{Im} H_V$ and so $H(K^- n \to \pi^- \Lambda)$ is purely real. In fact, $\pi^- p \to K^0 \Lambda$ and $K^- n \to \pi^- \Lambda$ should have (a) equal cross sections and (b) zero polarization if EXD were true and (c) equal but opposite values of $P \, d\sigma/dt$ if described by two Regge pole K^* and K^{**} exchange. Predictions (a), (b), and (c) are all false for $P_{\text{lab}} \lesssim 5$ GeV/c while at higher energies we can only say with conviction that (b) remains false. (Polarization in $\pi^+ p \to K^+ \Sigma^+$ is energy independent (136) and nonzero up to 14 GeV/c.) These deviations from simple Regge theory are not unexpected; it is a simple consequence of SU(3) which is confirmed by the data that these reactions have a large nonflip amplitude, and Figure 12 showed that Regge theory failed for such an amplitude.

Now our two reactions are determined by eight quantities Re, Im N, $F_{V,T}$ of which six are in principle determinable (each reaction has an unobservable overall phase). At the present time, only four observables have been measured, namely the polarization P and $d\sigma/dt$ for each reaction. Any amplitude analysis must therefore make additional assumptions. Now the results[16] of three analyses (132, 134, 135) are shown in Figure 17. There is not only little resemblance between the amplitudes from different analyses but even worse; each analysis predicts a large difference between $\operatorname{Im} N_V$ and $\operatorname{Im} N_T$ at $t = 0$. This is surprising because the flat total cross sections for $K^+ n$ and $K^+ p$ show that $\operatorname{Im} N_V = \operatorname{Im} N_T$ for both $V = \omega$, $T = P'$, and the pair $V = \rho$, $T = A_2$ exchange. The error in this assertion may be made quantitative by comparing at 5 GeV/c, the difference between $K^+ p$ and $K^- p$ total cross sections (measures $P' + \omega + A_2 + \rho$) to that between $K^+ p$ and its "flat" Pomeron value at 20 GeV/c (measures $P' - \omega + A_2 - \rho$). This procedure indicates at most a 10% breaking of exchange degeneracy.

---

[15] The Σ data are superior but FESR analysis has only been applied to the Λ reactions.

[16] We actually replotted the results of (132) to correspond to a conventional K^*, K^{**} Regge trajectory $\alpha(t) = 0.32 + 0.82t$.

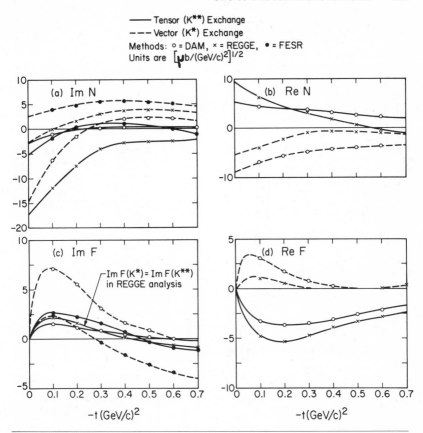

——— Tensor (K^{**}) Exchange
– – – Vector (K^*) Exchange
Methods: ○ = DAM, × = REGGE, ● = FESR
Units are $[\mu b/(GeV/c)^2]^{1/2}$

Figure 17   Results of amplitude analyses of $\pi^- p \to K^0 \Lambda$ and $K^- n \to \pi^- \Lambda$ at 4 GeV/c. The curves marked DAM, REGGE, and FESR come from (135, 134, 132, respectively). For the latter, we only take the imaginary part predictions. The amplitudes are split up into Vector (V) and Tensor (T) contributions in accordance with equation 4.5.

Further, using the quark model suggestion (34) that ϕ and f' exchange decouple (justified by the small cross sections for $\pi^- p \to \phi n, f' n$), we establish that V and T exchange have the same F/D value at the $N\bar{N}$ vertex. Then it is an immediate consequence of SU(3) that $\mathrm{Im}\, N_V = \mathrm{Im}\, N_T$ at $t = 0$ in our reactions and deviations should only be typical SU(3) breaking—say 20%—not the factor of five differences in Figure 17. This argument is strengthened by the fact that one gets a similar F/D ratio from the ratio of Λ to Σ reactions (131, 133–135) as one does from the above σ_{tot} argument. It is thus important to see whether one can find a set of amplitudes satisfying the constraint $\mathrm{Im}\, N_V \approx \mathrm{Im}\, N_T$ at $t = 0$. Here we merely detail the assumptions in the current work and indicate where they may be unreliable.

(*a*) The curves marked DAM (dual absorption model) in Figure 17 come from

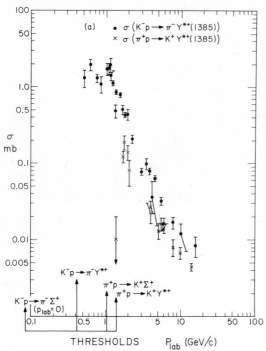

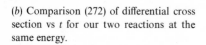

Figure 18 (*a*) Comparison of reaction cross sections for $K^-p \rightarrow \pi^- Y^{*+}(1385)$ and $\pi^+p \rightarrow K^+ Y^{*+}(1385)$. The data were taken from (272, 297, 305, 306).

(*b*) Comparison (272) of differential cross section vs *t* for our two reactions at the same energy.

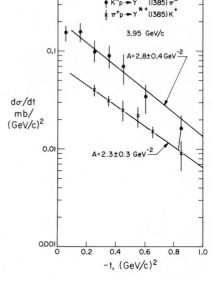

assuming that "imaginary parts are peripheral" while real parts are parameterized rather generally (135). In practice, this means $\text{Im}\,N_V$, $\text{Im}\,N_T$ and $\text{Im}\,F_V$, $\text{Im}\,F_T$ vanish together. Perhaps the bad results ($N_V \approx 6N_T$) could be overcome by just a small relaxation in this basic assumption. On the other hand, we will discuss in section 4F some evidence that the dual absorption model idea may itself be badly violated.

(b) The curves marked REGGE come from the assumption that the spinflip amplitudes are in perfect agreement with Regge theory (134). This is motivated by the nice agreement—illustrated in Figure 12c, d—of the ρ, A_2 spinflip amplitudes with Regge theory. However, it then predicts equal cross sections for spinflip K^*, K^{**} exchange processes—a suggestion violated by a factor of 2 by $\pi^+ p \to K^+ Y^{*+}_{1385}$ and $K^- p \to \pi^- Y^{*+}_{1385}$ at intermediate energies (137). This is demonstrated in Figure 18 which also marks the reaction thresholds. Naively one can interpret the suppression of $K^+ Y^{*+}$ compared with $\pi^- Y^{*+}$ as a consequence of the higher threshold for the former reaction. This may be true, but one would get exactly the same threshold suppression of $K^+ \Sigma^+$ vs $\pi^- \Sigma^+$ and can only deem 4 GeV/c too low an energy for asymptotic Regge assumptions.

Analyses using the above assumption have been presented in both (134) and (71).

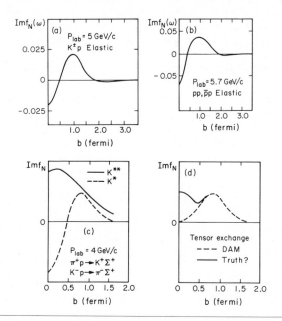

Figure 19 Some partial wave profiles for $\text{Im}\,N$ (imaginary part of nonflip amplitude). (a), (b) ω exchange from (139). (c) K^*, K^{**} exchange from (71). (d) Two possible tensor (P', A_2, K^{**}) profiles. That marked DAM is postulated by the dual absorptive model (31): that marked "truth" is suggested by the evidence discussed in section 4F. (a), (b) are normalized as in equation 4.1: (c), (d) are unnormalized.

For clarity, we only give curves from the former, although to play no favorites, in Figure 19 we take the partial wave analysis from (71).

(c) The curves marked FESR come from finite energy sum rule calculations of the type described in the last subsection (132). The good qualitative results in πN scattering suggest that the difficulties with this analysis are due to the poor quality of the low energy data on hypercharge exchange.

4F   Is Tensor Exchange Peripheral?

It is well established that the imaginary parts of vector exchange amplitudes are "peripheral"; in the b-plane this corresponds to a large partial wave amplitude near $b = 1$ fm, and in the t-plane to a zero in $\mathrm{Im}\,N$ near $t = -0.2$ $(\mathrm{GeV}/c)^2$. This was illustrated in Figures 12a and 14c for ρ exchange while for ω exchange it follows from the crossovers in $K^\pm p$ and $^{(\bar{p})}p$ scattering (see Figure 7). The latter allow estimation of $\mathrm{Im}\,N(\omega)$ by (138, 139)[17]

$$\mathrm{Im}\,N(\omega) = [\mathrm{d}\sigma/\mathrm{d}t(K^- p) - \mathrm{d}\sigma/\mathrm{d}t(K^+ p)]/4\sqrt{\mathrm{d}\sigma/\mathrm{d}t(K^+ p)}$$
$$\text{or} \ = [\mathrm{d}\sigma/\mathrm{d}t(\bar{p}p) - \mathrm{d}\sigma/\mathrm{d}t(pp)]/4\sqrt{\mathrm{d}\sigma/\mathrm{d}t(pp)} \qquad 4.6.$$

Expression 4.6 was partial wave analyzed in (139) and typical results[18] are shown in Figure 19a, b. In both KN and NN scattering ω exchange has the characteristic peak at 1 fm; note, however, that these figures and Figure 14c all show a significant negative peak at $b = 0$.

Now EXD would suggest that tensor exchange (P', A_2, K^{**}) has a similar partial wave profile, and this is indeed postulated in the dual absorption model. Nevertheless, there is no direct evidence for it and several papers have claimed that tensor exchange is nonperipheral (32, 33, 60). Let us discuss the pros and cons here.

(a) Typical of a "nonperipheral" tensor exchange is the solid line in Figure 19c. Note that the partial wave amplitude is still big at $b = 1$ fm; there is, however, a positive contribution at $b = 0$ and translated into the t-plane this moves the magic -0.2 zero out to ≈ -0.6 $(\mathrm{GeV}/c)^2$. On the other hand, the dotted line in Figure 19c, or in fact all our examples of vector exchange, has a negative contribution at small b which is destructive in the t-plane and moves the -0.2 zero in. The latter still looks "peripheral" in the t-plane even though its partial wave amplitude at $b = 0$ is no smaller in magnitude than the so-called "nonperipheral" tensor exchange. So the first point to emphasize is that by a uniform terminology, both vector and tensor exchanges appear nonperipheral in Figures 14c and 19. For vector exchange, the negative excursion at small b has been seen in essentially all analyses;[19] but one must admit it is sensitive to the large t behavior of the amplitude, and a common bias in the treatment of this could have produced a universally

---

[17] To be precise, this is really $\omega + \rho$, but the latter is only a 25% contaminant.

[18] Actually (139) uses $\frac{1}{2}[\mathrm{d}\sigma/\mathrm{d}t(K^+ p) + \mathrm{d}\sigma/\mathrm{d}t(K^- p)]$ rather than $\mathrm{d}\sigma/\mathrm{d}t(K^+ p)$ in denominator of equation 4.6. We will not debate the correct choice here.

[19] See (139) or Figure 9 of (16) for similar effects in $\bar{N}N$ scattering.

false conclusion. For tensor exchange, we will discuss whether the amplitude is positive at small b and the t-zero moved out; we follow convention and use the abused term "nonperipheral" for this situation.

(b) The study of P' exchange in πN, KN, and NN elastic scattering is complicated by the ubiquity of the Pomeron. As we showed in section 4D, low energy data suggest—unlike Figure 19c but consistent with either curve in 19d—a *peak* in the P' partial wave spectrum at $b \approx 1$ fm. However, these duality arguments can never be precise. For instance, the possible central component under discussion (i.e. the area between the dotted and solid curves in Figure 19d) will only contribute some 20% of the amplitude at $t = 0$. So even the best verified consequence of duality— the equality of the 1-fm peaks in P' and ω amplitude to within 20%— cannot settle the argument.

We should also note that the P' residue found by Harari & Zarmi (128) from resonance dominance of FESR was "nonperipheral."

(c) One can try to distinguish P' and Pomeron on the basis of their different energy dependence in $\pi^{\pm} p$ elastic scattering. If one assumes a flat Pomeron, the P' has no zero near $t = 0$; if one assumes a Pomeron with slope $\gtrsim 0.7$, the P' has the peripheral -0.2 $(\text{GeV}/c)^2$ zero. This was pointed out many years ago (125, 140), forgotten (141), and then rediscovered (32, 142). Anyhow, the first assumption is suggested by $\gamma p \to \phi p$, the second by *some* models[20] of pp and $K^+ p$ scattering. A choice is impossible at present.

(d) If one assumes a $t = -0.2$ $(\text{GeV}/c)^2$ zero in $\text{Im} N(A_2)$ and a central $\text{Re} N(A_2)$ [i.e. $\text{Re} N$ either has no zero or the zero occurs at $-t > 0.2$ $(\text{GeV}/c)^2$], then negative polarization is predicted (16) in $\pi^- p \to \eta^0 n$ at $t \simeq -0.2$ $(\text{GeV}/c)^2$. The data are meager (102, 144, 145) but they suggest positive polarization. Although there seems no compelling reason for the second assumption, better polarization data for $\pi^- p \to \eta^0 n$, $K^- p \to \bar{K}^0 n$ and $K^+ n \to K^0 p$ would be very helpful in pinning down the elusive tensor exchange.

(e) Modified absorption models (33, 60), using a large negative real to imaginary ratio in the absorbing amplitude, successfully fitted the ρ exchange amplitudes in Figure 12. They automatically predict less absorption for tensor exchange and produce partial wave profiles similar to the solid lines in Figure 19c, d.

(f) The amplitude analyses (71, 134) of hypercharge exchange reactions using a pure Regge spinflip amplitude produce a "nonperipheral" K^{**} exchange (see Figures 17a and 19c). Although we have expressed grave doubts about their validity, it appears that this particular conclusion is not sensitive to deficiencies in the analysis. The essential point is that the data show large (up to factors of 2) violations of line reversal symmetry with the "real" reaction ($K^- n \to \pi^- \Lambda$, $K^- p \to \pi^- \Sigma^+$) being larger than the "moving phase" reaction ($\pi^- p \to K^0 \Lambda$, $\pi^+ p \to K^+ \Sigma^+$) for $-t \lesssim 0.4$ $(\text{GeV}/c)^2$ (see Figure 20E1). In this region the spinflip amplitude is too small to explain the breaking. The flip amplitude can be estimated from isospin bounds (134), SU(3), and KN charge exchange (71) or the quark model

[20] The good fits of (143) found a Pomeron of low slope (0.3) could describe pp scattering up to ISR energies.

and $\pi N \to K Y^*$, $\bar{K} N \to \pi Y^*$. So, simply consider a nonflip amplitude and write its real and imaginary parts for vector and tensor as Re, Im V, T respectively. Then, taking Σ reactions, we can write

$$d\sigma/dt(K^- p \to \pi^- \Sigma^+) = \text{``Real''} = (\text{Re } T + \text{Re } V)^2 + (\text{Im } T - \text{Im } V)^2$$

$$d\sigma/dt(\pi^+ p \to K^+ \Sigma^+) = \text{``Moving''} = (\text{Re } T - \text{Re } V)^2 + (\text{Im } T + \text{Im } V)^2$$

4.7.

where we have chosen signs *un*conventionally so that all four amplitudes are positive at $t = 0$. Then it is easy to see that for *any* given ordering of zeros for the four amplitudes, we can predict the ratio of the two reactions at the zeros. Writing $x < y$ to denote that amplitude x vanishes before y, we force Im $V <$ Re V from πN amplitude analysis and Im $V <$ Im T from the sign of $K^- p \to \pi^- \Sigma^+$ polarization. (This is almost certainly implied by the negative excursion in vector amplitudes at $b = 0$.) Then there are eight possible zero orderings summarized in Table 3. Only two of these (Nos. 2, 8) predict $K^- p \to \pi^- \Sigma^+$ (real reaction) to be larger than $\pi^+ p \to K^+ \Sigma^+$ (moving) at all zeros and these are the "nonperipheral" Im T model of the effective absorption approach (33, 60). Options 1, 3, 5, and 6 have a possibly peripheral Im T but option 3 is ruled out at once by line reversal systematics; the remainder all predict the moving reaction to be come larger than the real at the third zero. For sensible zero positions, this is marginally consistent with the data while for 1, the line reversal data definitely imply a nonperipheral Im T: $\pi^+ p \to K^+ \Sigma^+$ favors 5 over 6 but, in fact, both of these predict the apparently wrong sign for the $\pi N \to \eta N$ polarization. Thus there is no satisfactory solution with a peripheral tensor exchange and current data clearly favor a nonperipheral Im $N(K^{**})$. Better data at higher energies are of obvious importance to test the hypothesis that the line reversal breaking is an asymptotic effect.

We have summarized a lot of rather questionable information. It appears to us that

Table 3   Possible zero orderings for real and imaginary parts of vector V and tensor T nonflip amplitudes

		Larger Reaction $R =$ Real, $M =$ Moving				Is Tensor Exchange Peripheral?
		Zero No.				
	Zero Ordering	1	2	3	4	
1	Im $V <$ Re $T <$ Im $T <$ Re V	R	R	M	M	Maybe
2	Im $V <$ Re $T <$ Re $V <$ Im T	R	R	R	R	No
3	Re $T <$ Im $V <$ Im $T <$ Re V	M	M	M	M	Maybe
4	Re $T <$ Im $V <$ Re $V <$ Im T	M	M	R	R	No
5	Im $V <$ Im $T <$ Re $T <$ Re V	R	R	M	M	Maybe
6	Im $V <$ Im $T <$ Re $V <$ Re T	R	R	M	M	Maybe
7	Im $V <$ Re $V <$ Im $T <$ Re T	R	R	M	M	No
8	Im $V <$ Re $V <$ Re $T <$ Im T	R	R	R	R	No

a nonperipheral $\text{Im}\,N$ is suggested for both vector and tensor exchange. These amplitudes appear to have opposite signs at small impact parameters so as to cancel in moving phase reactions—this is the content of the weak SCAP hypothesis of Chiu (71).

5   REACTIONLAND

Here we briefly document some of the important theoretical and phenomenological features that have been discovered in *meson* exchange processes. We stress particularly characteristics that appear to be common to several different reactions. We do not cover backward scattering because our current theoretical and experimental knowledge is too fragmentary to make such a compilation of ideas useful. The reactions discussed in sections 5A–5M are listed—with illustrative pictures— in Figures 20A–20M respectively. The latter are organized by exchanged quantum numbers, labeled A1 to M5, and list possible exchanges in ρ, A_2, π, and B nonets. For clarity, we omit possible exchanges of the A_1 nonet and its EXD partners; the evidence for the presence of these exchanges is detailed in section 5L(*iie*). Further we only record Pomeron exchange when it obeys vacuum selection rules— including the so-called Gribov-Morrison rule (146) which says the value of τP, signature × parity, is unchanged by diffraction. The experiment evidence against this is discussed in section 5I(*vi*). In section 6, we bring together some of the theoretical points raised in the following subsections.

Although all the reactions listed in Figure 20 have been measured somewhere, sometime, it is obviously impractical to give the original references here. The reader is referred to the many review articles and compilations—some of which we have cited.

5A   *Elastic Reactions*

(*i*) Total cross sections for reactions A1–5 measure $\text{Im}\,N(s,0)$: the imaginary part of the nonflip amplitude at $t = 0$ (cf. equation 3.3).

(*a*) All can be fitted from $p_{\text{lab}} = 5$ to 30 GeV/c as $\sigma_{\text{tot}} = a + bs^{-\frac{1}{2}}$ (see Figures 20A1, A2) where a is Pomeron and b is secondary (P', ω, ρ, A_2) Regge trajectory exchange (120). The value of a shows the Pomeron is an SU(3) singlet with around 20% octet admixture. The measurement of b for reactions A1, A2 allows extraction of the D/F ratio for $B\bar{B}$ coupling of V and T mesons. This turns out to be $D/F \approx -0.2$ (Figure 21*a*) in reasonable agreement with the quark model prediction $D/F = 0(147–149)$. In Figure 21*a*, we calculate D/F from comparison of ρ and ω exchanges by the formula

$$D/F = 1 - 2[\sigma_{\text{tot}}(K^-n) - \sigma_{\text{tot}}(K^+n)]/[\sigma_{\text{tot}}(K^-p) - \sigma_{\text{tot}}(K^+p)] \qquad 5.1.$$

One can also test the validity of SU(3) by comparing ρ exchange contributions to KN and πN; the ratio R should be 1 if SU(3) is exact where

$$R = \frac{[\sigma_{\text{tot}}(K^-p) - \sigma_{\text{tot}}(K^+p) - \sigma_{\text{tot}}(K^-n) + \sigma_{\text{tot}}(K^+n)]}{[\sigma_{\text{tot}}(\pi^-p) - \sigma_{\text{tot}}(\pi^+p)]} \qquad 5.2.$$

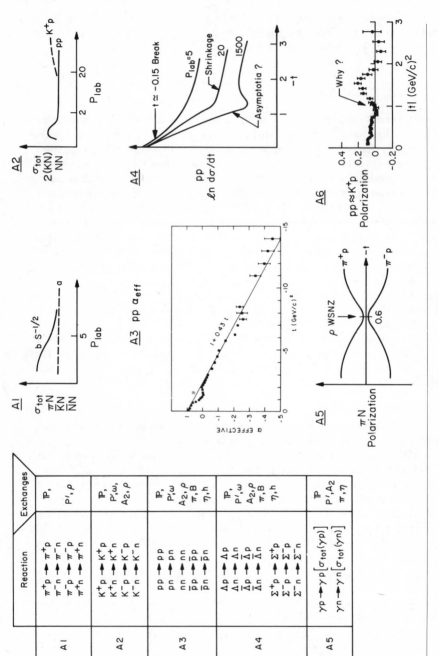

Figure 20A   Elastic reactions: (*A1, A2*) Sketch of total cross sections. (*A3*) $\alpha_{eff}(t)$ for *pp* elastic scattering for $8 < P_{lab} < 30$ GeV/c (162). (*A4*) Sketch of *pp* differential cross sections. (*A5*) Sketch of $\pi^{\pm}p$ elastic polarizations. (*A6*) *pp* polarization (176).

The data in Figure 21b are not very conclusive either way; it would be nice to improve them.

The experimentally observed value $b \approx 0$ for $K^+ p$ and pp implies EXD for P', ω and ρ, A_2. The error in this is at most 10% (section 4E) in KN but is as high as 25% in NN scattering (120).

(b) The comparison of reactions A5 with A1 and A2 successfully checks vector dominance plus quark model ideas of the similarity of $(\rho, \omega, \phi)N$ scattering to πN scattering (150).

(c) The comparison of reactions A3 with A1 and A2 (120, 148, 149, 151) checks to 20% the quark prediction $a(NN) = 3/2 \, a(\pi N)$ (Figure 21d); further the comparison checks ω universality $\omega \to N\bar{N} = 3 \, (\omega \to K\bar{K})$ amazingly well[21] (Figure 21c). ρ universality $(\rho \to NN = \rho \to K\bar{K})$ is not as easy to check due to poor neutron data. (Anyhow it is implied by ω universality plus $D/F = 0$.)

(d) Reactions A4 [latest data (152–154); theory (155)] allow a test of singlet Pomeron plus SU(3) and factorization for secondary trajectories. ($D/F \approx -0.2$ plus data on A3 reactions implies A4.)

(e) The rise in the $K^+ p$ total cross section (156) at Serpukhov energies (above 20 GeV/c) and the contrast with the otherwise similar pp total cross section, which does not rise (if at all) until ISR energies (156a), is not understood.

(f) Also the difference between the $\pi^- p$ and $\pi^+ p$ total cross sections (pure ρ exchange) falls surprisingly slowly with energy (151). It violates (157) unsubtracted dispersion relations when compared to the Serpukhov (158) $\pi^- p \to \pi^0 n$ data (reactions D1).

(ii) Coulomb interference measurements give Re $N(s, 0)$ and allow tests of dispersion relations and Regge phase. [See, for instance, (159) for πp, (160) for pp.] The former are particularly impressive in a recent Compton scattering $(\gamma p \to \gamma p)$ measurement (161). Here interference with the known Bethe-Heitler amplitude in principle allows measurement of the real part away from $t = 0$.

(iii) Striking features of the elastic differential cross sections are (12):

(a) pp, $K^+ p$ $d\sigma/dt$ shrink up to 30 GeV/c. As EXD says secondary trajectories should be real and not interfere with Pomeron, these should be our best glimpse of the pure Pomeron. The effective slope, found for the Pomeron this way (Figure 20A3) varies between 0.5 and 1 depending on reaction and energy range used (32, 162).

(b) pp elastic $d\sigma/dt$ develops a break at p_{lab} of 20 GeV/c for $t \approx -1.2$ (GeV/c)^2. This break becomes a dip at ISR energies.[22] The shrinkage, mentioned in (a), definitely slows down and may have essentially stopped for smaller [e.g. $-t \lesssim 2$ (GeV/c)^2] momentum transfer (Figure 20A4). This behavior implies that the *effective* Pomeron slope mentioned in (a) gets smaller, the higher the energy of the data; it follows that values calculated from $p_{lab} \lesssim 30$ GeV/c are of little fundamental significance.

---

[21] A check of universality at nonzero t-values is contained in (139).

[22] Rumor—which will be fact by the time this paper is published.

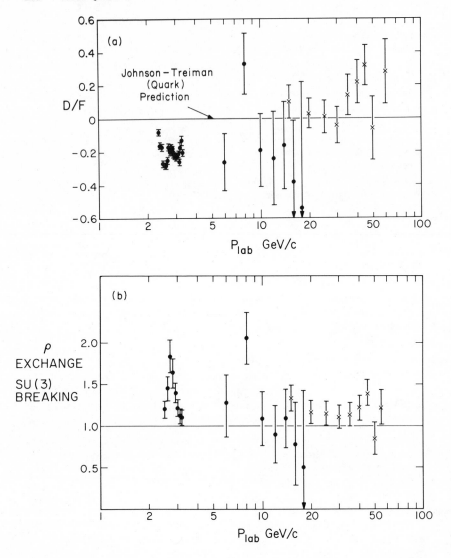

The dip and entire shape of $d\sigma/dt$ is nicely predicted by the Chou-Yang model (163–165).

(c) There is also a change in slope for pp $d\sigma/dt$ at ISR energies for $t \approx -0.15$ $(GeV/c)^2$ [i.e. in $d\sigma/dt \propto e^{At}$, A goes from 10.5 for $t < -0.15$ to 12.5 for $t > -0.15$ $(GeV/c)^2$]. This same effect is almost certainly present at lower energies (166). If there is a similar structure in πp scattering it occurs at smaller $-t$

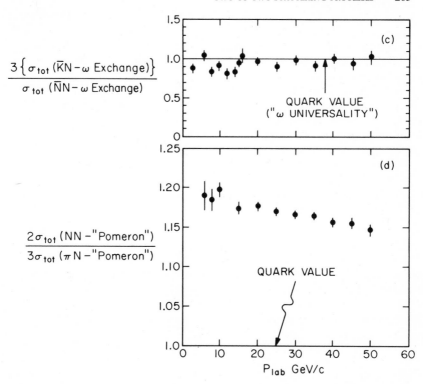

$$3\left\{\frac{\sigma_{tot}(\bar{K}N-\omega \text{ Exchange})}{\sigma_{tot}(\bar{N}N-\omega \text{ Exchange})}\right\}$$

$$\frac{2\sigma_{tot}(NN-\text{"Pomeron"})}{3\sigma_{tot}(\pi N-\text{"Pomeron"})}$$

Figure 21 (above and facing) (*a*) The D/F ratio for vector meson (ρ, ω) exchange in the nonflip amplitude at $t = 0$. It is calculated at each energy from total cross sections using equations 5.1 and as explained in section 5E(*vi*), a conventional interpretation of SU(3) suggests we should use the lower energy value $D/F \approx -0.2$ as our best estimate. The data are from (308, 309) and (151); the figure is adapted from the latter reference where the low energy points are incorrectly plotted. (*b*) The ratio R defined in equation 5.2 which should be 1 if SU(3) is exact. The data are from (308–311) and (151) and the figure is adapted from the latter reference. (*c*) 3 × (ratio of ω contributions to KN and NN scattering) which should be 1 if ω universality holds. The data are from (308, 312, 313, and 151), while the graph is taken from latter reference. (*d*) 2/3 of (ratio of Pomeron and P' contributions to NN and πN scattering) which should be 1 in quark model. The data are from (309, 311, 313, 314, and 151) while the graph is taken from latter reference.

(167, 168); an exact analysis is confused by uncertainties in experimental normalization (169).

Theoretically this slope change or curvature is given automatically by the Chou-Yang model. Alternately it may be due to the singularity at $t = 4m_\pi^2$ which must be present in the pp amplitude (170, 171).

(*d*) There is the notorious crossover between particle and antiparticle $d\sigma/dt$

(i.e. $\pi^- p$ vs $\pi^+ p$, $K^- p$ vs $K^+ p$, $\bar{p}p$ vs pp elastic) for $-t = 0.1$ to 0.2 $(\text{GeV}/c)^2$. This implies a zero in this t range for Im $N(s,t)$ of ρ and ω exchange, which disagrees with the Regge prediction of a zero at α_ρ or $\alpha_\omega = 0$. [$t \approx -0.5$ $(\text{GeV}/c)^2$.] These data on the other hand are splendid successes for absorption models (sections 2B, 4F, and Figure 7).

(e) $\pi^\pm p$ $d\sigma/dt$ do not shrink for $p_{\text{lab}} \lesssim 40$ GeV/c and $-t \lesssim 1$ $(\text{GeV}/c)^2$ (168); this can be understood in two ways. First (32, 125, 141, 142), interference of an "effective" Pomeron with slope 0.5 to 1 [as needed in pp or $K^+ p$—see (a)] with a P' exchange whose imaginary part changes sign at $t \approx -0.2$ $(\text{GeV}/c)^2$ [i.e. P' has peripheral zero required by EXD/duality and seen in (d) for ω/ρ exchange]. The second explanation is more natural but perhaps less consistent with other data; in this, the "effective" Pomeron has little slope and the P' has no structure in its t-dependence (cf. "peripheral" P' discussion in section 4F).

(iv) The larger $-t$ $d\sigma/dt$ data exhibit:

(a) A dip at $t \approx -3$ $(\text{GeV}/c)^2$ in $\pi^\pm p$; is this related to $t \approx -1.4$ dip in pp ISR data?

(b) All differential cross sections shrink very fast for $p_{\text{lab}} \lesssim 30$ GeV/c and $-t > 1$ $(\text{GeV}/c)^2$. The shrinkage disappears for pp at ISR energies and a similar fate no doubt awaits the other reactions (207) [cf. (iiib)].

(c) At $\cos\theta_{\text{cms}} = 0$, the shrinkage is translated into a fast fall of $d\sigma/dt$ with s. In agreement with theoretical prejudice, the fall can be fitted (162, 172–174) to s^{-n} with n lying between 7 and 12 for $\gamma p \to \pi p$, $\pi p \to \pi p$, and $pp \to pp$.

The origin of the shrinkage plus the transition from models for 90° scattering to small t(exchange) models is not understood. In particular, for $p_{\text{lab}} \lesssim 5$ GeV/c and, say, $1 < -t < 3$ $(\text{GeV}/c)^2$, the elastic shrinkage is similar to that in a host of quantum number exchange $(\pi^- p \to \pi^0 n, \; \pi^+ p \to K^+ \Sigma^+, \; \pi^- p \to p\pi^- \ldots)$ reactions (9, 16) and fittable in a Regge model.

(d) The $K^+ p$ $d\sigma/dt$ can be fitted to large $-t$ with a simple exponential form, e.g. at 5 GeV/c, this form fits up to $t = -2$ $(\text{GeV}/c)^2$. In particular, the break occurs at larger $-t$ than in otherwise similar pp elastic scattering; this can be interpreted in an absorption model as smaller multiple scattering in $K^+ p$ than in pp. However, it is then difficult to understand the earlier rise in σ_{tot} for $K^+ p$ than for pp.

(v) The polarization (Figures 20A5, A6) is sensitive to the spinflip amplitude of the Regge V, T exchange. The data for $-t \lesssim 1$ $(\text{GeV}/c)^2$ are in beautiful agreement with simple pole theory (15, 16, 175) in $\pi^\pm p$, $K^\pm p$, pp, and $\bar{p}p$ scattering. The pp polarization exhibits some rather surprising energy dependence (it falls a little too fast) and has ill-understood (141) structure at $t \approx -1$ $(\text{GeV}/c)^2$ (176). There are also difficulties in explaining the energy dependence of the (small) $I = 0$ spinflip exchange amplitude demanded by the $\pi^\pm p$ polarization data (177). Somebody should measure the pn polarization and check the theoretical prediction that it is equal and opposite to pp.

(vi) Comparison of $\pi^- n \to \pi^- n$ with $\pi^+ p \to \pi^+ p$, or $\Lambda n \to \Lambda n$ with $\Lambda p \to \Lambda p$ allows an unambiguous check of the Glauber technique for extracting neutron cross sections from deuterium data.

(*vii*) Reaction A5 with real photons demonstrated the similarity of real photon and hadron data (150). The inelastic electron data from SLAC measure off shell photon total cross sections, i.e. $\sigma_{tot}[\gamma(q^2)p,n]$, for (large) spacelike q^2 and the similarity disappears. (Scaling and point-like behavior sets in.) However, a Regge analysis is still possible for large "ν/q^2" and successful (178). In particular, roughly the same D/F ratio is applicable.[23]

It would be nice to measure $d\sigma/dt$ for $\gamma p \to \gamma p$ and $\gamma n \to \gamma n$ (real photons). A crossover at $t \simeq -0.2$ (GeV/c)^2 would directly indicate (179) a peripheral zero in Im $N(A_2)$ [cf. (*iiie*) and section 4F].

5B   Diffraction Dissociation

General reviews may be found in (13, 180, 181).

(*i*) Reaction B contains nonelastic processes whose quantum numbers allow vacuum (Pomeron) exchange (Figure 20B1). Their cross sections are roughly energy independent (Figure 20B2); although inelastic processes fall a little faster than elastic, the former still fall much slower than the typical $p_{lab}^{-1.5 \to 2.5}$ of a quantum number exchange reactions (e.g. $\pi^+ p \to K^+ \Sigma^+$, $\pi^- p \to \pi^0 n$).

(*ii*) It is not clear if bumps seen in reactions B1–B5 are true resonances. A_1, A_3, Q, L, and N^* (1400) are almost certainly mainly some sort of kinematic enhancement typified by Deck model (182, 183—Figure 20B3). In each case, one expects a real resonance roughly at the mass of the bump; however, the parameters of the bump are wrong and in the case of A_1, A_3, no resonant phase variation is found (92, Figure 20B4). N^*(1400) appears at different masses in πN and $\pi\pi N$ decays (expected in Deck model), but this is a general feature of multi-channel resonance decays (86) and need not be evidence against a resonance interpretation. It is not known if "diffractively" produced $N^*(\approx 1700)$ is $5/2^+ N^*(1688)$, $5/2^- N^*(1680)$, or just a kinematic bump! The former seems most reasonable (184, 185), but the A_2 data [see (*iv*)] might suggest the $5/2^-$ possibility.

(*iii*) There is theoretical (duality plus zero triple Pomeron coupling) and experimental evidence that background under "resonances" in $\pi N \to \pi N^*$ decreases with incident energy (186). This does not seem consistent with $\pi N \to (A_1, A_2, A_3)N$ data (92) while Einhorn et al (187) point out complications in the theory.

(*iv*) The A_2 is seen in the same $(\pi^- \pi^- \pi^+ p)$ final state as A_1 and A_3; the A_2 has the right (mass, width, phase) properties to be identified with a resonance but energy dependence of data (92) implies $\pi N \to A_2 N$ (reaction I1) is Pomeron exchange (Figure 20B2) contradicting the Gribov-Morrison rule (146).

(*v*) The cross sections for $\pi N \to \pi N^*$, $NN \to NN^*$, and $KN \to KN^*$ allow tests of Pomeron factorization (48). As reviewed in (13), the current data are consistent with this idea but the energy dependence difficulty [cf. (*i*)] and the uncertain identification

---

[23] E. D. Bloom, Private communication. D/F can be calculated from the proton/neutron ratio.

	Reaction	Exchanges
BI	$\pi^+ p \rightarrow A_1^+ p$ $\pi^- p \rightarrow A_1^- p$ $\pi^+ p \rightarrow A_3^+ p$ $\pi^- p \rightarrow A_3^- p$	$\mathbb{P}$ P', ρ η, B
B2	$\pi^+ p \rightarrow \pi^+ N^{*+}$ (many) $\pi^- p \rightarrow \pi^- N^{*+}$ (many) $\pi^- n \rightarrow \pi^- N^{*0}$ (many)	$\mathbb{P}$ P', ρ
B3	$K^+ p \rightarrow Q^+ p$ $K^0 p \rightarrow Q^0 p$ $K^- p \rightarrow Q^- p$ $\bar{K}^0 p \rightarrow \bar{Q}^0 p$ $K^+ p \rightarrow L^+ p$ $K^- p \rightarrow L^- p$	$\mathbb{P}$ P', ω A_2, ρ π, B η, h
B4	$K^+ p \rightarrow K^+ N^{*+}$ (many) $K^+ n \rightarrow K^+ N^{*0}$ (many) $K^- p \rightarrow K^- N^{*+}$ (many)	$\mathbb{P}$ P', ω A_2, ρ
B5	$pp \rightarrow p N^{*+}$ (many) $pn \rightarrow p N^{*0}$ (many) $\bar{p} p \rightarrow \bar{p} N^{*+}$ (many) $\bar{p} n \rightarrow N^{*+}$ (many) n $\bar{p} n \rightarrow \bar{p} N^{*0}$ (many)	$\mathbb{P}$ P', ω A_2, ρ π, B η, h
B6	$\gamma p \rightarrow \rho p$ $\gamma p \rightarrow \rho' p$ $\gamma p \rightarrow \omega p$ $\gamma p \rightarrow \phi p$	$\mathbb{P}$ P', A_2 π, η $\mathbb{P}$

Figure 20B (above and facing page) Diffraction dissociation: (*BI*) Typical Pomeron exchange reactions. Elastic, single, and double dissociation. (*B2*) Energy dependence (in roughly $5 \leq P_{lab} \leq 20$ GeV/c) of elastic and diffraction dissociation reactions [adapted from (13)]. (*B3*) Deck model. (*B4*) Mass dependence of cross section and relative $1^+(A_1)$ and 0^- (background) *s*-wave phases for $\pi^- p \rightarrow A_1^- p$ at 40 GeV/c (92). (*B5*) Ratio of $yp \rightarrow yN^*(1688)$ to $yp \rightarrow yp$ for $y = p$ (symbolized ●), $\pi(x)$ and electron (Δ) (189). (*B6*) *t*-dependence of $\pi^- p \rightarrow A_{1,2}^- p$ at 40 GeV/c (92).

of bumps with resonances confuses the interpretation. [Note that a Deck bump would factorize and this model would explain results in Tables II and III of (13)]. The reported failure for double diffraction processes (188) merely indicates large non-Pomeron components in high multiplicity states around 20 GeV/c.

Figure 20B5 shows that factorization works as a function of t for $N^*(1688)$ production (189).

(*vi*) More than one theoretical model predicts a zero at $t = 0$ for $d\sigma/dt$ of N^*

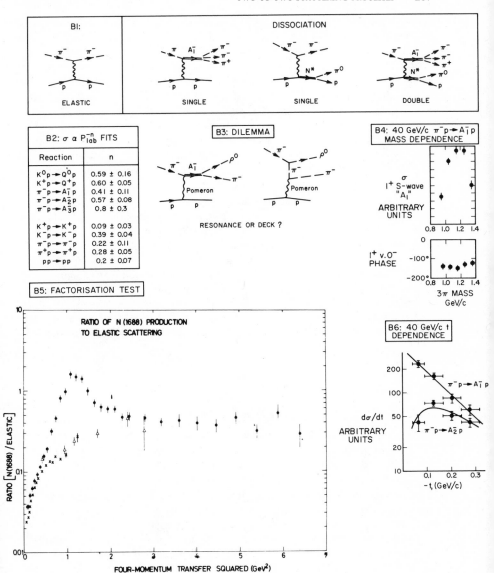

production. This is consistent with but not proved by the current data. Thus $d\sigma/dt$ is very sharp for $KN \to QN$, $\pi N \to A_1 N$, and $NN \to NN^*(1400)$ (Figure 20B6); however perhaps the Q, A_1 and $N^*(1400)$ are not resonances. $NN \to NN^*(1688)$ shows a flat $(e^{3t})\,d\sigma/dt$; taking a simple mass-cut to define the $N^*(1688)$ shows no $t = 0$ zero. However, a Breit-Wigner fit to the mass spectrum shows a decrease in signal/background as $t \to 0$ (189) and suggests a zero in the true resonance component.

The similarity in $eN \to eN^*$ and $NN \to NN^*$ after scaling by the elastic cross section, in Figure 20B5, supports models that make Pomeron couplings proportional to those of the photon (190).

(vii) $d\sigma/dt$ for $K^0 p \to Q^0 p$ vs $\bar{K}^0 p \to \bar{Q}^0 p$ show (191) a similar crossover at $t \approx -0.2$ (GeV/c)^2 to that in $K^+ p \to K^+ p$ vs $K^- p \to K^- p$. This suggests the universality and hence importance of such particle antiparticle crossovers. Secondly, the fact that $\bar{Q}^0 p$ is bigger at $t = 0$ than $Q^0 p$ contradicts the Deck model (showing it is a typical and not the only background amplitude).

A similar crossover is seen in $\pi^- p \to A_1^- p$ vs $\pi^+ p \to A_1^+ p$ but now the values at $t = 0$ are consistent with the Deck model (13).

(viii) The reactions in B6 are less confusing.

(a) The decay of the ρ shows s-channel helicity conservation (110) in agreement with πN elastic and Compton scattering. This helicity conservation was *not* seen in B1–B5 where t-channel helicity conservation is preferred (13).

(b) $\gamma p \to \phi p$ is useful as one expects zero coupling for secondary Regge (P') trajectories (as ϕ is purely strange quarks). This reaction is perfect for examining the Pomeron amplitude. The current data (13, 192) support a flat Pomeron (i.e. no shrinkage). Attempts to correlate this with the flat Pomeron observed for pp scattering at ISR are dubious because all $\gamma N \to VN$ reactions have t-dependences similar to elastic scattering in the $p_{\text{lab}} \lesssim 30$ GeV/c regime where the Pomeron is observed to shrink.

(c) $\gamma p \to \omega p$ also has sizable π exchange (see reactions G12).

(d) It is not known whether the ρ' observed (193) in $\gamma p \to \rho' p$ is a true resonance state or a kinematic enhancement of the type that plagued reactions (B1–B5).

5C   Elastic Deuteron and Nuclear Reactions

(i) We do not list the analogous reactions with nuclear targets. Neither the deuteron (23) (see also reactions K) or nuclear (194–197) processes has added much to our understanding of elementary particle theory. Their isolation of $I = 0$ exchange is, in practice, achieved also in the reactions of groups A and B.

(ii) Current data confirm (23) Glauber theory with interesting additional contributions (198, Figure 20C1) from inelastic states at high energy ($\gtrsim 10$ GeV/c). The latter affects extraction of neutron cross sections from data on deuteron targets.

There is also evidence for D-wave in the deuteron which obscures a multiple scattering dip (23, 199, Figure 20C2). The deuteron is strongly polarized at the t-value of the dip.

(iii) There is the d^* [$N\Delta(1234)$ state analogous to d as np state] which is an amusing example of quantum mechanics. We do not list associated reactions (e.g. $\pi d \to \rho d^*$) as their properties follow (in principle) straightforwardly from subreactions obtained by writing d and d^* in terms of their constituents. Typical is (200) which finds the curious sharing of decay π^+ between decays of L and d^* in $K^+ d \to K^+ \pi^- \pi^+ d$; a similar effect is seen in the "constituent" reactions $K^+ p \to K^+ \pi^- \pi^+ p$.

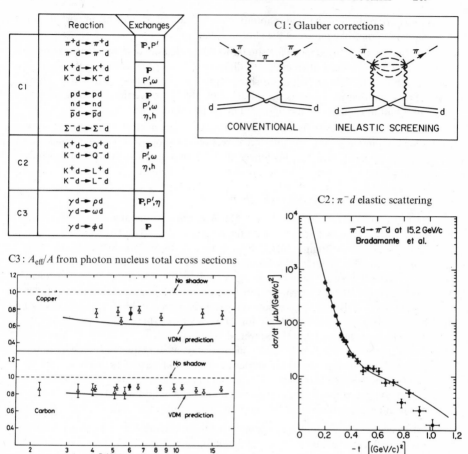

Figure 20C Elastic and diffractive deuteron and nuclear reactions: (*C1*) Diagrams calculated in Glauber theory—including Gribov's addition of inelastic contributions (198). (*C2*) Differential cross sections for $\pi^- d$ elastic scattering at 15.2 GeV/c, measured by the CERN-Trieste High Energy Group (307). The prominent features are the sharp forward peak, due mostly to single scattering, and a shallower slope at larger values of $|t|$, due to double scattering. The curve is a Glauber theory calculation by Sidhu & Quigg (199). (*C3*) Comparison of photon nucleus total cross section data with vector dominance (VDM) and point photon ("No Shadow") predictions (202).

(*iv*) The A (atomic number) dependence of γ Nucleus → (ρ, ω, ϕ) Nucleus lies between A^1(Volume) and $A^{2/3}$(Surface). This indicates that γ has both point-like and hadronic (vector-dominance) features (201). Typical is Figure 20C3, taken from (202), which analyzes data in terms of $A_{\rm eff}$ (201).

Electroproduction (virtual photon) data off nuclei would be important for clarifying point-like nature of the photon at large q^2.

(*v*) The reactions listed in C2 are similar to B3 [indicating that, say, structure on Q region (200) is indeed attributable to diffractive $I = 0$ production]. Data (203) on n Carbon $\rightarrow (N^{*0} \rightarrow p\pi^-)$ Carbon shows a $p\pi^-$ decay angular distribution which differs from that in reactions B2, B4, B5 (184, 185). It is presumably some nuclear effect but no explanation has appeared in the literature.

(*vi*) If you believe the relevant theory (204), coherent nuclear reactions $xA \rightarrow yA$ allow extraction of the yN total cross section. Sensible values are obtained for σ_{tot} (ϕp, ηp, etc.), but it is no longer thought that the $\pi A \rightarrow (3\pi)A$ data, which find $\sigma_{tot}[(3\pi)p] \approx \sigma_{tot}(\pi N)$, imply a resonant ($A_1$) 3π system.

5D $I = 0, 1$ Natural Parity Exchange

(*i*) The simplest of these reactions are D3. According to the quark model they are pure spinflip [this is tested by beautiful $K^+ p \rightarrow K^0 \Delta^{++}$ data (205) at 2.25 GeV/c which show $d\sigma/dt$ (Figure 20D2) dipping to 1/10 its maximum value at $t = 0$]. This model also predicts the Stodolsky-Sakurai decay distributions $\rho_{33} = 3/8$, $\rho_{3-1} = \sqrt{3}/8$, and $\rho_{31} = 0$, which agree well with data even at low incident energy (Figure 20D3). Current folklore (section 4) believes ρ, A_2 spinflip amplitudes should agree well with simple pole theory. This indeed holds for shape (WSNZ) and energy dependence of $d\sigma/dt$ for D3 reactions. There is, however, definite evidence for failure of EXD [line-reversal—see (*iv*)].

The picture is slightly muddied theoretically as the quark model naturally predicts a pure t-channel spinflip amplitude while folklore prefers s-channel spinflip as the perfect Regge amplitude. Rotation of the quark amplitude to the s-channel would give some non-Regge nonflip and double flip amplitudes. The effect is quite small and existing data are much too poor to test for its presence.

(*ii*) Reactions D1 are dominantly spinflip but $d\sigma/dt$ (Figure 20D1) shows a flattening near $t = 0$ corresponding to a small nonflip amplitude. The latter is isolated relatively exactly by the amplitude analyses and is definitely non-Regge (section 4). $d\sigma/dt$ is dominated by the spinflip amplitude and shows Regge energy dependence up to $-t \simeq 3$ (GeV/c)^2 (206) [cf. (207) which finds similar energy dependence for πp elastic]. The ρ, A_2 WSNZs at $-t = 0.6$ and 1.5 are seen but not the second ρ WSNZ at $-t \approx 2.5$ (GeV/c)^2 (206).

The quark model relates the spinflip amplitudes in reactions D1 and D3. This relation fails by factor of 1.5 to 2 in cross section—the Δ^{++} cross section is too big (208).

The polarization in $\pi^- p \rightarrow \pi^0 n$ first indicated the failure of pure Regge theory. This polarization was, however, correctly predicted at small $-t$ by absorption models (which correctly moved the zero of Im N from $\alpha_\rho = 0$ to smaller $-t$). The lack of a negative spike at $t \simeq -0.6$ (GeV/c)^2 then destroyed the old absorption model (which *incorrectly* moved the zero of Re N to smaller $-t$ than Im N). There is some controversy at present about the exact experimental values (101, 102, Figure 13).

The sign of the $\pi^- p \to \eta^0 n$ polarization at $-t \approx 0.2$ (GeV/c)^2 is an important test of models (section 4F); current data are poor (102, 144, 145).

(*iii*) This same marvelous Regge natural parity spinflip amplitude seen in reactions D1 and D3 can also be isolated in $K^\pm p \to K^{\pm *} p$ (reactions G5), $\pi^\pm p \to \rho^\pm p$ (G1), and $\pi^\pm p \to A_2^\pm p$ (I1). For these last reactions, we extract our amplitude using $(\rho_{11} + \rho_{1-1}) d\sigma/dt$ while polarized photon data isolate it in $\gamma p \to \pi^0 p$ and $\gamma n \to \pi^0 n$ (reactions I3). The amplitude always has the same phenomenological properties; universal shape, WSNZ where expected, shrinkage (not tested always) and line reversal breaking systematics [see (*iv*)]. The same amplitude also contributes to innumerable other processes but its exact isolation is not possible; for instance, it gives observed polarization in $\gamma p \to \pi^+ n$ (reaction H3) and has the usual line reversal breaking in $\pi^- p \to K^*(890)\Lambda$ vs $K^- p \to \omega \Lambda$ [section 5L(*v*)].

(*iv*) In spite of its many Regge beauties, this wondrous spinflip amplitude exhibits strong violation (209) of the expected line reversal symmetry [e.g. $d\sigma/dt(K^+ n \to K^0 p)$ $= d\sigma/dt(K^- p \to \bar{K}^0 n)$]. Always the "real" ($K^+ n \to K^0 p$, $K^+ p \to K^{+ *} p$) is larger than the "moving phase" ($K^- p \to \bar{K}^0 n$, $K^- p \to K^{- *} p$) reaction. For $K^+ n \to K^0 p$ the deviation is only present below 5 GeV/c; for $K^+ p \to K^0 \Delta^{++}$ vs $K^- n \to \bar{K}^0 \Delta^-$ or $K^- p \to \pi^- Y^{*+}(1385)$ vs $\pi^+ p \to K^+ Y^{*+}(1385)$ (reactions E4) current data suggest the effect persists to higher energy (Figures 20D4 and 18).

Examination of $d\sigma/dt$ reveals that $K^- p \to \bar{K}^0 n$ and $K^- n \to \bar{K}^0 \Delta^-$ have a break at $t \approx -0.6$ (GeV/c) (reminiscent of multiple scattering), but $K^+ n \to K^0 p$ and $K^+ p \to K^0 \Delta^{++}$ are quite smooth here.

We might as well note here that $K^+ n \to K^{+ *} n$ and $K^- n \to K^{- *} n$ do *not* show any line reversal breaking (210); this is believed to be because the spinflip amplitude is reduced in size as ρ and ω contribute with opposite signs: the observed $(\rho_{11} + \rho_{1-1}) d\sigma/dt$ is contaminated with double flip and nonflip amplitudes. For similar reasons, the difference (211) between $K^+ p \to K^{+ *} p$ and $K^- p \to K^{- *} p$ becomes more pronounced if one extracts the pure natural parity exchange contribution.[24]

(*v*) The reaction D2, $K_L^0 p \to K_S^0 p$, is important because it is dominantly nonflip— which amplitude disagrees with Regge theory. Current data (212) show a break at $t \approx -0.2$ (GeV/c)^2 [corresponding to zero in Im $N(\omega)$] and rapid shrinkage of $d\sigma/dt$ at larger $-t$ (even though Regge has wrong zero structure!). Detailed polarization (it should be negative at small $-t$) and $d\sigma/dt$ studies of this reaction will be rewarding.

(*vi*) The small $I = 0$ exchange polarization in $\pi^\pm p \to \pi^\pm p$ indicates that the f^0 (and hence by EXD ω) s-channel spinflip coupling is small. This implies an s-channel spinflip D/F value of 3 for V, T exchange. The quark model predicts $D/F = 3/2$ in the t-channel (16). Rotation of $\rho \to N\bar{N}$ s-channel spinflip coupling plus nonflip couplings from σ_{tot} shows this is consistent with current data.

(*vii*) It will be important to see whether there is a ρ exchange dip in $\pi^+ p \to \pi^0 \Delta(1940)$ or more generally in $\pi^+ p \to \pi^0 (\text{fast})(\pi^+ p)$ (reactions D4). The former and, for instance,

[24] B. Musgrave, Private communication.

$K^+ p \rightarrow K^0 \Delta(1940)$ are good places to look for possible decay symmetries analogous to Stodolsky-Sakurai for the $\Delta(1234)$. The ρ photon analogy (213) predicts a decay similar to that in photoproduction $\gamma N \rightarrow \Delta(1940) \rightarrow \pi N$.

(viii) Comparison of $\pi^+ p \rightarrow \eta \Delta^{++}$ and $\pi^+ p \rightarrow \eta' \Delta^{++}$ provides a direct measure of $\eta - \eta'$ mixing if one ignores the "disconnected quark graph" coupling. This gives (214, 215) a mixing angle of $-30°$ to $-40°$ in disagreement with $-10°$ expected for the favored quadratic mass formula. However, this large value should be disregarded as the analysis reported in section 5E(iii) renders the quark assumption invalid (216).

	Reaction	Exchanges
D1	$\pi^- p \rightarrow \pi^0 n$ $\pi^+ n \rightarrow \pi^0 p$	ρ
	$\pi^- p \rightarrow \eta n$ $\pi^+ n \rightarrow \eta p$ $\pi^- p \rightarrow \eta' n$ $\pi^+ n \rightarrow \eta' p$	A_2
	$K^+ n \rightarrow K^0 p$ $K^0 p \rightarrow K^+ n$ $K^- p \rightarrow \bar{K}^0 n$	A_2, ρ
D2	$K_L^0 p \rightarrow K_S^0 p$	ρ, ω
D3	$\pi^+ p \rightarrow \pi^0 \Delta^{++}$ $\pi^+ n \rightarrow \pi^+ \Delta^0$ $\pi^+ p \rightarrow \pi^+ \Delta^+$ $\pi^- p \rightarrow \pi^- \Delta^+$ $\pi^- p \rightarrow \pi^0 \Delta^0$	ρ
	$\pi^+ p \rightarrow \eta \Delta^{++}$ $\pi^+ p \rightarrow \eta' \Delta^{++}$	A_2
	$K^- n \rightarrow \bar{K}^0 \Delta^-$ $\bar{K}^0 p \rightarrow K^- \Delta^{++}$ $K^- n \rightarrow K^- \Delta^0$ $K^- p \rightarrow K^- \Delta^+$ $K^- p \rightarrow \bar{K}^0 \Delta^0$ $K^+ p \rightarrow K^0 \Delta^{++}$ $K^+ p \rightarrow K^+ \Delta^+$ $K^+ n \rightarrow K^0 \Delta^+$ $K^+ n \rightarrow K^+ \Delta^0$	ρ, A_2
D4	$\pi^- p \rightarrow \pi^0 N^{*0}$ (many) $\pi^+ p \rightarrow \pi^0 N^{*++}(I=3/2)$	ρ
	$K^+ p \rightarrow K^0 N^{*++}(I=3/2)$ $K^+ n \rightarrow K^0 N^{*+}$ (many) $K^- p \rightarrow \bar{K}^0 N^{*0}$ (many)	ρ, A_2

Figure 20D $I = 0, 1$ nondiffractive natural parity exchange: (D1) Typical differential cross sections at 5 GeV/c. (D2) $d\sigma/dt$; and (D3) Δ^{++} density matrix elements for $K^+ p \rightarrow K^0 \Delta^{++}$ at 2.5 GeV/c (205). (D4) Comparison of reaction cross sections for various line-reversed reactions. (Data is unpublished compilation of G. C. Fox.)

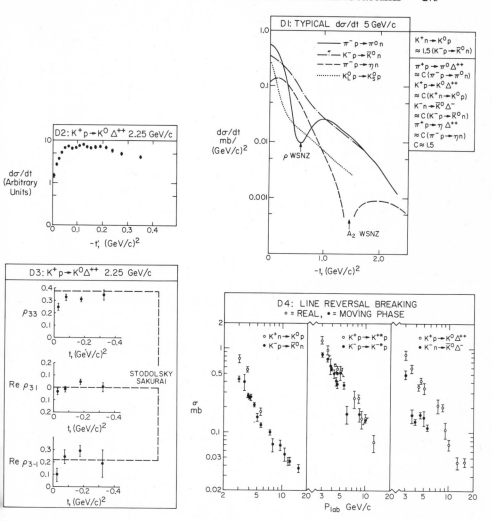

5E   Hypercharge and Natural Parity Exchange

(i) We have discussed reactions E1, E2 under amplitude analyses (sections 4E and 4F). The latter are still uncertain and require both R and A measurements (section 3) and clarification of line reversal systematics [see (v)]—e.g. careful measurement of $K^- p \rightarrow \Sigma^+$ and $\pi^+ p \rightarrow K^+ \Sigma^+$ with no relative normalization error. The most reasonable amplitude analyses suggest a surprising nonperipheral tensor (K^{**}) exchange (71, 134—Figure 19c).

(ii) As expected theoretically [SU(3)], reactions E1–E3 are mainly nonflip (cf.

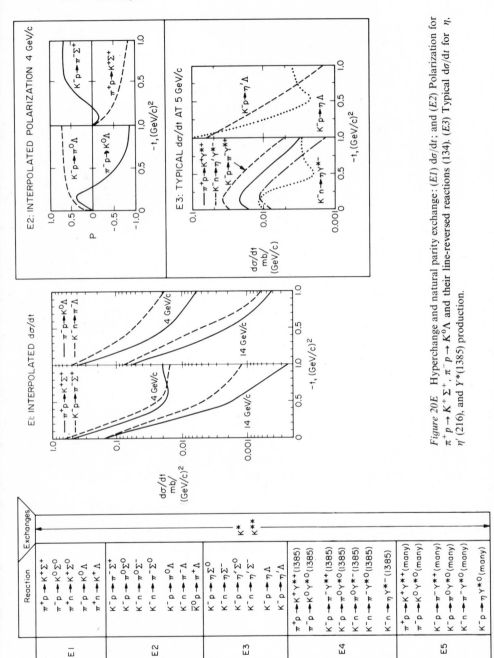

Figure 20E Hyperchange and natural parity exchange: (*E1*) $d\sigma/dt$; and (*E2*) Polarization for $\pi^+ p \to K^+ \Sigma^+$, $\pi^- p \to K^0 \Lambda$ and their line-reversed reactions (134). (*E3*) Typical $d\sigma/dt$ for η, η' (216), and Y^*(1385) production.

Figures 20 E1, E2) and so, like $K_L^0 p \to K_S^0 p$, should exhibit more deviations from Regge theory than reactions D1, D3. Comparison of Λ and Σ $d\sigma/dt$ at $t = 0$ (e.g. $\pi^- p \to K^0 \Lambda$ vs $\pi^+ p \to K^+ \Sigma^+$) gives directly a D/F value for the relevant K^* and K^{**} exchanges in the nonflip amplitude (10, 131, 133, 134). Although this qualitatively agrees with that obtained from σ_{tot}, there is a tendency for the Σ data to be higher than expected. The spinflip amplitude can only be bounded by polarization data (134) and extracted explicitly in models. This amplitude is also consistent with SU(3) and the D/F values discussed in section 5D. The polarization observed in "real" reactions $K^- n \to \pi^- \Lambda$, etc., is sizable (Figure 20E2). This suggests that K^* and K^{**} do not have the equal imaginary parts in nonflip amplitude N demanded by EXD. In the amplitude analyses of section 4E, these amplitudes indeed differed but by huge factors (Figure 17)—as we discussed there, this appears to violate SU(3) at $t = 0$.

There is an interesting dip at $t \approx -0.4$ $(\text{GeV/c})^2$ in $K^- p \to \eta \Lambda$ (Figure 20E3)— it is probably related to K^* WSNZ (in the spinflip amplitude?) for SU(3) predicts a suppression of K^{**} in this reaction (216).

The Y^* channel (reactions E4) is, like its SU(3) friends in reaction group D3, mainly spinflip (Figure 20E3). The expected dip in $K^- n \to \eta Y^{*-}$ has not yet been seen experimentally.

(iii) One can extract the $\eta - \eta'$ mixing angle by combining reactions in E3 with $\pi N \to (\eta, \eta') N$ (reactions D1). The analysis is more complicated than in $\pi N \to \eta \Delta$, $\eta' \Delta$ because of the different nonflip and spinflip D/F values at the $N \bar{N}$ vertex. However, the data are better and one can avoid the assumption of no "disconnected quark graphs" [this would also be possible for the Δ data if you also used $\pi N \to \pi \Delta$ and $KN \to K \Delta$ and did a full SU(3) analysis]. The analysis of (216) gives a mixing angle of $-11°$ in agreement with a quadratic mass formula.

(iv) The energy dependence of the data is odd. The best measured channel, $\pi^+ p \to K^+ \Sigma^+$, falls with an $\alpha_{\text{eff}}(0) \approx 0.7$ (0.35 expected) although it does exhibit canonical shrinkage. In particular, the break at $t \approx -0.4$ (GeV/c) disappears rapidly with energy (cf. Figure 20E1).

(v) The experimental information on line reversal breaking is inconsistent. It appears that the "real" reaction $(K^- p \to \pi^- \Sigma^+$, $K^- p \to \pi^0 \Lambda$, $K^- p \to \pi^- Y^{*+})$ is generally larger than its "moving phase" counterpart $(\pi^+ p \to K^+ \Sigma^+$, $\pi^- p \to K^0 \Lambda$, $\pi^+ p \to K^+ Y^{*+})$ but there are differences between Λ and Σ data (Figure 20E1), and it is not clear if the breaking vanishes at high energy (as it did for $K^- p \to \bar{K}^0 n$ vs $K^+ n \to K^0 p$—Figure 20D4).

Lai & Louie (209) first systematized these differences but since then the situation has been confused by fresh and contradictory data. The current Y^* data are summarized in Figure 18 while Figure 20E1 reproduces the interpolation (134) of Λ and Σ processes. New data of reliable normalization are urgently needed. We emphasize that current data does *not* support the idea (71, 134) that line reversal breaking is smaller in spinflip than in nonflip reactions (compare Figures 18 and 20E1).

(vi) The reactions E4 are related by SU(3) to D3 $(\pi^+ p \to K^+ Y^{*+}$ to $K^- n \to \bar{K}^0 \Delta^-$,

FULLY FLEDGED π $(1/(t-m_\pi^2)^2)$ GENERAL π EXCHANGE FIGURES

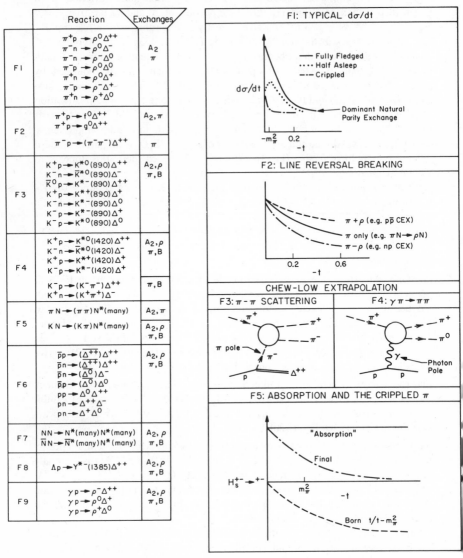

	Reaction	Exchanges
F1	$\pi^+p \rightarrow \rho^0\Delta^{++}$ $\pi^-n \rightarrow \rho^0\Delta^-$ $\pi^-n \rightarrow \rho^-\Delta^0$ $\pi^-p \rightarrow \rho^0\Delta^0$ $\pi^+n \rightarrow \rho^0\Delta^+$ $\pi^-p \rightarrow \rho^-\Delta^+$ $\pi^+n \rightarrow \rho^+\Delta^0$	A_2 π
F2	$\pi^+p \rightarrow f^0\Delta^{++}$ $\pi^+p \rightarrow g^0\Delta^{++}$	A_2,π
	$\pi^-p \rightarrow (\pi^-\pi^-)\Delta^{++}$	π
F3	$K^+p \rightarrow K^{*0}(890)\Delta^{++}$ $K^-n \rightarrow \overline{K}^{*0}(890)\Delta^-$ $\overline{K}^0p \rightarrow K^{*-}(890)\Delta^{++}$ $K^+p \rightarrow K^{*+}(890)\Delta^+$ $K^-n \rightarrow K^{*-}(890)\Delta^0$ $K^-p \rightarrow K^{*-}(890)\Delta^+$ $K^-p \rightarrow K^{*0}(890)\Delta^0$	A_2,ρ π,B
F4	$K^+p \rightarrow K^{*0}(1420)\Delta^{++}$ $K^-n \rightarrow \overline{K}^{*0}(1420)\Delta^-$ $K^+p \rightarrow K^{*+}(1420)\Delta^+$ $K^-p \rightarrow K^{*-}(1420)\Delta^+$	A_2,ρ π,B
	$K^-p \rightarrow (K^-\pi^-)\Delta^{++}$ $K^+n \rightarrow (K^+\pi^+)\Delta^-$	π,B
F5	$\pi N \rightarrow (\pi\pi)N^*(\text{many})$	A_2,π
	$K N \rightarrow (K\pi)N^*(\text{many})$	A_2,ρ π,B
F6	$\overline{p}p \rightarrow (\overline{\Delta^{++}})\Delta^{++}$ $\overline{p}n \rightarrow (\overline{\Delta^{++}})\Delta^+$ $\overline{p}n \rightarrow (\Delta^0)\Delta^-$ $\overline{p}p \rightarrow (\Delta^0)\Delta^0$ $pp \rightarrow \Delta^0\Delta^{++}$ $pn \rightarrow \Delta^{++}\Delta^-$ $pn \rightarrow \Delta^+\Delta^0$	A_2,ρ π,B
F7	$NN \rightarrow N^*(\text{many})N^*(\text{many})$ $\overline{N}N \rightarrow \overline{N}^*(\text{many})N^*(\text{many})$	A_2,ρ π,B
F8	$\Lambda p \rightarrow Y^{*-}(1385)\Delta^{++}$	A_2,ρ π,B
F9	$\gamma p \rightarrow \rho^-\Delta^{++}$ $\gamma p \rightarrow \rho^0\Delta^+$ $\gamma p \rightarrow \rho^+\Delta^0$	A_2,ρ π,B

Figure 20F Fully fledged π exchange reactions: Figures *F1–F5* are relevant for all π exchange reactions. (*F1*) Typical $d\sigma/dt$ from π exchange [section 5F(*i*)]. (*F2*) Typical line reversal breaking [section 5F(*iii*)]. (*F3*) Chew-Low extrapolation for π exchange [section 5F(*iv*)]. (*F4*) Chew-Low extrapolation for photon exchange [section 5F(*v*)]. (*F5*) Role of absorption for crippled π exchange amplitudes [section 5H(*xi*)].

G: HALF ASLEEP π $\left(-t/(t-m_\pi^2)^2\right)$

	Reaction	Exchanges
G1	$\pi^+p \rightarrow \rho^+p$ $\pi^-p \rightarrow \rho^-p$ $\pi^-p \rightarrow \rho^0n$ $\pi^+n \rightarrow \rho^0p$	ω, A_2 π, h π, A_2
G2	$\pi^-p \rightarrow f^0n$ $\pi^+n \rightarrow f^0p$	π, A_2
G3	$\pi^+p \rightarrow g^+p$ $\pi^-p \rightarrow g^-p$ $\pi^-p \rightarrow g^0n$ $\pi^+n \rightarrow g^0p$	ω, A_2 π, h π, A_2
G4	$\pi^+n \rightarrow (\pi^0\pi^0)p$ $\pi^-p \rightarrow (\pi^0\pi^0)n$ $\pi^+p \rightarrow (\pi^+\pi^+)n$ $\pi^-n \rightarrow (\pi^-\pi^-)p$	π
G5	$K^+p \rightarrow K^{*+}(890)p$ $K^+n \rightarrow K^{*+}(890)n$ $K^-n \rightarrow K^{*-}(890)n$ $K^-p \rightarrow K^{*-}(890)p$ $K^+n \rightarrow \underline{K}^{*0}(890)p$ $K^-p \rightarrow \bar{K}^{*0}(890)n$	P', ω A_2, ρ π, B η, h A_2, ρ π, B
G6	$K^+p \rightarrow K^{*+}(1420)p$ $K^+n \rightarrow K^{*+}(1420)n$ $K^-n \rightarrow K^{*-}(1420)n$ $K^-p \rightarrow K^{*-}(1420)p$ $K^+n \rightarrow \underline{K}^{*0}(1420)p$ $K^-p \rightarrow \bar{K}^{*0}(1420)n$	P', ω A_2, ρ π, B η, h A_2, ρ π, B
G7	$K^+p \rightarrow (K^+\pi^+)n$ $K^-n \rightarrow (K^-\pi^-)p$	π, B
G8	$\pi N \rightarrow (\pi\pi)N$ $KN \rightarrow (K\pi)N$	ω, A_2 π, h P', ω A_2, ρ π, B η, h
G9	$pp \rightarrow n \Delta^{++}$ $\bar{p}p \rightarrow (\overline{\Delta^{++}})n$ $\bar{p}p \rightarrow (\overline{\Delta^0})n$ $pp \rightarrow p \Delta^+$ $pn \rightarrow p \Delta^0$ $\bar{p}p \rightarrow (\overline{\Delta^+})p$	A_2, ρ π, B P', ω A_2, ρ π, B η, h

	Reaction	Exchanges
G10	$NN \rightarrow NN^*$ (many) $\bar{N}N \rightarrow N^*$ (many)N $\bar{N}N \rightarrow \bar{N}N^*$ (many)	P', ω A_2, ρ π, B η, h
G11	$\Lambda p \rightarrow \Sigma^- \Delta^{++}$	A_2, ρ π, B
G12	$\gamma p \rightarrow \rho^+n$ $\gamma n \rightarrow \rho^-p$ $\gamma p \rightarrow \omega p$	A_2, ρ π, B P P', A_2 π, η
G13	$\gamma p \rightarrow \pi^- \Delta^{++}$ $\gamma p \rightarrow \pi^+ \Delta^0$ $\gamma n \rightarrow \pi^+ \Delta^-$ $\gamma n \rightarrow \pi^- \Delta^+$	A_2, ρ π, B
G14	$\gamma p \rightarrow A_1^+ n$ $\gamma p \rightarrow A_2^+ n$ $\gamma p \rightarrow B^0 p$	A_2, ρ π, B P', A_2 π, η

H: CRIPPLED π $\left(t^2/(t-m_\pi^2)^2\right)$

	Reaction	Exchanges
H1	$np \rightarrow pn$ $\bar{p}p \rightarrow \bar{n}n$	A_2, ρ π, B
H2	$\Lambda p \rightarrow \Sigma^0 p$ $\Lambda p \rightarrow \Sigma^+ n$ $\Sigma^- p \rightarrow \Lambda n$ $\Sigma^- p \rightarrow \Sigma^0 n$	A_2, ρ π, B
H3	$\gamma p \rightarrow \pi^+ n$ $\gamma n \rightarrow \pi^- p$	A_2, ρ π, B

Figure 20G, H   G: Half-asleep π exchange reactions. H: Crippled π exchange reactions.

etc). The simple relation fails but this can be accounted for by the different Regge intercepts in $(s/s_0)^\alpha$ with $s_0 \approx 1$ $(GeV/c)^2$. A similar effect lowers reactions E1–E3 in the direct comparison with D1–D2. An unorthodox explanation of this suppression in terms of direct channel thresholds was advanced by Trilling (217).

The correctness of a low energy $s_0 \approx 1$ scale factor in this case suggests that in Figure 21a we should use the low energy value $D/F \approx -0.2$ value and not $D/F \approx 0$ suggested by the Serpukhov data (151).

(*vii*) The Y^* decays into π and Λ and the observed Λ polarization gives more information on the Y^* density matrix elements than the simple $\pi\Lambda$ decay angular distribution. This allows more stringent tests of the quark model predictions [section 5D(*i*), Figure 20D3] and seems to show some clear if small violations (218).

5F, G, H   General π Exchange

(*i*) All π exchange processes are described at small t $[-t$ is $O(m_\pi^2)]$ by the Poor Man's Absorption Model (PMA). The reader is referred to (67) for a full treatment—here we just note that PMA specifies that each s-channel helicity amplitude have the minimum t-dependence expected from analyticity $[(\sqrt{-t})^n$ with n the amount of spinflip] and the known residue at the π pole. Although this is a simple and successful prescription, it has never been understood theoretically. In the Regge pole language it corresponds to a complicated lack of factorization [conspiracy (219)] of the π Regge pole. Classical absorption models (50) have similar qualitative predictions to PMA but they never have been able naturally to get as precise a fit as the PMA. The model predicts that a reaction should be classified by the t-dependence of its Born term. This is the motivation for the three subdivisions F, G, and H (Figure 20F1).

(*ii*) Before specializing to the three divisions, note that almost all π exchange processes also allow natural parity exchange. Experimentally the natural parity exchange is observed to dominate the cross section for $-t \gtrsim 0.2$ $(GeV/c)^2$ (Figure 22). Theoretically the attribution of this to A_2 exchange is questionable because the PMA modifications of π exchange also produce a (large) natural parity component.

Experimentally the energy dependence of the data gives further information. The unnatural parity exchange component appears to shrink (220, 221) just as one expects for a π Regge pole of canonical slope ≈ 1 $(GeV/c)^{-2}$ (Figure 23a–c). Unfortunately, this conclusion is only statistically convincing if you use data below 5 GeV/c and there all processes known to man shrink (9, 16). (Simply because they are flat at threshold and peaked at higher energy!?) To be unambiguously identified with a Regge effect, the shrinkage must be confirmed in better data at higher energies. To further confuse the issue, some surprising effective π trajectories were reported (222) from studies of $KN \to K^*\Delta$, $NN \to \Delta\Delta$, and $\pi N \to \rho\Delta$.

The natural parity component exhibits little shrinkage up to large $-t \approx 3$ $(GeV/c)^2$ (Figure 23d–f). This contrasts with the reactions of groups A, D, and E which exhibit dramatic shrinkage at large $-t$. This difference is not understood theoretically. Note that this energy dependence implies that natural parity exchange dominance (hinted at) in Figure 22 will become more dominant as energy increases

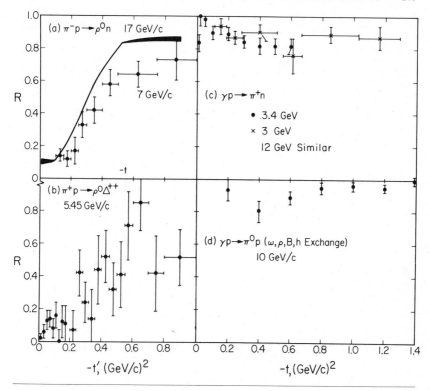

Figure 22 Fraction R of natural parity exchange discussed in section 5F(*ii*). The unnatural parity exchange is essentially π exchange for (*a*)–(*c*) and B exchange for (*d*). The data come from (*a*) (315, 316), (*b*) (317), (*c*) (223, 224, 318), and (*d*) (319).

(*iii*) There is also a universal structure of the line reversal breaking (67). This is generated from the interference between a π exchange which vanishes at $t = 0$ and the natural parity component of the PMA π (Figure 20F2). The effect is large because both exchanges are dominantly in the same flip-flip ($n = 0, 2$) amplitudes. The sign of interference is correctly predicted by the quark model in all reactions. The interference is destructive, giving a very sharp $d\sigma/dt$ and a reduced natural parity component in np charge exchange (CEX), $\gamma n \to \pi^- p$, $\gamma p \to \pi^- \Delta^{++}$, $K^+ n \to K^{*0} p$, $K^+ p \to K^{0*} \Delta^{++}$, $pp \to n\Delta^{++}$, $pn \to \Delta^{++}\Delta^-$. The interference is constructive in $\bar{p}p$ CEX, $\gamma p \to \pi^+ n$, $\gamma p \to \pi^+ \Delta^0$, $K^- p \to \bar{K}^{0*} n$, $K^- n \to \bar{K}^{0*}\Delta^-$, $\bar{p}n \to (\bar{\Delta}^{++})p$, $\bar{p}p \to (\bar{\Delta}^{++})\Delta^{++}$; these show a flat $d\sigma/dt$ (Figure 24) and an increased natural parity exchange. Thus this effect will alter the natural parity component at large $-t$ discussed in (*ii*) and reduce it in $\gamma n \to \pi^- p$ and $\gamma p \to \pi^- \Delta^{++}$. [This is in fact seen experimentally (223, 224).]

Note in particular that the effect is especially dramatic in np CEX where the

ρ exchange is three times bigger than in $\gamma n \to \pi^- p$ which has already a very sharp $d\sigma/dt$. It follows that the very sharp np CEX $d\sigma/dt$, which has evinced much interest, is an accidental combination of a π pole with a large destructive ρ exchange. Similarly the flat $d\sigma/dt$ for $p\bar{p}$ CEX (where ρ is constructive) is of no special significance.

The equality of $\gamma n \to \pi^- p$ and $\gamma p \to \pi^+ n$ at $t = 0$ is particularly interesting as direct evidence that, in accordance with factorization, ρ exchange truly vanishes at $t = 0$. [In an ancient language (219), the ρ is evasive or $M = 0$.] $p\bar{p}$ CEX and np CEX are also almost equal at $t = 0$; however, at low energies ($\lesssim 5$ GeV/c), the

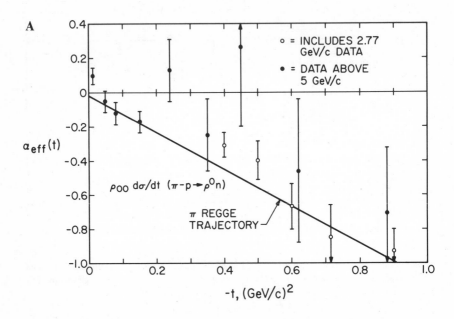

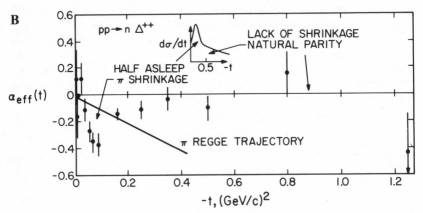

C

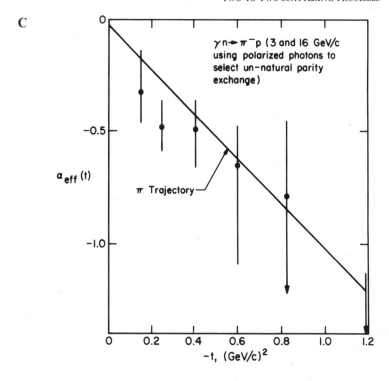

$\gamma n \rightarrow \pi^- p$ (3 and 16 GeV/c using polarized photons to select un-natural parity exchange)

π Trajectory

$\alpha_{eff}(t)$

$-t, (GeV/c)^2$

Figures 23a–c Shrinkage? Energy dependence of processes with large π-exchange discussed in section 5F(*ii*). Precisely (*a*) (data from 221) and (*c*) (220) select pure π exchange. Meanwhile, (*b*) (320) is π at small $-t$ and natural parity exchange at large $-t$.

resonance reactions (Figure 24*d–g*) differ at $t = 0$ (225–227). This inequality is not understood but maybe it disappears at higher energies.

For the resonance reaction this ρ interference effect occurs in the $I = 1$ "crippled π" $\rho_{11} \, d\sigma/dt$ amplitudes which in fact dominate at large $-t$ in the processes listed. One can also examine line reversal breaking in the other amplitudes contributing to resonance production. For instance, for $I = 0$ natural parity exchange seen in $(\rho_{11} + \rho_{1-1}) \, d\sigma/dt$ for $K^\pm p \rightarrow K^\pm{}^* p$, we have a single $n = 1$ spinflip amplitude and as explained in section 5D(*iv*), this exhibits the well-known "real" greater than "moving phase" systematic (Figure 20D4). The same systematic is also probably exhibited by the pure π exchange $\rho_{00} \, d\sigma/dt$ for the vector meson production processes; it has not been as carefully studied.

(*iv*) The reactions F1 to F7, G1 to G10 can be used to study $\pi\pi$, $K\pi$, and πN (this as a check on the technique) elastic scattering by Chew-Low extrapolation (Figure 20F3).

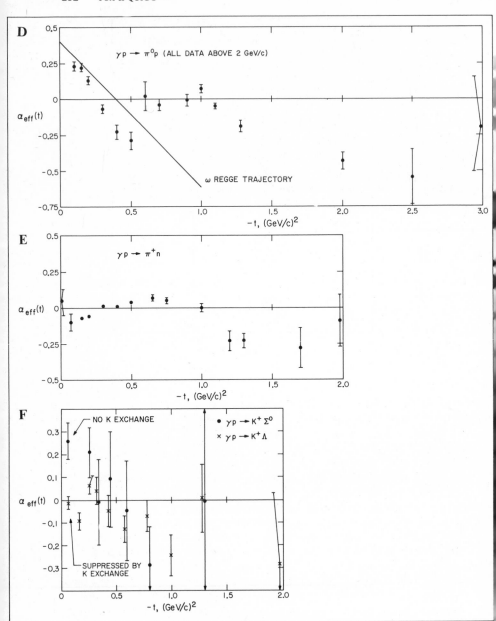

Figures 23d–f Shrinkage? Energy dependence of some dominantly natural parity exchange reactions. The graphs are updated from (14) and discussed in section 5F(*ii*).

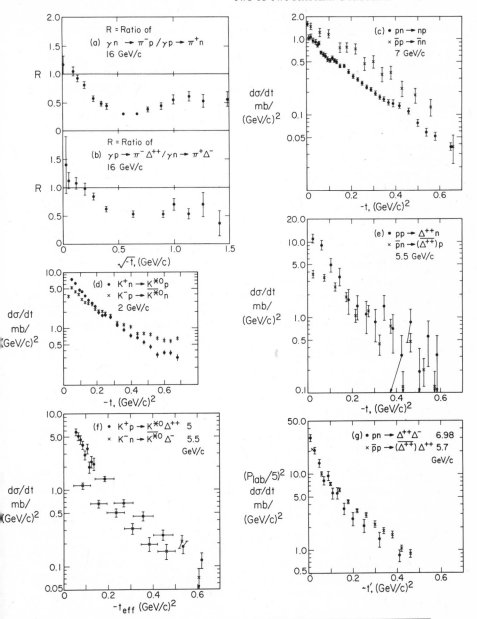

Figure 24   Line reversal breaking discussed in π exchange processes in section 5F(*iii*). The data are from (*a*) (321), (*b*) (243), (*c*) (322, 323), (*d*) (226), (*e*) (324, 325)—similar but less dramatic differences are seen in the 2.8-GeV/c data of (326), (*f*) (327, 328), and (*g*) (329, 330).

This has been generally very successful although it has been obscured at times by over-extravagant claims as to the reliability of the method. In principle, the fully fledged reactions F are best because the t in the Born Term (cf. Figure 20F1) suppresses the π exchange in group G reactions and renders it hard to distinguish from other (A_2, ρ) exchanges. However, this advantage (F vs G) can only be realized at high energies because the physical region boundary $(\propto 1/s$ in F, $1/s^2$ in G) keeps one away from the π pole. For instance, to study the spin 2 $[f^0, K^*(1420)]$ region well, one needs lab momentum of 50 GeV/c for group F but only 8 GeV/c for group G. So for established accelerators the latter are preferable (67).

It is important to note that the background (A_2, ρ) exchange *decreases* relative to π exchange as $(K\pi, \pi\pi)$ mass *increases*. This was demonstrated phenomenologically (213) for the 2^+ region (cf. ρ, B exchange $\pi^+p \to A_2^0\Delta^{++}$ in reaction group I1) and has been qualitatively explained theoretically in terms of duality and properties of the triple Regge coupling (228). It follows that increasing the energy in group F reaction *will* isolate π exchange by reducing t_{min} and *not* drown it by increasing the A_2 exchange with its higher intercept and hence asymptotic dominance over π exchange (Figure 17 of Ref. 213).

(v) It is convenient to mention here that a similar study of $\gamma\pi$ and γK scattering may be possible at high energy in reactions like $\pi^+ p \to (\pi^+\pi^0)p$ and $K^+ p \to (K^0\pi^+)p$ by "Chew-Low" extrapolation to the photon pole at $t = 0$ (Figure 20F4). The latter dominates at very high energies because its cross section is constant with energy while π exchange falls like $1/p_{lab}^2$ and ω exchange like $1/p_{lab}$ (229). Estimates suggest that the integrated photon exchange contribution to $\pi^+ p \to \rho^+ p$ will catch up with the strong interaction cross section at $p_{lab} = 200$ GeV/c. Further the photon cross section can be increased by using nuclear targets (230).

5F   Fully fledged π

(vi) The reactions F1 and F3 and $\pi^+ p \to \omega^0\Delta^{++}$ (reaction I1) have the imposing total of 20 independent observables to describe the simultaneous decay of the spin 1^- and $3/2^+$ particles. The quark model (34) relates the twelve independent (complex) amplitudes for such a reaction to only five amplitudes. This implies many relations between observables which are strikingly successful (16, 231). More generally, one need only supplement the experimental information by three extra assumptions (the quark model gives many more than this) to do an amplitude analysis for these reactions. This has been attempted (232, 233) but current data do not allow very precise conclusions. For instance, the simple Regge interpretation of (232) must be wrong because of the conspiracy necessary to give the nonzero amplitudes at $t = 0$ seen in the data and predicted by PMA (234). Similar quark tests are successful for the sparse data for F6 (233).

5G   Half-asleep π

(vii) Particularly neat phenomenology of half-asleep reactions was the use of data on $\pi^\pm p \to \rho^\pm p$, $\pi^- p \to \rho^0 n$ at the *same* energy to isolate the separate contributions of $I = 0$ (h, ω) and $I = 1$ $[\pi, A_2, A_1 (?)]$ exchanges plus their interference. Use of the

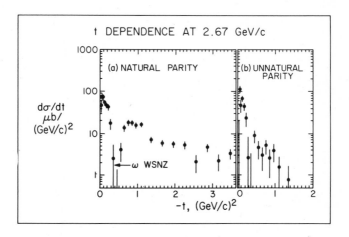

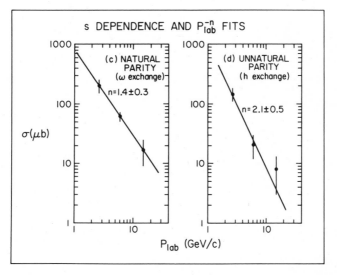

Figure 25 $I = 0$ exchange in $\pi N \to \rho N$ described in section 5G(*vii*). (*a*), (*b*) *t*-dependence of natural and unnatural parity exchange at 2.67 GeV/c (236). (*c*), (*d*) *s*-dependence of integrated cross sections (238). To be exact, (*b*) only shows $\rho_{00}^t \, d\sigma/dt$, not the full $(\rho_{00} + \rho_{11} - \rho_{1-1}) \, d\sigma/dt$ used in (*d*).

ρ density matrix elements (section 3B) allows further separation of each isospin channel into natural and unnatural parity exchange. This analysis (235–239) shows (Figure 25) the following:

    (*a*) A break occurs in $\rho_{00}^H \, d\sigma/dt$ for $I = 1$ unnatural parity (π) exchange at

$t \approx -0.4$ (GeV/c)^2—not too far from the canonical spinflip dip of the absorption model. This break is a dip in $\pi^+ p \to \rho^+ p$ at low energies (235).

(b) A dip occurs in $(\rho_{11} + \rho_{1-1}) d\sigma/dt$ at $t \approx -0.4$ (GeV/c)^2 for $I = 0$ natural parity (ω) exchange which has already been mentioned in section 5D(iii) and is interpreted as an ω WSNZ just as in the related reaction $\gamma p \to \pi^0 p$.

(c) The presence of $I = 0$ unnatural parity (h) exchange. The h is the SU(3) partner of the B and so the h amplitude in $\pi N \to \rho N$ can be related to that of the B in $\pi N \to \omega N$. For instance, in terms of the D/F ratio for (B, h) $N\bar{N}$ couplings

$$\frac{h \text{ part of } \pi^- p \to \rho^- p}{B \text{ part of } \pi^- p \to \omega^0 n} = \frac{(3F - D)^2}{2(F + D)^2} = 0.18 \qquad 5.3.$$

which is calculated for the accepted value $D/(F+D) = 0.6$ for π couplings. (The ratio for B should be the same by EXD.) At 2.7 GeV/c the experimental ratio is $(140 \pm 30) \mu b/(500 \pm 50) \mu b$ which is in fair agreement with equation 5.3. In fact, the latter assumed magic mixing for the B nonet—as suggested by duality (240). On the other hand, if we had used an unmixed B nonet, similar to the π nonet, 0.18 would be replaced by 0.06 with consequent disagreement with experiment. Further, the energy dependence of h exchange (238) is comparable to that of B (241)—as one would expect for a magically mixed nonet. (Both are p_{lab}^{-2} to $p_{\text{lab}}^{-2.5}$.)

It would be nice to have a similar analysis for the K^* reactions; partial results have been reported in (242).

(viii) The reactions G4, G7, and $\pi^- p \to (\pi^- \pi^-)\Delta^{++}$ are pure π exchange (to the extent that the S-wave dominates low mass $\pi\pi$, $K\pi$ scattering) and so provide a good place to examine it without confusion by natural parity exchange.

(ix) As pointed out in (213), reactions G14 are very favorable for meson spectroscopy of the 1^+ nonets because the π exchange should give much cleaner and (relatively) larger cross sections than the usual natural parity exchange (reaction group J).

5H   Crippled π

(x) We did not list in H the many crippled amplitudes (e.g. $2\rho_{11}^H d\sigma/dt$ for $\pi^- p \to \rho^0 n$) that can be isolated from half-asleep and fully fledged reactions. Also reactions H2 are only of the standard type if one neglects the $\Sigma - \Lambda$ mass difference (a similar proviso holds for G11).

(xi) For these reactions, the Born Term vanishes at $t = 0$; any absorption model will add an additional slowly varying and destructive amplitude to produce the resultant sharp forward peak seen in the data. This is sketched in Figures 20F1 and F5; the PMA model specifies the size of the absorption amplitude exactly; this value agrees with current data to around 10% (67).

(xii) It is appropriate to notice here that the reasonably successful vector dominance (VDM) relation between $\gamma N \to \pi N$ (reactions H3) and $\pi N \to \rho N$ (G1) can be understood in terms of the validity of VDM at the π pole and the universal dynamics of π exchange expressed through PMA (122). Similarly its failure to relate G13 to F1 correctly at current energies is a simple t_{min} effect which will disappear at higher energies (234, 243).

(*xiii*) The polarization in np CEX has been measured rather well (244, 245); it shows little energy dependence and is large ($\approx 50\%$) at $-t = 0.5$ (GeV/c)^2. It is of philosophical importance because the reaction is truly exotic and so EXD predicts all amplitudes to be real. Thus there can be no polarization. $K^- n \to \pi^- \Lambda$ is also exotic and sizable polarization is observed [see section 5E(*ii*)]; however, this reaction requires SU(3) to prove exoticity. Again $K^+ p \to p K^+$ is an exotic channel (not to be confused with exotic exchange) and polarization is in fact small (246); however, here, the measurements are nowhere near as detailed as in np CEX. Unfortunately, the many amplitudes discourage theoretical interpretation of the np CEX polarization. First, we need some hints from the polarization measurements in the simpler exotic reaction $K^+ n \to K^0 p$.

5I   B and Natural Parity Exchange

(*i*) Universal acclaim and rightful recognition was first enjoyed by the B meson upon discovery of the $\rho - \omega$ interference effect (201, 247, 248), illustrated in Figure 20I1) for the reaction $\pi^+ p \to \rho^0 \Delta^{++}$ where it was first seen convincingly (249). According to standard mass mixing theory (251), the effect depends on $\delta A(\omega)/A(\rho)$ where δ is the off-diagonal element of the ω, ρ electromagnetic mass matrix while $A(\omega)$, $A(\rho)$ are the ω, ρ production amplitudes. δ is best determined from the electromagnetic processes $e^+ e^- \to \pi^+ \pi^-$ which gives $\delta = 4 \pm 1$ MeV (250) or somewhat less reliably theoretically $\gamma A \to (\pi^+ \pi^-)A$ which gives $\delta = 1.8 \pm 0.2$ MeV [R. Marshall cited in (201)]. Figure 20I2 illustrates the various $\pi^+ \pi^-$ mass shapes possible with $\delta A(\omega)/A(\rho)$ of modulus 2 MeV and various phases β. The observed dip in $\pi^+ p \to \rho^0 \Delta^{++}$ indicated both $\beta \approx 90°$ and that ρ and ω production amplitudes were rather coherent. This phase is correctly predicted (251) by EXD between π and B, although the magnitude of the B amplitude is underestimated by a factor of 1.5 to 2 by the EXD argument (252).

EXD also makes precise predictions for $\pi^- p \to \rho^0 n$ ($\beta = 270°$ with a bump—see Figure 20I2) which is confirmed by data from 2 to 5 GeV/c (248, 253). At the higher energy of 13 GeV/c (254), interference is still seen but it is no longer simply explained by EXD. This failure is probably due to the increasing importance of other exchanges (ρ, A_2). This is not a disaster because the relative phases are all known from EXD while the overall sign of π, B compared with ρ, A_2 is known from the data discussed in section 5F(*iii*). Such an analysis even agrees with the 13-GeV/c experiment but precise statements require use of the density matrix elements of the produced ρ^0 meson to isolate definite exchanges (section 3B); this was attempted in (255) but limited statistics defeated the effort.

The $\rho - \omega$ interference phenomenon provides a unique opportunity to study strong interaction phases and will repay deeper investigation. Further good reactions with different theoretical phases β (e.g. $K^- p \to \rho^0 \Lambda$) where one can study the interference are discussed in (251).

(*ii*) One can also use EXD to relate B exchange in $\pi^+ p \to A_2^0 \Delta^{++}$, $\pi^+ n \to A_2^0 p$, $\pi^+ n \to \phi_N^0(1680)p$, and $\pi^- p \to \pi_N^-(980)p$ to π exchange in $\pi^+ p \to f^0 \Delta^{++}$, $\pi^+ n \to f^0 p$, $\pi^+ n \to g^0 p$, and $\pi^- p \to \varepsilon(\approx 700)n$, respectively. These relations are all good to within a factor of 2 in cross section (213). Reaction I2 ($\pi^- p \to \pi_N^- p$) is particularly

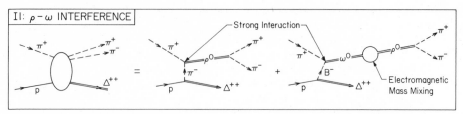

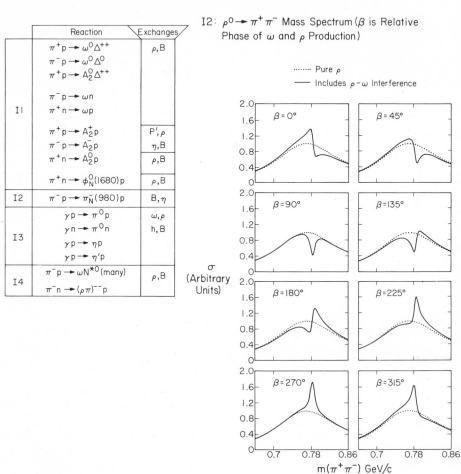

	Reaction	Exchanges
	$\pi^+ p \rightarrow \omega^0 \Delta^{++}$	ρ, B
	$\pi^- p \rightarrow \omega^0 \Delta^0$	
	$\pi^+ p \rightarrow A_2^0 \Delta^{++}$	
	$\pi^- p \rightarrow \omega n$	
I1	$\pi^+ n \rightarrow \omega p$	
	$\pi^+ p \rightarrow A_2^+ p$	P', ρ
	$\pi^- p \rightarrow A_2^- p$	η, B
	$\pi^+ n \rightarrow A_2^0 p$	ρ, B
	$\pi^+ n \rightarrow \phi_N^0 (1680) p$	ρ, B
I2	$\pi^- p \rightarrow \pi_N^-(980) p$	B, η
	$\gamma p \rightarrow \pi^0 p$	ω, ρ
I3	$\gamma n \rightarrow \pi^0 n$	h, B
	$\gamma p \rightarrow \eta p$	
	$\gamma p \rightarrow \eta' p$	
I4	$\pi^- p \rightarrow \omega N^{*0} (\text{many})$	ρ, B
	$\pi^- n \rightarrow (\rho \pi)^{--} p$	

I2: $\rho^0 \rightarrow \pi^+ \pi^-$ Mass Spectrum (β is Relative Phase of ω and ρ Production)

········ Pure ρ
——— Includes $\rho - \omega$ Interference

σ (Arbitrary Units)

$m(\pi^+ \pi^-)$ GeV/c

Figure 201   *B* and natural parity exchange: (*I1*) $\rho - \omega$ interference diagram. (*I2*) Shape of $\pi^+ \pi^-$ mass spectrum in the ρ region for various relative phases β for ω and ρ production [section 5I(*i*)].

interesting because it has no additional natural parity exchange (as π_N^- is spin 0^+).

(*iii*) We have already remarked in section 5F(*iv*) that the π/A_2 exchange ratio increases as we move from $1^-(\rho)$ to $2^+(f^0)$ production. By EXD, the same is true for the B/ρ ratio; correspondingly, $\pi^+ p \to A_2^0 \Delta^{++}$ and $\pi^+ n \to A_2^0 p$ are dominantly unnatural parity exchange (213).

(*iv*) B exchange, studied in $\rho_{00} \, d\sigma/dt$ for $\pi^+ n \to \omega^0 p$, has been shown to decrease with an $\alpha_{\text{eff}} \approx 0$ to -0.3 at $t = 0$ (241). This is not unreasonable for a B Regge trajectory either with canonical slope 0.9 $[\alpha(0) \approx -0.35]$ or passing through the $\pi [\alpha(0) \approx 0, \alpha'(0) \approx 0.65]$. This was discussed further in section 5G(*vii*).

(*v*) The data on reaction I3 are particularly good. ω exchange dominates (due to Clebsch-Gordon factors) $\gamma N \to \pi^0 N$ and has already been mentioned [section 5D(*iii*)]. According to naive expectation, this ω exchange reaction should be very similar to the ρ exchange reaction $\pi^- p \to \pi^0 n$. Both have a dip around $t \approx -0.5$ (GeV/c)^2; however, although $\gamma N \to \pi^0 N$ has some shrinkage, it is nowhere near as much as in $\pi^- p \to \pi^0 n$ especially for $-t > 1$ (GeV/c)^2. [See Figure 23*d*.] This difference cannot be explained in current theories (14, 123). In fact, the major theoretical difficulty with all photoproduction processes [see (256) for the latest review on this] is that although they show more or less canonical shrinkage for $-t \gtrsim 0.5$ (GeV/c)^2 (cf. Figure 23*d–f*) where two natural parity dominated reactions $\gamma p \to \pi^0 p$, $\gamma p \to K^+ \Sigma^0$, both shrink for small t; this disappears at large $-t$ where a "universal" $e^{3t} P_{\text{lab}}^{-2}$ behavior sets in. This contrasts (16, 67) with strong interactions which shrink at all t values.

$\gamma p \to (\eta, \eta')p$ are, on the other hand, dominantly ρ and B exchange. They are kinematically very similar to $\rho_{11} \, d\sigma/dt$ for $\pi N \to \omega N$; like the latter reaction they show no ρ WSNZ at $t \approx -0.6$ (GeV/c)^2 even if the natural parity component is extracted. According to folklore, this lack of a WSNZ [also seen in the nonequality of $\gamma n \to \pi^- p$ and $\gamma p \to \pi^+ n$ at $t \approx -0.6$ (GeV/c)^2] follows because ρ is dominantly in the non-Regge $n = 0$ and 2 amplitudes. There is some evidence that the usual P_{lab}^{-2} behavior does not apply to $\gamma N \to \eta N$ (256, 257).

(*vi*) Just as in $\pi N \to \rho N$ [section 5G(*vii*)], one can use data on $\pi^\pm p \to A_2^\pm p$, $\pi^+ n \to A_2^0 p$ to isolate the separate $I = 0$ and 1 exchanges (92). As described in section 5I(*ii*), $I = 1$ exchange is dominantly unnatural parity B. For $I = 0$, we have a reversal of roles; unnatural parity (η) exchange is still lost in the sea of statistical errors while a virile f^0 exchange dominates the reaction. At least that's what one would have thought; f^0 exchange lives in our fairyland spinflip amplitude [section 5D(*iii*)] where all Regge dreams come true. Indeed at 5 GeV/c, the $I = 0$ $\pi N \to A_2 N$ cross section has just the right size and shape for f^0 exchange (213); however, the energy dependence of the data is wrong and is suggestive of Pomeron dominance at higher energies (11, 92).

(*vii*) Finally we recall that it may be possible to study $f^0 - A_2$ interference in their mutual $K\bar{K}$ decay (258–260). The dominance of unnatural parity exchange in, say, $\pi^+ n \to (A_2^0, f^0)p$ and EXD should lead to even *more* amplitude coherence than seen in $\rho - \omega$ interference.

5J   Natural and Unnatural Parity Exchange without π, K Poles

These reactions are characterized by allowing both parities to be exchanged but not a 0⁻ particle. Thus unnatural parity exchange is both suppressed and in a segment of which we have essentially no theoretical understanding. The whole subject was reviewed very recently in (182) and (213), and we refer the reader there for more details and explicit references.

(i) There are good data on $\pi^+ p \to B^\pm p$ and these show a surprisingly small natural parity exchange cross section. Both the size and shape (it looks like a spinflip not the expected nonflip amplitude) are consistent with an extra t factor in the other-

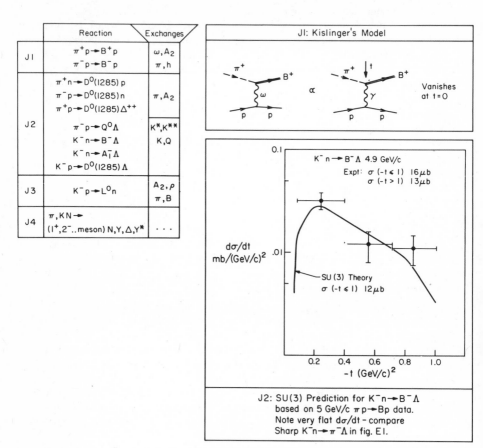

Figure 20J   Natural parity exchange plus unnatural parity with no π, K poles: (*J1*) Kislinger's Model (261). (*J2*) Typical successful SU(3) prediction (213) for $K^- n \to B^- \Lambda$ from $\pi^\pm p \to B^\pm p$ data at same energy.

wise dominant nonflip amplitude. This factor is predicted in Kislinger's (261) generalization of the photon-vector meson exchange model (Figure 20J1).

(*ii*) There are many SU(3) relations between reactions in J1 and J2 which work quite reasonably (Figure 20J2). However, many of the predicted values (in particular, those involving the elusive nondiffractive A_1 production) are too small to be reliably identified in existing experiments. The data and hence our knowledge of 1^+ meson spectroscopy—should improve quite soon. The hypercharge exchange $\pi N \to Q\Lambda$, $\bar{K}N \to (A, B, D, h)$ (Λ, Σ) reactions are probably best for such studies as there is little background from competing channels with larger cross sections.

5K   Nondiffractive Coherent Deuteron Reactions

This is a quiet backwater of conceptually beautiful reactions that limited statistics have prevented from telling us anything really new about reaction mechanisms. $I = 0$ exchange (ω, f^0, h, η) exchange is picked out and accordingly natural parity exchange dominates. The cross section is thus a curious combination of a t factor from the spinflip PV coupling and the sharp (e^{20t}) falloff of the deuteron wave function [see Figure 20K1 from (262)]. Best studied is $Kd \to K^{*\pm}(890)d$ which shows (263) an overall energy dependence, $\alpha_{\text{eff}}(0) \approx 0.33 \pm 0.04$ in agreement with $\omega - f^0$ exchange, plus a bit of unnatural parity exchange whose energy dependence cannot be studied with present statistics. The latter is presumably the h exchange we studied earlier in section 5G(*vii*) from $\pi N \to \rho N$ data. The latter study has

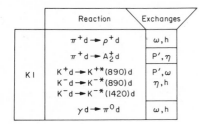

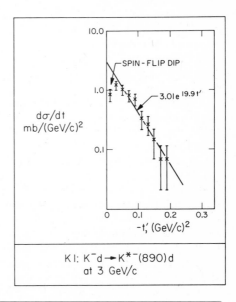

Figure 20K   Coherent nondiffractive deuteron reactions: (*K1*) $d\sigma/dt$ for $K^- d \to K^{*-}(890)$ d at 3 GeV/c (262).

intrinsic difficulties coming from the necessity of subtracting cross sections to isolate $I = 0$ exchange. The deuteron data are far cleaner but higher statistics are required to isolate the object of major theoretical interest, i.e. the energy dependence where possibilities range from P_{lab}^{-2} (as for π exchange) to a slope one trajectory through B mass which is $P_{\text{lab}}^{-2.8}$. h exchange can be isolated from density matrix elements of the vector meson in $\pi^{\pm}d \to \rho^{\pm}d$, $K^{\pm}d \to K^{*\pm}d$ (section 3B) or using polarized photons in $\gamma d \to \pi^0 d$.

The deuteron data can be successfully predicted (264) from proton and neutron data using standard (Glauber) theory.

5L   Natural and Unnatural Parity Exchange in Hypercharge Exchange Reactions

(*i*) A melting pot of sundry poorly measured processes which we should first categorize. Consider reactions L1–L9, which are arranged in order of decreasing importance of natural parity exchange. The reader can find the coupling structure using factorization; at small t, L1 is nonflip, L2–L5 single flip, and L6–L9 double flip; the increasing number of $\sqrt{-t}$ factors thereby decreasing the size of the natural parity exchange.

(*ii*) On the other hand, the importance of unnatural parity exchange increases as we move through the list. The rules governing K (and its EXD friend Q) exchange are:

(*a*) SU(3) implies a very small $p\bar{\Sigma}$ coupling. However, the $p\bar{\Lambda}$ and $p\bar{Y}^*$ couplings are both large. The suppression of unnatural parity exchange for Σ reactions agrees with experiment (265, 266, Figure 20L1).

(*b*) For reactions involving a $p\bar{Y}^*$ vertex, the unnatural parity exchange is particularly apparent because of $\sqrt{-t}$ suppression of the spinflip $p\bar{Y}^*$ natural parity coupling.

(*c*) As discussed in sections 5F(*iv*) and 5I(*iii*), unnatural parity exchange is additionally enhanced over natural parity for 2^+ nonet (reactions L10–L12) production.

(*d*) The reactions in L13 have *no* natural parity exchange. SU(3) and EXD relate the processes—successfully but with large errors (213)—to $\pi^- p \to \varepsilon^0(700)n$ [cf. section 5I(*ii*)].

(*e*) It is also possible to exchange the strange partner Q_{A_1} of the A_1 and its EXD friend. However, the quark model (has this ever been badly wrong?) predicts such an exchange to have the same D/F ratio as nonflip V, T exchange, i.e. such an exchange would have similar $\bar{p}\Sigma$ and $\bar{p}\Lambda$ couplings. To the extent that there is no evidence for unnatural parity exchange in any Σ reaction, it follows that there is also no evidence for A_1 nonet exchange in these reactions. This can be compared with studies (267) of $\pi^- p \to \pi^+ \pi^- n$ near $t = 0$ which showed no A_1 exchange and with polarization (268) in $\pi^- p \to (\pi^0 \pi^0)n$ which indicated some. The latter data deserves confirmation. Theoretical fits (269, 270) to reactions L2, L3 needed Q_{A_1} exchange but the expected D/F was not enforced; it is possible that the Q_{A_1} in these analyses just reflects an oversimplified parameterization of $K - Q$ exchange.

	Reaction	Exchanges
L1	$\bar{p}p \rightarrow (\bar{\Sigma^+})\Sigma^+$ $\bar{p}p \rightarrow (\bar{\Sigma^0})\Sigma^0$ $\bar{p}p \rightarrow \bar{\Lambda}\,\Lambda$ $\bar{p}p \rightarrow (\bar{\Sigma^0})\Lambda$ $\bar{p}n \rightarrow \bar{\Lambda}\,\Sigma^-$	
L2	$\pi^+p \rightarrow K^{*+}(890)\Sigma^+$ $\pi^-p \rightarrow K^{*0}(890)\Sigma^0$ $\pi^-p \rightarrow K^{*0}(890)\Lambda$	
L3	$K^-p \rightarrow \rho^-\,\Sigma^+$ $K^-p \rightarrow \rho^0\,\Sigma^0$ $K^-n \rightarrow \rho^0\,\Sigma^-$ $K^-n \rightarrow \omega\,\Sigma^-$ $K^-n \rightarrow \phi\,\Sigma^-$ $K^-p \rightarrow \phi\,\Sigma^0$ $K^-p \rightarrow \rho^0\,\Lambda$ $K^-n \rightarrow \rho^-\,\Lambda$ $K^-p \rightarrow \omega\,\Lambda$ $K^-p \rightarrow \phi\,\Lambda$	K^*,K^{**} K,Q
L4	$\gamma p \rightarrow K^+\Sigma^0$ $\gamma p \rightarrow K^0\Sigma^+$ $\gamma n \rightarrow K^+\Sigma^-$ $\gamma p \rightarrow K^+\Lambda$	
L5	$\bar{p}p \rightarrow \bar{Y}^{*+}(1385)\Sigma^+$ $\bar{p}p \rightarrow \bar{Y}^{*0}(1385)\Lambda$	
L6	$\pi^+p \rightarrow K^{*+}(890)Y^{*+}(1385)$ $\pi^-p \rightarrow K^{*0}(890)Y^{*0}(1385)$	
L7	$K^-p \rightarrow \rho^- Y^{*+}(1385)$ $K^-p \rightarrow \rho^0 Y^{*0}(1385)$ $K^-n \rightarrow \rho^- Y^{*0}(1385)$ $K^-n \rightarrow \rho^0 Y^{*-}(1385)$ $K^-p \rightarrow \omega\, Y^{*0}(1385)$ $K^-n \rightarrow \omega\, Y^{*-}(1385)$ $K^-p \rightarrow \phi\, Y^{*0}(1385)$ $K^-n \rightarrow \phi\, Y^{*-}(1385)$	

	Reaction	Exchanges
L8	$\gamma p \rightarrow K^+Y^{*0}(1385)$ $\gamma n \rightarrow K^+Y^{*-}(1385)$	
L9	$\bar{p}p \rightarrow \bar{Y}^{*+}(1385)Y^{*+}(1385)$	
L10	$\pi^-p \rightarrow K^{*0}(1420)\Lambda$ $\pi^-p \rightarrow K^{*0}(1420)\Sigma^0$	
L11	$K^-p \rightarrow A_2^-\Sigma^+$ $K^-n \rightarrow A_2^-\Sigma^0$ $K^-p \rightarrow f\,\Sigma^0$ $K^-n \rightarrow f\,\Sigma^-$ $K^-p \rightarrow f'\Sigma^0$ $K^-n \rightarrow f'\Sigma^-$ $K^-p \rightarrow A_2^0\,\Lambda$ $K^-n \rightarrow A_2^-\,\Lambda$ $K^-p \rightarrow f\,\Lambda$ $K^-p \rightarrow f'\,\Lambda$	K^*,K^{**} K,Q
L12	$K^-p \rightarrow f'Y^{*0}(1385)$	
L13	$K^-n \rightarrow \pi_N^-(980)\Lambda$ $K^-p \rightarrow \pi_N^-(980)Y^{*+}(1385)$	K,Q
L14	$\pi N \rightarrow K^*(\text{many})N^*(\text{many})$ $KN \rightarrow (\rho,\omega,\phi)\,Y^*(\text{many})$ $\gamma N \rightarrow KY^*(\text{many})$ $\bar{p}p \rightarrow \bar{Y}^*(\text{many})\,(\Lambda,\Sigma)$	K^*,K^{**} K,Q

Figure 20L Natural and unnatural parity exchange in hypercharge exchange reactions.

For instance, the observed Λ polarization in $K^-p \rightarrow \omega\Lambda$, after unnatural parity exchange has been selected using ω density matrix elements, only implies Q_{A_1} exchange if SU(3) is used (neglect $p-\Lambda$ splitting) to restrict $K-Q_B$ exchange to a single $\bar{p}\Lambda$ helicity state. Perhaps this is not reliable.

Experimentally it would be nice to study unnatural parity exchange in Σ reactions —particularly in the enhanced reaction L10–L12. Again it would be important to confirm the small unnatural parity exchange in $\pi^- p \to K^{*0}\Sigma^0$ which is in fact seen in the first bin of Figure 20L1.

(*iii*) There are polarization and even correlation data for the $\bar{p}$ reactions L1, L5, L9. However, the complicated amplitude structure and pauce data have discouraged

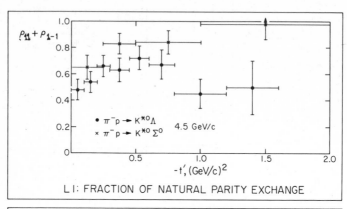

L1: FRACTION OF NATURAL PARITY EXCHANGE

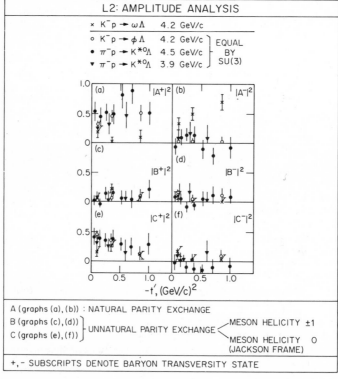

L2: AMPLITUDE ANALYSIS

A (graphs (a),(b)) : NATURAL PARITY EXCHANGE

B (graphs (c),(d))
C (graphs (e),(f)) } UNNATURAL PARITY EXCHANGE < MESON HELICITY ±1
MESON HELICITY 0 (JACKSON FRAME)

+,– SUBSCRIPTS DENOTE BARYON TRANSVERSITY STATE

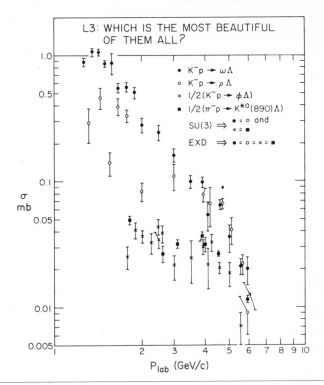

Figure 20L 1–3 Natural and unnatural parity exchange in hypercharge exchange reactions: (*L1*) Fraction of natural parity exchange in $\pi^- p \to K^{*0}(890)$ (Λ, Σ^0) at 4.5 GeV/c (266) discussed in section 5L(*ii*). (*L2*) Amplitude analysis (271) for $\pi^- p \to K^{*0}(890)$ Λ, $K^- p \to (\omega, \phi) \Lambda$ discussed in section 5L(*iv*). (*L3*) Comparison of reaction cross sections for $K^- p \to (\rho, \omega, \phi) \Lambda$, $\pi^- p \to K^{*0}(890) \Lambda$ discussed in sections 5L(*iv*) and (*v*) with data from (265, 266, 272, 297).

authoritative theoretical statements. An attempt to relate Λ polarization in $\pi^- p \to K^0\Lambda$ and $\bar{p}p \to \bar{\Lambda}\Lambda$ was given by Michael (11). Michael also suggests qualitative relations between, say, polarization in $\pi^- p \to K^{*0}\Lambda$ and $\pi^- p \to K^0\Lambda$ by assuming that natural parity exchanges have the same modifications of the Regge phase in each reaction. These predictions disagree with experiment but offhand this is not surprising as the basic assumption isn't valid in any absorption model. The situation is muddied somewhat as the assumption *does* give a correct relation between $\pi^- p \to \pi^0 n$ and $\gamma p \to \pi^0 p$ polarization (11).

(*iv*) Observation of the joint decay distribution is possible in reactions L2, L3 using the weak decay of the hyperon. This gives all but one of a complete set of observables. The work of Field and collaborators (271) is very encouraging. SU(3) predicts the equality $K^- p \to \phi\Lambda = \pi^- p \to K^{*0}\Lambda$ while EXD further states that both are equal to

$2(K^- p \to \omega \Lambda)$. The amplitude analysis shown in Figure 20L2 showed that the first equality is valid amplitude by amplitude; the last equality is, however, badly violated not only in magnitude, $d\sigma/dt$, but also in amplitude structure. This argues that the EXD violations seen [sections 4E, 4F, 5E(v)] in hypercharge exchange reactions are not simply due to SU(3) violations (e.g. to predict $K^- n \to \pi^- \Lambda$ to be real needs SU(3)+EXD; the observed polarization is then a consequence of EXD not SU(3) violation). This is spectacularly illustrated in Figure 20L3. Supplementing the experimental observables by additional assumptions, they were able (271) to find the individual amplitudes; in particular, they suggest that the $n = 2$ amplitude in $K^- p \to \omega \Lambda$ is indeed real and so suffered little absorption. This important conclusion [natural in most models (2) except the dual absorption model (31)] deserves confirmation by higher statistics data. It is apparent from Figure 20L2 that the most difference between $K^- p \to \omega \Lambda$ and $\phi \Lambda$ occurs in the natural parity exchange segment; this is also a bit surprising as one would have expected it and not the low-lying (in J plane) unnatural parity exchange to agree best with theoretical expectations (EXD).

(v) In fact, the reader will by now be bored to hear that these reactions exhibit exactly the same systematic in line reversal breaking; "real" reactions $[K^- p \to (\omega, \rho)\Lambda]$ are greater than "moving phase" $(K^- p \to \phi \Lambda,\ \pi^- p \to K^{*0} \Lambda)$ reactions as found in section 5D(iv). The production cross section comparison is summarized in Figure 20L3; if you select definite exchange, then just as in $K^\pm p \to K^{\pm *} p$ [see section 5D(iv)], one finds more breaking for natural than unnatural parity exchange (213). We can note again, just as in Figure 18, that the suppressed reactions ($\phi \Lambda$, $K^{*0}\Lambda$) have a higher threshold than $\omega \Lambda$, $\rho \Lambda$ and so must be lower at low energy. Perhaps the observed violation at 4 GeV/c is connected with this.[25]

(vi) It is possible to use SU(3) and vector dominance $\gamma = \rho + \omega/3 - \sqrt{2}\,\phi/3$ to predict the amplitudes for $\gamma p \to K^+ (\Lambda, \Sigma)$ from those for vector meson production $K^- p \to (\omega, \phi)(\Lambda, \Sigma)$; this unfortunately predicts the photon data to be a mixture of real and rotating phases and one can only say that the sign of the polarization and the size of $d\sigma/dt$ for $\gamma p \to K^+ (\Lambda, \Sigma^0)$ are in reasonable accord with expectations (11). Note that K exchange is important at small t in $\gamma p \to K^+ \Lambda$ but its effect falls with energy compared with the dominant K^*, K^{**} exchange. This produces a noticeable effect on $\alpha_{\rm eff}(t \approx 0)$ for the two processes, i.e. that for Λ lies lower than that for Σ^0 (see Figure 23f).

(vii) There are of course many quark model relations (34) for the class L reactions. First, we have the double correlation relations for $\bar{p}p \to \bar{Y}^* Y^*$, $K^- p \to \rho Y^*$, etc.; the SU(3) analogs of those for $\bar{p}p \to \overline{\Delta^{++}} \Delta^{++}$ and $\pi^+ p \to \rho^0 \Delta^{++}$. Current data do not provide stringent tests, but the nice data of (272) showed pleasing agreement between theory and experiment.

Secondly, there are the quark model relations following from the equality of the

---

[25] Such ideas cannot directly explain difference between $K^+ n \to K^0 p$ and $K^- p \to \bar{K}^0 n$ which have the same threshold. Nevertheless, there is a lot of difference between low energy $K^- p$ and $K^+ n$ scattering, i.e. in unitarity effects.

PV (e.g. πK^*), $B\bar{B}$ spinflip (e.g. $N\bar{\Lambda}$), and $B\bar{D}$ (e.g. $P\bar{Y}^*$) vertices. These only worked to a factor of 1.5 to 2 for the similar $B\bar{B}$ to $B\bar{D}$ nonstrange exchange (ρ, A_2) relation coupling [section 5D(ii)]. A similarly precise statement has not been extracted from the hypercharge exchange data.

Thirdly, there are simple predictions of the quark model D/F values. That for unnatural parity exchange is OK experimentally (i.e. the coupling is roughly zero in Σ reactions as in section 5L(iia). There are two D/Fs for natural parity exchange and so it is not possible to relate Λ to Σ reactions directly. However, adding the spinflip $B\bar{B}$ to $B\bar{D}$ relation enables one to write the sum rule (273, 274).

$$\sigma(AN \rightarrow B\Lambda) = 3\big(\sigma(AN \rightarrow B\Sigma) + \sigma(AN \rightarrow BY^*[1385])\big) \qquad 5.4.$$

for nucleon N and any hadrons A, B.

This is a cavalier approach because it is sensitive to kinematic factors and breaking of the $B\bar{B}$ to $B\bar{D}$ relation. Anyhow, however you do it, the Σ data is much larger than expected (272, 273). The reason for this is twofold. First, the nonflip $B\bar{B}$ amplitude has a $D/F \approx -0.2$ which does enhance the Σ reactions compared with the quark $D/F = 0$ (265, 275). Secondly, the quark model predicts Σ to have essentially zero t-channel spinflip coupling at $t = 0$ (taking, for definiteness, the t-channel as the additivity frame). Rotating this prediction, it corresponds to comparable Σ and Λ s-channel spinflip amplitudes [section 5D(vi)]. Folklore suggests that it is sensible to take the ratio of s-channel (and *not* t-channel) amplitudes at $t = 0$ as a measure of their ratio away from $t = 0$ (i.e. we assume s-channel amplitudes have a universal shape). This simple dynamical effect completely invalidates the application of equation 5.4 away from $t = 0$; nearly all claimed violations of equation 5.4 come from comparing it with data integrated over t and so are irrelevant. An alternate view on the violation of equation 5.4 was given in (274).

The application of equation 5.2, exactly at $t = 0$, is extremely interesting because all amplitudes for reactions L2–L13 vanish there for factorizable Regge poles. So examination of the quark relations at $t = 0$ then enables one to check if they are more general than the pole approximation or if they are just symmetries of pole residues. This point is best studied in $\gamma p \rightarrow K^+(\Lambda, \Sigma^0)$ where the data (276, 277) are good and $t_{\min}$ effects small. The Σ^0/Λ ratio does decrease towards $t = 0$ but, *miserabile dictu*, it is not enough for the quark model. Rather it is in rough agreement with absorption calculations[26] based on poles whose residues obey quark relations.

5M   Exotic Exchange

(i) In this category, we can place all reactions whose quantum numbers forbid the exchange of all known single particle states. As anticipated in section 2, the cross sections are indeed small and fall fast with energy. A representative sample of the data (266, 278–280) is shown in Figures 20M2–M5. First consider the energy dependence of the data: below 2 GeV/c, there is a rapid ($\approx s^{-10}$) fall which is in fact shared by many processes—exotic and nonexotic. Figure 20M2 shows the

[26] G. C. Fox, Unpublished fits and (275).

similarity of $K^{\pm}p$ backward scattering at low energy. From 2 to 5 GeV/c, the rate of fall slows—s^{-5} seems typical while there is at present no data above this energy. Now consider the t-dependence: at low energy, we found the Odorico zeros [Figure 8, (45); see also (281) for $K^-p \to \pi^+\Sigma^-$]. These have not been reported above 2 GeV/c. Rather, around 5 GeV/c, the data exhibit a definite peak with a rather flat (e^{4t}) behavior. It is worth noting explicitly that although $K^-p \to K^-p$ has a backward peak at 5 GeV/c (280 and Figure 20M2), it has a backward dip below 2.5 GeV/c (282).

(*ii*) Theoretical models (Figure 20M1) include double scattering or equivalently Regge-Regge cuts (64, 65, 283–289), exotic exchange, s-channel effects (290), and reflection mechanism (291).

(*iii*) The Regge-Regge cut model naturally reproduces the t-dependence observed at 5 GeV/c: it predicts energy dependence[27] $\sigma \propto s^{2\alpha - 2}$ where Quigg (65) finds $\alpha \approx \alpha_{\text{cut}} - \frac{1}{2}$ where $\alpha_{\text{cut}}(0) = \alpha_1(0) + \alpha_2(0) - 1$, for exchange of trajectories α_1 and α_2, is the tip of the cut. It is clear from Figures 20M3 and 20M4 that data below 5 GeV/c must be due to some other mechanism. It is worth noting that (even at 5 GeV/c) $\pi^-p \to K^+\Sigma^-$, $K^-p \to K^-p$, $\bar{p}p \to \bar{p}p$, and $\bar{p}p \to K^+K^-$ all have similar size exotic peaks. This equality is rather surprising theoretically. For instance, one naively expects

$$\frac{\sigma(\bar{p}p \to p\bar{p})}{\sigma(\pi^-p \to K^+\Sigma^-)} \approx \frac{\sigma(\bar{p}p \to \pi^0\pi^0)^2}{\sigma(\pi^-p \to \pi^0 n) \times \sigma(\pi^0 n \to K^+\Sigma^-)} \ll 1$$

References (65) and (289) emphasize that this argument ignores important spin effects. However, Figure 20M4 shows that the model is unable to account for the experimental difference between $K^-p \to \pi^+\Sigma^-$ and $\pi^-p \to K^+\Sigma^-$.

(*iv*) The resonance reactions M5 exhibit large exotic exchange. $\pi^-p \to \pi^+\Delta^-$ is ≈ 45 μb at 4 GeV/c (292) and $\pi^-p \to K^+Y^{*-}$ (1385) roughly $\frac{1}{2}$ μb at 5 GeV/c (266). These large values may be due to a reflection mechanism proposed by Berger (291, 293) and calculated in the latter case by Sivers & Von Hippel (294). Note that if one includes final state interactions (between π^- and n in Figure 20M1) the reflection model becomes equivalent to the double scattering model plus the remark that cross sections involving π exchange are particularly large.

(*v*) Figure 20M3 indicates that the statistical calculation of Frautschi (290) qualitatively reproduces backward K^-p scattering. It cannot explain the similarity of $K^{\pm}p$ at low energy because K^+p is predicted to be zero as there are no resonances.

(*vi*) Exotic exchange can also be found by comparing values of nonexotic reactions (295). For instance, the difference between 2 $(\pi^-p \to K^0\Sigma^0)$ and $\pi^+p \to K^+\Sigma^+$ (reactions E1) measures an $I = 1/2, 3/2$ interference. In principle this is more sensitive than the direct measurement of the $I = 3/2$ amplitude squared in $\pi^-p \to K^+\Sigma^-$. It is more difficult, however, because of normalization and systematic errors in

---

[27] Such a power law energy dependence with s also given by exchange of an exotic trajectory $\alpha(t)$.

the cross sections being compared. The comparison method, outlined above, used for $\gamma p \to \pi^+ \Delta^0$ vs $\gamma n \to \pi^+ \Delta^-$ (reactions G13) (243) and $\gamma p \to K^+ \Sigma^0$ vs $\gamma n \to K^+ \Sigma^-$ (reactions L4) (277) gave a large (10% of nonexotic) exotic exchange amplitude at 11–16 GeV/c. The SLAC group checked that their extraction of neutron cross sections from deuterium was correct by comparison of $\gamma p \to K^+ \Lambda$ and $\gamma p \to \pi^+ n$ for free protons and those in deuterium.

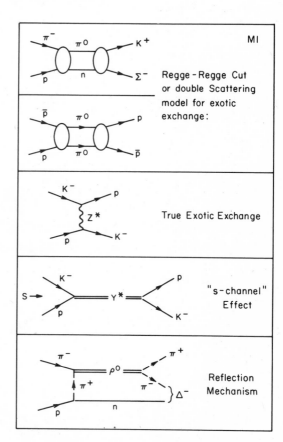

Figures 20M, 20M1   Exotic exchange: (M1) Theories for exotic exchange discussed in section 5M(ii). (M2) Elastic differential cross sections over the whole angular range at 5 GeV/c (280). (M3) Comparison of energy dependence for $K^\pm p$ backward scattering (280) with Regge cut (64) and statistical model (290) calculations for $K^- p$. The latter is calculated at $\theta_{cms} = 180°$ not $u = 0$ like the data but this is not a large effect. (M4) Comparison of Regge-cut calculations (65) with $d\sigma/dt$ data on $K^- p \to \pi^+ \Sigma^-$ and $\pi^- p \to K^+ \Sigma^-$ at $t = 0$. (M5) Compilation (279) of cross sections σ for some exotic reactions with results of $\sigma \alpha P_{lab}^{-n}$" fits.

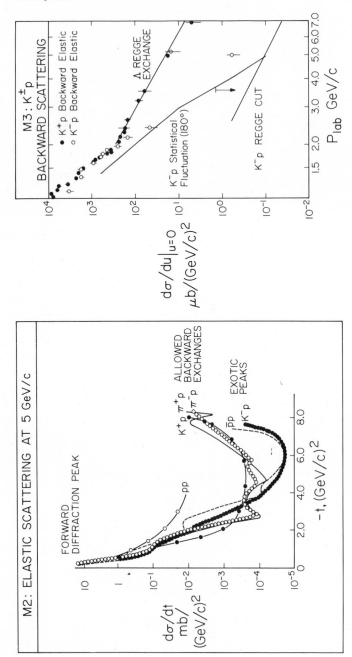

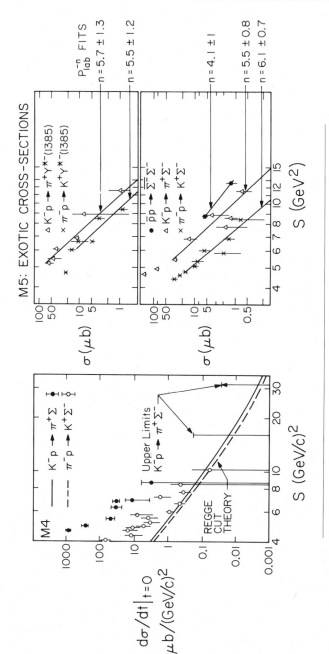

Figures 20M 2–5   Exotic exchange (see legend to Figures 20M, 20M1).

(*vii*) Reactions M2 are only exotic for canonical quark rules (34). A current limit (214) is

$$\frac{\sigma(\pi^+ p \to \phi^0 \Delta^{++})}{\sigma(\pi^+ p \to \omega^0 \Delta^{++})} \lesssim 1/330$$

at 3.7 GeV/c.

6 KEY CONCEPTS AND INTRIGUING EXPERIMENTS

The spellbound reader of section 5 will have gained the impression (through repetition) that a number of theoretical ideas stand out as especially important organizers of the experimental systematics. We describe them briefly again, for the particular benefit of the nonspecialist who has not waded through the details of sections 3–5. In addition, we point the way to a modest number of future experiments at intermediate energies which may be expected to clarify several of the questions raised in section 5.

6A   Particularly Useful Ideas

As is our custom, we shall neglect to mention Regge theory in our catalog of constructs because acceptance of Reggeism is *sine qua non* for trying to make sense of two-body scattering. Therefore the schemes we outline here are some which are useful adjuncts to the basic Regge pole hypothesis.

Predictive power is what is lacking in the unadorned Regge pole framework. We are rarely able to compute anything from what we might hope to call "first principles," but we can, with great success, relate amplitudes for one reaction to amplitudes for another by invoking internal symmetries, specifically SU(3) and its more detailed incarnation, the quark model. There are no important violations of the conjecture that Regge residues satisfy SU(3) symmetry. The source of the $\approx 20\%$ violations which are common has never been traced satisfactorily. For resonance decays, all deviations of partial widths from the predictions of SU(3) are understood in terms of kinematic effects. (In other words, apart from mass splittings within SU(3) multiplets, there are no sizable deviations from SU(3) symmetric behavior.) Attempts to make similar kinematic excuses for the deviations from SU(3) for scattering amplitudes have been too casual to settle the issue. The quark model has no qualitative failures, neither for the hadron spectrum nor for selection rules, nor indeed for the relative magnitudes of couplings. At the same level, predictions for the spin structure of amplitudes deduced from the quark model are successful as well. The quark model thus poses a two-pronged challenge; it cries out for experimental obstacles and for theoretical understanding.

Exchange degeneracy provides line reversal relations, predictions for *t*-dependence of cross sections (by the occurrence of nonsense wrong signature zeros), and, especially when coupled with the quark model in duality diagrams, predictions for relative and absolute phases of scattering amplitudes. In most cases line reversal predictions are roughly satisfied and in the pair of reactions for which the data seem most credible (*KN* charge exchange), line reversal symmetry seems well-satisfied above 10 GeV/c. Nonsense dips seem to occur, as prescribed, in all but (s-channel) nonflip amplitudes. The phase predictions are in general successful: the amplitude for $K^+ n \to K^0 p$ does seem to be real, π exchange and B exchange are $90°$ out of phase in $\pi N \to (\rho, \omega)\Delta$, the polarization in $K^- p$ elastic scattering suggests a Regge pole contribution to the flip amplitude proportional to $\exp[-i\pi\alpha_\rho(t)]$, etc. Much less successful are predictions of the absence of polarization in inelastic reactions which are based on the requirement that all spin amplitudes should have exactly the same phase. (Inability to account for polarization naturally in the archetypal Regge reaction, $\pi^- p \to \pi^0 n$, was a chink in the Regge armor long before exchange degeneracy was invoked.)

The physical basis of absorption models seems almost unimpeachable, and examples exist (see section 4B, F; section 5F, G, H) of strong circumstantial evidence for absorptive effects, but there is no successful quantitative implementation of the absorption scheme. Four years ago, absorptive corrections were heralded as solutions to all the woes of Regge phenomenology. The early enthusiasm has gone unrewarded by spectacular results, but the idea on which hopes were based retains its appeal. In fact, there has been renewed theoretical interest[28] in calculating Regge cuts (absorption effects) using the triple Regge techniques[29] developed to study inclusive reactions. So far the only practical application has been to the evaluation of the Glauber correction (198), but one can anticipate that the union of the Gribov calculus[28] with triple Regge phenomenology will lead to a better understanding of cuts.

The study of inclusive reactions as a function of s and x may be likened to total cross section phenomenology.[29] In the same analogy, triple Regge studies correspond to $d\sigma/dt$, density matrix, and polarization. Independent of theoretical breakthroughs, we must expect the triple Regge couplings, extracted from detailed analysis of inclusive data, to exhibit the same wealth of structure section 5 detailed for two-body scattering.

6B   Experiments of Future Interest

Although we display the obligatory reviewer's hubris by enumerating our favorite list of essentially impossible experiments which will tell us everything we need to know,[30] we shall keep the shopping list mercifully short. Many specific suggestions will be found midst the flood of information in sections 4 and 5, so here we concentrate on slightly bizarre experiments which lie just beyond present-day experimental capabilities.

Let us contemplate the possibilities for experiments with a polarized HD target.[31]

[28] There is no comprehensive review: see the many talks at the 1972 Batavia Conference (296).

[29] For the nonexpert, we remark that the triple Regge coupling is that between three Regge poles whereas two-to-two scattering processes are analyzed in terms of couplings of one Regge pole and two particles.

[30] Or at least everything we know we need to know.

[31] Such targets exist in the low-temperature lab (E. Graf, Private communication) and will someday soon (we hope) find their way onto the accelerator floor.

Beyond improving polarization measurements for the two-body reactions already studied (by improving the dilution factor), targets free of complex nuclei open new experimental vistas. For example, the information to be gained from now impractical two-body reactions such as[32] forward

$$n + \vec{p} \rightarrow n + p \qquad\qquad 6.1.$$

could be extracted from

$$p + \vec{d} \rightarrow p + n + p_s \qquad\qquad 6.2.$$

One can also hope to make precise measurements of the polarization in reactions for which, according to duality diagrams, the scattering amplitudes should be purely real, such as

$$K^+ \vec{n} \rightarrow K^0 p \qquad\qquad 6.3.$$

In the absence of complex target nuclei, it will also be possible to study (107) quasi two-body or authentic multibody final states, among which

$$\pi^- \vec{p} \rightarrow \pi^0 \pi^0 n \qquad\qquad 6.4.$$

and

$$\pi^- \vec{p} \rightarrow \rho^0 n \qquad\qquad 6.5.$$

can clarify the role of A_1 exchange. In reactions such as

$$\pi \vec{N} \rightarrow K Y^* \qquad\qquad 6.6.$$
$$\quad\ \ \llcorner\!\!\rightarrow \pi Y$$
$$\qquad\quad\ \ \llcorner\!\!\rightarrow \pi N$$

one can go far toward measuring directly *all* of the participating amplitudes.

Recently constructed hyperon beams may be used to perform diffraction dissociation experiments in which background problems are substantially reduced. Good signal-to-background ratios seem to make these experiments appealing in spite of the relatively low flux of hyperons. Since Σ^- beams exist at Brookhaven and CERN, we will give examples appropriate to them, but it will be obvious that a Λ beam, for example, would serve equally well.

The reactions

$$\Sigma^- p \rightarrow Y_1^{*-} p \qquad\qquad 6.7.$$

are SU(3) analogs of the already studied reactions

$$pp \rightarrow N^{*+} p \qquad\qquad 6.8.$$

Corresponding to the prominent N^* peaks

$N^*(1518)$ $3/2^-$ $\Gamma \approx 105$ to 150 MeV
$N^*(1688)$ $5/2^+$ $\Gamma \approx 105$ to 180 MeV

---

[32] We denote polarized particles by a superior arrow.

in reactions 6.8, reaction 6.7 should exhibit

$\Sigma(1670)$ $3/2^-$ $\Gamma \approx 50$ MeV
$\Sigma(1915)$ $5/2^+$ $\Gamma \approx 70$ MeV

which should be relatively easier to establish, due to their narrower widths. Even in a missing mass experiment, therefore, it may be possible to obtain more reliable production angular distributions for equation 6.7 than for equation 6.8. The important potential advantage of channel (6.7) lies in the measurement of decay angular distributions of the Y^*s. Both Y^*s (probably) decay appreciably into $\bar{K}N$, as well as into the $\pi\Lambda$, $\pi\Sigma$, and $\pi\pi\Sigma$ channels which are likely to be complicated by Deck background. In particular, the sequence

$$\Sigma^- p \to Y_1^{*-} p \qquad\qquad 6.9.$$
$$\phantom{\Sigma^- p \to Y_1^{*-}} \llcorner\!\!\to \pi^- \Lambda$$
$$\phantom{\Sigma^- p \to Y_1^{*-} \pi^-} \llcorner\!\!\to p\pi^-$$

which, for narrow Y_1^{*-}, should be distinguishable from nonresonant (Deck) background, contains very detailed information about the Y_1^* decay angular distribution which in turn has much to say about the diffractive production process. If reaction 6.9 is feasible, it should be an exceptionally clean experiment by which to elucidate the detailed dynamics of diffraction dissociation.

How large will the cross sections be? If the Pomeranchuk trajectory were a pure SU(3) singlet, we would have

$$(d\sigma/dt)[\Sigma^- p \to \Sigma(1670)p]_{\text{Pomeranchuk}} = (d\sigma/dt)[pp \to N(1518)p]_{\text{Pomeranchuk}}$$

and

$$(d\sigma/dt)[\Sigma^- p \to \Sigma(1915)p]_{\text{Pomeranchuk}} = (d\sigma/dt)[pp \to N(1688)p]_{\text{Pomeranchuk}}$$

Similar remarks apply to the meson (e.g. ω) exchange contributions. These crude estimates suggest that at 10 GeV/c the forward differential cross section for equation 6.7 should be comparable to those (189) for the corresponding N^* reactions $[0.5–5$ mb/$(\text{GeV/c})^2]$.

7 OUTLOOK

We see no immediate danger that two-body reaction mechanisms will become completely understood and relegated to a Handbuch on a dusty shelf. Developments of the past three years convince us that a description of two-body reactions at intermediate energies is unlikely to be simple in detail, however neat the broad outlines of a future theory seem. More than a sign of resignation, amplitude analysis is a recognition that nature is less simple (and perhaps more intricately fascinating) than had been hoped in earlier days. Progress is therefore likely to follow detailed and difficult experiments which map out amplitude structure and present us with (we can only hope) more systematic systematics.

Looking ahead to studies of two-body reactions at the higher energies of the Serpukhov and Batavia machines, we anticipate the disappearance of canonical

Regge behavior. J-plane singularities more complicated than Regge poles must surely exist, and some of these ineluctably will resemble absorptive Regge cuts. Although these may be weak, they lie higher in the J-plane than the poles which generate them, and give rise to amplitudes with milder energy dependence than those corresponding to Regge pole exchange. At some point Regge cuts will have to dominate the reaction mechanisms, and we guess on the basis of model fits that they will do so by 100 GeV/c.

Finally, we deem it unlikely that new facts discovered at higher energies will invalidate the basic picture of two-body reactions that we hold today. The framework we described in section 2 is simply too firmly grounded in experimental results for us to get off so easily. The elementary facts are in our possession; the challenge is to build a theory from them.

ACKNOWLEDGMENTS

We would like to thank Nancy Clark and Pat Lee for their excellent and cheerful job of typing and drawing our herculean effort.

Literature Cited

1. Jackson, J. D. 1972. *Particle Physics,* ed. M. Bander, G. L. Shaw, D. Y. Wong, p. 164. New York: Am. Inst. Phys.
2. Chiu, C. B. 1972. *Ann. Rev. Nucl. Sci.* 22:255
3. Barger, V., Cline, D. B. 1969. *Phenomenological Theories of High Energy Scattering.* New York: Benjamin
4. Jacob, M. 1968. *Resonances and Exchange Processes.* Presented at L'Ecole Int. Phys. Particules Elem., *Herceg-Novi, Yugoslavia*
5. Collins, P. D. B., Squires, E. J. 1968. *Springer Tracts in Modern Physics.* Vol. 45. Berlin: Springer-Verlag
6. Jackson, J. D. 1970. *Rev. Mod. Phys.* 42:12
7. Harari, H. 1969. *Theories of Strong Interactions at High Energies,* ed. R. F. Peierls, p. 385. *Brookhaven Nat. Lab. Rep. BNL50212*
8. Schmid, C. 1971. *Proceedings of the Amsterdam International Conference on Elementary Particles,* ed. A. G. Tenner, M. J. G. Veltman, p. 275. Amsterdam: North-Holland
9. Schmid, C. 1971. *Phenomenology in Particle Physics,* ed. C. B. Chiu, G. C. Fox, A. J. G. Hey, p. 629. Pasadena: Caltech
10. Michael, C. 1972. *Proc. Fourth Int. Conf. High Energy Collisions,* ed. J. R. Smith. *Rutherford Lab. Rep. RHEL-72-001*
11. Michael, C. 1972. *Proc. XVI Int. Conf. High Energy Physics, Batavia*
12. Giacomelli, G. 1972. *Proc. XVI Int. Conf. High Energy Physics, Batavia*
13. Leith, D. W. G. S. 1972. *Proc. XVI Int. Conf. High Energy Physics, Batavia*
14. Fox, G. C. 1970. *High Energy Collisions,* ed. C. N. Yang et al, p. 367. New York: Gordon and Breach
15. Lovelace, C. 1971. *Phenomenology in Particle Physics,* ed. C. B. Chiu, G. C. Fox, A. J. G. Hey, p. 668. Pasadena: Caltech
16. Fox, G. C. 1971. *Phenomenology in Particle Physics,* ed. C. B. Chiu, G. C. Fox, A. J. G. Hey, p. 703. Pasadena: Caltech
17. Chiu, C. B. 1971. *Proc. Workshop Particle Physics at Intermediate Energies,* ed. R. D. Field, Jr., p. 396. *Lawrence Radiation Lab. Rep. UCRL-20655*
18. Guisan, O. 1971. *High Energy Phenomenology,* ed. J. Tran Thanh Van, p. 105. Paris: Société Polygraphique Mang.
19. Harari, H. 1971. *Proc. Int. Conf. Duality and Symmetry in Hadron Physics,* ed. E. Gotsman, p. 148. Jerusalem: Weizmann Sci.
20. Phillips, R. J. N. 1971. *Proceedings of the Amsterdam International Conference on Elementary Particles,* ed. A. G. Tenner, M. J. G. Veltman, p. 110.

Amsterdam: North-Holland
21. Lovelace, C. 1971. *Proceedings of the Amsterdam International Conference on Elementary Particles*, ed. A. G. Tenner, M. J. G. Veltman, p. 141. Amsterdam: North-Holland
22. Phillips, R. J. N., Ringland, G. A. 1972. *High Energy Physics*, ed. E. H. S. Burhop, 5:187. New York: Academic
23. Joachain, C. J., Quigg, C. 1972. *Nat. Accelerator Lab. Rep. NAL-THY-99; Rev. Mod. Phys.* In press
24. Sonderegger, P. 1972. *Two-Body Collisions*, ed. J. Tran Thanh Van, p. 518. Paris: Société Polygraphique Mang.
25. Quigg, C. 1971. *Particles and Fields, 1971*, ed. A. C. Melissinos, P. F. Slattery, p. 40. New York: Am. Inst. Phys.
26. Chan, Hong-Mo. 1972. *Proc. Fourth Int. Conf. High Energy Collisions*, ed. J. R. Smith. *Rutherford High Energy Lab. Rep. RHEL-72-001*
27. Mueller, A. H. 1972. *Proc. XVI Int. Conf. High Energy Physics, Batavia*
28. Abarbanel, H. D. I. 1971. *Phys. Lett. B* 34:69
29. Chan, Hong-Mo, Hsue, C. S., Quigg, C., Wang, J.-M. 1971. *Phys. Rev. Lett.* 26:672
30. Mueller, A. H. 1970. *Phys. Rev. D* 2:2963
31. Harari, H. 1971. *Phys. Rev. Lett.* 26:1400
32. Barger, V., Geer, K., Halzen, F. 1972. *Nucl. Phys. B* 49:302
33. Martin, A., Stevens, P. R. 1972. *Tensor Meson Exchange Amplitudes*, Preprint
34. Bialas, A., Zalewski, K. 1968. *Nucl. Phys. B* 6:449, 465, 478, 483
35. Evans, R. D. 1955. *The Atomic Nucleus*. New York: McGraw
36. Benecke, J., Chou, T. T., Yang, C. N., Yen, E. 1969. *Phys. Rev.* 188:2159
37. Jackson, J. D. 1966. *Proceedings of the XIII International Conference on High Energy Physics*, ed. M. Alston-Garnjost, p. 148. Berkeley: Univ. Calif.
38. Hearn, A. C., Drell, S. D. 1967. *High Energy Physics*, ed. E. H. S. Burhop, 2:219. New York: Academic
39. Kokkedee, J. J. J. 1969. *The Quark Model*. New York: Benjamin
40. Chew, G. F. 1966. *The Analytic S-Matrix*. New York: Benjamin
41. Sivers, D., Yellin, J. 1971. *Rev. Mod. Phys.* 43:125
42. Arnold, R. C. 1965. *Phys. Rev. Lett.* 14:657
43. Dolen, R., Horn, D., Schmid, C. 1968. *Phys. Rev.* 166:1768
44. Veneziano, G. 1968. *Nuovo Cimento A*

57:190
45. Odorico, R. 1971. *Phys. Lett. B* 34:65
46. Berger, E. L. 1971. *Phenomenology in Particle Physics*, ed. C. B. Chiu, G. C. Fox, A. J. G. Hey, p. 83. Pasadena: Caltech
47. Finkelstein, J. 1969. *Phys. Rev. Lett.* 22:362
48. Freund, P. G. O. 1968. *Phys. Rev. Lett.* 21:1375
49. Chen, M.-S. et al 1971. *Phys. Rev. Lett.* 26:1585
50. Ross, M., Henyey, F. S., Kane, G. L. 1970. *Nucl. Phys. B* 23:269
51. Dar, A., Kugler, M., Dothan, Y., Nussinov, S. 1964. *Phys. Rev. Lett.* 12:82
52. Roy, D. P. 1971. *High Energy Phenomenology*, ed. J. Tran Thanh Van, p. 309. Paris: Société Polygraphique Mang.
53. Tran Thanh Van, J. 1972. *Two Body Collisions*, ed. J. Tran Thanh Van, p. 1. Paris: Société Polygraphique Mang.
54. Martin, A., Stevens, P. R. 1972. *Phys. Rev. D* 5:147
55. Martin, A., Stevens, P. R. 1972. *Comparative Evaluation of Absorption Models for ρ Exchange*, Preprint
56. Ross, M. 1972. *Phys. Lett. B* 38:321
57. Johnson, R. C. 1972. *Phys. Lett. B* 38:325
58. Ringland, G. A., Roberts, R. G., Roy, D. P., Tran Thanh Van, J. 1972. *Nucl. Phys. B* 44:395
59. Hartley, B. J., Kane, G. L. 1972. *Phys. Lett. B* 39:531
60. Hartley, B. J., Kane, G. L. 1973. *Towards a General Description of Two-Body Hadron Reactions*, Preprint
61. Worden, R. 1972. *Duality and Regge Absorption Models*, Michigan preprint
62. Finkelstein, J. 1971. *Nuovo Cimento A* 5:413
63. Worden, R. 1972. *Phys. Lett. B* 40:260
64. Michael, C. 1969. *Phys. Lett. B* 29:230; *Nucl. Phys. B* 13:644
65. Quigg, C. 1971. *Nucl. Phys. B* 34:77
66. Williams, P. K. 1970. *Phys. Rev. D* 1:1312
67. Fox, G. C. 1972. *π-Exchange*. In *ANL/HEP 7208*, Vol. II
68. Gottfried, K., Jackson, J. D. 1964. *Nuovo Cimento* 34:735
69. Harari, H. 1969. *Phys. Rev. Lett.* 22:562
70. Rosner, J. L. 1969. *Phys. Rev. Lett.* 22:689
71. Chiu, C. B., Ugaz, E. 1973. *Phys. Lett. B* 43:327
72. Fox, G. C., Rosenfeld, A. H. 1971. Unpublished Particle Data Group Memo
73. Jacob, M., Wick, G. C. 1959. *Ann.*

Phys. New York 7:404
74. Trueman, T. L., Wick, G. C. 1964. *Ann. Phys. New York* 26:322
75. Wolfenstein, L. 1956. *Ann. Rev. Nucl. Sci.* 6:43
76. Cronin, J. W., Overseth, O. E. 1963. *Phys. Rev.* 129:1795
77. Gottfried, K., Jackson, J. D. 1964. *Phys. Lett.* 8:144
78. Gottfried, K., Jackson, J. D. 1964. *Nuovo Cimento* 33:309
79. Jackson, J. D. 1964. *Nuovo Cimento* 34:1644
80. Jackson, J. D. 1965. *High Energy Physics*, ed. C. DeWitt, M. Jacob, p. 325. New York: Gordon and Breach
81. Berman, S. M., Jacob, M. 1965. *Phys. Rev. B* 139:1023
82. Pilkuhn, H., Svensson, B. E. Y. 1965. *Nuovo Cimento* 38:518
83. Kotanski, A. 1966. *Acta Phys. Pol.* 29:699; 30:629
84. Barger, V., Phillips, R. J. N. 1972. *Phys. Lett. B* 42:479
85. de Lesquen, A. et al 1972. *Phys. Lett. B* 40:277
86. von Hippel, F., Quigg, C. 1972. *Phys. Rev. D* 5:624
87. Eberhard, P., Pripstein, M. 1963. *Phys. Rev. Lett.* 10:351
88. Wolters, G. F. 1968. *Kinematics and Multiparticle Systems*, ed. M. Nikolić, p. 129. New York: Gordon and Breach
89. Jongejans, B. 1969. *Methods in Subnuclear Physics*, ed. M. Nikolić, 4:349. New York: Gordon and Breach
90. Brau, J. E., Dao, F. T., Hodous, M. F., Pless, I. A., Singer, R. A. 1971. *Phys. Rev. Lett.* 27:1481
91. Ascoli, G. 1972. *Experimental Meson Spectroscopy-1972*, ed. A. H. Rosenfeld, K.-W. Lai, p. 185. New York: Am. Inst. Phys.
92. Ascoli, G. 1972. *Summary of the Experimental Situation Regarding A_1, A_2, A_3 Production.* Submitted to *XVI Int. Conf. High Energy Physics, Batavia*
93. Berger, E. L., Fox, G. C. 1970. *Phys. Rev. Lett.* 25:1783
94. Halzen, F., Michael, C. 1971. *Phys. Lett. B* 36:367 [Erratum in footnote 2 of Ref. (10)]
95. Kelly, R. L. 1972. *Phys. Lett. B* 39:635
96. Cozzika, G. et al 1972. *Phys. Lett. B* 40:281
97. Johnson, P., Lassila, K. E., Koehler, P., Miller, R., Yokosawa, A. 1973. *Phys. Rev. Lett.* 30:242
98. Barger, V., Phillips, R. J. N. 1969. *Phys. Rev.* 187:2210
99. McClure, J. A., Pitts, L. E. 1972. *Phys. Rev. D* 5:109

100. Höhler, G., Jakob, H. P. 1971. *Nuovo Cimento Lett.* 2:485
101. Hill, D. et al 1973. *Phys. Rev. Lett.* 30:239
102. Bonamy, P. et al 1973. *Nucl. Phys. B* 52:392
103. Ambats, I. et al 1972. *Phys. Rev. Lett.* 29:1415
104. Pietarinen, E. 1972. *Nucl. Phys. B* 49:315
105. Elvekjaer, F., Pietarinen, E. 1973. *Optimized FESR Applied to πN Charge Exchange Amplitudes (Preprint RPP/T/36)*
106. Lovelace, C. 1972. *XVI Int. Conf. High Energy Physics, Batavia*
107. Fox, G. C. 1971. *Proceedings of the II International Conference on Polarized Targets*, ed. G. Shapiro. Berkeley: Lawrence Berkeley Laboratory
108. Högaasen, H., Michael, C. 1972. *Nucl. Phys. B* 44:214
109. Barger, V., Halzen, F. 1972. *Phys. Rev. D* 6:1918
110. Gilman, F. J., Pumplin, J., Schwimmer, A., Stodolsky, L. 1970. *Phys. Lett. B* 31:387
111. Zarmi, Y. 1971. *Phys. Rev. D* 4:3455
112. Dronkers, J., Kroll, P. 1972. *Nucl. Phys. B* 47:291
113. Jakob, H. P., Kroll, P. 1972. *The Slope of the πN Charge Exchange Amplitude at $t = 0$ (Preprint TKP 20/72)*
114. Alcock, J. W., Cottingham, W. N. 1971. *Nucl. Phys. B* 31:443
115. Shirkov, D. V., Serebryakov, V. V., Mescheryakov, V. A. 1969. *Dispersion Theories of Strong Interactions at Low Energy*. New York: Wiley Interscience
116. Donnachie, A., Kirsopp, R., Lovelace, C. 1968. *Phys. Lett. B* 26:161
117. Ayed, R. et al 1972. *Zero Strangeness Baryon States Below 2.5 GeV Mass*, Saclay preprint
118. Almehed, S., Lovelace, C. 1972. *Nucl. Phys. B* 40:157
119. Höhler, G., Strauss, R. 1970. *Z. Phys.* 232:205
120. Barger, V., Phillips, R. J. N. 1971. *Nucl. Phys. B* 32:93
121. Jackson, J. D., Quigg, C. 1969. *Phys. Lett. B* 29:236; Ibid. 1970. *Nucl. Phys. B* 22:301
122. Cho, C. F., Sakurai, J. J. 1970. *Phys. Rev. D* 2:517
123. Worden, R. 1972. *Nucl. Phys. B* 37:253
124. Dass, G. V., Michael, C. 1968. *Phys. Rev.* 175:1774
125. Berger, E. L., Fox, G. C. 1969. *Phys. Rev.* 188:2120
126. Harari, H. 1968. *Phys. Rev. Lett.* 20:1395

127. Freund, P. G. O. 1968. *Phys. Rev. Lett.* 20:235
128. Harari, H., Zarmi, Y. 1969. *Phys. Rev.* 187:2230
129. Fukugita, M., Inami, T. 1972. *Impact Parameter Analysis of $\bar{K}N$ and πN Scattering in the Intermediate Energy Region (Preprint Rutherford Lab. RPP/T/32)*
130. Irving, A. C., Martin, A. D., Michael, C. 1971. *Nucl. Phys. B* 32:1
131. Michael, C., Odorico, R. 1971. *Phys. Lett. B* 34:422
132. Field, R. D., Jackson, J. D. 1971. *Phys. Rev. D* 4:693
133. Martin, A. D., Michael, C., Phillips, R. J. N. 1972. *Nucl. Phys. B* 43:13
134. Irving, A. C., Martin, A. D., Barger, V. 1972. *Analysis of Data for Hypercharge Exchange Reactions (Preprint CERN TH-1585)*
135. Loos, J. S., Matthews, J. A. J. 1972. *Phys. Rev. D* 6:2463
136. Bashian, A. et al 1972. *Phys. Rev. D* 4:2667
137. Mott, J. 1969. *Nucl. Phys. B* 13:565
138. Davier, M., Harari, H. 1971. *Phys. Lett. B* 35:239
139. Barger, V., Geer, K., Halzen, F. 1972. *Nucl. Phys. B* 44:214
140. Dikmen, F. N. 1969. *Phys. Rev. Lett.* 22:622
141. Barger, V., Halzen, F. 1972. *Nucl. Phys. B* 43:62
142. Davier, M. 1972. *Phys. Lett. B* 40:369
143. Austin, D. M., Greiman, W. H., Rarita, W. 1970. *Phys. Rev. D* 2:2613; Ibid. 1971. *Phys. Rev. D* 4:3507
144. Drobnis, D. D. et al 1968. *Phys. Rev. Lett.* 20:274
145. Bonamy, P. et al 1970. *Nucl. Phys. B* 16:335
146. Morrison, D. R. O. 1968. *Phys. Rev.* 165:1699
147. Johnson, K., Treiman, S. B. 1965. *Phys. Rev. Lett.* 14:189
148. Lipkin, H. J., Scheck, F. 1966. *Phys. Rev. Lett.* 16:71
149. Levinson, C. A., Wall, N. S., Lipkin, H. J. 1966. *Phys. Rev. Lett.* 17:1122
150. Wolf, G. 1971. *Proceedings of the 1971 International Symposium on Electron and Photon Interactions at High Energies,* ed. N. Mistry. Ithaca: Cornell
151. Denisov, S. P. et al 1972. *Differences of Total Cross Sections for Momenta up to 65 GeV/c. Submitted to XVI Int. Conf. High Energy Physics, Batavia*
152. Gjesdal, S. et al 1972. *Phys. Lett. B* 40:152
153. Badier, J. et al 1972. *Phys. Lett. B* 41:387
154. Kadyk, J. A., Hauptman, J. M., Trilling, G. H. 1972. *Λp Interactions in the Momentum Range 0.5 → 8 GeV/c (Preprint LBL-1064)*
155. Lipkin, H. J. 1971. *Quark Model Predictions for Reactions with Hyperon Beams (Preprint NAL-THY-81)*
156. Denisov, S. P. et al 1971. *Phys. Lett. B* 36:415
156a. Amaldi, U. et al 1973. *Phys. Lett. B* 44:112; Amendolia, S. R. et al 1973. *Phys. Lett. B* 44:119
157. Soloviev, L. D., Shelkacher, A. V. 1972. *XVI Int. Conf. High Energy Physics, Batavia*
158. Bolotov, V. N. et al 1972. *A Study of $\pi^- p \to \pi^0 n$ Charge Exchange in the Momentum Range 20 to 50 GeV/c.* Preprint submitted to *XVI Int. Conf. High Energy Physics, Batavia*
159. Foley, K. J. et al 1969. *Phys. Rev.* 181:1775
160. Beznogikh, G. G. et al 1972. *Phys. Lett. B* 39:411
161. Alvensleben, H. et al 1973. *Phys. Rev. Lett.* 30:328
162. Barger, V., Halzen, F., Lutke, J. 1972. *Phys. Lett. B* 42:428
163. Chou, T. T., Yang, C. N. 1968. *Phys. Rev. Lett.* 20:1213
164. Durand, L., Lipes, R. 1968. *Phys. Rev. Lett.* 20:637
165. White, J. N. J. 1973. *Nucl. Phys. B* 51:23
166. Carrigan, R. A. 1970. *Phys. Rev. Lett.* 24:168
167. Dzierba, A. R. et al 1973. *Phys. Rev. D* 7:725
168. Antipov, Yu. M. et al 1972. *Elastic Scattering of $\pi^- p$, $K^- p$ and $\bar{p}p$ at 25 and 40 GeV/c.* Submitted to *XVI Int. Conf. High Energy Physics, Batavia*
169. Höhler, G., Staudenmaier, H. M. 1972. *Is There a Break in the $\pi^- p$ Diffraction Peak? (Preprint TK19/72)*
170. Anselm, A. A., Gribov, V. N. 1972. *Phys. Lett. B* 40:487
171. Barshay, S., Chao, Y.-A. 1972. *Structure in Diffraction Scattering Inside $|t| \sim 4m_\pi^2$.* Preprint submitted to *XVI Int. Conf. High Energy Physics, Batavia*
172. Gunion, J. F., Brodsky, S. J., Blankenbecler, R. 1972. *Phys. Lett. B* 39:649
173. Theis, W. R. 1972. *Phys. Lett. B* 42:246
174. Horn, D., Moshe, M. 1972. *Nucl. Phys. B* 48:557
175. Phillips, R. J. N., Ringland, G. A. 1971. *Nucl Phys. B* 32:131
176. Borghini, M. et al 1971. *Phys. Lett. B* 36:501

177. Odorico, R. 1973. *Nucl. Phys. B* 52:248
178. Bloom, E. D., Gilman, F. J. 1971. *Phys. Rev. D* 4:2901
179. Harari, H. 1972. *Proceedings of the 1971 International Symposium on Electron and Photon Interactions at High Energies,* ed. N. B. Mistry. Ithaca: Cornell
180. Morrison, D. R. O. 1970. *Proc. XV Int. Conf. High Energy Physics, Kiev*
181. Zachiariasen, F. 1971. *Phys. Rep. C* 2:1
182. Fox, G. C. 1972. *Experimental Meson Spectroscopy-1972,* ed. A. H. Rosenfeld, K.-W. Lai. New York: Am. Inst. Phys.
183. Berger, E. L. 1969. Review talk at Irvine (*Preprint ANL/HEP 6927*)
184. Oh, Y. T. et al 1972. *Phys. Lett. B* 42:497
185. Lissauer, D. et al 1972. *Phys. Rev. D* 6:1852
186. Wang, J. M., Wang, L.-L. 1971. *Phys. Rev. Lett.* 26:1287
187. Einhorn, M. B. et al 1973. *Phys. Rev. D* 7:102
188. Ljung, S. 1972. *Diffraction Dissociation in the Reaction* $pp \rightarrow pp\pi^+\pi^-\pi^0$ *at 19 GeV/c.* Preprint submitted to *XVI Int. Conf. High Energy Physics, Batavia*
189. Allaby, J. V. et al 1973. *Nucl. Phys. B* 52:316
190. Ravndal, F. 1971. *Phys. Lett. B* 37:300
191. Brandenburg, G. et al 1972. *Nucl. Phys. B* 45:397
192. Anderson, R. L. et al 1973. *Phys. Rev. Lett.* 30:149
193. Bingham, H. H. et al 1972. *Phys. Lett. B* 41:635
194. Beusch, W. 1971. *Production of Particles and Resonances on Nuclei, Dubna Conf.*
195. Lubatti, H. J. 1971. *Hadron Nucleus Interactions—An Experimental Study.* Preprint
196. Binon, F. 1971. *Pion-Nucleus Scattering: Experiment.* Rapporteur's Talk at *Int. Seminar π-Meson Nucleus Interactions, Strasbourg*
197. Bingham, H. H. 1970. *Review of Coherent Multiparticle Production Reactions from Nuclei* (*Preprint CERN/PHYS/70-60*)
198. Quigg, C., Wang, L.-L. 1973. *Phys. Lett. B* 43:314
199. Sidhu, D. P., Quigg, C. 1973. *Phys. Rev. D* 7:755
200. Firestone, A., Goldhaber, G., Lissauer, D., Trilling, G. H. 1972. *Phys. Rev. D* 5:505
201. Gottfried, K. 1972. *Proceedings of the 1971 International Symposium on Electron and Photon Interactions at High Energies,* ed. N. B. Mistry. Ithaca: Cornell
202. Heynen, V. et al 1972. *Total Photoproduction Cross Section for Hadrons on Carbon and Copper at* 6 *GeV/c.* Preprint submitted to *XVI Int. Conf. on High Energy Physics, Batavia*
203. Vander Velde, J. C. et al 1972. *Nucl. Phys. B* 45:1
204. Goldhaber, A. S. 1973. *Phys. Rev. D* 7:765
205. Neuhofer, G. et al 1972. *Phys. Lett. B* 39:271
206. Brockett, W. S. et al 1971. *Phys. Rev. Lett.* 26:527
207. Cronillon, P. et al 1973. *Phys. Rev. Lett.* 30:403
208. Chiu, C. B. 1971. *Nucl. Phys. B* 30:477
209. Lai, K. W., Louie, J. 1970. *Nucl. Phys. B* 19:205
210. Brody, A. D. et al 1970. *Preprint SLAC-PUB-823.* Unpublished
211. Carney, J. N. et al 1972. *Phys. Lett. B* 42:124
212. Brody, A. D. et al 1971. *Phys. Rev. Lett.* 26:1050
213. Fox, G. C., Hey, A. J. G. 1972. *Nondiffraction Production of Meson Resonances* (*Preprint CALT-68-373*)
214. Butler, W. R. 1970. *UCRL-19845.* Ph.D. thesis. Univ. Calif., Berkeley. Unpublished; Butler, W. R. et al 1969. Contribution to the *Lund Int. Conf. on Elementary Particles* (*UCRL-19225*); Abrams, G. S. et al 1970. Contribution to *XV Int. Conf. High Energy Physics, Kiev*
215. Bloodworth, I. J., Jackson, W. C., Prentice, J. D., Yoon, T. S. 1972. *Nucl. Phys. B* 39:525
216. Martin, A. D., Michael, C. 1971. *Phys. Lett. B* 37:513
217. Trilling, G. H. 1972. *Nucl. Phys. B* 40:13
218. Aguilar-Benitez, M., Chung, S. U., Eisner, R. L., Field, R. D. 1972. *Phys. Rev. Lett.* 29:749, 1201
219. Leader, E. 1968. *Phys. Rev.* 166:1599
220. Sherden, D. J. et al 1973. *Asymmetries in Charged Pion Photoproduction on Nucleons by 16 GeV Polarized Photons* (*Preprint SLAC-PUB-1206*)
221. Estabrooks, P., Martin, A. D. 1972. *Phys. Lett. B* 42:229
222. Firestone, A., Colton, E. 1972. *Determination of the Effective Trajectory in* Δ(*1238*)-*Producing Reactions* (*Preprint CALT-68-353*)
223. Schwitters, R. F. et al 1971. *Phys. Rev. Lett.* 27:120
224. Bar-Yam, Z. et al 1970. *Phys. Rev. Lett.* 25:1053
225. Fox, G. C. et al 1971. *Phys. Rev. D* 4:2647

226. Poster, R. et al 1971. *Comparison of $K^+n \to K^{*0}p$ and $K^-p \to K^{*0}n$ at 2 GeV/c.* Submitted to *1971 Amsterdam Conf,* Preprint

227. Alexander, G., Fridman, A., Gotsman, E. 1972. *Nucl. Phys. B* 40:1

228. Hoyer, P., Roberts, R. G., Roy, D. P. 1973. *New Relations for Two-Body Reactions from Inclusive Finite Mass Sum Rules (Preprint Rutherford Lab. RPP/T/35)*

229. Dar, A. 1971. *High Energy Phenomenology,* ed. J. Tran Thanh Van, p. 17. Paris: Société Polygraphique Mang.

230. Stodolsky, L. 1971. *Phys. Rev. Lett.* 26:404

231. Abrams, G. S., Barnham, K. W. J. 1972. *Quark Model Comparisons with $\rho^0\Delta^{++}$ and $\omega^0\Delta^{++}$ Data at 3.7 GeV/c (Preprint LBL-961)*; Barnham, K. W. J. 1972. *Two Body Collisions,* ed. J. Tran Thanh Van, p. 459. Paris: Société Polygraphique Mang.

232. Barnham, K. W. J. et al 1972. *$\rho^0\Delta^{++}$ and $\omega\Delta^{++}$ Joint Decay Correlations at 3.7 GeV/c (Preprint LBL-960)*

233. Kotanski, A., Zalewski, K. 1969. *Nucl. Phys. B* 13:119

234. Carnegie, R. K. et al 1972. *Study of Reactions $\pi^- p \to \pi^+\pi^-\Delta^0$ and $\pi^+ p \to \pi^+\pi^-\Delta^{++}$ at 15 GeV/c.* Preprint submitted to *XVI Int. Conf. High Energy Physics, Batavia*

235. Williamson, Y. et al 1972. *Dip Structures in $\pi^+ p \to \rho^+\rho$ at 1.55–1.84 GeV/c.* Preprint submitted to *XVI Int. Conf. High Energy Physics, Batavia*

236. Michael, W., Gidal, G. 1972. *Phys. Rev. Lett.* 28:1475

237. Crennell, D. J. et al 1971. *Phys. Rev. Lett.* 27:1674; Scarr, J. M., Lai, K.-W. 1972. *Phys. Rev. Lett.* 29:310

238. Pratt, J. C. et al 1972. *Phys. Lett. B* 41:383; Ibid. *Isolation of Isoscalar Exchange in $\pi^\pm p \to \rho^\pm p$ at 15 GeV/c,* SLAC preprint

239. A-B-B-C-C-H-W collaboration. 1972. *Nucl. Phys. B* 46:46

240. Mandula, J., Weyers, J., Zweig, G. 1970. *Ann. Rev. Nucl. Sci.* 20:289

241. Holloway, L. E. et al 1971. *Phys. Rev. Lett.* 27:1671

242. Aguilar-Benitez, M., Eisner, R. L., Kinson, J. B., Samios, N. P., Scarr, J. M. 1971. *Observation of Structure in Production and Decay of $K^{*-}(890)$ (Preprint BNL-15707).* Unpublished

243. Boyarski, A. M. et al 1970. *Phys. Rev. Lett.* 25:695

244. Robrish, P. R. et al 1970. *Phys. Lett. B* 31:617

245. Abolins, M. A. et al 1972. Quoted by A. Diddens in *Proc. 4th Int. Conf. High Energy Collisions,* ed. J. R. Smith. *Rutherford Lab. Report RHEL 72-001*

246. Yokosawa, A. 1971. *Phenomenology in Particle Physics,* ed. C. B. Chiu, G. C. Fox, A. J. G. Hey. Pasadena: Caltech

247. Donnachie, A., Gabathuler, E., Eds. 1970. *Vector Meson Production and $\omega-\rho$ Interference, Daresbury Study Weekend No. 1*

248. Goldhaber, G. 1970. *Experimental Meson Spectroscopy,* ed. C. Baltay, A. H. Rosenfeld. New York: Columbia

249. Goldhaber, G. et al 1969. *Phys. Rev. Lett.* 23:1351

250. Le Francois, J. 1972. *Proceedings of the 1971 International Symposium on Electron and Photon Interactions at High Energies,* ed. N. B. Mistry. Ithaca: Cornell

251. Goldhaber, A. S., Fox, G. C., Quigg, C. 1969. *Phys. Lett. B* 30:249

252. Abrams, G. S., Maor, U. 1970. *Phys. Rev. Lett.* 25:617

253. Hagopian, S. et al 1970. *Phys. Rev. Lett.* 25:1050

254. Ratcliff, B. N. et al 1972. *Phys. Lett. B* 38:345

255. Bloodworth, I. J. et al 1971. *Nucl. Phys. B* 35:133

256. Wiik, B. 1971. *Proceedings of the 1971 International Symposium on Electron and Photon Interactions at High Energies,* ed. N. B. Mistry. Ithaca: Cornell

257. De Wire, J. et al 1971. *Photoproduction of η Mesons from Hydrogen (Preprint CLNS-170)*

258. Rosner, J. L. 1971. *Phenomenology in Particle Physics,* ed. C. B. Chiu, G. C. Fox, A. J. G. Hey. Pasadena: Caltech

259. Michael, C., Ruuskanen, P. V. 1971. *Phys. Lett. B* 35:65

260. Biswas, N. N. et al 1972. *Phys. Rev. D* 5:1564

261. Kislinger, M. 1971. *Vector, Tensor and Pomeron Regge Couplings (Preprint Caltech CALT-68-341).* Unpublished

262. Hoogland, W. et al 1969. *Nucl. Phys. B* 11:309

263. Firestone, A. et al 1973. *Nucl Phys. B* 52:403

264. Harrison, D. et al 1972. *Phys. Rev. D* 5:2730

265. Abramovich, M. et al 1972. *Nucl. Phys. B* 39:189

266. Crennell, D. J., Gordon, H. A., Lai, K.-W., Scarr, J. M. 1972. *Phys. Rev. D* 6:1220

267. Estabrooks, P., Martin, A. D. 1972. *Phys. Lett. B* 41:350

268. Sonderegger, P., Bonamy, P. 1969. Sub-

mitted to *1969 Lund Conf.*, Preprint. Results quoted in (24)

269. Field, R. D. et al 1972. *Phys. Rev. D* 6: 1863
270. Abramovich, M., Irving, A. C., Martin, A. D., Michael, C. 1972. *Phys. Lett. B* 39: 353
271. Field, R. D., Eisner, R. L., Chung, S. U., Aguilar-Benitez, M. 1972. *Transversity Amplitude Analysis of the Reactions* $K^- p \to (\omega, \phi)\Lambda$. Brookhaven preprint
272. Aguilar-Benitez, M., Chung, S. U., Eisner, R. L., Samios, N. P. 1972. *Phys. Rev. D* 6: 29
273. Hirsch, E. et al 1971. *Phys. Lett. B* 36: 139
274. Hirsch, E., Karshon, U., Lipkin, H. J. 1971. *Phys. Lett. B* 36: 385
275. Michael, C., Odorico, R. 1971. *Phys. Lett. B* 34: 422
276. Boyarski, A. M. et al 1969. *Phys. Rev. Lett.* 22: 1131
277. Boyarski, A. M. et al 1971. *Phys. Lett. B* 34: 547
278. Akerlof, C. W. 1970. *Double Charge Exchange Reactions,* Michigan preprint
279. Atherton, H. W. et al 1972. *Phys. Lett. B* 42: 522
280. Chabaud, V. et al 1972. *Phys. Lett. B* 38: 441, 445, 449
281. Langbein, W., Wagner, F. 1972. *Nucl. Phys. B* 47: 477
282. Carroll, A. S. et al 1969. *Phys. Rev. Lett.* 23: 887
283. Rivers, R. J. 1968. *Nuovo Cimento A* 57: 174
284. Dean, N. 1968. *Nucl. Phys. B* 7: 311
285. Chiu, C. B., Finkelstein, J. 1969. *Nuovo Cimento A* 59: 92
286. Quigg, C. 1970. PhD thesis. Univ. Calif., Berkeley; *Preprint UCRL-20032*
287. Quigg, C. 1971. *Nucl. Phys. B* 29: 67
288. Harari, H. 1971. *Phys. Rev. Lett.* 26: 1079
289. Henyey, F. S., Kane, G. L., Scanio, J. J. G. 1971. *Phys. Rev. Lett.* 27: 350
290. Frautschi, S. 1972. *Ericson Fluctuations and the Bohr Model in Hadron Physics (Preprint CERN-TH-1463)*
291. Berger, E. L. 1969. *Phys. Rev. Lett.* 23: 1139
292. Dauber, P. M. et al 1969. *Phys. Lett. B* 29: 609
293. Berger, E. L., Morrow, R. A. 1970. *Phys. Rev. Lett.* 25: 1136
294. Sivers, D., von Hippel, F. 1972. *Phys. Rev. D* 6: 874
295. Halzen, F., Mandula, J., Weyers, J., Zweig, G. 1971. *Tests of the Absence of Exotic Exchanges (CERN-TH-1371)*
296. Gribov, V. N., Low, F. E. 1972. *Proc.*

XVI Int. Conf. High Energy Physics, Batavia
297. Bracci, E., Droulez, J. P., Flaminio, E., Hansen, J. D., Morrison, D. R. O. 1972. *CERN/HERA 72-1*
298. Carroll, A. S. et al 1966. *Phys. Rev. Lett.* 16: 289
299. Yvert, M. 1969. PhD thesis, Orsay. Unpublished
300. Schneider, J. et al 1969. *Phys. Rev. Lett.* 23: 1068
301. Litchfield, P. J. et al 1971. *Nucl. Phys. B* 30: 125
302. Lovelace, C., Wagner, F. 1971. *Nucl. Phys. B* 28: 141
303. Cutkosky, R. E. et al 1973. *$K^+ p$ Phaseshift Analysis Using the Accelerated Convergence Expansion,* Preprint
304. Davies, A. 1971. *Nucl. Phys. B* 21: 359
305. Flaminio, E., Hansen, J. D., Morrison, D. R. O., Tovey, N. 1970. *CERN/HERA 70-7*
306. Birnbaum, D. et al 1970. *Phys. Lett. B* 31: 484
307. Bradamante, F. et al 1970. *Phys. Lett. B* 31: 87
308. Abrams, R. J. et al 1970. *Phys. Rev. D* 1: 1917
309. Galbraith, W. et al 1965. *Phys. Rev. B* 138: 913
310. Citron, A. et al 1966. *Phys. Rev. B* 144: 1101
311. Foley, K. J. et al 1967. *Phys. Rev. Lett.* 19: 330
312. Bugg, D. V. et al 1966. *Phys. Rev.* 146: 980
313. Foley, K. J. et al 1967. *Phys. Rev. Lett.* 19: 857
314. Jones, L. W. et al 1971. *Phys. Lett.* 36: 509
315. Matthews, J. A. J. et al 1971. *Nucl. Phys. B* 32: 366
316. Grayer, G. 1972. *Nucl. Phys. B* 50: 29
317. Bloodworth, I. J., Jackson, W. C., Prentice, J. D., Yoon, T. S. 1971. *Nucl. Phys. B* 35: 79
318. Geweniger, C. et al 1969. *Phys. Lett. B* 29: 41
319. Anderson, R. L. et al 1971. *Phys. Rev. D* 4: 1937
320. Grayer, G. et al 1972. *Measurement of* $pp \to (p\pi^+)n$ *at 12.5 and 16.9 GeV/c.* Preprint submitted to *XVI Int. Conf. High Energy Physics, Batavia*
321. Boyarski, A. M. et al 1968. *Phys. Rev. Lett.* 21: 1767
322. Astbury, P. et al 1966. *Phys. Lett. B* 23: 160
323. Miller, E. L. et al 1971. *Phys. Rev. Lett.* 26: 984
324. Braun, H. et al 1970. *Phys. Rev. D* 2: 488

325. Alexander, G. et al 1967. *Phys. Rev. B* 154:1284
326. Bacon, T. C. et al 1966. *Phys. Rev.* 162:1320
327. Werner, B. et al 1970. *Nucl. Phys. B* 23:37
328. De Baere, W. et al 1969. *Nuovo Cimento* A 61:397
329. Atherton, H. W. et al 1972. *Preliminary Results on $\bar{p}p \rightarrow (\overline{\Delta^{++}})\Delta^{++}$ at 5.7 GeV/c.* Preprint submitted to *XVI Int. Conf. High Energy Physics, Batavia*
330. Shapira, A. et al 1970. *Nucl. Phys. B* 23:583

THE RADIATION ENVIRONMENT OF ⋇ 5540
HIGH-ENERGY ACCELERATORS

A. Rindi and R. H. Thomas
Lawrence Berkeley Laboratory, University of California, Berkeley, California

Dedicated to the memory of Burton J. Moyer

CONTENTS

1   INTRODUCTION

High-energy[1] particle accelerators have been used primarily as physics research instruments. Due, perhaps, to the haste at the beginning for exploiting them, very

---

[1] Any definition of the term "high energy" is necessarily arbitrary. For this review, we may use the term "high-energy accelerator" to mean those accelerators capable of generating π-mesons. This will include most synchrotrons and synchrocyclotrons.

315

little thought was given to considering safety measures, and as a result, the radiation environments of these accelerators were initially unknown. The ignorance of accelerator radiation protection in the late forties and early fifties, as described by Rotblat (1), led to underground construction of many of the early synchro-cyclotrons (2), avoiding, but inhibiting any fundamental understanding of, radiation problems (3, 4). This solution could not be continued indefinitely. As accelerators grew in physical size, energy, and intensity, and were applied more widely to the problems of medicine, industry, and research, the concern for personnel safety and the pressure of economic considerations made it necessary that accelerator radiation studies be placed on a systematic basis.

Concern for personnel safety was given impetus toward the end of 1948 when it became known that several nuclear physicists in France and the United States, who had been exposed to radiation produced by a cyclotron, had manifested incipient cataract (5). Review of these cases drew attention to the poor status of radiation dosimetry, particularly that of neutrons, at high-energy accelerators. At some cyclotron laboratories there immediately followed an extensive effort to improve dosimetric techniques (6, 7): effort further stimulated by the need to reduce radiation levels around accelerators because the successful performance of experimental research often demands radiation intensities one or two orders of magnitude below the maximum level required for personnel safety. The orderly investigation of accelerator-produced radiation required more emphasis on general solutions than was usual in health physics. Over the past decade our knowledge of high-energy radiation environments has made significant advances (8–16) with the operation of several accelerators in the GeV energy region and the design, construction, and operation of several accelerators in the energy region of 10 to 400 GeV.

At high-energy accelerators a large variety of particles may be produced, extending over a wide range of energies, and their measurement presents many novel problems. It was, therefore, necessary to investigate in some detail the production and transmission through shielding of accelerator-produced radiation. Radiation detectors initially designed for nuclear physics research are the natural choice for such investigations. With the understanding provided by such detectors, *all* the requirements of a radiation protection program may be undertaken (7): possible radiation hazards may be anticipated and their magnitude estimated; protective shielding may be designed and operational procedures selected which permit efficient operation under safe conditions; the response of any radiation detector may be correctly interpreted, and, finally, radiation survey instruments with response approximately proportional to dose equivalent may be designed for use in a limited range of environments.

The lessons learned from the development of techniques of measurement in mixed radiation fields and their interpretation are of general interest because they bear directly on the problem of developing a general, self-consistent scheme of dosimetry in radiation protection (17–20). Furthermore, experience has shown that the radiation environments of high-energy accelerators are in many respects similar to those produced by lower-energy accelerators or even nuclear reactors: neutrons and photons are the dominant components. So, the techniques of measurement developed for their radiation fields may be applied equally well to both high- and low-energy accelerators.

Experience with accelerator-produced radiation has, until recently, necessarily been limited to a relatively small number of research institutions. However, exposure to accelerator-like radiation environments is no longer only of academic interest. The uses of ionizing radiation in research, industry, and medicine have increased dramatically over the past decade. This has been made possible by impressive developments in accelerator design.

A large variety of particle accelerators capable of accelerating a wide range of particles to high energy with high beam intensities is now commercially available. We anticipate a rapid increase in their industrial application to a host of diverse tasks, which has only begun (21, 22). Burill (23) has documented the increasing uses of accelerators in industry and medicine and shows the number of accelerators in use to be increasing at a rate of roughly 10% per year. Many of the electron accelerators presently being installed are of sufficiently high energy to produce neutrons. Techniques of radiotherapy using fast neutrons, π-mesons, or energetic heavy ions are being extensively investigated (24–26). If widely adopted, considerable numbers of hospital personnel may be occupationally exposed to mixed radiation fields.

There are, in addition, applications of technology that result in exposures to accelerator-like radiation environments. For example, the use of aircraft for mass transportation will expose large numbers of the general population to a radiation environment similar to that produced by high-energy accelerators (27).

There has recently been much speculation and some controversy concerning the possible biological impact on man of the increasing uses of radiation (28). The effects of low levels on man are not yet fully understood, but what does seem probable is that radiation effects due to densely ionizing radiations (high LET)[2] will be greater than those due to lightly ionizing radiations (low LET) (29). As we shall see, the radiation environments of many accelerators are particularly rich in high-LET radiation; so, as the uses of accelerators increase, more people may be exposed to high-LET radiations.

In a recent report published by the International Commission on Radiological Protection (ICRP), it is suggested that, in the foreseeable future, high-LET radiations will contribute only a small fraction of the total general exposure (30). Nevertheless, from our present understanding of radiation effects, it is entirely possible that high-LET radiations may have a greater biological impact than this relatively small contribution to exposure might suggest. The possible biological consequences of exposure to high-LET radiation certainly merit continuing study.

In a short review such as this it is not feasible to attempt an encyclopedic coverage of the entire field of high-energy radiations. Rather, we have attempted to set the development of the subject over the past decade in some perspective by emphasizing those experimental and theoretical advances we believe to be of the greatest importance.

[2] LET is an abbreviation for Linear Energy Transfer.

2   THE RADIATION ENVIRONMENT OF ACCELERATORS

Despite the large variety of high-energy particle accelerators, both with respect to beam characteristics and utilization, their external radiation environments are often quite similar, and are dominated by photons and neutrons.

In many branches of health physics it has been customary to quantify radiation fields solely in terms of gross properties such as exposure, absorbed dose, and dose equivalent (see section 3 on radiation protection). This procedure is inadequate at accelerators. In order properly to perform the tasks required of a health physicist at an accelerator (such as personal dosimetry, the design and construction of radiation-measuring instruments, general radiation and particle beam dosimetry, shielding design or determination of induced activity), it is vital that the detailed composition of the radiation environment be understood in terms of the constituent particles. The study of these environments in terms of the energy spectra of their separate components is still in its early stages; techniques of measurement and data analysis are still being developed, and more extensive measurements are required. Consequently, the limited information that has been published only describes neutron spectra. But, when supplemented by information from cosmic-ray experiments and neutron transport theory, some general conclusions can be made concerning radiation fields produced by proton accelerators.

Shielding studies have shown that the radiation field reaches an equilibrium condition within a few mean-free paths inside an accelerator shield (see section 4 on shielding). The shape of the neutron spectrum observed at a shield/air interface is very close to that which exists within the shield, but may be perturbed at the low-energy end, due to the scattering and leakage through holes in the shielding.

A   High-Energy Proton Accelerators

In the latter part of the fifties, experience at the 184-inch Synchrocyclotron and Bevatron (Lawrence Berkeley Laboratory) and at the Cosmotron (Brookhaven) allowed estimation of the qualitative nature of their radiation environments outside thick shielding (31–34). Although detailed spectra were not obtained, a general rule emerged for proton accelerators showing that neutrons between 0.1 and 10 MeV contributed more than 50% to the total dose equivalent in the radiation field; γ rays and low-energy neutrons contributed about 10–20%, with the balance made up by neutrons greater than 10 MeV in energy. Patterson et al (35) explained this observation by pointing out that the equilibrium neutron spectrum in the lower atmosphere that is produced by the interaction of the primary galactic cosmic radiation (mainly protons) must be very similar to that generated in the shield of a high-energy proton accelerator. Somewhat later Tardy-Joubert (36, 37) noted that at neutron energies above 50 MeV the neutron spectrum outside the 3-GeV proton synchrotron "Saturne" shield, deduced from an analysis of the prong-number distribution of stars produced in nuclear emulsion, was consistent with the cosmic-ray neutron spectrum measured by Hess et al (38). The relative unimportance of protons is also explained by analogy with the cosmic-ray radiation (36, 39). At energies greater than a few hundred MeV, protons are present in numbers comparable with neutrons; at lower energies, however, protons are depleted by ionization losses.

By 1965 there was sufficient experience at many high-energy proton accelerators around the world to confirm that the neutron spectra outside accelerator shields and the cosmic-ray spectrum were, in general, quite similar (32–34, 36, 40–42); any attempts to measure proton spectra have not been reported.

There were, however, apparent discrepancies in some data. Table 1 (43–45) gives

Table 1   Composition of radiation fields above thick shields at the CPS

	Percentage of dose equivalent	
Radiation component	Above concrete shield bridge[a]	Above target through earth shield[b]
Thermal neutrons	11–12	<1–3
Fast neutrons (0.1 MeV $< E <$ 20 MeV)	50–70	10–37
High-energy particles ($E >$ 20 MeV)	2–25	52–89
γ rays and ionization from charged particles	2–19	1–13

[a] Reference 43.
[b] References 44, 45.

a typical example. The relative composition of dose equivalent measured through thick shielding above an accelerator target is given for a concrete shield and for an earth shield at the CERN proton synchrotron (CPS). The data measured above the concrete shield are very similar to those reported at other accelerators, such as the British 7-GeV proton synchrotron (41), and suggested a neutron spectrum similar to that produced by cosmic rays (35), while the data measured above an earth shield indicated a relatively large contribution to the dose equivalent by high-energy neutrons. Relative data, as in Table 1, are not adequate to determine whether the high fraction of dose equivalent contributed by high-energy particles was due to a deficit of low-energy neutrons or to a surfeit of high-energy neutrons. For this, more specific information on the neutron spectrum would be necessary. In the past ten years more specific information on the neutron spectra found around accelerators has been obtained by the use of nuclear emulsions and activation detectors (46).

Very little data currently exist on the radiation environments of heavy-ion accelerators but it seems probable that there will be little qualitative difference from the features exhibited by proton accelerators.

1   NEUTRON SPECTROMETRY TECHNIQUES

a   Nuclear emulsions   Lehman & Fekula (47) have summarized the neutron spectra determined at the Bevatron from the measurement of recoil protons in thick nuclear emulsions by saying that the general form of the stray neutron spectra (measured between 0.7 and 20 MeV) at eight locations near the Bevatron is a broad peak in the 0.5- to 2-MeV region, followed by a smooth 100-fold drop in value between the peak and 12 MeV. Unfortunately, proton recoil measurements in emulsions are unreliable above about 20 MeV, because track loss corrections become difficult. Nuclear emulsions may be used at higher energies, however, to give some indication of the slope of a smooth neutron spectrum assumed to be of the form E^{-n}, if the average number of grey prongs per star is determined. This was first done for cosmic rays (48), but the technique has been refined and used outside shielding of accelerators at Berkeley (49–51). Values of spectral slope, n, ranging between 1.5 and 2.0 were obtained at the Bevatron and the CPS and are consistent with threshold detector data.

b *Threshold detectors* The use of threshold detectors is a well-understood and universally accepted technique in neutron detection. This technique has found widespread application at most high-energy particle accelerators and has been extensively described in the literature (36, 41, 52–55). Detectors of high sensitivity over the entire energy range normally of interest at accelerators (0.1–100 MeV) are available. For radiation protection purposes, detailed knowledge of the neutron spectrum in the energy range from thermal up to about 10 keV is rarely required because, as we have seen, intermediate-energy neutrons are usually of little importance. The dose equivalent per unit fluence is independent of neutron energy below 10 keV, and usually a simple measurement of flux density is sufficient. However, detailed spectral information in this energy range may be obtained, if required, by the use of several Bonner spheres of different sizes (56, 57).

c *Spectrum determination* Measurements with several threshold detectors, whose excitation functions are known, provide information on the energy distribution of the neutron flux density. Specifically, a solution for the neutron spectrum $\phi(E)$ is sought from a set of activation equations of the form:

$$A_j = C_j \int_{E_{\min}}^{E_{\max}} \sigma_j(E)\phi(E)\,\mathrm{d}E \quad \text{for} \quad j = 1, 2, \ldots m \qquad 1.$$

where A_j is the saturation activity of the jth detector, $\sigma_j(E)$ is the cross section for the appropriate reaction at energy E, C_j is a normalizing constant between activity and flux density, and $E_{\min}$, $E_{\max}$ are the minimum and maximum neutron energies in the spectrum.

Equation 1 is a degenerate case of a Fredholm integral equation of the first kind. Formal methods of solution are not applicable when, as is the case with activation detectors, the A_j's *and* σ_j's are known only as a set of discrete points (58). Early attempts to obtain neutron spectra from activation detector data were frustrated by difficulties such as nonuniqueness or an oscillatory (and even negative) character to the solutions to the Fredholm equations. Some of these problems arise from the mathematical characteristics of the equations to be solved, while others are related to the specific method of solution adopted. Routti (58) has critically reviewed the numerical techniques commonly used for solution of such first-order Fredholm equations, and the interested reader is referred to his paper for a detailed account.

Routti suggests that a suitable method of solution must be able to combine the information contained in the measured data with existing information of the neutron spectrum. Thus, for example, the solution must be non-negative and zero beyond a given maximum energy. In addition, the spectrum of radiation penetrating thick shields constructed of a complex material such as concrete is assumed to be smooth. Some information on intensity or shape may be available from previous measurements. However, care must be taken to ensure that the consequent additional constraints imposed on the spectrum do not prevent it from matching the measured responses or from assuming any physically acceptable shape.

It is important that any method of solution be tested to ensure that it meets all these requirements. This is most conveniently done by computing how the system responds to test spectra. The resolution of the system and the influence of experimental errors or uncertainties in the detector response functions may then be systematically studied.

Routti has applied a generalized least-squares method to solve the activation equations in a computer program LOUHI (59). In his technique, the solution is forced to be non-negative, and prior information on the spectrum can be incorporated. Considerable experience has now been obtained and LOUHI has proved to be extremely reliable and capable of calculating neutron spectra with accuracy adequate for radiation protection purposes.

2   TYPICAL NEUTRON SPECTRA MEASURED AT PROTON ACCELERATORS   The application of threshold detectors to accelerator radiation environments at several laboratories simultaneously, rapidly expanded our understanding.

Figure 1 shows several typical unnormalized neutron spectra obtained outside thick shields at proton synchrotrons, where $E\phi(E)$ is plotted as a function of neutron energy [$\phi(E)$ is the differential energy spectrum]. In such a plot, a $1/E$ spectrum becomes a horizontal line (Figure 1a). This representation of the Hess cosmic-ray spectrum (Figure 1b) clearly shows the large excess of neutrons in the MeV region (due to evaporation processes) in comparison with a $1/E$ spectrum. At lower energies the spectrum is $1/E$ in character, but there is a noticeable dearth of thermal neutrons.

The neutron spectrum obtained above the concrete shielding around targets at the CPS is shown in Figure 1c. (Compare with Table 1.) The spectrum is seen to be $1/E$ in character from about 1 MeV down to thermal energies. This would be expected from neutron slowing-down theory in a hydrogenous medium, such as concrete. At about 1 MeV the evaporation peak, also evident in the Hess spectrum, is clearly seen, and the spectrum shows a rapid decline at energies above 50 MeV.

Figure 1d shows the neutron spectrum measured above the earth shield of the CPS. (Compare with Table 1.) This spectrum is depleted of neutrons below about 1 MeV, but in other respects is similar to the spectrum shown in Figure 1c. The water content of the earth shield through which the neutrons penetrated was very high (approximately 15% by weight) compared to concrete (few percent by weight), and so this paucity of low-energy neutrons is to be expected.

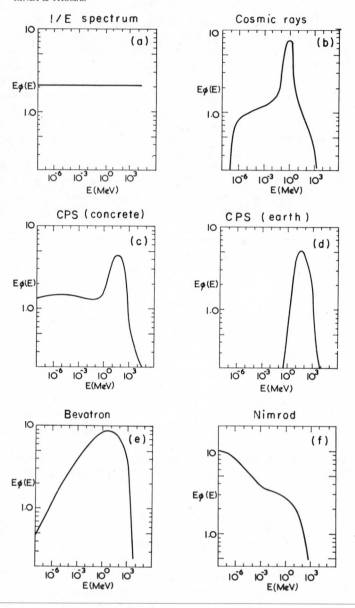

Figure 1   Some typical high-energy neutron spectra (see Ref. 17). (*a*) $1/E$ spectrum (for comparison). (*b*) Cosmic-ray spectrum. (*c*) Spectrum at concrete shielding bridge at CPS. (*d*) Spectrum on earth shield of CPS. (*e*) Spectrum outside Bevatron shielding. (*f*) Spectrum outside steel shielding of Nimrod external proton beam.

The neutron spectrum outside the Bevatron shielding (Figure 1e) is intermediate in character between the two spectra measured at the CPS and suggests that the hydrogen content of the concrete at Berkeley is higher than that at CERN. (To the authors' knowledge this speculation has never been tested.)

Finally, Figure 1f shows the spectrum around a steel-shielded proton beam of the British 7-GeV synchrotron. Compared with the other spectra shown, a large buildup of neutrons below 1 keV is seen, and is attributed to leakage of low-energy neutrons through holes in the shielding (60). It is unlikely, however, that this is the entire explanation because such a buildup is frequently observed outside steel shields. For example, measurements of the neutron spectrum emerging from the main Bevatron magnet identified a very large component near 100 keV (34), while Perry & Shaw (41) observed large increases in radiation levels when steel replaced concrete in shield construction. However, recent theoretical calculations of the neutron spectrum produced in steel by the interaction of 200-GeV protons do not indicate a buildup (61). Such a discrepancy shows that although we now have a fair understanding of high-energy environments, more needs to be done.

It will be shown later how such neutron spectral data may be used to calculate dose equivalent. Gilbert et al (52) have given the distribution of dose equivalent as a function of energy for several of the neutron spectra shown in Figure 1.

B   Electron Accelerators

Early measurements at high-energy electron accelerators were principally concerned with the development and transmission of the electromagnetic cascade through the shield (62–66). These studies confirmed that there was good theoretical understanding of these processes (67). Photon spectra at accelerators up to now have not been measured, but a β-γ spectrometer and a NaI(Tl) anticoincidence γ-ray spectrometer have been used to measure dose rate for space missions (68, 69). The application of such instruments to accelerator radiation fields may prove illuminating.

Bathow et al (63, 70) measured significant neutron production at the DESY 4-GeV electron synchrotron. De Staebler has shown that at high energies and intensities the radiation environments of electron and proton accelerators will be quite similar outside their shields (71). Increasing attention has been given to the measurement of neutrons in recent years. Thus, for example, measurements outside thick shielding at the Stanford Mark III 1-GeV electron linac showed that neutrons were the dominant component of the radiation field; in addition, a significant flux density of neutrons above 20 MeV was identified (72–74). Neutrons are a major component of the radiation field (75) in the earth shield at the 4-GeV electron synchrotron NINA, and are the only significant radiation component surviving at large distances from the SLAC 20-GeV electron linac.[3] Recently Pszona et al (76) have demonstrated the dominating presence of neutrons around the 1-GeV Frascati synchrotron by measurements with ionization chambers.

In comparing the radiation environments of electron and proton accelerators it is interesting to note that close to the primary proton beam, very high photon fluxes

[3] D. Busick, SLAC, Private communication.

have been observed at the 7-GeV synchrotron Nimrod (77) and the CPS (52). The source of these photons has not been established, but has been tentatively attributed to the decay of π^0 mesons produced by proton interactions.

C   Accelerators with Energies > 10 GeV

At high energies (greater than about 10 GeV) the production of energetic muons can be sufficiently great to pose a serious shielding problem at both electron and proton accelerators. Cowan (78) has reported that substantial muon intensities were observed downstream from targets when the BNL 33-GeV AGS first came into operation. Several authors including Keef (12, 79, 80), Bertel et al (81), Theriot, Awschalom & Lee (82), and Kang et al (83) have shown that for the new generation of accelerators above 100 GeV muons will dominate shielding requirements in regions downstream of beam targets.

At these high energies we need more measurements of neutron spectra outside of various shielding materials in order to study the influence of shield construction. In particular we need to extend our knowledge of these spectra above 100 MeV. At these higher energies it may prove to be technically more feasible to detect the equilibrium proton spectrum. Penfold & Stevenson (84) have reported the use of a proton telescope to detect intense sources of radiation inside thick shields along an external proton beam. The application of spark chambers to this problem should prove extremely helpful; Hajnal et al (85) have reported the use of an optical spark chamber to study the secondary-neutron energy spectra emerging from a 40-cm-thick iron shield bombarded by 2.9-GeV protons. Rindi (86) and Lim (87) have described the construction of an instrument that utilizes multiwire spark chambers with magnetostrictive readout and that may be used for measuring neutron and proton spectra up to energies of about 300 MeV in low-intensity fields.

3   RADIATION PROTECTION

A   The Dose Equivalent

The numerical scale used in radiation protection is expressed in terms of the parameter dose equivalent whose unit is the rem. Conceptually, dose equivalent is a measure of radiation used in radiation protection, based upon its ability to induce disease (somatic and genetic injury) in humans who are chronically exposed to low intensities of ionizing radiations (88). (A complete definition of dose equivalent would more adequately define the terms "disease," "chronically exposed," and "low intensities." However, with our present limited understanding of the biological effects of ionizing radiations in humans, such a definition can only be approximated.) Recent discussions on the methods of evaluating the dose equivalent in high-energy radiation fields have clarified the concept of dose equivalent. So, we believe it is useful to review this development.

Early observations in radiology and radiobiology suggested that the dominant parameter that largely determined subsequent injury to irradiated tissue was the quantity of energy absorbed per unit mass of tissue. (Absorbed dose is usually measured in units of rads where 1 rad = 100 ergs g^{-1}.) More sophisticated experiments showed that absorbed dose was not an entirely adequate parameter, and that to better express biological damage absorbed dose had to be weighted by other parameters that depended upon the characteristics of the radiation. This problem was empirically solved in radiobiology by expressing exposures to different radiations in terms of absorbed dose of some standard radiation (usually X or γ rays of specified energy). Thus the biological effects of irradiation by all different types of radiation would be identical to that from χ rads of standard radiation where

$$\chi = \sum_{i=1}^{n} R_i D_i \qquad \qquad 2.$$

and R_i is the relative biological effectiveness (RBE) of the ith radiation defined by $R_i = D_x/D_i$, and D_x, D_i are, respectively, the absorbed doses of the standard radiation and the ith radiation required to produce the same biological effect.

The quantity defined in equation 2, referred to in the literature as the RBE dose, is clearly an equivalent dose of standard radiation and has the same physical dimensions as absorbed dose [as does dose equivalent (89)]. Radiobiologists have measured many RBEs, even for a specific type of radiation, depending upon the biological system, the biological effect considered, the dose rate and distribution, and many other biological and physical factors. One parameter found to have an important influence on the RBE is the average LET, or collision stopping power of the ionizing radiation. [LET still continues to play an important role in the thinking of radiobiologists although recently some have suggested it has only limited value in specifying radiation quality (90–92).]

For radiation protection purposes, the appropriate "RBEs" required would be those for chronic low-level exposures of humans. The biological effects of low-level exposures are not entirely known but probably include carcinogenesis, leukemia induction, life-span shortening, and deleterious mutations. There are no data on these biological effects in humans exposed at sufficiently low doses and dose rates, and furthermore it seems unlikely that data will be directly obtained in the foreseeable future, since such human experiments are not feasible. Nor does it seem likely that epidemiological studies will greatly alleviate this situation if the risks of somatic injury are of the magnitude estimated by the International Commission on Radiological Protection (ICRP) (93, 94). Any values of RBE currently used in radiation protection are, therefore, extrapolations from epidemiological studies of humans acutely exposed or from animal experiments, and are essentially administrative in character.

The solution adopted by the ICRU/ICRP was to express the "RBE used in radiation protection" as the product of several modifying factors. Provision was made for several such factors including those which take account of LET (the Quality Factor), the nonuniform spatial distribution of absorbed dose, and differences in the absorbed dose rate (95). For external radiation exposure, however, only the Quality Factor (Q), which accounts for the difference in LET of ionizing radiations at the locations of interest, is defined. When ionizing radiation of

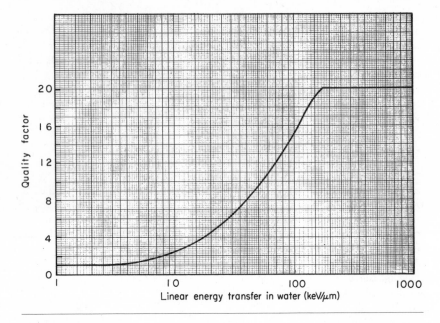

Figure 2   Quality factor as a function of linear energy transfer in water, interpolated from ICRP recommendations (98).

more than one LET, L, is present at the point of interest, the dose equivalent at that point may be expressed by a modification of equation 2 as (96)

$$H = \sum_{i=1}^{n} Q_i D_i \qquad\qquad 3.$$

In practice, the ionizing particles producing the absorbed dose have a continuous distribution in L, and equation 3 becomes (97)

$$H = \int_0^{L_{max}} Q(L)D(L)\,dL \qquad\qquad 4.$$

where $D(L)$ is the absorbed dose per unit interval of LET due to particles with LET between L and $L+dL$. L_{max} is the maximum value of LET at the point of interest. Dose equivalent in high-energy environments is evaluated from a knowledge of the parameters of the radiation environment by calculating the $D(L)$ distribution as a function of depth and using the relationship between $Q(L)$ and L defined by the ICRP (98) (Figure 2).

B   Dose-Equivalent—Depth Distributions

It is the current practice of regulatory organizations to set maximum permissible limits for the dose equivalent (MPD) in certain so-called "critical organs" such as the gonads, red bone marrow, thyroid, etc. For radiation protection purposes the dose equivalent in these critical organs must be calculated to determine whether those MPDs have been exceeded.

The quantity H, as defined by equation 4, in principle may be calculated as a function of position in the human body, under any irradiation conditions. In practice, however, such detailed calculations, involving as they do complex details of geometry and nuclear interactions, require extensive computing facilities for their execution. Furthermore, even with the aid of large digital computers, certain simplifications have been necessary to make the calculations tractable.

At present most calculations have been made under limited radiation conditions for uniform, semi-infinite slabs of tissue-like material (e.g. water, polystyrene, "standard-tissue"), but an increasing number of calculations are being made for finite phantoms (parallelepipeds, cylinders, elliptical cylinders). In addition, attention is being given to the effects of nonuniform body compositions (99). In the case of irradiation by neutrons, several summaries of these calculations have been published (100); comparison with experimental measurements indicates good agreement (101).

C   Conversion Factors

In selecting a single set of particle-flux-density to dose-equivalent rate conversion factors as a function of particle energy, it is conventional to choose those irradiation

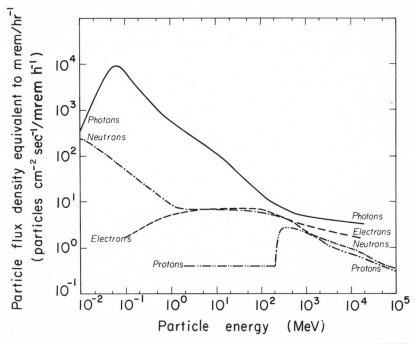

Figure 3   Conversion factors as a function of energy for different particles.

conditions that maximize the dose equivalent in the body. These generally occur for unilateral irradiation by a normally incident beam of particles. In addition, such conversion factors are derived from the maximum in the calculated dose-equivalent —depth distributions. Figure 3 shows conversion factors for electrons, neutrons, photons, and protons derived in this way by ICRP. In practice it is usually necessary to evaluate dose equivalent due to particles distributed over a range of energies.

The dose-equivalent rate $\dot{H}$ may be approximated by the equation

$$\dot{H} = \int_{E_{min}}^{E_{max}} \phi(E) \, dE / g(E) \qquad\qquad 5.$$

where $g(E)$ is the appropriate conversion factor for particles of energy E, and E_{min}, E_{max} are the appropriate energy limits.

Because the conversion factors $g(E)$ are derived from irradiation conditions which maximize the dose *at each energy*, the use of equation 5 may overestimate the dose equivalent due to a continuous particle spectrum. Equation 5 expresses the sum of the maxima of the dose-equivalent depth curves at each energy rather than the maximum of the sum of the dose-equivalent distributions from each component of the spectrum (102).

For irradiation by particles extending over a wide energy range, Shaw et al (17) have suggested that the dose equivalent should be obtained by calculating the dose-equivalent distribution in the body due to the entire spectrum. The maximum dose equivalent in the body (or the dose equivalent in the internal organs) may then be evaluated. They have reported such calculations for some typical accelerator neutron spectra (Figure 1) and showed that the use of equation 5 with these spectra was accurate enough for practical purposes. This may be seen from Table 2, which compares effective conversion factors averaged over the entire energy range. Column 1 shows approximate values obtained using equation 5 [reported by Gilbert et al (52)]; column 2 gives more precise values reported by Shaw et al (17). There is essential agreement between these two sets of values.

Table 2   Effective conversion factors for neutron spectra

Spectrum	Effective neutron conversion factors $\left(\dfrac{n/cm^2 \ sec}{mrem/h} \right)$	
	Gilbert et al (52)	Shaw et al (17)
Cosmic ray	12.1	14.1
Bevatron	8.8	11.9
CERN synchrotron bridge	7.3	12.1
CERN ringtop	4.3	5.1
$1/E$	4.7	6.4

4   SHIELDING

As recently as 1960, Jaeger described the design of high-energy radiation shielding as "more an art than a science" (103). During the period 1960–1970, several experiments at high-energy electron and proton accelerators as well as theoretical studies of the propagation of radiations through shielding have radically changed the state of the art. A critical review of many of these experiments has recently been given by Patterson & Thomas (3, 4), and Ranft (104) has compared experimental data with Monte Carlo calculations.

The immediate aims of the early experiments were to obtain a qualitative understanding of the transmission of accelerator-produced radiation, to express this understanding in some physically plausible, but empirical manner, and, finally, to make this empirical formulation quantitative. As our understanding of radiation transport increased, experiments became more sophisticated in an attempt to understand the development of electromagnetic and hadronic cascades in matter. Now, as theoretical techniques become increasingly reliable, experiments are designed with a view of testing (and improving) theoretical models of transport phenomena.

A   Phenomenological Models

Consider an effective point source produced by protons interacting in a thick target (Figure 4). The radiation level on the outside surface of a shield may be written by analogy with the corresponding photon shielding problem as

$$H = r^{-2} \int F(T)B(T) \exp\left[-d(\theta)/\lambda(T)\right] \cdot \mathrm{d}^2 n(T, \theta)/(\mathrm{d}T \, \mathrm{d}\Omega) \, \mathrm{d}T \qquad 6.$$

where r is the distance from the source, T is the neutron energy, F is a factor which

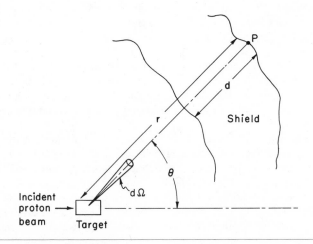

Figure 4   Schematic diagram of typical shielding geometry.

converts fluence to dose equivalent, d is the shield thickness, λ is the effective removal mean-free path, B is a buildup factor, and $d^2n/(dT\,d\Omega)$ is the yield of neutrons per unit solid angle between T and $T+dT$ at angle θ (see Figure 4).

De Staebler (8) wrote equation 6 as:

$$H = r^{-2} \sum_i B_i F_i \exp(-d/\lambda_i) \cdot (dn/d\Omega)_i \qquad\qquad 7.$$

where the subscript i denotes a range of neutron energies for which B, F, and λ are fairly constant and the definition of $(dn/d\Omega)$ is obvious.

Moyer (105, 106) made an extremely important contribution when he recognized that equation 7 may be approximated by a single-energy group because the nature of the radiation field outside the shield of a high-energy proton accelerator will be determined by neutrons with energy greater than about 150 MeV. Neutron attenuation lengths above 150 MeV are roughly independent of energy, but diminish rapidly with energy below about 100 MeV. Consequently, the greater yields of low-energy as compared to high-energy neutrons, at the primary interaction, will be more than compensated for by the greater attenuating action of the shield for these neutrons. Deep in the shield, high-energy ($E > 150$ MeV) neutrons regenerate the cascade but are present in relatively small numbers. At a shield interface the radiation field observed consists of these "propagators," born close to the primary radiation source, accompanied by many particles of much lower energy born near the interface. Equation 7 therefore becomes

$$H \propto [Ng(\theta)/r^2] \cdot \exp(-d/\lambda) \qquad\qquad 8.$$

where N is the proton intensity incident on the target, θ is the angle subtended to the beam direction, $g(\theta)$ is the angular distribution of high-energy particles at the source, d is the shield thickness, and λ is the effective attenuation length of high-energy neutrons. The total neutron flux density (and consequently the dose-equivalent rate) will be proportional to the high-energy neutron flux density. Because the low-energy components are produced by interaction of the high-energy propagators, their intensity decreases through the shield in an exponential manner with effectively the same attenuation length for all directions through the shield.

Moyer (105, 106) generated appropriate parameters to be used in equation 8 in calculating shielding for the Bevatron. Smith (54) has described the excellent agreement between measured radiation levels outside the Bevatron shield and those predicted by Moyer.

Many shielding experiments have subsequently confirmed Moyer's basic assumptions. For example, Smith et al (55) used threshold detectors to measure the spatial variation of flux density produced in concrete bombarded by 6-GeV protons. Figure 5 shows the relative flux density distribution, measured by the ^{27}Al $\rightarrow$ ^{24}Na reaction (threshold 6 MeV) along paths drawn at several angles to the incident beam direction. The transmission curves are seen to be exponential and essentially parallel, within the limits of experimental accuracy. Similar results were obtained with detectors utilizing the ^{12}C $\rightarrow$ ^{11}C reaction (threshold 20 MeV). In addition, Smith et al demonstrated the existence of an equilibrium spectrum by

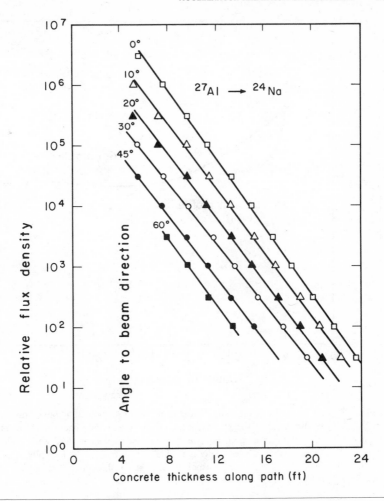

Figure 5 Relative flux-density distribution measurements along paths drawn at several angles to the point of incidence of the proton beam on a concrete shield. Measurements made with the ^{27}Al $\rightarrow$ ^{24}Na reaction. Incident proton energy 6 GeV (55).

calculating the ratio of the response of the carbon and aluminum activation detectors. Figure 6 shows that this ratio becomes constant both in the beam direction and transverse to it. Equilibrium is evidently much more rapidly attained in the transverse direction than in the beam direction.

In the past five years effort has been devoted to obtaining optimum values of λ and $g(\theta)$ for use in equation 8.

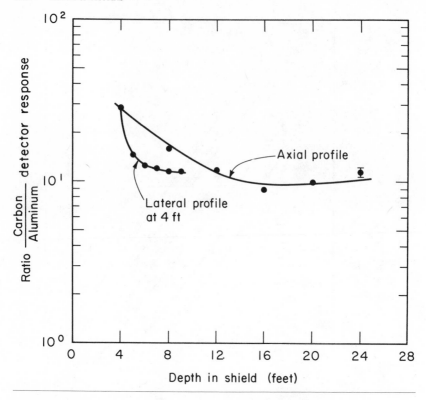

Figure 6 Typical example of the ratio of detector response as a function of distance from the point of incidence of the proton beam on the shield. The figure demonstrates the existence of an equilibrium spectrum. The curve labeled "axial profile" was obtained in the beam direction; that labeled "lateral profile" was obtained at a depth of four feet into the shield in a direction normal to the beam direction. Incident proton energy 6 GeV (55).

1 HIGH-ENERGY ATTENUATION LENGTH   It is readily seen that the results of calculations using the Moyer model are most sensitive to the value of attenuation length used. At high energies, particle attenuation is essentially determined by inelastic interactions and so we might expect the appropriate value of λ to be given by

$$\lambda = 1/N\sigma_{in} \qquad\qquad 9.$$

where N is the number of atoms/cm^3 and σ_{in} is the inelastic cross section.

Measurements of nucleon-nucleus inelastic cross sections as a function of mass number are quite well represented by the formula

$$\sigma_{in} = 43.1A^{0.70} \text{ mb} \qquad\qquad 10.$$

for $A \geqq 3$, $E \geqq 150$ MeV irrespective of whether the incident particle is a proton or neutron (107, 108). Substitution into equation 9 gives

$$\rho\lambda = 38.5A^{0.30} \text{ g cm}^{-2} \qquad\qquad 11.$$

Over the past fifteen years many shielding experiments have been performed (3, 4), but accurate data are limited. The earlier experiments, in particular, were subject to many sources of error, especially with regard to an accurate knowledge of the density of the shielding material. Furthermore, in much of the earlier literature there are conflicting interpretations of the term "attenuation length" (3, 4). Of the later experiments, that reported by Gilbert et al (52) at the CPS, which analyzed the experimental data in terms of the Moyer model, obtained a value of $\rho\lambda = 117\pm2$ g/cm^2 in earth, close to that predicted by equation 11. Use of attenuation lengths calculated from equation 11 is certainly consistent with the available experimental determination of attenuation length.

2 ANGULAR DISTRIBUTION OF SECONDARY PARTICLES   The exact nature of the angular distribution function $g(\theta)$ that should be used in equation 8 is not immediately obvious. One approach is to deduce the angular distribution from measurements of particle flux density within the shield around the radiation source. Using such an approach, Gilbert et al (52) found that an angular distribution of the form

$$g(\theta) = a\exp(-b\theta) \quad 60° \leqq \theta \leqq 120° \qquad\qquad 12.$$

well represented the flux density data measured in the earth shield of the CPS. In these measurements a thin Be-Al target was bombarded by 14.6- or 26.4-GeV/c protons. In their experiment the parameter b did not seem to be strongly dependent upon primary proton energy. Values of b in the range 2.1–2.4 radian^{-1} were reported by Gilbert et al consistent with values of b around 2.5 reported by Stevenson et al (109), using a similar technique, for a primary proton energy of 7 GeV.

The angular distribution of secondary hadrons determined from measurements around fairly thin targets is of more fundamental interest. Such data are needed to test the validity of Monte Carlo and other transport model calculations, which are increasingly used to estimate the magnitude of a variety of radiation phenomena such as radiation damage, induced radioactivity, and radiation intensity. Measurements of momentum-integrated secondary-particle yields around internal targets are difficult because of poorly defined source geometry (52, 110). Recently some careful measurements of the angular dependence of hadron yields from various target materials bombarded by 3-GeV (111), 7-GeV, and 23-GeV (112) protons have been reported. Levine et al (112) conclude from their measurements that the shape of the angular distribution measured with any particular detector is independent of primary proton energy and, within the range $60° \leqq \theta \leqq 120°$, is consistent with the form suggested by Gilbert et al (52) (equation 12). Table 3 summarizes values of the parameter b obtained at 7 GeV, from which it may be concluded that b is strongly dependent upon the energy threshold of the radiation

Table 3 Values of relaxation parameters

Detector	Assumed reaction threshold	Relaxation parameter b, radians^{-1}				
		7-GeV data[a]			3-GeV data[b]	
		"W"	Cu	Al	Pb	Fe
HPD[c] & TLD		1.65 ± 0.1	1.36 ± 0.05	1.25 ± 0.05		
$^{32}S\text{-}^{32}P$	3.0 ± 0.5 MeV	0.29 ± 0.03	0.39 ± 0.04	0.50 ± 0.06	0.23 ± 0.07	0.30 ± 0.08
$^{27}Al\text{-}^{24}Na$	6.0 ± 0.5 MeV	0.51 ± 0.02	0.65 ± 0.04	0.71 ± 0.04		
$^{19}F\text{-}^{18}F$	11 ± 1 MeV	0.73 ± 0.05	0.90 ± 0.05	1.05 ± 0.07		
$^{12}C\text{-}^{11}C$	22 ± 3 MeV	1.28 ± 0.05	1.34 ± 0.03	1.32 ± 0.05	1.10 ± 0.12	1.35 ± 0.16
$^{12}C\text{-}^{7}Be$	35 ± 5 MeV	1.6 ± 0.1	1.7 ± 0.1	1.4 ± 0.2		
$^{27}Al\text{-}^{18}F$	35 ± 5 MeV	1.6 ± 0.1	1.7 ± 0.1	1.4 ± 0.2	0.84 ± 0.14	1.07 ± 0.13
Au fission	90 ± 10 MeV	2.1 ± 0.3	2.1 ± 0.3	2.1 ± 0.3		

[a] Reference 112.
[b] Reference 111.
[c] Hydrogen pressure dosimeter.

detector. Comparison with the 3-GeV data of Awschalom & Schimmerling (111) indicates no strong dependence of b upon primary proton energy. Figure 7 shows the data of Table 3. (A range for threshold energy is indicated because different hadrons may produce the radioactive species observed.)

In using the Moyer model to calculate transverse shielding for proton accelerators, the appropriate angular distribution $g(\theta)$ is assumed to be that of particles with energies greater than about 150 MeV (105). Extrapolation of the data of Figure 7 gives a value of b of 2.3 ± 0.3 at 150 MeV. This value is in surprisingly (and perhaps fortuitously) good agreement with the values of b in the range 2.1–2.5 extrapolated from measurements deep in the shield.

The absolute yield of secondary hadrons depends both upon target material and primary proton energy. At large angles the yields appear to be dominated by contributions from the intranuclear cascade and are not inconsistent with a variation proportional to $A^{1/3}$ (112). If the yield y is expressed in the form

$$y = \text{constant} \cdot E^n \cdot g(\theta) \qquad 13.$$

n lies in the range 0–0.5, depending upon the detector used, over the angular range $30° \leq \theta \leq 80°$.

Comparison of the experimental with the integrated momentum spectra of secondary particles predicted by a modification of the semi-empirical Trilling production formula (113, 114) indicates good absolute agreement at angles less than 30°. At larger angles there is a divergence between the experimental data and theoretical predictions for two reasons: Firstly, this Trilling formula does not correctly describe the production of particles with high transverse momentum. Secondly, the production of particles in the intranuclear cascade and by evapora-

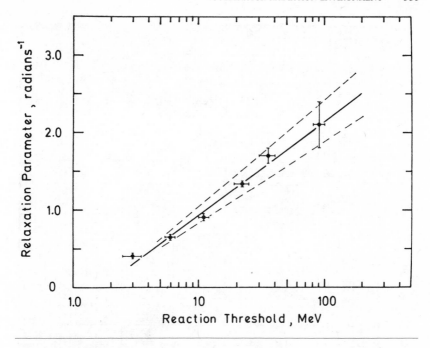

Figure 7 Angular relaxation parameter, *b*, as a function of reaction threshold energy (after Levine et al, 112).

tion processes must be correctly accounted for. Recently Ranft & Routti have described suitable empirical formulae which predict angular distributions in good agreement with available experimental data at all angles (115).

3 ACCURACY OF THE MOYER MODEL Use of the Moyer model with appropriate input data, and under fairly simple geometrical conditions, leads to estimates of radiation levels usually accurate to better than a factor of two. Figure 8 indicates the accuracy possible when experimental data are fitted to a Moyer-type equation. Calculated and measured neutron flux densities in the earth shield of the CPS are shown (116). Fluxes are plotted as a function of longitudinal distance from an internal target for five different depths in the shield. Flux densities were measured utilizing the ^{27}Al$(n, \alpha)^{24}$Na reaction in aluminium. In this particular example, flux densities are predicted to about 20%, over a range of five orders of magnitude. Estimates of dose-equivalent rate follow from a knowledge of neutron flux density and spectrum.

For the calculation of shield thicknesses transverse to a proton beam, for uniform beam loss, the Moyer model takes a particularly simple form. Substituting for $g(\theta)$ using equation 13, and using experimental data from the CPS it may be shown that

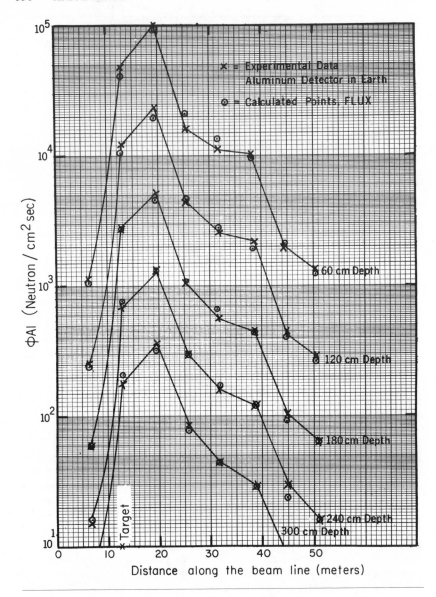

Figure 8 Comparison of measured and calculated flux density as a function of position in the earth shield of the CPS. The abscissa gives the distance along the beam line measured from an arbitrary point. The 25-GeV proton beam interacted with a beryllium target at T (12.5 m from the origin). The ordinate shows the neutron flux density measured with an Al detector (116).

$$\dot{H} = [0.11L/(a+d)] \int_0^\pi \exp(-2.3\,\theta)\exp[-(d/\lambda)\csc\theta]\,d\theta \qquad 14.$$

where $\dot{H}$ is measured in mrem/h, L is the beam loss in units of GeV/cm sec, a is the accelerator tunnel radius in meters, d is the shield thickness in meters, and λ is the attenuation length. Integrals of the form appearing in equation 14 have been tabulated in the region of physical interest by Routti & Thomas (117).

Phenomenological models permit simple, rapid, and fairly accurate shield estimates. Furthermore, they provide a valuable physical insight into the problems of shielding. Such models are, however, necessarily limited by operational experience.

B Monte Carlo Calculations

One of the most important advances in the study of accelerator radiation environments over the past ten years has been the development of Monte Carlo techniques to calculate electromagnetic and hadronic cascade phenomena. These calculations have recently been reviewed by Ranft (104), and space does not permit a complete discussion here. Accurate and reliable calculations of radiation phenomena at accelerators have required development of an understanding not only of the interaction of primary particles with internal targets and machine components, but also of particle production resulting from primary particle interaction, the transport of these primary and secondary particles together with their interaction products through matter and, finally, the conversion from the calculation of particles transported to observable phenomena. Ranft has reported good agreement with experimental data in such diverse areas as induced radioactivity, radiation doses, radiation heating, and shielding.

A good example of the agreement between theoretical and experimental data is the recent calculation of the neutron spectrum in the earth's atmosphere by Armstrong et al (118). These workers used a Monte Carlo code to compute the production of protons, charged pions, and neutrons by the incident galactic protons, and the subsequent transport of these particles down to energies of 12 MeV. The production of neutrons of energy ≤ 12 MeV as calculated by the Monte Carlo code was used as input to a discrete-ordinates code to obtain the low-energy neutron spectrum. Figure 9 shows the results of these calculations and an *absolute* comparison with the experimental data of Hess et al (38) at atmospheric depths of 200 and 1033 g/cm². The calculated and measured spectra differ somewhat at lower energies but are in good agreement at high energies. The increasing number of such examples of good agreement between calculated and experimental data is extremely encouraging.

5 INDUCED RADIOACTIVITY

The development and transport of the electromagnetic and hadronic cascades (sections 3 and 4) also result in the production of radioactivity in accelerators and their surroundings. Accelerator shielding and accelerator components such as

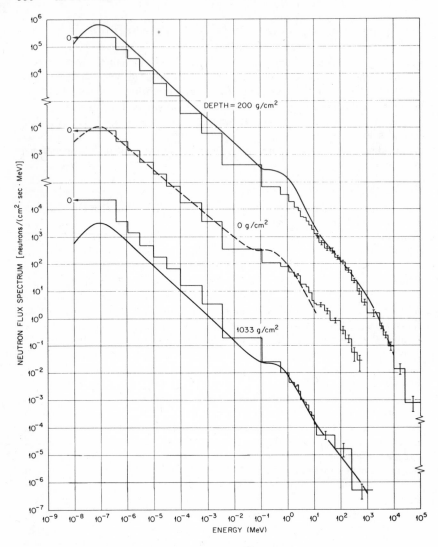

Figure 9 Histograms show calculated values of the cosmic-ray neutron spectra due to Armstrong et al (118), which are compared with the measurements of Hess et al (38) (solid line) at depths in the atmosphere of 200 g/cm² and 1033 g/cm², and are also compared with calculations of Lingenfelter—quoted by Armstrong et al (118)—at the top of the atmosphere (dotted lines). The calculations and measurements are made in the range of geomagnetic latitudes 40–44°.

targets, vacuum chamber, magnets and rf cavities, cooling water or ground water close to the accelerator buildings, and air in the accelerator room may all become radioactive to some degree.

Barbier (119) has summarized the mechanism for the production of radioactivity at high-energy accelerators. In principle, all the nuclides that have atomic mass and atomic number equal to, or less than, the sum of the numbers of the target plus projectile nuclei can be produced. Many of the radionuclides that can be produced have half-lives so short that they need not be considered in protection problems.

A   Radioactivity of Accelerator Components and Other Solids

The number of radionuclides that might be produced is potentially very large. Fortunately the materials used in accelerator construction are limited in number, the most important being iron, several stainless steels, copper, aluminum, aluminum alloys, and several plastics. Charalambus & Rindi (120) have reported a table of all the main radionuclides that can be produced at a typical proton accelerator. They considered only radionuclides with a half-life longer than one hour and show that about 70% of them are γ-emitters. However, even shorter half-lives may be of concern for protection purposes if they are produced in large quantities.

Table 4 summarizes the radionuclides commonly identified in materials used in accelerators; those with half-lives of less than 10 minutes are excluded. Most of the radionuclides listed are produced by simple nuclear reactions such as (n, xn), (p, xn), (p, pn) etc, but some result from spallation, fragmentation, or capture reactions.

Table 4   Radionuclides commonly identified in solid materials irradiated around accelerators

Irradiated material	Radionuclides
Plastics, oils	^{7}Be, ^{11}C
Concrete, aluminum	As above, plus ^{22}Na, ^{24}Na, ^{32}P, ^{42}K, ^{45}Ca
Iron, steel	As above, plus ^{44}Sc, ^{44m}Sc, ^{46}Sc, ^{47}Sc, ^{48}V, ^{51}Cr, ^{52}Mn, ^{52m}Mn, ^{54}Mn, ^{56}Mn, ^{57}Co, ^{58}Co, ^{60}Co, ^{57}Ni, ^{55}Fe, ^{59}Fe
Copper	As above, plus ^{65}Ni, ^{61}Cu, ^{64}Cu, ^{63}Zn, ^{65}Zn

Several measured cross sections for high-energy reactions have been reported by Bruninx (121–123). Rudstam (124) has proposed a very useful empirical formula for their calculation, while Bertini (125) has reported intranuclear cascade calculations of these cross sections.

Because the number of radionuclides produced in accelerator components is large and accelerator operation often variable, the production and decay of gross radioactivity is a complex function of time. Notwithstanding, for radiation protection purposes it may be necessary to have some estimate of the dose rate, and its variation with time.

The decay of dose rate near the 600-MeV CERN synchrocyclotron has been reported by Baarli (126) and Rindi (127). Reliable experimental data of this type are few because of the difficulty of obtaining them at most accelerators. During periods of accelerator shutdown, gross changes in the remnant radiation field may result from structural changes in the accelerator and its shielding. What data are available, however, show that beginning a few minutes after the shutdown, the dose rate decays by about a factor of two in the first two hours and by about another factor of two within the next 50 hours. This is in agreement with measurements at all the accelerators at the Lawrence Berkeley Laboratory (128) and elsewhere (129). Indeed it seems confirmed by general experience that the gross features of the decay of induced activity near accelerators that have been in operation for several years are nearly independent of the type of particles accelerated and their maximum energy.

Sullivan & Overton (130) have shown that the dose-rate decay may be approximated by an equation of the form

$$D(t) = B\phi \ln\left[(t_i + t)/t\right] \qquad\qquad 15.$$

where $D(t)$ is the dose rate at time t after irradiation ceases, ϕ is the flux density of high-energy primary particles, t_i is the irradiation time, and B is a parameter which depends on several variables but is a constant for any given set of irradiation, target, and geometrical conditions. Equation 15 is in good qualitative agreement with the form of the buildup and decay of dose rates observed in an accelerator environment, and in a recent paper Sullivan (131) has reported values of B for heavy materials that give reasonably good absolute agreement with observation. More accurate calculations required detailed Monte Carlo techniques of the type used in shielding calculations (see section 4). Armstrong et al (132, 133) have calculated the dose rate resulting from the irradiation of steel by 200-MeV, 3-GeV, and 200-GeV protons. For long irradiation times they find that [52m]Mn (21 min), [56]Mn (2.6 h), [52]Mn (5.6 d), [48]V (16 d), [51]Cr (27.8 d), and [54]Mn (280 d) are the dominant radionuclides (Figure 10). These calculations are supported by recent observations at the 76-GeV proton synchrotron in Serpukhov (134).

At electron accelerators, too, only few nuclides are dominant. For example, Saxon (135) reports that at the 4-GeV electron synchrotron NINA, [56]Mn, [52]Mn, and [48]V are dominant in steel. Similar results have been reported by Wyckoff (136) from exposure to the 100-MeV bremsstrahlung beam of the NBS linac. De Staebler (137) has estimated the gross production of radioactivity by a high-energy electron accelerator as some 34 Ci at saturation per kW of beam power.

B   Radioactivity of Air

Radioactive gases are produced by the interaction of primary and secondary particles with the nitrogen, oxygen, argon, and carbon nuclei of air circulating in the accelerator vaults. In Table 5 we show the radionuclides that have been found in

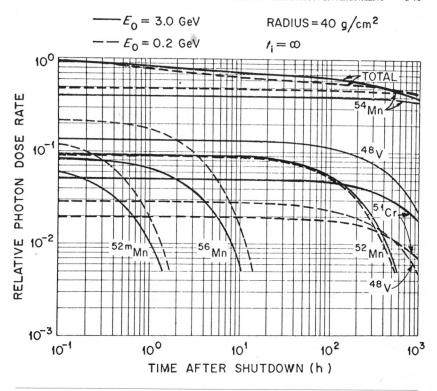

Figure 10 Relative contribution to the photon dose rate, due to six radionuclides, at the surface of an iron cylinder (diameter 80 g/cm²) irradiated axially by 200-MeV and 3-GeV protons for an infinite time (from Armstrong & Barish, 132).

the air at different accelerators. Radionuclides with half-lives less than one minute are of no concern, decaying to negligible activities before personnel can enter the accelerator room or before the air can reach populated areas around the accelerator. Long-lived activities, on the other hand, may be discounted because of their low production rate. Such arguments, supported by the measurements cited in Table 5, suggest that at existing accelerators only four radionuclides need be considered: ^{15}O, ^{13}N, ^{11}C, and ^{41}A. A further increase in the energy or in the intensity of the accelerators, however, could cause the production of amounts of ^{7}Be and ^{3}H which may be important.

Presently, the concentrations of radioactive gases measured in the accelerator room a few minutes after shutdown may range between 10 and 30 times the MPC for continuous inhalation (139). However, the air is quickly mixed with inert air and the radioactivity decays rapidly so that the associated dose rate is negligible compared to that from the solid machine parts.

Table 5   Radionuclides identified in the air of different accelerators

Radionuclide	Where identified	Explanation of previous symbol and reference
^{7}Be	A	A = Saclay 560-MeV electron linac (138)
^{11}C	A, B, C, D, E, F	
^{13}N	A, B, C, D, E, F, G, H	B = CERN 600-MeV proton synchrocyclotron (139)
^{14}O	D	C = CERN 28-GeV proton synchrotron (140)
^{15}O	A, C, D, E, G, H	
^{16}N	E	D = PPA 3-GeV proton synchrotron (141)
^{24}Na	A, D	
^{37}S	D	E = RHEL 7-GeV proton synchrotron (142)
^{38}Cl	A, H	
^{39}Cl	A	F = BNL 30-GeV proton synchrotron (143)
^{41}A	A, B, C, D, F	
^{34m}Cl	D	G = RPI 50-MeV electron linac (144)
		H = Frascati 300-MeV electron linac (145)

C   Radioactivity of Water

Radioactivity induced in cooling water circuits of high-intensity accelerators is potentially of concern for the following reasons: high dose rates around pipes carrying this water, radioactive contamination resulting from spills, and disposal problems. Rose et al (146) reported that external radiation levels as high as 100 mrem/h were found at various regions close to the cooling system of the Harwell 150-MeV cyclotron when it was operated with an internal beam of about 1 μA. Warren et al (147) have reported dose rates of between 0.5 and 4 mrem/h from cooling water circuits along the accelerator structure of the Stanford 20-GeV electron linear accelerator. Considerably higher levels are found from heat exchangers for high-power beam dumps—rates up to 120 mrem/h being observed.

Distenfeld (143) has concluded from measurements at the Brookhaven AGS that with a proton beam intensity of 10^{13} protons/sec the external radiation hazard from induced activity in cooling water would be trivial. However, the dose rate from large volumes of water, such as heat exchangers or storage tanks, would be measurable during accelerator operation. Some rough experimental studies of the production of radionuclides in water from typical high-energy neutron spectra (148) have confirmed ^{11}C, ^{7}Be, and ^{3}H as the most important ones produced. The ratio of the specific activities of tritium and ^{7}Be extrapolated at saturation in samples of water irradiated under several different conditions varied between 1.3 and 5.8. Disposal of irradiated water to streams would generally be controlled by the tritium content, since ^{7}Be is strongly absorbed in the mixed bed resins used for demineralizing the waters (149). Careful studies of the radioactivity produced in water irradiated by high-energy electrons (147) have identified ^{15}O, ^{11}C, and ^{7}Be as the most important radionuclides.

D   Environmental Impact of Radioactivity

The environmental impact of accelerators has been given some study over the past few years. Large accelerators are capable of producing many thousands of curies of radioactivity, a small fraction of which may be released to the environment. The transportation of these radionuclides, induced directly in ground water or leached from irradiated earth in the accelerator shield, has been studied around proton synchrotrons with energies up to several hundred GeV (150–153). These studies show that the inventory of long-lived radionuclides produced in earth shields of such accelerators is of the order of tens of curies, or less. Chemical sorption plays an extremely important role in preventing the migration of many of the nuclides. From these studies the levels of radioactivity likely to appear in ground water systems will be minimal and may not even be detectable.

6 CONCLUSIONS

We hope that in this brief review we have been able to indicate the advances in the qualitative and quantitative understanding of accelerator radiation problems that have been achieved in the past fifteen years. From almost complete ignorance of these problems we have reached the point where accurate estimates of shielding, induced radioactivity, radiation damage, and radiation levels are generally possible.

The increasing use of accelerators in industry and medicine necessitates the wider dissemination of our expanding knowledge of accelerator-produced radiation. There is a need to educate sufficient operating personnel as to the special problems of shielding and dosimetry at particle accelerators if adequate safety standards are to be maintained. This need is especially important at accelerators used for radiation therapy, and is not fulfilled by specialized conferences open only to a limited audience. A number of recent monographs (4, 119, 154, 155) specifically discussing accelerator radiation problems will help alleviate this situation. In the United States a start has been made to provide more formal instruction through short training courses.

The radiation problems of very high energy accelerators (> 10 GeV) continue to be of rather specialized interest, but the systematic solution of these problems will be of general interest to other disciplines as well.

Despite the significant advances reported here, much remains to be done. The problems of personal dosimetry for high-energy neutrons ($E > 20$ MeV) has, to date, been given little attention. Shield computations still need to be refined and this must be done by comparison with observation. Improvements are still needed in instrumentation for the measurement of fast neutrons "in the field." We suggest that the problems of induced radioactivity will need increasing study if the application of accelerators in industry is not to be inhibited. Radiation damage studies, too, are of increasing importance so that accelerators may be economically and efficiently designed.

ACKNOWLEDGMENTS

It is a pleasure to thank H. Wade Patterson (Lawrence Berkeley Laboratory) and H. de Staebler (Stanford Linear Accelerator Center) for helpful advice in the preparation of this article. We are grateful to our colleagues of the several high-energy research centers for providing us with updated information on their researches.

Thanks are also due to LBL editor Mary Wildensten, to Mary Long of the Health Physics group, and to Carmen Hubbard for typing the manuscript.

This work was done under the auspices of the U.S. Atomic Energy Commission.

Literature Cited

1. Rotblat, J. 1950. *The Acceleration of Particles to High Energies,* Sect. 1, 1–13. London: Inst. Phys.
2. Livingston, M. S. 1952. *Ann. Rev. Nucl. Sci.* 1:167
3. Patterson, H. W., Thomas, R. H. 1971. *Particle Accel.* 2:77
4. Patterson, H. W., Thomas, R. H. 1973. *Accelerator Health Physics.* New York: Academic. In press
5. Ham, W. T. 1953. *Arch. Opthalmol.* 50:618
6. Moyer, B. J. 1954. *Radiat. Res.* 1:10
7. Moyer, B. J. 1958. *Ann. Rev. Nucl. Sci.* 8:327
8. De Staebler, H. 1962. *Transverse Radiation Shielding for the Stanford Two-Mile Accelerator (USAEC Rep., SLAC 9)*
9. Los Alamos Scientific Laboratory 1964. *A Proposal for a High Flux Meson Facility*
10. Yale University Design Study Staff 1964. *Final Rep. on the Design of Very High-Intensity Proton Linear Accel. as a Meson Factory of 750 MeV (Yale Rep. Y-12)*
11. CERN Laboratory 1964. *Rep. on the Design Study of a 300-GeV Proton Synchrotron [CERN Rep. 563 (2 vols.)]*
12. Lawrence Radiation Laboratory 1965. *200-GeV Accel. Design Study Rep. [UCRL-16000 (2 vols.)]*
13. National Accelerator Laboratory 1968. *Design Report* (Revised version)
14. Awschalom, M. 1969. In *Proc. Conf. Radiat. Prot. in Accel. Environ.* England: Rutherford
15. Awschalom, M. 1971. *Proc. Int. Congr. Radiat. Prot. Accel. and Space Radiat., CERN Lab.,Geneva(CERN 71-16: 1050)*
16. Goebel, K., Ed. 1971. *Radiat. Problems Encountered in the Design of Multi-GeV Res. Facilities (CERN 71-12)*
17. Shaw, K. B., Stevenson, G. R., Thomas, R. H. 1969. *Health Phys.* 17:459
18. Patterson, H. W., Routti, J. T., Thomas, R. H. 1971. *Health Phys.* 20:517
19. Burlin, T. E., Wheatley, B. M. 1971. *Phys. Med. Biol.* 16:47
20. Rindi, A., Thomas, R. H. 1972. *Health Phys.* 23:715
21. Rosen, L. 1971. *IEEE Trans. Nucl. Sci.* NS-18:29
22. Morgan, I. L. 1973. *IEEE Trans. Nucl. Sci.* NS20, No. 2, p. 36
23. Burill, E. A. 1969. In *Proc. Int. Conf. Accel. Dosimetry, 2nd, Stanford, Calif. (CONF-691101)*
24. Boom, M. L. M., Wiley, A. L. 1971. See Ref. 21, p. 36
25. Raju, M. R., Richman, C. 1971. *Current Topics in Radiation Research,* eds. M. Ebert, A. Howard. Amsterdam: North-Holland
26. Tobias, C. A., Lyman, J. T., Lawrence, J. H. 1971. *Recent Advan. Nucl. Med.* Vol. 3, Chap. 6, ed. J. H. Lawrence. New York: Grune and Stratton
27. See for example: Schaefer, H. J. 1971. *Science* 173:780
28. Holcomb, R. W. 1970. *Science* 167:853
29. Rossi, H. H., Kellerer, A. M. 1972. *Science* 175:200
30. ICRP 1972. *Publication No. 18.* Oxford: Pergamon
31. Patterson, H. W. 1957. *Proc. Conf. Shielding High-Energy Accel., New York (USAEC Rep. TID-7545),* p. 3
32. Smith, A. R. 1958. *LBL Rep. UCRL-8377*
33. Smith, A. R. 1962. *Proc. Premier Colloq. Int. Prot. auprès des Grands Accel.* Paris: Presses Univ. de France
34. Patterson, H. W. 1965. *Proc. Symp. Accel. Radiat. Dosimetry, 1st, BNL, Brookhaven (CONF-651109),* p. 3
35. Patterson, H. W., Hess, W. N., Moyer, B. J., Wallace, R. 1959. *Health Phys.* 2:69
36. Tardy-Joubert, P. 1965. See Ref. 34, p. 117
37. Tardy-Joubert, P. 1967. *Progr. Radiology,* Vol. 2. Amsterdam: Excerpta Medica
38. Hess, W. N., Patterson, H. W., Wallace, R. 1959. *Phys. Rev.* 116:445
39. Puppi, G., Dallaporta, N. 1952. *Progr. Cosmic Ray Phys.* 1:317
40. Baarli, J., Sullivan, A. H. 1965. *Health Phys.* 11:353
41. Perry, D. R. 1966. *Proc. Symp. Neutron Monitoring for Radiological Prot.,* p. 355. Vienna: IAEA; See also Perry, D. R.,

Shaw, K. B. in Ref. 34, p. 20
42. Komochkov, M. M. 1970. *Eng. Compend. Radiat. Shielding,* 3:171. New York: Springer-Verlag
43. Baarli, J. 1964. Private communication to Middlekoop, W. C. Reported in *CERN Intern. Rep. AR/INT/SG 64-6*
44. Capone, T. et al 1965. *CERN Intern. Rep. DI/HP/71*
45. Baarli, J., Sullivan, A. H. 1965. See Ref. 34, p. 103
46. Thomas, R. H. 1972. *Proc. IAEA Symp. Neutron Dosimetry.* Vienna: IAEA. In press
47. Lehman, R. L., Fekula, O. M. 1964. *Nucleonics* 22(11):35
48. Rossi, B. 1954. *High-Energy Particles.* New York: Prentice-Hall
49. Remy, R. 1965. *UCRL Rep. 16325*
50. Omberg, R. P., Patterson, H. W. 1967. *UCRL Rep. 17063*
51. Patterson, H. W., Heckman, H. H., Routti, J. T. 1969. See Ref. 23, p. 750
52. Gilbert, W. S. et al 1968. *UCRL Rep. 17941*
53. Sullivan, A. H. 1969. See Ref. 23, p. 625
54. Smith, A. R. 1965. See Ref. 34, p. 224
55. Smith, A. R. et al 1965. See Ref. 34, p. 365
56. Bramblett, R. L., Ewing, R. I., Bonner, T. W. 1960. *Nucl. Instrum. Methods* 9:1
57. Aleinkov, V. E., Gerat, V. P., Komochkov, M. M. 1972. See Ref. 46
58. Routti, J. T. 1969. See Ref. 23, p. 494
59. Routti, J. T. 1969. PhD thesis. Univ. Calif., Berkeley. *UCRL Rep. 18514*
60. Stevenson, G. R. 1972. See Ref. 46
61. Gabriel, T. A., Santoro, R. T. 1972. *ORNL-TM-3945*
62. Bathow, G., Freytag, E., Tesch, K. 1965. *Nucl. Instrum. Methods* 33:261
63. Bathow, G., Freytag, E., Tesch, K. 1967. *Nucl. Phys.* 82:669
64. Tesch, K. 1969. See Ref. 23, p. 595
65. Busick, D. et al 1969. See Ref. 23, p. 782
66. Ladu, M. 1969. *Progr. Nucl. Energ. Series XII,* Part 1, p. 365. New York: Pergamon
67. Bathow, G., Freytag, E., Kajikawa, R., Köbberling, M. 1969. See Ref. 24, p. 222
68. Farmer, B. J., Rainwater, W. J. 1968. *Rep. 0-71100/8R-S (NASA Contract NAS 9-7565)*
69. Farmer, B. J., Johnson, J. H., Bagwell, R. G. 1972. *Proc. Nat. Symp. Natural and Man-Made Radiat. in Space,* p. 142; *NASA-TM-X-2440*
70. Bathow, G., Freytag, E., Tesch, K. 1967. *Nucl. Instrum. Methods* 51:56
71. De Staebler, H. 1965. See Ref. 34, p. 429
72. Carter, T. G., Thomas, R. H. 1969. *Stanford Univ. Intern. Rep. RHT/TN/ 69-11*

73. Chakalian, V. M., Thomas, R. H. 1969. *Stanford Univ. Intern. Rep. RHT/TN/ 69-11*
74. Thomas, R. H. 1969. *Stanford Univ. Intern. Rep. RHT/TN/69-18*
75. Coleman, F. J., Thomas, D. C., Saxon, G. 1972. *Daresbury Nucl. Phys. Lab. Rep. DNPL/P72*
76. Pszona, S. et al 1972. *INR Rep. 1415, Warsaw*
77. Shaw, K. B., Thomas, R. H. 1967. *Health Phys.* 13:1127
78. Cowan, F. P. 1962. See Ref. 33, p. 143
79. Keefe, D. 1964. *LBL Intern. Rep. UCID-10018*
80. Keefe, D., Noble, C. M. 1968. *Nucl. Instrum. Methods* 64:173
81. Bertel, E., de Sereville, B., Freytag, E., Wachsmuth, E. 1971. See Ref. 16, p. 79
82. Theriot, D., Awschalom, M., Lee, K. 1971. See Ref. 16, p. 641
83. Kang, Y. et al 1972. *Particle Accel.* 4:31
84. Penfold, J., Stevenson, G. R. 1968. *Rutherford Lab. Rep. RP/PN/28*
85. Hajnal, F. et al 1969. *Nucl. Instrum. Methods* 69:245
86. Rindi, A. 1969. See Ref. 23, p. 660
87. Lim, C. B. 1973. PhD thesis. Univ. Calif., Berkeley; *LBL-1719*
88. Rindi, A., Thomas, R. H. 1971. *Proc. Ann. Health Phys. Topical Symp., 6th, Richland, Washington,* p. 465
89. Boag, J. W. et al 1972. *Brit. J. Radiol.* 45:314
90. Rossi, H. H. 1969. In *Proc. Symp. Neutrons in Radiobiol., Oak Ridge,* (*CONF-691106*)
91. Katz, R. 1971. *Proc. Biophysical Aspects of Radiat. Quality,* p. 11. Vienna: IAEA
92. Katz, R., Sharma, S. C., Homayoonfar, M. 1972. *Health Phys.* 23:740
93. ICRP 1969. *Publication No. 14.* Oxford: Pergamon
94. Patterson, H. W., Thomas, R. H. 1970. *Proc. Berkeley Symp. Math. Statistics and Probability, 6th,* 6:313. Berkeley: Univ. Calif.
95. RBE Committee 1963. *Health Phys.* 9:357
96. Neary, G. J. 1963. *Phys. Med. Biol.* 7:419
97. ICRP 1970. *Publication No. 15,* Para. 13. Oxford: Pergamon
98. ICRP 1972. *Publication No. 22.* Oxford: Pergamon
99. Figerio, N. A., Coley, R. F., Branson, M. H. 1973. *Phys. Med. Biol.* 18:53
100. See for example: Auxier, J. A., Snyder, W. S., Jones, T. D. 1968. *Radiation Dosimetry,* eds. F. H. Attix, W. C. Roesch, Chap. 6. New York: Academic
101. See for example: Fuller, E. W., Eustace, R. C. in Ref. 15, p. 344; or Armstrong,

T. W., Bishop, B. L. 1971. *ORNL-TM-3304*

102. Patterson, H. W., Thomas, R. H. 1973. See Ref. 4, Chap. 2
103. Jaeger, T. 1960. *Grundzeuge der Strahlenschutz*. Berlin: Springer-Verlag
104. Ranft, J. 1972. *Particle Accel.* 3:129
105. Moyer, B. J. 1961. *UCRL Rep. 9769*
106. Moyer, B. J. 1962. See Ref. 33, p. 65
107. Keefe, D., Scolnick, M. 1966. *LBL Intern. Rep. AS/EXPER/01*
108. Schimmerling, W., Devlin, T. J., Johnson, W., Vosburgh, K. G., Mishke, R. F. 1973. *Phys. Rev.* 67:248
109. Stevenson, G. R., Shaw, K. B., Hargreaves, D. M., Lister, L. P., Moth, D. A. 1969. *Rutherford Lab. Rep. RHEL/M/48*
110. Charalambus, S., Goebel, K., Nachtigall, D. 1967. *CERN Lab. Rep. DI/HP/97*
111. Awschalom, M., Schimmerling, W. 1969. *Nuovo Cimento A* 64:871
112. Levine, G. S. et al 1972. *Particle Accel.* 3:91
113. Ranft, J. 1967. *Nucl. Instrum. Methods* 48:133, 261
114. Ranft, J., Borak, T. 1969. *Nat. Accel. Lab. Rep. NAL-FN-193*
115. Ranft, J., Routti, J. T. 1972. *Particle Accel.* 4:101
116. Gilbert, W. S. 1967. *Proc. Int. Conf. High Energ. Accel., 6th, Cambridge, Mass*, p. 221
117. Routti, J. T., Thomas, R. H. 1969. *Nucl. Instrum. Methods* 72:157
118. Armstrong, T. W., Chandler, K. C., Barish, J. 1972. *Neutron Phys. Div. Ann. Rep., ORNL-4800*, p. 63
119. Barbier, M. 1969. *Induced Radioactivity*. Amsterdam: North-Holland
120. Charalambus, S., Rindi, A. 1967. *Nucl. Instrum. Methods* 56:125
121. Bruninx, E. 1961. *CERN Rep. 61-1*
122. Bruninx, E. 1962. *CERN Rep. 62-9*
123. Bruninx, E. 1964. *CERN Rep. 64-17*
124. Rudstam, G. 1966. *Z. Naturforsch. A* 21:1027
125. Bertini, H. W. 1969. See Ref. 23, p. 42
126. Baarli, J. 1962. See Ref. 23, p. 123
127. Rindi, A. 1964. *Colloq. Intern. Dosimetrie Irradiat. Sources Externes, Paris.* 2:153
128. Boom, R. W., Toth, K. S., Zucker, A. 1961. *ORNL Rep. 3158*
129. Awschalom, M., Larsen, F. L., Sass, R. E. 1965. See Ref. 34, p. 57
130. Sullivan, A. H., Overton, T. R. 1965.

Health Phys. 11:1101
131. Sullivan, A. H. 1972. *Health Phys.* 23:253
132. Armstrong, T. W., Barish, J. 1969. *ORNL Rep. TM-2383*
133. Armstrong, T. W., Alsmiller, R. J. 1969. *Nucl. Sci. Eng.* 38:53
134. Golovachik, V. T., Britvich, G. I., Lebedev, V. N. 1969. *IFVE-69-76.* Transl. *ORNL-tr-2328*
135. Saxon, G. 1969. *DNPL Rep. P8.* Daresbury, England
136. Wyckoff, J. 1967. *IEEE Trans. Nucl. Sci.* 14(3):990
137. De Staebler, H. 1963. *SLAC Rep. TN-63-92*, Stanford, Calif.
138. Vialettes, H. 1969. See Ref. 23, p. 121
139. Rindi, A., Charalambus, S. 1967. *Nucl. Instrum. Methods* 47:227
140. Höfert, M. 1969. See Ref. 23, p. 111
141. Awschalom, M., Larsen, F. L., Schimmerling, W. 1968. *Health Phys.* 14:345; Idem. 1969. *Nucl. Instrum. Methods* 75:93
142. Shaw, K. B., Thomas, R. H. 1967. *Health Phys.* 13:1127
143. Brookhaven Nat. Lab. 1964. *BNL Rep. 7956*, p. 225
144. George, A. C., Breslin, A. T., Haskins, T. W., Ryan, R. M. 1965. See Ref. 34, p. 513
145. Ladu, M., Pelliccioni, M., Roccella, M. 1967. *Giorn. Fis. San. Minerva Fis. Nucl.* 11(2):112
146. Rose, B. et al 1958. *AERE NP/R-2768.* Harwell, England
147. Warren, G. J., Busick, D. D., McCall, R. C. 1969. See Ref. 23, p. 99
148. Stapleton, G. B., Thomas, R. H. 1967. *Rutherford Lab. Rep. RHEL RP/PN/45*
149. Busick, D. D., Warren, G. T. 1969. See Ref. 23, p. 139
150. Stapleton, G. B., Thomas, R. H. 1972. *Health Phys.* 23:689
151. Thomas, R. H. 1970. *Proc. Symp. Health Phys. Aspects of Nucl. Facility Siting, Idaho Falls, Idaho*, p. 93
152. Stapleton, G. B., Thomas, R. H. 1973. *Water Research.* In press
153. Borak, T. D. et al 1972. *Health Phys.* 23:679
154. Zaitsev, L. N., Komochkov, M. M., Sychev, B. S. 1971. *Fundamentals of Shielding Around Accelerators.* Moscow: Atomizdat (In Russian)
155. Freytag, E. 1972. *Strahlenschutz an hochenergiebeschleunigern.* Karlsruhe: Braun (In German)

RADIOACTIVE HALOS[1]

✕ 5541

Robert V. Gentry

Chemistry Division, Oak Ridge National Laboratory, Oak Ridge, Tennessee

CONTENTS

INTRODUCTION

In some thin samples of certain minerals, notably mica, there can be observed tiny aureoles of discoloration which, on microscopic examination, prove to be concentric dark and light circles with diameters between about 10 and 40 μm and centered on a tiny inclusion. The origin of these halos (first reported between 1880 and 1890) was a mystery until the discovery of radioactivity and its powers of coloration; in 1907 Joly (1) and Mügge (2) independently suggested that the central inclusion was radioactive and that the α-emissions from it produced the concentric shells of coloration (the circular patterns observed in this sections are, of course, simply plane sections through concentric spheres). The earliest name, "pleochroic halos," was partially a misnomer, since halos were later found in isotropic minerals which do not show pleochroism.

Aside from their interest as attractive mineralogical oddities, halos command attention because they are an integral record of radioactive decay in minerals that constitute the most ancient rocks. Most importantly, this thermal-resistant record is detailed enough to allow estimation of the decay energies involved and

[1] Research sponsored by the US Atomic Energy Commission under contract with Union Carbide Corporation. RVG is a guest scientist at ORNL from Columbia Union College, Takoma Park, Maryland.

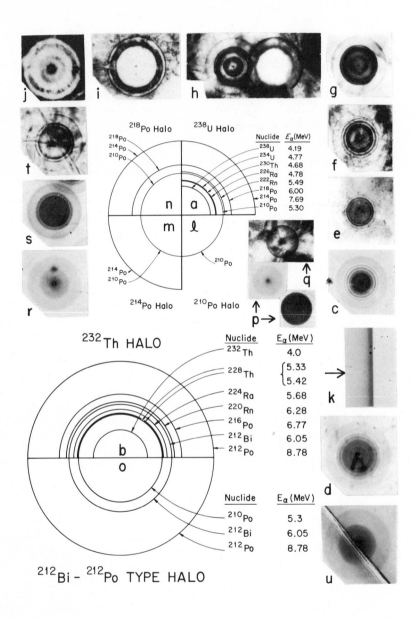

²¹⁸Po Halo   **²³⁸U Halo**

²¹⁸Po
²¹⁴Po
²¹⁰Po

Nuclide	E_α (MeV)
²³⁸U	4.19
²³⁴U	4.77
²³⁰Th	4.68
²²⁶Ra	4.78
²²²Rn	5.49
²¹⁸Po	6.00
²¹⁴Po	7.69
²¹⁰Po	5.30

n   a

m   l

²¹⁴Po
²¹⁰Po

²¹⁰Po

²¹⁴Po Halo   **²¹⁰Po Halo**

²³²Th HALO

Nuclide	E_α (MeV)
²³²Th	4.0
²²⁸Th	{5.33 / 5.42}
²²⁴Ra	5.68
²²⁰Rn	6.28
²¹⁶Po	6.77
²¹²Bi	6.05
²¹²Po	8.78

b
o

Nuclide	E_α (MeV)
²¹⁰Po	5.3
²¹²Bi	6.05
²¹²Po	8.78

²¹²Bi - ²¹²Po TYPE HALO

to identify the nuclides decaying (through the energies and through genetic connections). This latter possibility is particularly exciting because classes of halos exist which correspond to no known radionuclide. Barring the possibility of a nonradioactive origin, these are evidence for hitherto undiscovered or presently extinct radionuclides. Joly, a geology professor at Dublin, lost nearly all his halo evidence for an element he called "hibernium" in the Easter uprising of 1916.

This review of halos and their bearing on nuclear science is organized into three main sections covering: the radioactive origin and provenance of halos, evidence for constancy of radioactive decay energies over geologic time, and variant halos and evidence for unknown radionuclides.

THE RADIOACTIVE ORIGIN AND PROVENANCE OF HALOS

Evidence to support the Joly-Mügge hypothesis of a radioactive origin of halos is both analytical and synthetic. The correspondence between halo dimensions and the ranges of typical alpha particles (extrapolated from air to mica) is suggestive of a causative connection, and an even stronger qualitative argument is the correspondence between the number of α-emitters in the ^{238}U (or ^{232}Th) series and the number of rings in a corresponding halo. Schematic drawings (based on air α-ranges) of a ^{238}U halo (Figure 1a) and a ^{232}Th halo (Figure 1b) may be compared with halo photomicrographs in Figures 1c and 1d respectively. This illustrates rather well the compelling evidence of the patterns. Quantitative evidence about the patterns will be described after noting the synthetic evidence.

Rutherford (3) and Joly & Rutherford (4) provided synthetic evidence by showing that α-radioactivity could produce halo-like coloration in solids. Rutherford (3) in 1910 reported that radon enclosed in a soda-lime glass capillary tube produced a colored annulus whose depth approximated the range of radon alphas in glass. In 1913, Joly & Rutherford (4) reported that a dose of 1.5×10^{13} α-particles/cm^2 produced a halo-like coloration in biotite (an Fe-rich mica).

Quantitative evidence results from using newer methods that permit observation

---

←

Figure 1   The scale for the photomicrographs is 1 cm equivalent to 45 μm. (a) Schematic drawing of ^{238}U halo with radii proportional to α-ranges in air. (b) Schematic drawing of ^{232}Th halo. (c) ^{238}U halo in biotite formed by sequential α-decay of the ^{238}U decay series. (d) ^{232}Th halo in biotite. (e) Embryonic ^{238}U halo in fluorite. (f) Well-developed ^{238}U halo in fluorite. (g) Overexposed ^{238}U halo in fluorite showing inner ring diminution. (h) Two overexposed ^{238}U halos in fluorite showing inner ring radius diminution in one halo and obliteration in the other. (i) Further overexposed ^{238}U halo in fluorite showing outer ring reversal effects. (j) Second stage reversal ^{238}U halo in fluorite. Rings not related to ^{238}U α-emitters. (k) Coloration band formed in mica by 7.7 MeV ^{4}He ions. Arrow shows direction of beam penetration. (l) Schematic of ^{210}Po halo. (m) Schematic of ^{214}Po halo. (n) Schematic of ^{218}Po halo. (o) Schematic of ^{212}Bi-^{212}Po halo. (p) Two ^{210}Po halos in biotite, one light and one very dense. (q) ^{210}Po halo in fluorite. (r) ^{214}Po halo in biotite. (s) ^{218}Po halo in biotite. (t) ^{218}Po halo in fluorite. (u) ^{212}Bi-^{212}Po type halo in biotite.

of single fission tracks in biotite (Fleischer, Price & Walker 5). After appropriate etching, an embryonic U halo (only first ring visible) exhibits a cluster of 20 to 30 ^{238}U fossil fission tracks (Gentry 6), implying that $\sim 5 \times 10^7$ ^{238}U atoms have decayed ($\lambda_\alpha/\lambda_f = 2 \times 10^6$ for ^{238}U). Thus at the first U ring radius of 13 μm the natural α-dose approximates that determined by Joly & Rutherford (4). This limited agreement in diverse samples is accidental, for, as Picciotto & Deutsch (7) have shown, different biotites may show wide coloration threshold responses.

By irradiating halo-containing biotites with ^{4}He ions Gentry (6, 8) found the natural and induced coloration thresholds compare favorably enough to suggest that for some biotites reciprocity is a valid principle (i.e. discoloration is approximately independent of dose rate up to an α flux of 10 nA/mm^2). Furthermore, coloration reversal effects arising from and occurring adjacent to highly radioactive inclusions (Figure 2c) have been duplicated by high-dose exposures (7). However, other halo reversal effects suggest reciprocity is not uniformly applicable. For example, in certain samples coloration inversion occurs immediately exterior to the halo ring terminus, thus accentuating the halo ring structure (Figures 1c, d). Generally this effect is not observed in briefly irradiated (0.1 to 30 min) samples of biotite or fluorite. This experimental approach to halo coloration as pursued by Hövermann (9), Poole (10), Przibram (11), Ramdohr (12), and others (3, 4, 6–8), informative as it has been, has not been paralleled by an equivalent amount of theoretical work.

Now halos generally occur in an igneous environment and even in some igneous fragments that occasionally occur interspersed in sedimentary rocks (Stark 13). An exception is their occurrence in coal and asphaltized wood (Jedwab 14). While mineralogists have found over 40 minerals in which halos occur, their distribution within these minerals is very erratic (12, 13). Further, halos in opaque minerals, such as ilmenite, are of course visible only in reflected light and exhibit poorly defined ring structure, partly due to smearing effects produced by large ($\simeq 10$ μm) halo inclusions (Ramdohr 12). Only those minerals that are transparent in 30-μm thin section are suitable for detailed ring analysis. Even in these considerable microscopic scanning time is required to locate good halo specimens.

Ease of thin section preparation and fairly good registration properties have made biotite an ideal choice for several halo investigations (namely Joly 15, van der Lingen 16, Wiman 17, Iimori & Yoshimura 18, Kerr-Lawson 19, Henderson et al 20, and Gentry 6, 8). The three-dimensional nature of a halo is demonstrated when a biotite specimen is prepared for microscopy. Because of its perfect cleavage properties, the leaves of a book of mica are easily separated with Scotch tape, each successive section revealing a ring pattern of increasing size until the diametral section is obtained. Ring sizes must be measured with diametral sections, and most accurate measurements result from specimens having exceptionally small nuclei.

As good as some halos in certain biotite specimens are, the investigations of Hirschi (21), Gudden (22), Schilling (23), and Gentry (24) have shown that fluorite halos afforded even better microscopic resolution of the halo rings. Mahadevan (25) has shown that halos in cordierite may also yield suitable ring structure for accurate radii measurement.

The halo inclusions occur as accessory minerals and have been identified as zircon, xenotime, monazite, allanite, sphene, apatite, uraninite, and thorite (Joly 15, Ramdohr 12, Hutton 26, Snetsinger 27, and Gentry 8, 28). However, their small size sometimes prevents use of petrologic identification techniques.

Uranium, Th, and other specific halo types have been observed mainly in Precambrian rocks (6, 8, 15, 17–20, 28), but much yet remains to be learned about their occurrence in rocks from other geological periods. Mineralogical studies have shown halos do exist in rocks stretching from the Precambrian to the Tertiary Periods (13, Holmes 29), but unfortunately the halo type existing in these formations was not identified by ring structure.

EVIDENCE FOR CONSTANCY OF RADIOACTIVE DECAY ENERGIES OVER GEOLOGIC TIME

Because radiohalos are integral α-radiographs, an early interest was to ascertain whether U and Th halo radii were of the same size in rocks from different geological periods. Any unaccounted for nonuniformity would imply different decay energies and perhaps suggest a different value of λ, the decay constant. Such investigations were based on the premise that well-defined halo rings could be associated with specified α-energies.

Joly, perhaps not realizing all the subtle factors which may influence halo radii sizes, announced in 1923 that U halo radii were indeed variable in rocks from different geological periods and suggested this was evidence for a change in λ. In 1926–1927, other investigators (18, 19, 23) reported that U and Th halo radii in biotite, cordierite, and fluorite approximated the expected values. However, these measurements were made on halos within the same host rock and did not in themselves counter Joly's claim of variable halo radii in rocks from different epochs.

Even so, Kerr-Lawson (19) made an important observation. In very thin biotite flakes, he saw the inner ^{238}U ring resolved from the subsequent U halo rings (see Figure 1a, c). Since Joly (15) had not reported this inner ring, Kerr-Lawson inferred Joly's halo mineral sections were too thick to permit good ring resolution and attributed Joly's variable halo radii to a confusion in measuring the inner U halo rings. Personal examination of some of Joly's thin sections has confirmed this point.

In the 1930s Henderson and co-workers (20) performed a classic series of experiments using a halo photometer for direct recording of halo ring patterns. With little theoretical justification Henderson et al (20) compared [as had others (15, 18, 19)] halo ring sizes with the minima of the composite curve formed by superimposing all the ^{238}U decay chain Bragg ionization curves. There were inherent problems because a single composite curve could at best represent only one of the many halo growth stages. Yet this approach seemed partially successful, for the composite curve (uncorrected for inverse square effect) did resemble the photometric scan of a U halo which showed all rings (Figure 1c). However, when corrected for the inverse square effect, the peaks and valleys in the curve previously associated with the halo rings became almost nonexistent. Clearly it is not possible to obtain reliable standards for radii comparison using this procedure.

Further attempts to compare halo radii with equivalent mineral ranges derived from α-ranges in air (20) are also unsatisfactory because of uncertainties in the air-mineral conversion factor. Comparison of halo radii with α-ranges calculated from the Bragg-Kleeman (or a similar) relation is possible, but necessarily assumes that ring coloration extends the full α-range. Now newer techniques do allow appropriate standards to be developed, but before discussing these we first digress to examine in more detail halos in different growth stages.

Radiohalos in fluorite will be used as examples. Figure 1*e* shows an embryonic U halo wherein only the first two rings are prominent. Figure 1*f* shows the normal or intermediate stage U halo wherein all rings may be detected. The U halo in Figure 1*g* is overexposed in the inner halo region, resulting in coloration reversal and the formation of a diminutive ghost ring in the center. Figure 1*h* shows two other partially reversed U halos, one of which shows the diminutive inner ring, while the other has experienced complete obliteration of all the inner rings. The U halo in Figure 1*i* is even more overexposed, and encroaching reversal effects have given rise to another ghost ring just inside the outermost periphery. Figure 1*j* shows a still more highly overexposed U halo in which second-stage reversal effects have produced spurious ghost rings that are unrelated to the terminal α-ranges.

Tracing out the above pattern of U halo development in fluorite is no straightforward task. Only by observing differential growth increments in thousands of halos (produced of course by differential amounts of U in the halo inclusions) is it possible to construct the sequence shown. Earlier investigators (12, 23) as well as a later one (Gentry 28, 30) at one time erred in inventing new α-activity to account for some of the aforementioned ghost rings. Clearly a one-to-one correspondence between halo radius and α-energy is not always valid.

We return to the discussion of the constancy of α-decay energies and the problem of developing reliable standards for halo radii comparisons. The most direct technique is to irradiate halo-containing minerals with a sufficient dose of monoenergetic ^{4}He ions until halo-like coloration develops (Gentry 8, 24). If reciprocity holds, the ^{4}He ions will produce a coloration band (see Figure 1*k*) that in theory is equivalent in depth or size to a halo radius produced by the same energy α-particles. These induced coloration band (CB) sizes then form the standards against which halo radii may be compared. [Interestingly a densitometer profile of the CB in Figure 1*k* shows a marked resemblance to the shape of the Bragg ionization curve, thus providing some basis for the superposition procedure used by Henderson (20) and Joly (15).] However, the actual comparison procedure requires that certain additional factors be considered.

First, in some minerals, especially biotite, this reviewer has found that halo radii are somewhat dose dependent; darker rings show slightly higher radii than faint rings (24). (CB sizes show a similar effect, implying coloration intensities in natural and induced specimens should be matched before size comparisons are made.) This dose effect is often masked by subtle halo radius variations produced by attenuation of the αs emitted within the inclusion itself. For example, when the inclusion (e.g. zircon) is more dense than the host mineral (e.g. biotite), slightly smaller radii result in embryonic halos; extreme values are reached only after a heavy dose. However, a heavy dose means a dark halo which tends to obscure the inner halo rings, making measurements difficult.

Although the finite inclusion size renders all radii measurements uncertain to a degree, there are two cases when this correction is minimized. For *densely colored* halos surrounding large inclusions (e.g. see Figure 2c), the only reasonable radius measurements are those made from the inclusion edge. In the other case for halos in biotite with only tiny 1-μm inclusions, this reviewer arbitrarily makes no correction for inclusion size.

In fluorite the situation is different. Here the effect of finite inclusion size is minimized because some halos exist with nuclei only 0.5 μm or less in diameter. Yet in contrast to halos in biotite, there is evidence that in some cases radioactivity is surficially distributed on the inclusion, implying halo radii be measured from the inclusion edge. Fortunately, the halo ring size in this mineral is only very minimally dependent on α-dose so that even embryonic halo rings form at practically maximum size (22–24).

Table 1 shows U halo radii measurements in biotite, fluorite, and cordierite (19, 20, 23–25) as well as CB sizes produced by accelerator ^{4}He ion beams of varying energy in the same minerals (Gentry 24). Note that the comparison between fluorite CB sizes and halo radii is quite good, while in biotite the halo radii are somewhat smaller than the equivalent CB sizes. As discussed earlier, this deviation in biotite is likely due to the effects of a finite inclusion size plus the dose-dependent radius effect rather than a change in α-energy.[2] Note also the comparison between CB sizes and halo radii in cordierite. Here the agreement is fair even though the U halos in cordierite possessed large inclusions, necessitating that halo radii be measured from the inclusion edge (Mahadevan 25).

Interestingly densitometer profiles of some of my U halos show the same small halo rings that Henderson (20) noted, but these appear to be only artifacts of the coloration process and are not included in Table 1.

With respect to the decay rate question, Spector (31) has argued that the differences between Henderson et al (20) halo radii measurements and equivalent air mineral ranges present a case for a variable λ. In the light of the above experimental uncertainties, this conclusion is not necessarily valid. On the other hand, Gentry (24) has shown that even exact agreement between halo radii and corresponding CB sizes does not necessarily imply an invariant λ and in fact uncertainties in radius measurements alone preclude establishing the stability of λ for ^{238}U to more than 35%.

VARIANT HALOS AND EVIDENCE FOR UNKNOWN RADIONUCLIDES

Variant halos exhibit ring patterns different from either U or Th halos. Some are due to α-decay from known radionuclides while others present diverse ring structure

---

[2] My U halo radii in mica shown in the G column of Table 1 give the range of radii sizes in this particular mica.

Table 1   Halo radii and induced coloration band (CB) size measurements[a]

^{4}He Induced Coloration Band Sizes (μm) Mica $G_L\to$	Mica $G_M\to$	Mica $G_D\to$	Fluorite $G\to$	Cordierite $G\to$	Nuclide	E (MeV)	U Halo Mica $K-L\to$	Mica $H\to$	Mica $G\to$	Fluorite $S\to$	Fluorite $G\to$	Cordierite $M\to$	Cordierite $H\to$	Po Mica ^{210}Po $G_{L-D}\to$	^{214}Po $H\to$	^{214}Po $G_L\to$	^{214}Po $H\to$	^{218}Po $H\to$	^{218}Po $G_{L-M}\to$	Po Fluorite ^{210}Po $G\to$	^{218}Po $G\to$
13.4	13.8	14.2	14.1	16.2	^{238}U →	4.2	12.3	12.7	12.2→13.0	14.0	14.2	16	NP[d]	NP	NP	NP	NP	NP	NP	NP	NP
NM[b]	NM	NM	17.3	19.2	^{226}Ra →	4.77	15.4	15.3	14.9→15.6	16.9	17.1	19	NP	NP	NP	NP	NP	NP	NP	NP	NP
NM	NM	NM	NM	NM	^{230}Th →	4.66	NR	NR	NR	NR	NR	NR	NP	NP	NP	NP	NP	NP	NP	NP	NP
NM	NM	NM	17.3	19.2	^{234}U →	4.78	15.4	15.3	14.9→15.6	16.9	17.1	19	NP	NP	NP	NP	NP	NP	NP	NP	NP
NM	19.3	20.0	19.6	22.5	^{210}Po →	5.3	NR[c]	NR	NR	19.3	19.5	NR	19.8	18.2→19.9	20.0	18.3	19.9	18.1→19.3	NP	19.8	19.8
NM	20.5	21.1	NM	NM	^{222}Rn →	5.49	18.6	19.2	17.9→18.8	20.5	20.5	23.5	NP	NP	NP	NP	NP	NP	NP	NP	NP
NM	23.0	23.9	23.6	26.7	^{218}Po →	6.0	22.0	23.0	21.7→22.2	23.5	23.5	26.5	NP	NP	NP	NP	24.0	22.5→23.3	NP	NP	23.7
33.1	33.9	34.4	34.6	38.7	^{214}Po →	7.69	33.0	34.1	31.0→32.9	34.5	34.7	38.5	NP	NP	34.5	33.3	34.0	33.2→34.0	NP	NP	34.9

[a] The symbols $K-L$, H, S, M, and G represent halo measurements by Kerr-Lawson, Henderson, Schilling, Mahadevan, and Gentry; G_L, G_M, and G_D represent Gentry's measurements on light (L), medium (M) [dose $\simeq$ 10 to 20 times coloration threshold], or dark (D) [$\simeq$ 50 times coloration threshold] induced coloration bands. G_{L-D} and G_{L-M} represent Gentry's measurements on light-to-dark and light-to-medium coloration halos respectively; halos with these designations were visually determined. Gentry's measurements were made with a filar micrometer that could be read to 0.07 μm. The estimated overall uncertainty was ±0.3 μm.

[b] NM = not measured.

[c] NR = not resolved.

[d] NP = not present.

unlike any known α-decay sequence. In this reviewer's opinion unusual size halos present a most interesting challenge for geochemistry, nuclear science, and cosmology.

Polonium Halos

PHOTOGRAPHIC EVIDENCE   Figure 1*l, m, n, o* shows schematic drawings of several Po halo types. Note that those in Figure 1*l, m, n* exhibit only Po rings and are designated by the first (or only) Po α-emitter in the decay sequence. Note also these Po α-emitters are end members in the ^{238}U decay chain. The similarity between these schematics and the actual photographs in Figure 1 is easily seen; i.e. compare Figure 1*l* with Figure 1*p, q*; Figure 1*m* with 1*r*; and Figure 1*n* with 1*s, t*. Figure 1*o* shows a combination ^{212}Bi-^{212}Po halo, which also has a ^{210}Po ring that is difficult to see in the photograph in Figure 1*u*. (Note this inner ring is not due to ^{228}Th α-decay because sequential decay rings are missing.)

Figure 1*q* shows a ^{210}Po halo in fluorite. The two slightly different size ^{210}Po halos in Figure 1*p* well illustrate the dose effect on halo radius in biotite. Figure 1*r* shows a ^{214}Po halo in biotite that approximately matches the coloration of the light ^{210}Po halo in 1*p*. In these cases the ^{210}Po halo radii agree well. Likewise the dark inner ^{210}Po ring in the ^{218}Po halo shown in Figure 1*s* compares well in size with the dark ^{210}Po halo in 1*p*. Figure 1*t* shows a ^{218}Po halo in fluorite. Here an important observation can be made. Note that the annulus between the ^{210}Po and ^{218}Po rings in Figure 1*t* (cf Figure 1*n*) is distinctly wider than the same annulus in the ^{238}U halo in 1*f*. This is clear evidence that the ^{222}Rn ring in Figure 1*f* is missing from the halo in 1*t*. In other words, the halo in Figure 1*t* indeed originated with ^{218}Po rather than ^{222}Rn α-decay.

COMPARISON OF PO HALO RADII AND CB SIZES   Table 1 gives measurements of Po halos in biotite and fluorite by Gentry (24) and Henderson et al (32). Henderson's measurements were made using the halo photometer, while the rest were made with a micrometer ocular. For comparison purposes I include measurements of some Po halos in biotite that exhibit radius variations due to dose effects. As Table 1 shows, the correspondence between Po halo radii and induced CB sizes leaves little doubt the Po halo ring structure does indeed match the respective Po α-emitters. Note the very good agreement between Po and ^{238}U halo radii in fluorite. In biotite the ^{238}U halo radii must be compared with very light Po halos in order to minimize radii differences from dose effects. Although not shown in Table 1, Po halo radii measurements in cordierite (25) also agree with CB sizes.

GENESIS, GEOPHYSICAL IMPLICATIONS, AND QUESTION OF ISOMER PRECURSORS
Because Schilling (23) saw Po halos located only along cracks in his Wolsendorf fluorite sample, he suggested they originated from preferential deposition of Po from U-bearing solutions. [I have also found them separated from conduits (Figure 1*q*).] Later Henderson (32) invoked a similar but more quantitative hypothesis to explain Po type halos along conduits in biotite. Those Po halos he found occurring apart from conduits (similar to those found by Gentry in Figure

1p, r, s) were more difficult to account for. Here a very qualitative laminar flow hypothesis was proposed. The intervention of World War II and his untimely demise soon thereafter prevented Henderson from determining whether these explanations were valid. [Although Henderson (32) suggested that such types existed, the halo in Figure 1u was discovered only recently by Gentry and cannot be explained on the basis of U daughter α-activity alone.]

Now the reason for the various attempts to account for Po halos by some sort of secondary process is quite simple; the half-lives of the respective Po isotopes are far too short to be reconciled with slow magmatic cooling rates for Po-bearing rocks such as granites ($T_{\frac{1}{2}} = 3$ min for ^{218}Po).

Yet Gentry (6), by using fission-track and α-recoil techniques, found no evidence for a secondary origin of those Po halos in biotite, which occurred apart from conduits (cf Figure 1p, r, s). Consistent with the ring structure, fission-track analysis of the Po halo inclusions showed very little, if any, U. Further, the α-recoil technique (Huang & Walker 33), which permits the observation of a single α-recoil pit in biotite, was employed to measure the distribution of decayed α-radioactivity in regions both adjacent to and far removed from Po halo inclusions. No differences in α-recoil density were noted in the two areas. If U daughter α-activity had fed the Po inclusions, a significantly higher α-recoil density would have been in evidence.

One solution to this dilemma is the suggestion (24) that the parent nuclides of the Po in the halo inclusions were long half-life β-decaying isomers that may yet exist. Halo ring structure allows this, for β-radiation produces no coloration in halos in biotite, fluorite, or cordierite. [Laemmlein (34) reported β-halos in a certain quartz sample, but this is unconfirmed.] If such isomers existed, Po, U, and Th halos could all be explained by assuming that minute quantities of the respective nuclides were incorporated into separate halo inclusions either before or coincident with the host rock crystallization.

Dwarf Halos

Joly (15) reported the existence of dwarf halos with radii ~ 5.2 and 8.5 μm in the black micas from the pegmatite quarry at Ytterby near Stockholm. A rather unsatisfactory attempt has been made (7) to identify the 5.2-μm dwarf halo with ^{147}Sm α-decay ($E_\alpha = 2.24$ MeV). Dwarf halos are extremely rare and erratically distributed even in the few mica samples which contain them. Joly (15) considered their radioactive origin beyond question and attributed their bleached appearance to radiation overexposure or to elevated temperatures.

Gentry (28) has also found dwarf halos in a mica specimen from Ytterby, and the sizes are difficult to correlate with α-decay systematics of known radioactive nuclides. The smallest dwarf halos range from only 1.5 to about 2.5 μm with associated α-energies in the 1-MeV range (Figure 2a). The half-lives of known radionuclides are in excess of 10^{13} years for α-decay energies of 2 MeV or less. Such weakly active nuclides almost escape detection and would hardly be expected to produce a halo. Other dwarf halos with radii from 3 to 11 μm (Figure 2b) correspond to α-energies of ~ 1.1 to 3.4 MeV. Some of these dwarf halos reflect

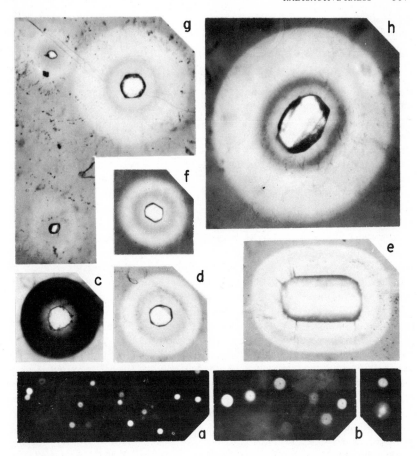

Figure 2   The scale for all photographs is 1 cm = 50 μm. (*a*) Dwarf halos ($\simeq$ 2-μm radius) in Ytterby mica. (*b*) Dwarf halos (3 $\mu m \leqq r \leqq$ 9 μm) in Ytterby mica. (*c*) Overexposed Th halo in ordinary biotite. (*d*) Th halo in Madagascar mica. (*e*) Th halo in Madagascar mica with a larger inclusion. (*f*) U halo in Madagascar mica. (*g*) Giant halo of $\simeq$ 65-μm radius, and two light Th halos (Madagascar mica). (*h*) Giant halo of $\simeq$ 90-μm radius, Madagascar mica.

coloration differences possibly due to varying amounts of parent radionuclides in the inclusions. In this respect the occurrence of dual-ring dwarf halos also suggests a radioactive origin. Although Gentry has not yet seen dwarf halos in cordierite, Mahadevan's (25) report of equivalent dwarf sizes in this mineral lends much credence to the radioactive origin of such halos. Whether there exists a causal relation between the dwarf halos and the previously reported, though unconfirmed, low-energy α-activity found by Brukl, Hernegger & Hilbert (35) is presently open to question. As noted earlier, the dwarf rings which Henderson (20) reported in some U halos appear to be only artifacts of the coloration process.

X Halos and Other Intermediate Size Varieties

Still rarer than the dwarf halos are the X halos first reported by Joly (15) in the micas from Ytterby and Arendal (Norway). Later van der Lingen (16) reported halos of similar dimensions occurred in a granite near Capetown. According to Joly (15) the inside ring of the X halo may be somewhat diffuse and measures about 8.5 to 9.8 μm in radius, corresponding to an E_α of about 2.9 to 3.2 MeV. The bleached rings extend out to a radius of about 14 to 15 μm, and sometimes an adjacent dark ring is evident at about 17 μm ($E_\alpha = 4.4$ to 5.0 MeV). The outer wide band extends to approximately 28 μm corresponding to an E_α of 6.7 to 6.9 MeV. Despite some similarities with the Th halo there is no known α-decay sequence corresponding to these energies. Although well documented (and even photographed) by Joly, Gentry has yet to find any X halos in scanning thousands of sections of biotite from Ytterby, Arendal, and Capetown. Above all others, the search for this halo has taxed my eyesight. In this respect the much earlier (and quite obscure) reports (Schintlmeister et al 36) of genetically related α-decay of 3 MeV and 4.5 MeV are interesting, but they remain unconfirmed. Therefore any association with the very elusive X halo is only speculation.

Gentry (37) has reported the existence of a halo with rings apparently due to α-energies of about 4.4 to 5.4 MeV. However, the relatively large size of the inclusion of this halo (6 μm in diameter) necessitates a re-examination of this halo with other techniques. Possibly the inner disc is a ghost ring resulting from α-particle attenuation within the inclusion. If so, the halo may be the ^{210}Po variety. I have also reported a halo possibly due to ^{211}Bi α-decay (30). Thus far I found only two of these halos and more specimens are needed to confirm the identity of this type.

Another unusual halo was the so-called D halo reported by Henderson (32) to exhibit a diffuse boundary of radius ~ 16 μm. He tentatively attributed this halo to ^{226}Ra α-decay because the radius approximated the size of the ^{226}Ra ring in the U halo. The absence of other rings, which should have appeared from daughter product α-activity, was explained by assuming ^{222}Rn diffused from the inclusion before it decayed. This, of course, is contrary to the situation observed in the normal ^{238}U halo. Gentry has examined such halos, both in Henderson's original thin sections and in other biotites. They generally possess inclusions several microns in diameter and are without detailed ring structure. While they cannot be explained on the basis of ^{226}Ra α-decay, the presence of fossil fission tracks indicates that U series α-emitters produced part of the coloration in this halo type.

In addition to U and Th halos Iimori & Yoshimura (18) reported three halo sizes they designated as Z_1, Z_2, and Z_3 halos. These halos were attributed to actinium series α-emitters. Gentry has examined some of the original slides as well as separate Japanese biotite samples in which such halos were reported. Gentry has observed that many halos in this biotite are very dark, so that it is necessary to prepare very thin sections in order that the halo inclusion can be seen. Some of the original sections containing Z halos were simply too thick to permit accurate halo radii measurements. In my opinion the Z_1 and Z_2 halos can be explained as U and/or combination U-Th halos without postulating the actinium α-emitters. The Z_3 halo is actually a ^{210}Po halo. In this context it should be noted that Joly's "emanation" halo was also a ^{210}Po halo.

Giant Halos and Unknown α-Radioactivity

Although Hirschi (21) was apparently the first to report giant halos, Wiman's (17) report of biotite halos with radii of ~ 55 μm and 67 μm was better documented. Even so, on the basis of a rather cursory examination, Höppe (38) was unable to confirm Wiman's results. Gentry (8), however, later found, after examining more than 1000 thin sections, that giant halos do exist in the rocks described by Wiman.

A more abundant source of giant halos was found in a mica from Madagascar. In this specimen Gentry (8) reported seven different groups of giant halos ranging from 45 μm to 110 μm. An unanswered question is whether these halos originate with high energy α-emitters in the range from 9.5 MeV to 15 MeV. One group in particular with radii between 50 and 58 μm was tentatively attributed to the low-abundance (1 : 5500), high energy (10.55 MeV) α-particles of ^{212}Po, the last α-emitter in the Th decay chain. Tentatively, this low-abundance group was associated with the 55-μm radius giant halo in the granites that Wiman studied. No other known nuclides occur with sufficient energy and/or abundance to produce the other groups of giant halos.

Therefore, seven other possibilities were considered (8) as explanations for the giant halos. These were: (a) Variations in α-particle range due to structural changes in mica, (b) Diffusion of a pigmenting agent from the inclusion into the matrix, (c) Diffusion of radioactivity from the inclusion to the matrix, (d) Channeling, (e) β-radiation instead of α-emission, (f) Long-range α-particles from spontaneous fission, and (g) α-particles or protons from (n, α) or (α, p) reactions.

Gentry (8) has shown that none of the above alternatives are very probable, implying that the giant halos may well represent unknown α-radioactivity. In this respect the giant halos in cordierite reported (and photographed) by Krishnan & Mahadevan (39) are very significant. The "Th" giant halos they report may be due to the low abundance αs from ^{212}Po. In contrast the "U" giant halos correspond to an E_α of about 9.5 MeV and cannot be explained on the basis of a low-abundance ^{238}U decay series α-emitter.

It has been suggested (24) that high spin or shape isomers that exhibit high energy α-decay may be responsible for the giant halos (8), a hypothesis that is presently under investigation (Gentry et al 40). There is practically nothing to suggest a superheavy element connection (8).

Figure 2c shows a dark, overexposed and partially reversed (near the inclusion) Th halo in a sample of ordinary biotite. Figure 2d, e shows Th halos in the Madagascar mica. Here the α-dose has created a lighter rather than a darker coloration. Figure 2f shows a U halo in the same mica. The ring structure in Figure 2d, e, f is missing, of course, due to the large inclusions. Figure 2g shows a 65-μm giant halo close to two very light Th halos. Figure 2h exhibits a giant halo of about 90-μm radius. If produced by high energy αs, the halos in Figure 2g, h correspond to an E_α of ~ 11.6 and 14 MeV respectively.

Mass Analysis of Halo Inclusions

The preceding sections have presented data on unusual size halos that may well have important implications for nuclear science and cosmology. If, for example, Po halos did originate with isomers that were long-lived β-decaying precursors of Po, then some of these isomers may represent extinct natural radioactivity (Kohman 41) and hence will be of cosmological significance. Possibly the isomer hypothesis for the giant halos may bear similar implications. Now the above inferences have been deduced solely on the basis of halo ring structure. Clearly the variant halo inclusions themselves contain another important source of information about the radionuclides which generated such halos. However, the small inclusion size effectively prevents use of ordinary mass spectrometric techniques.

To circumvent these problems Gentry has utilized the recently developed ion microprobe mass spectrometer (Anderson & Hinthorne 42) for in situ mass analysis of even the smallest halo inclusions (8, 28, 40). Many U and Th inclusions were analyzed to obtain isotopic data against which the results from the variant halos may be compared. The most important specific application thus far has been the analysis of Po halo inclusions.

Because the Po halos shown in Figure 1p, q, r, s, t all initiate with Po isotopes that terminate with ^{206}Pb, these inclusions should reflect an excess of this Pb isotope; further because ring structure and fission track analysis show only small amounts, if any, of U, mass analysis of such Po halo inclusions should be consistent with these observations as well. A mixed type Po halo such as in Figure 1u (or Figure 1o) would be expected to exhibit excesses of both ^{206}Pb (from ^{210}Po decay) and ^{208}Pb (from ^{212}Po decay). If ^{211}Bi halos have been properly identified (30), an excess of the decay product ^{207}Pb should exist in these inclusions. Generally then, while any given Po halo inclusion might have initially contained varying amounts of the different Po isotopes (or their β-decaying precursors), halo rings would develop only if $\sim 10^8$ atoms of a specific nuclide were present. This implies Po isotope ratios may be variable even when examining different Po halos of the same general type. (In principle this is similar to the case where varying amounts of Th are found in halo inclusions around which only U rings appear.)

Specifically ion probe analyses have shown one ^{210}Po halo inclusion that contained Pb without any detectable ^{204}Pb, U, or Th; the ^{206}Pb/^{207}Pb ratio was $\geqq 20$. Absence of U and Th means this Pb was not radiogenically derived from these elements; absence of ^{204}Pb means this Pb cannot be primordial or common Pb as those terms are usually defined. Briefly, while these values are inexplainable on the basis of any heretofore known types of Pb, they definitely agree with a Pb type derived from Po decay independent of U or Th.

In other Po inclusions the ^{206}Pb/^{207}Pb ratio has been determined as ca 10, 12, 18, 22, 25, 40, 62, and 80. The theoretical maximum possible radiogenic ^{206}Pb/^{207}Pb ratio, based on an instantaneous production of Pb from normal isotopic

U decay, is 21.8. Therefore, ^{206}Pb/^{207}Pb ratios greater than 21.8 not only reflect the existence of a unique type of Pb, but in a different way confirm the existence of Pb derived from Po decay independent of the normal U decay chain (28, 40).

Several of the Po halo inclusions described above were concurrently analyzed for the presence of the postulated Po isomers but with negative results so far (40). Gentry is continuing the search using different types of Po halos taken from rocks from several geological epochs. In closing we note successful application in obtaining the Pb isotope ratios gives promise that the mysteries of the other variant halos will not long remain unsolved.

Literature Cited

1. Joly, J. 1907. *Phil. Mag.* 13:381–83
2. Mügge, O. 1907. *Zentralbl. Mineral.* 1907:397–99 (See also Oak Ridge Nat. Lab. transl. ORNL-tr-757)
3. Rutherford, E. 1910. *Phil. Mag.* 19:192–94
4. Joly, J., Rutherford, E. 1913. *Phil. Mag.* 25:644–57
5. Fleischer, R. L., Price, P. B., Walker, R. M. 1965. *Science* 149:383–93
6. Gentry, R. V. 1968. *Science* 160:1228–30
7. Picciotto, E. E., Deutsch, S. 1960. *Pleochroic Halos.* Rome: Comitato Nazionale per l'Energia Nucleare
8. Gentry, R. V. 1970. *Science* 169:670–73
9. Hövermann, G. 1912. *Neues Jahrb. Mineral. Geol. Palaeontol. Beilageband Bd.* 34:321–400 (See also ORNL-tr-923)
10. Poole, J. H. J. 1928. *Phil. Mag.* 5:132–41; Idem. 1928. *Phil. Mag.* 5:444
11. Przibram, K. 1956. *Irradiation Colors and Luminescence.* London: Pergamon
12. Ramdohr, P. 1933. *Neues Jahrb. Mineral. Beilageband Abt. A* 67:53–65 (See also ORNL-tr-701); Ramdohr, P. 1957. *Abh. Deut. Akad. Wiss. Berlin. Kl. Chem. Geol. Biol.* 2:1–17 (See also ORNL-tr-755); Ramdohr, P. 1960. *Geol. Rundsch.* 49:253–63 (See also ORNL-tr-758)
13. Stark, M. 1936. *Chemie Erde* X:566–630 (See also ORNL-tr-700)
14. Jedwab, J. 1966. *Advan. Chem. Ser.* 55:119–30
15. Joly, J. 1917. *Phil. Trans. Roy. Soc. London Ser. A* 217:51–79; Idem. 1917. *Nature* 99:456–58, 476–78; Idem. 1923. *Proc. Roy. Soc. London Ser. A* 102:682–705; Idem. 1924. *Nature* 114:160–64
16. Lingen, J. S. van der 1926. *Zentralbl. Mineral. Abt. A,* 177–83 (See also ORNL-tr-699)
17. Wiman, E. 1930. *Bull. Geol. Inst. Univ. Uppsala* 23:1–170
18. Iimori, S., Yoshimura, J. 1926. *Sci. Pap. Inst. Phys. Chem. Res.* 5:11–23
19. Kerr-Lawson, D. E. 1927. *Univ. Toronto Stud. Geol. Ser.* 24:54–71; Idem. 1928. *Univ. Toronto Stud. Geol. Ser.* 27:15–27
20. Henderson, G. H., Bateson, S. 1934. *Proc. Roy. Soc. London Ser. A* 145:563–81; Henderson, G. H., Turnbull, L. G. 1934. *Proc. Roy. Soc. London Ser. A* 145:582–98; Henderson, G. H., Mushkat, C. M., Crawford, D. P. 1934. *Proc. Roy. Soc. London Ser. A* 158:199–211
21. Hirschi, H. 1920. *Vierteljahresschr. Naturforsch. Ges. Zuerich* 65:209–47 (See also ORNL-tr-702); Idem. 1924. *Naturwissenschaften* 12:939–40 (See also ORNL-tr-708)
22. Gudden, B. 1924. *Z. Phys.* 26:110–16
23. Schilling, A. 1926. *Neues Jahrb. Mineral. Abh. A* 53:241–65 (See also ORNL-tr-697)
24. Gentry, R. V. Unpublished
25. Mahadevan, C. 1927. *Indian J. Phys.* 1:445–55
26. Hutton, C. O. 1947. *Am. J. Sci.* 245:154–57
27. Snetsinger, K. G. 1967. *Am. Mineral.* 52:1901–3
28. Gentry, R. V. 1971. *Science* 173:727–31
29. Holmes, A. 1931. *Bull. Nat. Res. Counc.* 80:124–460
30. Gentry, R. V. 1967. *Nature* 213:487–89
31. Spector, R. M. 1971. *Phys. Rev. A* 5:1323–27
32. Henderson, G. H., Sparks, F. W. 1939. *Proc. Roy. Soc. London Ser. A* 173:238–49; Henderson, G. H. 1939. *Proc. Roy. Soc. London Ser. A* 173:250–64
33. Huang, W. H., Walker, R. M. 1967. *Science* 155:1103–6
34. Laemmlein, G. G. 1945. *Nature* 155:724–25
35. Brukl, A., Hernegger, F., Hilbert, H. 1951. *Oesterr. Akad. Wiss. Math. Naturwiss. Kl. Sitzungsber. Abt. IIA* 160:129–46

36. Schintlmeister, J., Hernegger, F. 1940. *Ger. Rep. G-55, Part I*; Idem. 1941. *Ger. Rep. G-112, Part II*; Schintlmeister, J. 1941. *Ger. Rep. G-111, Part III*; Schintlmeister, J. 1942. *Ger. Rep. G-186*
37. Gentry, R. V. 1967. *Earth Planetary Sci. Lett.* 1:453–54
38. Höppe, G. 1959. *Geol. Fören. Stockholm Förh.* 81:485–94 (See also ORNL-tr-756)
39. Krishnan, M. S., Mahadevan, C. 1930. *Indian J. Sci.* 5:669–80
40. Gentry, R. V., Cristy, S. S., McLaughlin, J. F., McHugh, J. A. 1973. *Nature* 244:282
41. Kohman, T. P. 1961. *J. Chem. Educ.* 38:73–82
42. Anderson, C. A., Hinthorne, J. R. 1972. *Science* 175:853–60

THE MANY FACETS OF NUCLEAR STRUCTURE[1]

✳ 5542

Aage Bohr and Ben R. Mottelson
Niels Bohr Institute and NORDITA, Copenhagen, Denmark

CONTENTS

EARLY DEVELOPMENTS

Fifty years ago, the nucleus of the atom had just been discovered. Its very existence had provided a decisive clue to the unraveling of the structure of atoms, a development that was rapidly leading to the establishment of quantal mechanics, but the internal structure of the nucleus remained unexplored. However, already the variety of radiations emanating from the radioactive substances had given the first indication of the richness of the nuclear phenomena. The distinction between the α, β, and γ rays established by the early pioneers was gradually to be recognized as a manifestation of the hierarchy of the strong, weak, and electromagnetic interactions, and the nucleus was to become a laboratory for exploring the symmetries and structure of the novel interactions that go beyond the classical world of gravity and electromagnetism.

[1] The present review originates from a request by the International Union of Pure and Applied Physics for a survey of the development and perspectives in the field of nuclear physics to be presented on the occasion of the 50th anniversary of IUPAP. The article represents an expanded version of the talk (delivered by A. Bohr) at the International Symposium organized on this occasion in Washington, D.C. in September 1972.

As the nucleus was subjected to more extensive probes, a wealth of different, often contrasting facets was revealed, and the resulting picture of the nuclear structure underwent profound developments from decade to decade. The discoveries of the thirties gave a broad view of the basic nuclear processes and revealed the complexity and subtlety of the strong interactions. During the first decade after the war, the nuclear physicists found themselves exploring a quite novel type of many-body system, in which the motion of the nucleons gives rise to a shell structure as well as to vibrational and rotational modes. The subsequent period brought into focus the great variety of elementary modes in the nucleus involving shape oscillations, pair correlations, isospin, etc, and decisive progress was made toward understanding the occurrence of these modes in terms of the interplay of the individual particles. Current developments involve an expansion of the field along many frontiers and continue to reveal entirely new facets of the nuclear structure.

THE NUCLEUS AS A QUANTAL MANY-BODY SYSTEM

The developing understanding of the nuclear structure has been part of the broad development of quantal concepts appropriate to the description of systems with many degrees of freedom. The early phases of quantal theory emphasized phenomena that could be described in terms of a few simple degrees of freedom or, as in electrodynamics, in terms of a perturbation in the motion of free quanta. Such phenomena could be directly mastered by a solution of the basic equations of motion. However, in the study of the many-body systems, ranging from the macro-molecules and matter in bulk to the strongly interacting elementary particles, the situation has been different. The structural possibilities and the variety of correlation effects that may occur in a system such as the nucleus are so vast that a frontal attack on the Schroedinger equation in multidimensional configuration space has provided only limited guidance. The crucial problem has been the identification of appropriate concepts and degrees of freedom to describe the observed phenomena. Progress in this direction has been achieved by a combination of many different approaches, including theoretical studies of model systems and the establishment of general relations following considerations of symmetry. Above all, the development has repeatedly been given entirely new directions by clues provided by startling experimental discoveries.

The many-faceted nature of the nuclear systems makes it impossible to give a one-dimensional presentation, but a few examples of phenomena encountered on some of the active frontiers of nuclear research may perhaps convey a feeling for the flavor and perspective of the development.

INTERPLAY OF SHELL STRUCTURE AND NUCLEAR DEFORMATIONS

In the exploration of nuclear dynamics, the efforts to achieve a proper balance between independent-particle and collective degrees of freedom has been a recurring and central theme. This may be true of all many-body systems, but, in the nucleus, because of the possibility of detailed studies of individual quantum states, this issue was encountered in an especially concrete form.

The dual aspects of nuclear structure manifest themselves in a very direct manner in the energy of the nucleus considered as a function of its shape. While the general features of this "potential energy function" can be described in terms of bulk properties of the nuclear matter such as surface tension and electrostatic energy, the specific geometry of the quantized orbits of the individual nucleons contribute important anisotropic effects; a striking consequence is the occurrence of nuclear equilibrium shapes deviating strongly from spherical symmetry.

The effect of the shell structure on the nuclear potential energy has come into new perspective as a result of the recent discovery of metastable states of the heavy nuclei that decay by spontaneous fission. The potential energy function of such a nucleus is illustrated schematically in Figure 1. The dashed curve represents the estimate of this function as given by the liquid drop model; the occurrence of a maximum, the fission barrier, results, as is well known, from the competition between the surface tension and the electrostatic repulsion. The quantized energies of the individual particles have a more specific dependence on the shape and symmetry of the potential, and give rise to rather large modifications of the potential energy function. In particular, the shell-structure effect implies a ground state equilibrium with an eccentricity of about 20% and a second minimum at much larger deformations, in the region of the fission barrier. The occurrence of the second minimum is responsible for the existence of the isomeric states that decay by spontaneous fission with lifetimes shorter than those of the ground state by factors

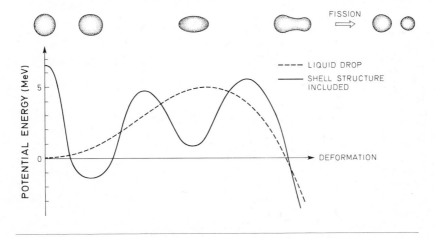

Figure 1   Potential energy function for deformations leading to fission. The schematic figure illustrates qualitative features in the structure of the fission barrier. The deformation parameter labels the path towards fission, as indicated by the shapes at the top of the figure. The shapes at the first and second minima correspond to those observed in the ground states and in the shape isomers in nuclei in the region of uranium.

of 10^{20} or more. This feature of the potential energy function is also revealed by many striking phenomena in the fission process. An example is shown in Figure 2, which exhibits a resonance structure in the cross section for fast neutron fission of ^{230}Th. The resonance occurring in the threshold region can be interpreted in terms

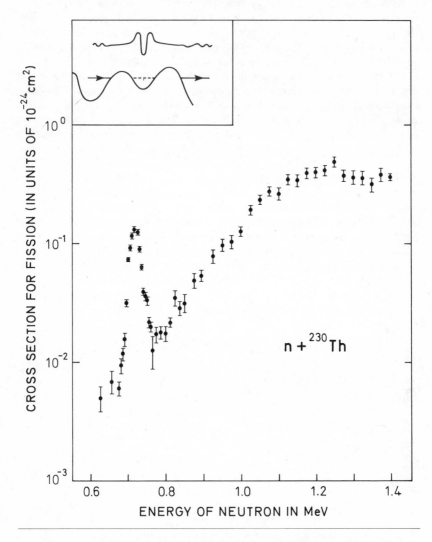

Figure 2   Resonance in threshold function for neutron-induced fission of ^{231}Th. This figure shows the cross section for fission as a function of the neutron energy (1). The insert illustrates the interpretation of the resonance phenomenon in terms of a semistationary state of vibration in the second minimum of the potential energy function.

of standing waves of vibrational motion in the second minimum, not unlike the resonance transmission in a Fabry-Perot interferometer. A wide range of ingenious approaches are currently being brought to bear in the developing spectroscopy of excitations in this new phase of the nuclear systems.

The occurrence of a highly deformed metastable state of the nucleus, referred to as a shape isomer, reflects a special stability associated with the shell structure. The nature of the new shells can be understood in a simple manner by reference to one-particle motion in a spheroidally deformed harmonic oscillator potential. As illustrated in Figure 3, the degeneracies of the isotropic oscillator are removed by the deformation, but new major shells (degeneracies) reappear when the oscillator frequencies in the different directions have rational ratios. Especially large effects occur for a deformation with the frequency ratio $\omega_\perp : \omega_3 = 2:1$, and the associated nucleon numbers for closed shells are $N = \ldots 110, 140, \ldots$ The nuclear potential differs from the harmonic oscillator in radial dependence as well as in the occurrence of a rather large spin-orbit coupling; as shown in Figure 4, the inclusion of these effects leaves intact the main features of the oscillator shell structure in the $2:1$ potential, but modifies the closed-shell numbers to $N = \ldots 116, 148, \ldots$ The number $N = 148$ corresponds with the region of neutron numbers for which the shape isomers are observed to be especially stable.

The discovery of the shape isomers has given new scope to the concept of shell structure in quantal many-body systems, and has raised the question of the general conditions on the potential that are necessary for the occurrence of significant deviations from uniformity in the eigenvalue spectrum. In particular, the development has focused attention on the intimate connection between the occurrence of shell structure in the quantal spectrum and the occurrence of degenerate families of periodic orbits in the corresponding classical motion.

INTERACTION OF NUCLEAR MATTER IN BULK (HEAVY ION REACTIONS)

The study of the fission isomers is part of a much broader program exploring the stability of nuclear matter as a function of neutron and proton numbers as well as of the deformation parameters. One of the exciting perspectives is the possible occurrence of islands of metastable nuclei with mass numbers much greater than those hitherto encountered, or with very different ratios between the numbers of neutrons and protons.

The part of the multidimensional particle-number and deformation space accessible to experimental study is being greatly expanded by the possibility of studying reactions initiated by the impact of heavy ions. As already indicated by pioneering experiments like those shown in Figure 5, such reactions lead with high probability to the transfer of many particles between the interacting nuclei. The current discussion is concerned with the physical conditions that determine the numbers of nucleons transferred, the nature of the flow involved, and the extent to which a statistical equilibrium is established during the reaction.

Reactions between heavy nuclei are also found to provide new possibilities for

$$V = \tfrac{1}{2}M(\omega_\perp^2(x_1^2 + x_2^2) + \omega_3^2 x_3^2)$$
$$E = \hbar\omega_\perp(n_\perp + 1) + \hbar\omega_3(n_3 + \tfrac{1}{2})$$

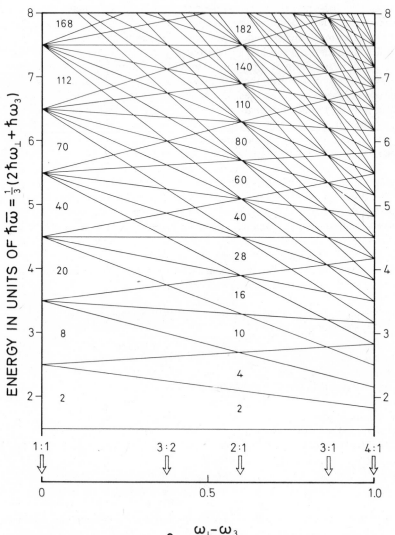

$$\varepsilon = \frac{\omega_\perp - \omega_3}{\bar{\omega}}$$

exploring such features of nuclear matter as the texture of the surface. The example in Figure 6 shows the elastic cross section as well as the inelastic excitation of a shape oscillation, considered as a function of incident energy. The semiclassical nature of the collision process makes it possible to relate the incident energy to the distance of closest approach, and the observed oscillations in the cross sections are a sensitive measure of the interactions that come into play in a grazing collision.

Studies have already been initiated with projectiles much heavier than those in the above examples. It is apparent that we are on the threshold of a whole new dimension of nuclear studies involving nuclear matter under pressure and strains that may only be rivaled in the extreme phases of cosmic evolution, such as in the formation of the neutron stars.

ELEMENTARY MODES OF EXCITATION; UNIFIED DESCRIPTION OF NUCLEAR DYNAMICS

The great sophistication that has been attained in the experimental study of nuclear spectra is making possible a penetrating analysis of the concept of the elementary modes of excitation. Examples of elementary excitations based on the closed-shell configuration of $^{208}_{82}$Pb are shown in Figure 7 (upper part). The single-particle and single-hole states are especially identified in reaction processes in which a nucleon is added to or removed from the ^{208}Pb ground state, and the figure shows the observed proton and proton-hole states in $^{209}_{83}$Bi ($\Delta Z = +1$) and $^{207}_{81}$Tl ($\Delta Z = -1$), respectively. The states are labeled by the conventional notation inherited from the early atomic spectroscopists. Collective modes corresponding to shape oscillations in ^{208}Pb have been identified by their large excitation probabilities in electromagnetic processes and inelastic scattering. While the shape oscillations can be resolved into particle-hole excitations, another type of collective excitation involves the addition or removal of correlated pairs of particles. The figure shows identified quanta of this type involving pairs of protons. [See, for example, the 0^+ states in $^{206}_{80}$Hg ($\Delta Z = -2$) and $^{210}_{84}$Po ($\Delta Z = +2$)].

The elementary excitations provide the building blocks in terms of which one

---

←

Figure 3   Shell structure in anisotropic harmonic oscillator potential. This figure shows the single-particle energy levels, as a function of deformation, in a prolate axially symmetric oscillator potential. The frequencies ω_3 and $\omega_\perp$ refer to motion parallel and perpendicular to the symmetry axis, while $\bar{\omega}$ is the mean frequency. The single-particle states can be specified by the number of quanta n_3 and $n_\perp$, and each energy level has a degeneracy $2 (n_\perp + 1)$, due to the spin and the degeneracy in the motion perpendicular to the axis. Additional degeneracies leading to the formation of major shells may occur when the ratio of the frequencies $\omega_\perp : \omega_3$ is equal to the ratio between integers. The deformations corresponding to the most prominent shell-structure effects are indicated by the arrows labeled by the corresponding frequency ratio. For the shells with frequency ratio 1:1 (spherical shape) and 2:1, the figure gives the particle numbers for closed-shell configurations.

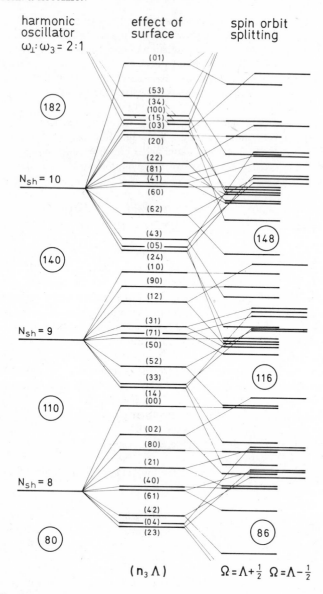

Figure 5 Transfer of many nucleons in collisions between heavy ions. This figure shows the intensities of the reactions involving the transfer of ΔN neutrons and ΔZ protons from the target nucleus $^{232}_{92}$Th to the impinging $^{22}_{10}$Ne nucleus (2). The estimated limits of particle stability (3) are indicated by the drawn lines. The most probable transfers lead to an increase of the mass to charge ratio for the projectile, which can be understood in terms of a tendency towards a more uniform distribution of the neutron excess during the collision.

←

Figure 4 Shell structure in nuclear potential with 2:1 symmetry. The single-particle spectrum to the left is identical with that of Figure 3, for the frequency ratio $\omega_\perp : \omega_3 = 2:1$. The shells are labeled by the shell quantum number $N_{sh} = n_3 + 2n_\perp$. The spectrum in the center shows the spreading of the shells that results from the deviation of the nuclear potential with its rather well-defined surface from that of a harmonic oscillator. The states are labeled by the component Λ of orbital angular momentum along the symmetry axis and by the quantum number n_3, which remains an approximate constant of the motion. Each level is fourfold degenerate, since the energy is independent of the sign of Λ and of the direction of the spin.

Finally, the spectrum to the right includes the effect of the rather strong spin-orbit coupling in the nuclear potential, which splits the states with component of total angular momentum $\Omega = \Lambda \pm \frac{1}{2}$, in such a way as to favor parallel spin and orbit $(\Omega = \Lambda + \frac{1}{2})$. The resulting spectrum is seen to preserve the existence of the major shells obtained from the oscillator potential though with a modification of the closed-shell numbers (given in circles) resulting from the fact that a few of the orbits with largest Λ and $\Omega = \Lambda + \frac{1}{2}$ have been shifted into the next lower shell.

The effects illustrated in Figure 4 are quite similar to those that govern the shell structure in spherical nuclei and lead to the closed-shell numbers..., 82, 126..., replacing the sequence..., 70, 112,... for the oscillator potential (see Figure 3).

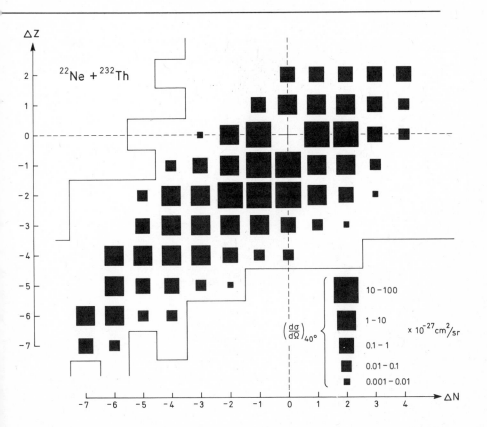

may attempt an analysis of the total excitation spectra. As an example, the spectrum of ^{209}Bi, in the lower part of Figure 7, shows the occurrence of one-particle states as well as states corresponding to a single particle or a single hole combined with the boson-like excitations of ^{208}Pb.

The description in terms of independent elementary modes is an approximation that is limited by the interrelation of the different quanta. In the nucleus, the analysis of these interrelations can be based on the average potential fields that are generated by the collective motion; these dynamic fields represent a generalization

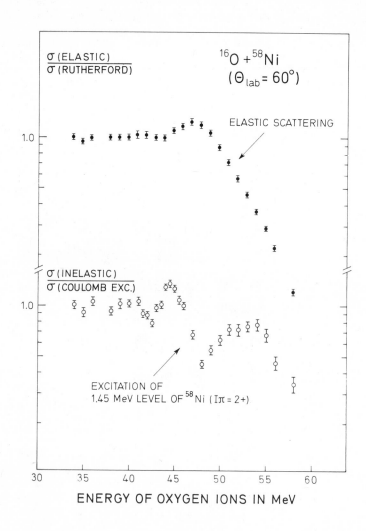

of the familiar static self-consistent potential. The dynamic fields provide a coupling between the motion of the individual particles and the collective modes (see Figure 8), which plays a similar role in the nucleus as the particle-phonon or particle-plasmon couplings in condensed media.

A systematic treatment of the particle-vibration coupling leads to a nuclear field theory, which is being actively explored as a basis for interpreting the rapidly accumulating evidence on the many interaction effects between the elementary modes including the anharmonicity in the collective motion and polarization effects leading to a renormalization of charge and moments of the particles.

Such a field theory also appears to provide a consistent treatment of the problem of the redundance in the total degrees of freedom, which has been felt as a paradoxical feature in the treatment of nuclear dynamics. Such a redundance is inherent in the description of a system, like the nucleus, where the collective modes are manifestations of the same underlying degrees of freedom as the particle modes of excitation. As an illustration of this point, Figure 9 shows the Feynman diagrams for the leading order interaction energy between a single-particle and a collective shape oscillation, as in the $(h_{9/2}, 3^-)$ septuplet in ^{209}Bi (see Figure 7). The fact that the particle configuration also appears as a component in the particle-hole expansion of the vibrational excitation finds expression in the exchange interaction represented by the last diagram.

←

Figure 6   Elastic and inelastic scattering of ^{16}O on ^{58}Ni. The cross section for elastic scattering is given in units of the Rutherford cross section for scattering of two point charges. The inelastic scattering involves the first excited state in ^{58}Ni, which can be approximately described as a quadrupole surface oscillation; the cross section for this excitation process is given in units of the cross section for excitation by the electric field of the oxygen nucleus, assuming that the projectile does not penetrate into the target nucleus. The experimental data is from (4).

The cross sections in the figure refer to a fixed scattering angle of 60°, in the laboratory system, and are shown as a function of the energy of the incident ion. Because of the small wave length of the heavy colliding particles, the process can be approximately described by considering the particles as moving along classical trajectories, and the incident energy and scattering angle therefore specify a rather sharply defined distance of closest approach. For energies below 45 MeV, this distance exceeds the range at which the nuclear interactions become effective, and the cross sections in the figure are close to unity, in the scale employed. The energy range from about 45 to 50 MeV corresponds to collisions in which the nuclear interactions are beginning to have a significant role. There is still some ambiguity in the analysis of the observed oscillations, but a possible interpretation involves the interference between the attractive nuclear forces and the repulsive electrostatic interaction acting in a grazing collision. This interference leads to a decrease in deflection angle (resulting in an increase of the cross section per unit solid angle of scattering) and a decrease in the probability for excitation. For still smaller impact parameters, the strong nuclear interactions lead to more violent reactions, and the probability that an oxygen nucleus emerges intact decreases rapidly with increasing energy (for fixed scattering angle).

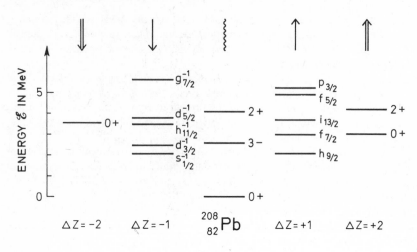

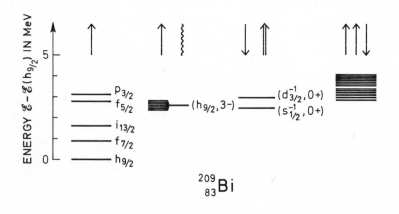

Figure 7 Elementary excitations based on the ground state of ^{208}Pb. The spectra of the nuclei around $^{208}_{82}$Pb can be described in terms of elementary excitations based on the ground state of ^{208}Pb, which has closed shells of neutrons and protons and acts as the "vacuum state" for the excitations.

The upper part of the figure shows fermion-like excitations involving the addition or removal of a single proton ($\Delta Z = +1$ or $\Delta Z = -1$), and boson-like excitations involving correlated pairs of protons ($\Delta Z = \pm 2$) as well as collective excitations in ^{208}Pb itself. The latter type of quanta can be expressed in terms of coherent particle-hole excitations with a resulting density oscillation approximately corresponding to that of a surface vibration. The energy scale employed in the figure involves a linear term in ΔZ, so chosen that the

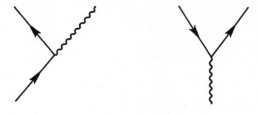

Figure 8   Particle-vibration coupling. The Feynman diagrams illustrate the basic coupling between particle and collective motion arising from the average one-particle potential generated by the collective vibrational motion. The first diagram represents the scattering of a particle with emission of a vibrational quantum (phonon); the second diagram represents the transition of a phonon into a particle-hole pair.

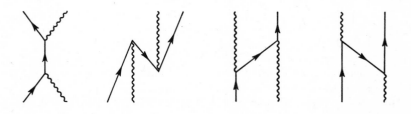

Figure 9   Interaction energy of a particle and a phonon. The basic particle-vibration coupling (see Figure 8) acting in second order gives rise to an interaction energy between a particle and a vibrational quantum, as illustrated by the four Feynman diagrams in the figure. This interaction energy contributes, for example, to the splitting of the $(h_{9/2}, 3^-)$ multiplet in the spectrum of ^{209}Bi (see Figure 7). The effects represented by the diagrams in Figure 9 include the renormalization of the phonon in the presence of the particle as well as the consequences of the identity of the particles and of the boson degrees of freedom.

←

lowest one-particle and one-hole excitations ($h_{9/2}$ and $s_{1/2}^{-1}$) have the same ordinate. Additional elementary excitations not shown in the figure involve changes in neutron number ($\Delta N = \pm 1, \pm 2$).

The lower part of the figure represents the low-energy spectrum of $^{209}_{83}$Bi. In addition to the one-particle excitations shown to the left, the spectrum exhibits excitations involving a single particle or a single hole combined with a collective excitation. The configuration $(h_{9/2}, 3^-)$ gives rise to a multiplet of states with total angular momentum $3/2, 5/2, \ldots$ $15/2$ that have all been identified within an energy region of a few hundred keV. The configurations involving a hole and a pair quantum with $I\pi = 0^+$ each give rise to only a single state. At an excitation energy of about 3 MeV, a rather dense spectrum of two-particle one-hole states sets in, as indicated to the right in the figure.

It is seen that the four diagrams in Figure 9 are completely equivalent to those describing the Compton scattering. Indeed, it is a remarkable feature of the spectrum of the nuclear many-body system that the vibrational quanta, though ultimately composed of particle excitations, appear just as elementary as the photon in electrodynamics.

THE FINE-GRAIN STRUCTURE OF NUCLEAR MATTER (HIGH-ENERGY PROBES)

The interpretation of the equilibrium density and binding energy of nuclear matter in terms of the forces acting between the nucleons has been a challenge since the early days of nuclear physics. The problem has turned out to require a much deeper analysis than could have been anticipated, not only because of the complexity of the strong interactions, but also because of the many subtle correlation effects that can be involved in collective properties of many-body systems.

A vast expansion of this field is resulting partly from the discovery of the great variety of collective modes generated by average fields of different symmetries and partly by the development of experimental probes capable of resolving the fine-grain structure of nuclear matter. Such resolution demands momentum transfers comparable with or larger than the Fermi momentum, as can be achieved by the scattering of high-energy particles from the nucleus.

An example illustrating some of the potentialities of the high-energy probes is shown in Figure 10, which gives the cross section for inclusive electron scattering representing the total yield per unit energy loss for a fixed scattering angle. The main features of the cross section can be understood in terms of scattering of individual nucleons and give a value of the Fermi momentum in agreement with that deduced from the nuclear size. The experiments also provide evidence that the effective mass for nucleons in the nucleus is modified as a result of the interactions (velocity-dependence of the average potential). In experiments of this type, it may become possible to identify short-range correlations between the nucleons as well as modifications in the structure of the nucleons themselves resulting from their binding in nuclear matter.

The high-energy probes that are becoming available for the study of the nucleus can also excite the internal degrees of freedom of the nucleons and thus a new field of strong interaction physics is opening up. Investigations on this frontier will illuminate the nuclear dynamics from new points of view and may reveal novel aspects in elementary quanta occurring as components in a strongly interacting system.

The question of elementarity appears in another guise when one considers the nuclei in the broader context of the hadronic spectra. It appears as a capricious feature of the strong interactions that they lead to bound systems of multiple baryonic number, but with binding energies so small that it is useful to consider these systems as composed of a definite number of neutrons and protons. In the corresponding problem of quantum electrodynamics, the existence of atoms and condensed matter describable as nonrelativistic many-body systems reflects the

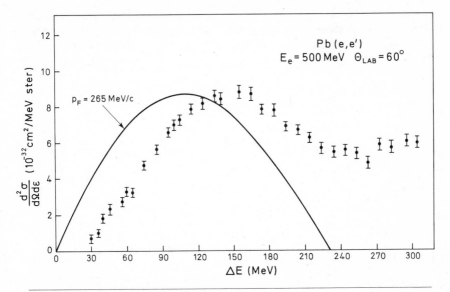

Figure 10   Inelastic electron scattering on ^{208}Pb. This figure shows the differential cross section per unit solid angle and unit energy loss for the scattering of 500-MeV electrons on a target of ^{208}Pb (5). The solid curve is obtained by considering the scattering of individual nucleons with a momentum distribution corresponding to that of a Fermi gas (quasi-elastic scattering). The width of the distribution is determined by the Fermi momentum, which is taken to be $p_F = 265$ MeV/c, corresponding to a bulk nuclear density of 0.17 nucleons fm^{-3}, as determined from the radius of the charge distribution.

Comparison of the calculated and observed cross sections lends support to the assumed value of the Fermi momentum, but reveals a shift in the maximum of the distribution by about 40 MeV. Such a shift can be understood in terms of the velocity-dependence of the nuclear binding field, which implies that the potential acting on the fast outgoing nucleon is weaker than the potential acting on the more slowly moving bound nucleon. The velocity-dependence obtained from the present experiment is of a similar magnitude to that deduced from the elastic scattering of nucleons on nuclei (optical potential).

The large cross sections observed for energy losses $\Delta E \gtrsim 240$ MeV cannot be accounted for in terms of quasi-elastic scattering from a Fermi gas. In this energy region important contributions arise from meson production as well as from the presence of nucleons with high momenta, resulting from violent short-range interactions.

smallness of the fine-structure constant, but one may ask: Where is there a small number in the structure that underlies the strong interactions?

COMPOUND NUCLEUS; STATISTICS OF QUANTAL STATES

Most of the tools available for probing the nuclear spectra can only resolve the individual levels in the low-energy region, where relatively few excitation quanta are involved. However, the whole development of nuclear physics has been decisively influenced by the existence of a small window, in the region of the neutron binding energy, within which the slow neutron reactions provide a probe of enormously

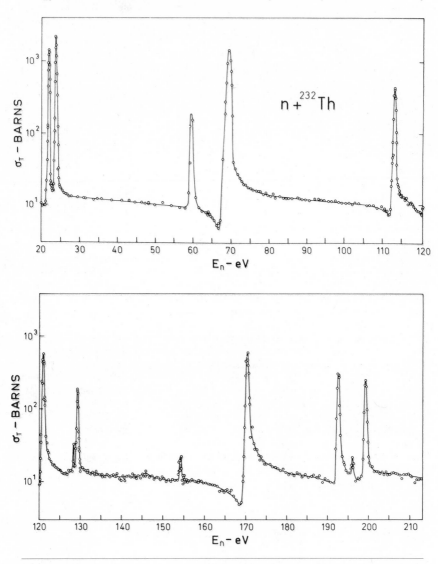

Figure 11   Resonances in slow neutron scattering. This figure shows the total cross section for neutrons incident on ^{232}Th (6); the resonances correspond to metastable states in the compound nucleus ^{232}Th that all have total angular momentum $I = 1/2$ and positive parity.

greater resolving power. Already the earliest experiments with slow neutrons revealed, very unexpectedly, a dense spectrum of resonances. This discovery led to the recognition of a strong coupling between the motion of the incident neutron and many degrees of freedom of the target. The coupling gives rise to the formation of a compound system with a lifetime very long compared with the one-particle periods.

The refinement that has been achieved in the study of the neutron resonance spectra is illustrated in Figure 10, which shows the total cross section for neutrons incident on ^{232}Th, in the energy range up to about 200 eV. The information provided by data of this type has led to significant new developments in the characterization of statistical equilibrium and the formulation of the concept of

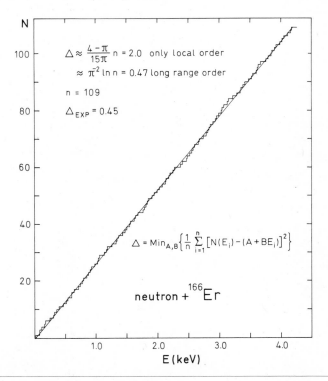

Figure 12   Long-range order in the energy spectrum of the compound nucleus. This figure shows the number of levels N of spin and parity $I\pi = 1/2^+$ that have been observed for neutron energies up to a given value E (7). The mean square deviation Δ from a uniform level spacing (corresponding to a straight line in the figure) is compared with the predictions based on a model (local order) that only includes the effect of repulsion between neighboring levels and a model (long-range order) that includes the repulsive effect of more distant levels as given by a random matrix obeying the time reversal and Hermiticity symmetries.

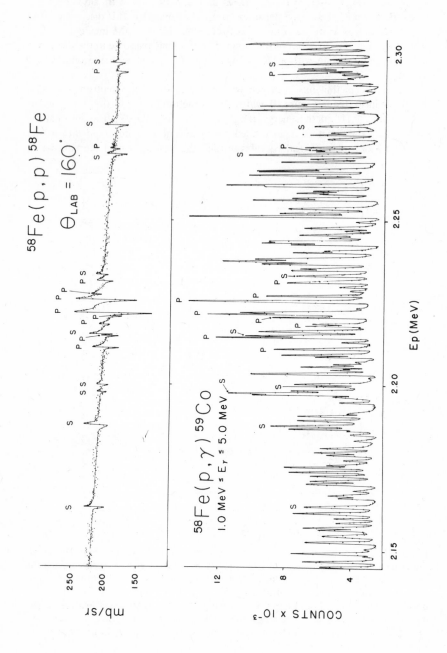

Figure 13   Gross structure in the resonance scattering of protons associated with the occurrence of isobaric analogue states. The upper part of the figure shows the elastic scattering cross section for protons on $^{58}_{26}$Fe as a function of the energy of the incident proton. The background represents nonresonant scattering primarily resulting from the Coulomb interaction between projectile and target nucleus; the resonances correspond to metastable states in the compound nucleus $^{59}_{27}$Co. For the proton energies considered, the resonances associated with s-wave scattering ($l = 0, I\pi = 1/2^+$) are clearly resolved from the background. The p-wave scattering ($l = 1, I\pi = 1/2^-$, $3/2^-$) is expected to give rise to three times as many resonances as the s-wave scattering, corresponding to the relative statistical weights, but the p-wave resonances are weakened by the reduced penetration through the Coulomb barrier that results from the centrifugal effect. Hence, most of the p-wave resonances remain undetected in this experiment. However, strongly enhanced p-wave resonances are found to occur in a narrow energy interval around $E_p = 2.22$ MeV; this gross structure effect can be associated with the approximate conservation of the isospin quantum number.

The target nucleus $^{58}_{26}$Fe$_{32}$ has total isospin $T = 3$ [the smallest value compatible with the isospin component $M_T = \frac{1}{2}(N-Z) = 3$]. Similarly, all the low-lying states in the compound nucleus $^{59}_{27}$Co$_{32}$ have $T = 5/2$. The first $T = 7/2$ state belongs to the same T-multiplet as does the ground state of $^{59}_{26}$Fe$_{33}$, which has $I\pi = 3/2^-$; the energy of the corresponding state in ^{59}Co (isobaric analogue state) can be obtained by adding the extra Coulomb energy (≈ 8 MeV) associated with the transformation of a neutron into a proton inside the nucleus. The resulting energy just corresponds to the incident proton energy in Figure 12, for which the strong p-wave resonances are observed.

If the isospin quantum number were exactly conserved, the isobaric analogue state would give rise to a single $I\pi = 3/2^-$ resonance, with a very large strength reflecting the simple internal structure of this state; in fact, the ground state of ^{59}Fe is known to be qualitatively described in terms of the ^{58}Fe core with the addition of a particle in a $p_{3/2}$ orbit. The proton width of the isobaric analogue state is therefore of a magnitude corresponding to a single-particle p-wave resonance in a potential. In contrast, the $T = 5/2$ states have the complexity of a compound nucleus with many strongly coupled degrees of freedom resulting in a high level density and a small decay width for particle emission. The violation of the isospin symmetry by the electromagnetic interactions (especially the electrostatic forces) implies that the properties of the $T = 7/2$ isobaric analogue state in the spectrum of ^{57}Co are shared among adjacent compound states with $T = 5/2$, each of which thereby acquires an enhanced p-wave resonance strength.

The lower part of Figure 12 shows the yield of the proton capture process with the emission of γ rays. Because of the absence of a strong nonresonant background, this process is an even more sensitive detector of the compound resonances than the elastic scattering. The labels s and p exhibit the $l = 0$ and $l = 1$ resonances that are also detected in the elastic scattering process (upper part of figure); as expected, the capture process is not sensitive to the isobaric analogue structure. The total density of levels observed in the (p, γ) process corresponds approximately to that estimated for s- and p-wave resonances, on the basis of the observed spacing of the s-wave resonances in the (p, p) process.

The experimental data in Figure 13 is from W. C. Peters, G. E. Mitchell & E. G. Bilpuch, to be published. We are indebted to Dr. Bilpuch for communication of these results prior to publication.

randomness at the level of individual quantal states of a many-body system. A model that has especially been explored expresses the randomness in the couplings of the different degrees of freedom in terms of an ensemble of matrix elements that is invariant with respect to transformations of the chosen basis. This formulation leads to predictions concerning the distributions of eigenvalues and decay widths that have been quantitatively tested. Thus, the repulsion between neighboring levels of the same spin and parity gives rise to a low probability for finding very small level spacings, and this short range order has been well established for over a decade. The theory also predicts a long-range order of a rather subtle type, which only very recently has been confirmed by experiments such as those illustrated in Figure 11. The test of this long range order is shown in Figure 12, which gives the extent of the deviations of the observed level sequence from that of a completely ordered uniform distribution. While a level distribution with only short-range order implies mean square deviations increasing linearly with the number of levels, the effect of the interactions between levels, represented by random couplings, implies a much higher degree of ordering resulting in a mean square deviation that is only logarithmic in the number of levels. Present efforts are directed towards identifying the limitations on the concept of randomness, as formulated above, that result from a more detailed analysis of the degrees of freedom involved in the nuclear spectra.

STRENGTH FUNCTION; DIRECT INTERACTIONS

The establishment of statistical equilibrium leading to the quantal randomness of the fine-structure resonances requires a time that is sufficiently long so that the couplings between the elementary modes of excitation can be effective. This time scale is revealed by the widths of the strength functions (or gross structure resonances), which show that the lifetimes τ_{coupl} of the elementary excitations correspond to energies that are typically of the order of MeV. Such a lifetime, though many order of magnitude smaller than the periods τ_{comp} of the compound nucleus, is large compared with the time τ_{sp} for a nucleon to traverse the nucleus

$$\tau_{sp} \quad \ll \quad \tau_{coupl} \quad \ll \quad \tau_{comp}$$
$$(\Delta E \sim 10 \text{ MeV}) \quad (\Delta E \sim 1 \text{ MeV}) \quad (\Delta E \sim 10 \text{ eV})$$

The nuclear reactions, therefore, also exhibit important effects involving only a single or few degrees of freedom ("direct" interactions). The possibility of studying processes extending over six orders of magnitude in the time or energy dimension (varying from the single-particle to the compound periods) provides the opportunity to explore in considerable detail the different levels of complexity and order of which the concepts of direct interactions and compound processes represent two extremes.

ISOBARIC ANALOGUE RESONANCE

Strength functions with especially long lifetimes may result from symmetries or other dynamical features that lead to approximately conserved quantum numbers. Figure

13 shows the cross section for elastic scattering of protons on $^{58}_{26}Fe_{32}$. The remarkable resolution achieved in such experiments makes it possible to study the dense spectrum of resonances interfering with the Coulomb scattering. Superposed on the individual compound levels is a gross structure that can be attributed to the charge independence of the strong interactions, as expressed in the conservation of the total isospin. The compound levels have isospin $T = 5/2$, as in the ground state of $^{59}_{27}Co$ formed by adding a proton to $^{58}_{26}Fe$, but, in the energy region studied, there is in addition a single state with $T = 7/2$, which has the same (relatively simple) internal structure as the ground state of $^{59}_{26}Fe$ (isobaric analogue state). If there were no interactions in the nucleus distinguishing between neutrons and protons, the isospin would be exactly conserved, and the $T = 7/2$ state would appear as a single sharp level in the spectrum of $^{59}_{27}Co$. However the Coulomb forces that act between the protons give rise to a spreading of the properties of the $T = 7/2$ state into the adjacent compound states with $T = 5/2$. The lower part of Figure 13 shows the cross section for radiative capture of protons on ^{58}Fe, a process that is even more sensitive than the proton scattering in revealing compound states in several angular momentum and parity channels, but which does not respond specifically to the isobaric analogue state.

The distribution of the strength of the isobaric analogue state as revealed in the proton scattering has a width of only about 20 keV, which reflects the smallness of the electromagnetic interactions as compared with the strong interactions. The analysis of this strength function involves a symmetry-breaking produced by a well-defined term in the Hamiltonian (the Coulomb energy); the problem, however, is not a trivial one since one is studying the effects of the perturbation in the actual physical states of the many-body system. Indeed, the total Coulomb potential in a heavy nucleus is not small compared with the nuclear potential, on account of the long range of the electric forces, and the weakness of its symmetry-breaking effect was only recognized after the discovery of the sharp isobaric analogue states. This unexpected discovery added a powerful new tool to the arsenal of reactions available to the nuclear spectroscopists.

COLLECTIVE MODES IN THE NUCLEAR PAIR FIELD (TWO-PARTICLE TRANSFER REACTIONS)

Collective modes associated with variations in the shape and density are familiar from classical systems. The more specifically quantal degrees of freedom associated with spin, isospin, and nucleon number give new dimensions to the nuclear dynamics. The exploration of the great variety of collective modes that can occur in these dimensions has only recently been initiated but appears to be a frontier of considerable perspective.

Collective motion with quanta carrying nucleon number is related to the nuclear pairing effect, which was identified in the nuclear systematics at a very early stage. The striking difference in the binding energies of nuclei with even and odd numbers of nucleons finds a rather dramatic expression in the different fissility of the even and odd isotopes of uranium. However, the collective significance of the pair correla-

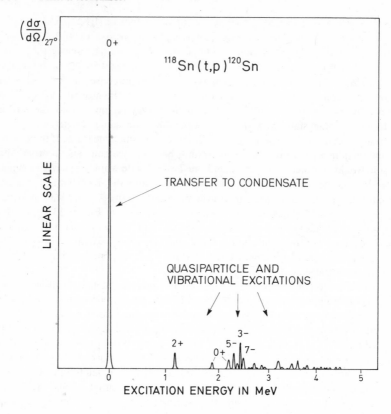

Figure 14   Two-particle transfer process in Sn nuclei. This figure shows the cross section for transfer of two neutrons from an incident triton to a target of $^{118}_{50}$Sn (8). The transfer takes place as a "direct interaction" with an angular distribution characteristic of the transferred angular momentum. The cross sections in the figure refer to protons emerging at an angle of 27° from the direction of the incident beam; this angle corresponds to a maximum in the angular distribution for zero angular momentum transfer.

The transition to the ground state involves a transfer of a pair of neutrons in a correlated state with angular momentum and parity 0^+. The process receives a further strong enhancement by the presence, in the target nucleus, of a rather large number of such correlated pairs formed out of neutrons in the partially filled shell. (Closed-shell configurations correspond to neutron numbers 50 and 82). This enhancement is a quantal effect associated with the identity of the quanta (Boson factor; see caption to Figure 15). The occurrence of many Bosons in a definite quantal state is referred to as a condensate, and the pair correlations in the nucleus have basic properties in common with the condensates of electron pairs in superconductors and of ^{4}He particles in superfluid helium. On account of the finite size of the nucleus, which is small compared with the correlation length for the pairs, supercurrents do not occur in nuclei (but may occur in superfluid nuclear matter in neutron stars).

tion effect and its far-reaching consequences for many nuclear properties were only gradually recognized.

A powerful tool for the study of the nuclear pair correlations has become available in the two-nucleon transfer reactions. Figure 14 shows an example of a (t, p) process on ^{118}Sn by which a pair of neutrons is added to the target nucleus. The spectrum is dominated by the transition to the ground state of ^{120}Sn, for which the cross section is found to be between one and two orders of magnitude greater than would correspond to the transfer of two neutrons each into a definite orbit in the nucleus. The enhancement reflects the fact that many neutrons in the partially filled shell

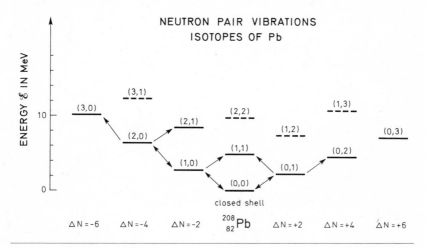

Figure 15 Neutron pair vibrations based on ^{208}Pb. This figure shows the spectrum of excitations involving quanta corresponding to the addition or removal of correlated pairs of neutrons with angular momentum and parity 0^+ (cf Figure 7). The levels in the pair-vibrational spectrum are labeled by the number of pair-removal and pair-addition quanta. The levels $(n, 0)$ and $(0, n)$ correspond to the ground states of the even Pb isotopes. The energy scale includes a term linear in ΔN, which has been chosen in a similar way as for the excitations with different ΔZ in Figure 7.

The observed levels are indicated by solid lines. In the harmonic approximation (neglect of interaction between quanta), the energy would be a linear function of the number of quanta, but the observed level positions reveal interaction effects that can be approximately described in terms of a quadratic form in the number of quanta (pair-wise interactions of the quanta). The dashed lines indicate predicted pair-vibrational excitations with energies that have been calculated with the inclusion of the interactions derived from the observed levels.

The pair vibrations are characterized by large cross sections in the two-neutron transfer processes. The enhancement reflects the correlation of the neutrons in the individual quanta. When several identical quanta are present, the transition probability for the processes $n \leftrightarrow n-1$ involves an additional factor n (Boson factor). The observed (t, p) and (p, t) transitions, indicated by the arrows in the figure, confirm the collective character of the pair-vibrational excitations.

form a condensate consisting of correlated neutron pairs, similar to the condensate of the paired electrons in a superconductor. In fact, the transfer process may be compared with the transmission of electron pairs between superconductors, as in the Josephson junction.

In a closed-shell nucleus (as in an insulator), there is no possibility of pair correlation and therefore no condensate. However, additional particles or holes can form correlated pairs that constitute elementary modes of excitation, as referred to above. The successive addition or removal of pairs leads to a vibrational-like spectrum, as illustrated in Figure 15, which shows the neutron pair vibrations with angular momentum zero, based on the closed-shell nucleus, ^{208}Pb. There are two

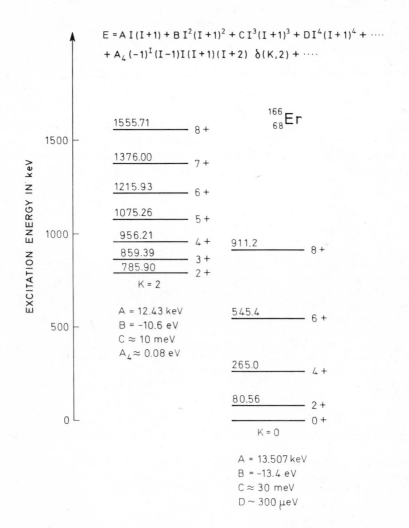

distinct quanta in the spectrum corresponding to the addition or removal of a pair of particles with respect to the closed-shell configuration, and the quantum numbers on the levels in Figure 15 give the numbers of quanta of each kind. The arrows in the figure indicate experimentally observed strong transitions in two-neutron transfer processes populating ground states of the even Pb isotopes $[(n, 0)$ and $(0, n)]$, as well as some of the excited states of the type $(n+1, 1)$, occurring at excitation energies of 5–6 MeV.

The vibrational spectrum in Figure 15 is an example of a collective family involving states in different nuclei. The vibrational motion is associated with the oscillation of a field, the pair field, that creates two nucleons. These oscillations occur not in the usual space but involve other dimensions including a so-called gauge space. The nucleon number operator, which appears as the angular momentum in gauge space, is usually assigned a rather passive role as an overall constant of the motion, a superselection rule that divides the phenomena into sharply separated compartments, each with a definite value of nucleon number. In the nuclear pair correlation effects, however, we are dealing with phenomena that relate states with

---

←

Figure 16   Rotational bands in ^{166}Er. This figure shows the observed energy levels belonging to the two lowest rotational bands in ^{166}Er, as studied in inelastic excitations and radioactive decay processes (9). Each energy level is labeled by the total angular momentum and parity $I\pi$, and the bands are further labeled by the quantum number K, which represents the component of the total angular momentum on the intrinsic symmetry axis.

The collective deformation of the nucleus is directly revealed by the strong transitions between members of a band. For these transitions the electric quadrupole strength is several hundred times larger than for a single-particle transition. Moreover, the $E2$ transition probabilities for the various transitions within a band follow intensity relations with a simple geometrical basis (similar to those known from the band spectra of axially symmetric molecules).

The nuclear shape is approximately spheroidal, with axial symmetry and reflection invariance. Such a deformation represents an only partially developed anisotropy (intrinsic breaking of rotational symmetry), and the rotational degrees of freedom are correspondingly reduced. Thus, the axial symmetry implies the absence of collective rotations with respect to the symmetry axis, and each rotational family of quantum states therefore involves only a single sequence with a fixed value of K. (The $K = 2$ band in ^{166}Er appears to represent a vibrational rather than a rotational mode of excitation). The invariance of the nuclear shape with respect to a rotation of 180° about an axis perpendicular to the symmetry axis implies that this operation is a property of the intrinsic motion, and leads to the selection rule $I = 0, 2, 4 \ldots$, for the $K = 0$ band in ^{166}Er.

The observed energy levels can be expressed in terms of a power series in the rotational angular momentum. The form of this expansion, given in the upper part of the figure, follows from the symmetry of the deformation. The coefficients in the expansion, derived from the empirical energies, are given for each band. It is seen that successive coefficients decrease by a factor of order 10^3 and hence the convergence is rather rapid, for the range of angular momenta in the figure. (The rate of convergence is similar to that observed in the H_2 molecule).

different nucleon number and that therefore involve operators, such as the orientation of the pair field in gauge space, that are complementary to the nucleon number. In the transfer processes, these operators are directly measured and the new dimensions are therefore experienced in a very real manner.

ROTATIONAL SPECTRA

The response to rotational motion is a basic property of physical systems that has played a prominent role in the development of dynamical concepts, ranging from celestial mechanics to the spectra of elementary particles. The question of whether nuclei possess rotational spectra was raised already in the early days of nuclear spectroscopy. Quantized rotational motion was known from molecular spectra, but atoms provide examples of quantal systems that do not rotate collectively. The early discussion of this issue was, however, hampered by the expectation that rotational motion would either be a property of all nuclei or would be generally excluded, and by the expectation that the moment of inertia would have the classical value, as for rigid rotation.

The establishment of the nuclear shell structure provided a description in terms of single-particle motion that might seem to preclude the occurrence of collective rotation. However, a new situation was created by the recognition that the shell structure may lead to equilibrium shapes that deviate from spherical symmetry. It was evident that such a collective deformation that defines an orientation of the system as a whole would imply rotational degrees of freedom, but one was faced with the need for a generalized treatment of rotations applicable to quantal systems that do not have a rigid or semirigid structure like that of molecules.

An example of rotational band structure in a nuclear spectrum is shown in Figure 16, which gives the two lowest bands observed in ^{166}Er. The angular momentum and parity quantum numbers of the states occurring in the rotational bands imply a deformation with axial symmetry and invariance with respect to space and time reflection. The energies can be represented by a power series expansion in the angular momentum, which converges rather rapidly for the range of angular momentum values included in the figure.

Similar expansions can be given for the matrix elements of tensor operators characterizing electromagnetic transitions, β decay, particle transfer, etc. As an example, Figure 17 shows the electric quadrupole ($E2$) matrix elements between the two bands of ^{166}Er in Figure 16. The analysis exemplified by Figures 16 and 17 is based only on the symmetry of the deformation, and it is seen that such an analysis provides an appropriate framework for interpreting the detailed body of evidence on the nuclear rotational spectra.

GENERALIZED ROTATIONAL MOTION

The recognition of the deformation and its degree of symmetry breaking as the central element in defining rotational degrees of freedom opens new perspectives for generalized rotational spectra associated with deformations in many different dimensions including isospace, gauge spaces, as well as orbital space. The resulting rotational band structure may involve comprehensive families of states labeled by the different quantum numbers of the internally broken symmetries, and there may be relations between quantum numbers referring to different spaces.

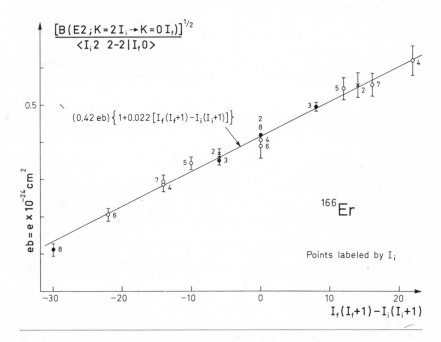

Figure 17 Intensity relation for electric quadrupole transitions between rotational bands. The figure shows the measured reduced electric quadrupole transition probabilities $B(E2)$ for transitions between members of the $K = 2$ and $K = 0$ band in ^{166}Er shown in Figure 16 (10). An expansion similar to that of the energies in Figure 16, but taking into account the tensor properties of the $E2$ operator, leads to an expression for $[B(E2)]^{\frac{1}{2}}$ which involves a Clebsch-Gordan coefficient (geometrical factor) multiplied by a power series in the angular momenta I_i and I_f of the initial and final states. The leading order term in this expansion is a constant and the next term is linear in $I_f(I_f + 1) - I_i(I_i + 1)$. The experimental data are seen to be rather well represented by these two terms.

The angular momentum dependent terms in the expansion of energies and matrix elements can be interpreted in terms of the coupling between bands produced by the Coriolis and centrifugal forces. Thus, a term in the $E2$ matrix element proportional to $I_f(I_f + 1) - I_i(I_i + 1)$ arises from a coupling between the $K = 2$ and $K = 0$ bands; even very small admixed amplitudes resulting from this coupling may contribute significantly to the $E2$ transitions between the bands, since these admixed amplitudes involve the strongly enhanced $E2$ matrix elements characteristic of transitions within a band. This circumstance accounts for the relatively large value of the term proportional to $I_f(I_f + 1) - I_i(I_i + 1)$ in Figure 17; higher order terms do not have a similar enhancement and are therefore expected to be much smaller.

The Regge trajectories that play such a prominent role in the current study of the structure of hadrons have features reminiscent of rotational spectra, but as yet there appears to be no definite evidence concerning the nature of the deformation that would define an orientation of the intrinsic structure of a hadron.

The condensates in superfluid Fermion systems involve a static deformation of the pair field, and the processes of addition or removal of pairs from the condensate constitute a rotational mode in gauge space. An example of such a rotational excitation is provided by the two-neutron transfer process linking the ground states of the even Sn-isotopes, as illustrated in Figure 14. The intensity of this transition provides a measure of the deformation of the pair field, also referred to as the order parameter.

PHASE TRANSITIONS INDUCED BY ROTATIONAL PERTURBATION; YRAST REGION

The very detailed studies of the nuclear rotational motion in the three-dimensional space of our daily experience have provided a large body of information concerning the response of the nuclear structure to the rotation of the average deformed field in which the nucleons move. In particular, these studies have revealed the major effect of the superfluidity on the collective rotational flow and the resulting moments of inertia.

The perturbations of the intrinsic structure produced by the rotation increase rapidly with angular momentum, and an exciting new frontier is opening as a result of the possibility of transferring to the nucleus such large amounts of angular momentum that the structure undergoes major modifications.

The nuclear spectrum as a function of angular momentum is illustrated schematically in Figure 18. The states with lowest energy for given angular momentum are referred to as the yrast line (literally dizziest, from Swedish yr); in this region of the spectrum, the nucleus though highly excited may be considered as "cold," since almost the entire excitation energy is expended in generating the angular momentum. The available beams of heavy ions provide powerful tools for producing nuclear states with 20, 50, perhaps 100 units of angular momentum. In this angular momentum range, one can envisage a number of different phase changes in the nuclear structure including the disappearance of the pair condensate (in analogy to the break-down of superconductivity by a magnetic field), the loss of axial symmetry (an effect related to instability phenomena in rotating stars), as well as a variety of different fission processes engendered by the centrifugal forces. In other situations, the quantal effects associated with the shell structure may lead to discontinuities in the yrast line, associated with a maximal alignment of the angular momenta of the particles within each shell.

Already the first glimpse into the region $I \approx 15\hbar$ to $20\hbar$ has revealed striking new phenomena as illustrated in Figure 19, which shows the moment of inertia as a function of the rotational frequency along the yrast line of ^{162}Er. The interpretation of the pronounced structure in Figure 19 is not yet definitely established, but it is possible that we are encountering the disappearance of superfluidity, which is expected to lead to a normal phase rotating with a moment of inertia equal to that for rigid rotation. We here face the challenging task of analyzing a phase transition as revealed by the properties of the individual quantum states of the system.

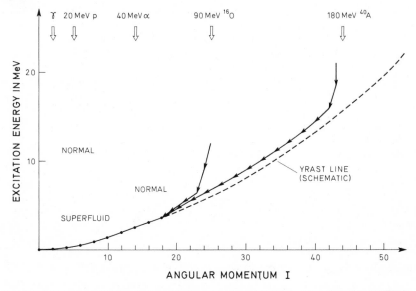

Figure 18   Nuclear spectrum as a function of angular momentum. The figure is a schematic representation of some of the important landmarks in the nuclear spectrum as a function of angular momentum I. (The magnitudes of the energies are chosen to correspond to a nucleus with mass number $A \approx 160$). The lowest state of given angular momentum is referred to as the "yrast" level. The low angular momentum portion of the yrast line ($I \lesssim 18$) corresponds to the known levels of ^{162}Er (see Figure 19); the tentative continuation of the yrast line is based on the assumption that, for high angular momentum, the nucleus will rotate with a moment of inertia corresponding to rigid rotation. In the region above the yrast line, the density of levels increases rapidly with excitation energy (see, for example, Figure 11).

The low energy and low angular momentum states exhibit strong pair correlations of the type that are characteristic of a superfluid phase. With increasing angular momentum and energy it is expected that these correlations will become less effective and that the system will go over to a "normal" phase.

The arrows at the top of the figure indicate the magnitude of the angular momentum that is brought into the nucleus in a grazing collision with a number of different projectiles currently available. In a typical reaction, the initially formed compound nucleus is in a highly excited state, which "cools" through a sequence of neutron emissions; the evaporated neutrons are light and of low energy (the nuclear temperature is of the order of one MeV) and thus carry away very little of the angular momentum. When the system arrives at an excitation energy with respect to the yrast line that is less than the neutron separation energy (≈ 5 MeV), the subsequent decay takes place by a cascade of γ rays. The figure shows two illustrative (hypothetical) γ cascades leading eventually to the yrast levels in the region of $I \approx 20$.

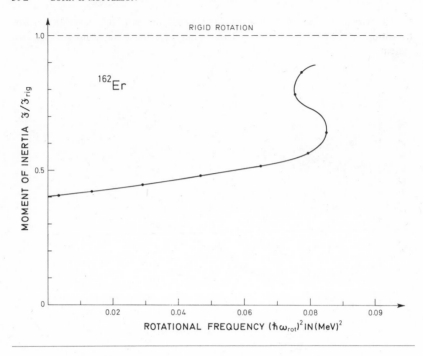

Figure 19   Moment of inertia of ^{162}Er as a function of rotational frequency (11). In a rotational sequence, the energy is a smooth function of the angular momentum; from the observed energies, it is thus possible to derive the rotational frequency, which is the derivative of the energy with respect to the angular momentum. In the figure, the frequency ω_{rot} has been obtained by interpolating linearly in the variable $I(I+1)$ between successive members of the rotational band. The moment of inertia $\mathscr{I}$ is obtained in a similar manner from the canonical definition, as the ratio of angular momentum to rotational frequency.

CONNECTIONS TO OTHER FIELDS

In the present report it has only been possible to touch upon a few of the areas of activity in present-day nuclear structure research, but the examples may give an impression of the depth of penetration into the nuclear world that has been achieved by the ingenuity and resourcefulness of the experimental effort, and illustrate the richness of the phenomena that have been revealed to the explorers. These examples may also indicate some of the perspectives that can be seen at the various frontiers, and the opportunities that lie ahead for vastly expanding the range of phenomena accessible to investigation. With the tools that are now becoming available or are technically feasible, we already appear to be on the threshold of a major increase in the scope of the field; the heavy ion beams will make it possible to study nuclear matter in quite new regimes, and the high-energy probes are providing microscopes for exploring the texture of this matter at new orders of resolution. On the "inner frontier," the greatly increased precision and flexibility of nuclear spectroscopy will constitute a challenge for sharpening the concepts employed in the description of the nuclear many-body system.

At many points we have also alluded to the intimate connection between the nuclear phenomena and those encountered in other domains of quantal physics. These connections that stem from the ubiquitous role of many-body problems in the present stage of physics have become increasingly apparent in recent years and have been of inspiration not least to the nuclear physicists, who find themselves at an intermediate position on the "quantum ladder." The efforts to view the developments in the various domains of quantal physics as a unified whole may in the future, to an even greater extent, become a stimulus for continued progress over the broad front.

It has not been possible within the framework of this report to deal with the vast network of applications of nuclear science and techniques. The access to energy sources hitherto confined to the sun and stars has opened opportunities and provided challenges of truly revolutionary dimensions. Less dramatic, but perhaps of no less profound significance, is the increasing extent to which nuclear processes and techniques provide tools for probing and transforming both living and inanimate matter at the various levels of organization. Such applications have become indispensable in almost all the modern fields of natural science and in many areas of industry, and indeed contribute a major component to the arsenal of tools available for attacking problems in our modern society.

Literature Cited

1. Preliminary data of Earwaker & James, quoted in Lynn, E. 1969. *Physics and Chemistry of Fission,* p. 249. Vienna: IAEA
2. Artukh, A. G., Gridnev, G. F., Mikheev, V. L., Volkov, V. V., Wilczynsky, J. Unpublished
3. Garvey, G. T., Gerace, W. J., Jaffe, R. L., Talmi, I., Kelson, I. 1969. *Rev. Mod. Phys.* 41:S1
4. Videbaek, F., Chernov, I., Christensen, P. R., Gross, E. E. 1972. *Phys. Rev. Lett.* 28:1072
5. Moniz, E. J. et al 1971. *Phys. Rev. Lett.* 26:445
6. Neutron Cross Sections. 1964. *BNL 325,* Suppl. 2. Brookhaven, NY: Brookhaven

Natl. Lab.; See also the more recent data by Rahn, F. et al 1972. *Phys. Rev. C* 6:1854
7. Liou, H. I. et al 1972. *Phys. Rev. C* 5:974
8. Bjerregaard, J. H. et al 1968. *Nucl. Phys. A* 110:1
9. Reich, C. W., Cline, J. E. 1970. *Nucl. Phys. A* 159:181; Lederer, C. M., Hollander, J. M., Perlman, I. 1967. *Table of Isotopes.* New York: Wiley
10. Gallagher, C. J. Jr., Nielsen, O. B., Sunyar, A. W. 1965. *Phys. Lett.* 16:298; Günther, C., Parsignault, D. R. 1967. *Phys. Rev.* 153:1297
11. Johnson, A., Ryde, H., Hjorth, S. A. 1972. *Nucl. Phys. A* 179:753

MOLECULAR STRUCTURE EFFECTS ON ATOMIC AND NUCLEAR CAPTURE OF MESONS

✲ 5543

L. I. Ponomarev

Laboratory of Theoretical Physics, Joint Institute for Nuclear Research, Dubna, USSR

CONTENTS

INTRODUCTION

Once negative mesons (μ^-, π^-, K^-) penetrate into some substance they slow down and, after stopping, form with the atomic nuclei atom-like systems: mesic atoms and mesic molecules. These systems are tens and hundreds of times smaller than the corresponding atoms and as a rule are entirely inside their K-shells. Therefore the electron shells of atoms are usually neglected when mesoatomic processes are studied. For the same reason one does not expect a priori that the capture of mesons into mesoatomic orbits and their reactions with the nuclei may depend on the peculiarities of chemical bond. The chemical properties of substances are defined by the outer electron shells whose size ($\sim 10^{-8}$ cm) much exceeds the range of action of nuclear forces ($\sim 10^{-13}$ cm) and the size of mesic atoms (10^{-11}–10^{-10} cm) from whose levels the nuclear capture of mesons proceeds.

Nevertheless the experiments carried out in the past ten years have provided convincing evidence that the molecular structure of matter affects the processes of atomic and nuclear capture of mesons. The observed effects were first established most clearly in the study of the charge-exchange reaction of π^- mesons on the nuclei of chemically bonded hydrogen and then in the study of the structure of muon X-ray series of elements in chemical compounds and mixtures.

Attempts to explain these facts have led to significant improvement in the picture of negative meson capture in matter. In particular, it was established that in the process of slowing down and stopping in a chemical compound a large fraction of mesons is captured into the orbits of the molecule as a whole rather than into the orbits of individual atoms. In such excited complexes, called "large mesic molecules," the mesoatomic orbits are comparable to the size of the molecules of matter and exceed by hundreds of times the characteristic mesoatomic distances.

The discovery and study of these phenomena may, in principle, lead to new methods for the investigation of the electronic structure of molecules and for application to a variety of applied problems of the qualitative and quantitative analysis of matter. The production of intense meson beams by meson factories makes these hopes more realistic.

The present paper is a brief survey of the experimental studies in which the effects of the chemical structure of matter on the meson capture processes are detected and analyzed. Some theoretical ideas currently used for the analysis of the experimental results are also given. A detailed bibliography of these problems is presented in (1–3). Different related topics are discussed in (4–7).

SUCCESSIVE STAGES OF ABSORPTION OF MESONS IN MATTER

In 1947 Fermi & Teller (8) formulated the so-called Z-law, according to which, in the process of deceleration of negative mesons (μ^-, π^-, K^-) in a chemical compound $Z'_k Z_m$, the probabilities $W(Z)$ and $W(Z')$ for the nuclear capture of mesons are proportional to the charges of their nuclei Z and Z':

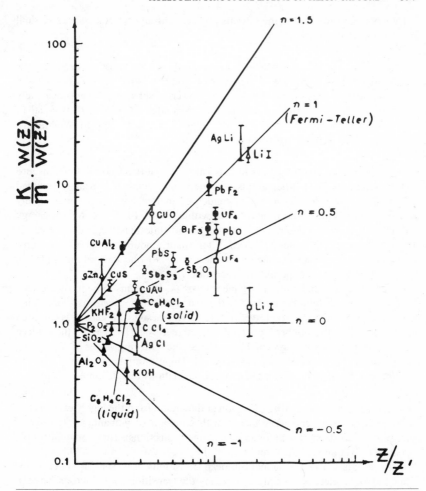

Figure 1   The results of checking the Z-law, following Baijal et al (9).

$$W(Z)/W(Z') = mZ/kZ' \hspace{4cm} 1.$$

The results of the verification of the Z-law validity are given by Baijal et al (9) in Figure 1. The straight lines correspond to the formula

$$kW(Z)/mW(Z') = (Z/Z')^n \hspace{4cm} 2.$$

for the different exponents n. The value $n = 1$ corresponds to the Z-law. On the whole, the experimental facts are shown to testify against the Z-law. Then there arises the question of what causes its violation.

The capture of mesons by the nucleus proceeds through several stages; we shall note the principal ones.

SLOWING DOWN OF THE MESONS FROM VELOCITIES $v \sim c$ TO VELOCITIES $v \sim \alpha c$ Up to velocities $v \sim \alpha c$, which are of the order of atomic electron velocities ($\alpha = 1/137$ is the fine-structure constant), the stopping power of substances is approximately proportional to Z. The time of deceleration in condensed matter (τ), is equal to 10^{-9}–10^{-10} sec, while in gases it is 10^3 times larger. According to Wightman's calculations (10), in liquid hydrogen $\tau = 10^{-9}$ sec.

ATOMIC CAPTURE OF MESONS   Very little is known about the capture of mesons into highly excited levels of mesic atoms and mesic molecules. It is usually assumed that mesons with velocities $v \sim \alpha c$ in condensed matter slow down quite rapidly and are captured into mesoatomic orbits, within 10^{-13}–10^{-14} sec. Meson energy loss at $0 < v < \alpha c$ and the transition to the bound state must depend significantly on the properties of the target material, for small amounts of energy are transferred in this case from the meson to the atom. No one has yet succeeded in taking into account these peculiarities accurately.

In hydrogen the adiabatic capture mechanism (8, 10) seems to be most effective. The meson approaching the nucleus of the hydrogen atom screens its charge, and, as a consequence, the electron leaves the atom. The meson with mass m is captured into the orbit with the number $N_0 \approx 0.8\,(\mu^*)^{\frac{1}{2}}$, where $\mu^* = m(1 + m/M_p)$ is the reduced mass of the meson-proton system. Hence we get the following values of N_0: 14, 15, and 26 for μ^-, π^-, and K^-, respectively (10). These estimates were employed by Eisenberg & Kessler (11, 12) for the cascade calculations in different mesic atoms.

DEEXCITATION PROCESSES   The transition of the meson from the highly excited state N to states with lower energy occurs with emission of γ-quanta or transfer of energy to the electron of the atomic shell (Auger transitions). The probability of the radiative transition is $\sim(\Delta E)^3$ and that of the Auger transition is $\sim(\Delta E)^{-\frac{1}{2}}$; therefore the γ-radiation is predominant in deep transitions of mesons in heavy atoms while the Auger electrons mainly accompany the transitions of the meson between highly excited levels of light mesic atoms with small transition energy ΔE. Both types of transitions obey the selection rules with respect to the angular momentum l and the projection m for dipole transitions: $\Delta l = \pm 1$ and $\Delta m = 0, \pm 1$. The total time of the cascade is greatly lengthened in an isolated atom, because these selection rules only enable the meson to descend from the state (Nlm) to a state not lower than $n' > l - 1$, therefore reaching the ground state only by successive transitions.

NUCLEAR CAPTURE OF MESONS   After having reached the orbit $n' = 1$ of the mesic atom $Z\mu$, the negative muon either decays in accordance with the reaction

$$\mu^- \to e + v_e + \tilde{v}_\mu \qquad\qquad 3.$$

or interacts with the proton of the nucleus

$$\mu^- + p \rightarrow n + \tilde{\nu}_\mu \qquad\qquad\qquad 3a.$$

The rates of these processes are not large (10^6–10^7 sec^{-1}).

The rates of nuclear capture of π^- and K^- mesons from the ns-states of the $Z\pi$ and ZK mesic atoms are rather large ($\sim Z^4 n^{-3}$), resulting in the absorption of mesons from orbits as low as $n \sim 4$. The π meson capture by Z nuclei leads to the disintegration of the nuclei, while for hydrogen atoms the charge exchange reaction of π^- mesons on protons occurs. The absolute rate of this reaction from the ns-state of the $p\pi$ atom, $\Gamma = 1.6 \times 10^{15}\ n^{-3}$ sec^{-1} (13), exceeds the decay rate for the free π^- meson ($\sim 10^8$ sec^{-1}) by many orders of magnitude.

COLLISIONS OF HYDROGEN MESIC ATOMS Collisions of hydrogen mesic atoms precede the nuclear capture of mesons by protons and in some cases can significantly affect its probability. The time of the fastest transition, 2p-1s, in the isolated mesic atom $p\pi$ is 6×10^{-12} sec, but the total cascade time from the level (nlm) to the level 1s is much larger ($\sim 10^{-9}$ sec) due to the selection rules $\Delta l = \pm 1$. This latter value much exceeds the measured one, $\tau_H = (2.3 \pm 0.6)\ 10^{-12}$ sec (14–16). This means that in the process of $p\pi$ atom deexcitation the radiative transitions are unimportant, and the collisions of $p\pi$ mesic atoms with the atomic nuclei play the main role.[1] Due to electroneutrality and small dimensions (10^{-10}–10^{-11} cm) the $p\pi$ atoms penetrate the electron shell of—and give their excitation energy to the electrons of another atom [the so-called external Auger effect (17)].

In addition, the mixing of highly excited states of the $p\pi$ atom with different l in the projectile Coulomb field results in permanent enrichment of the ns-states from which the mesons are intensively captured by nuclei. This leads to the nuclear capture of mesons from the states with $l \neq 0$, escaping the cascade stage [the Day-Snow-Sucher mechanism (18, 19)]. Consideration of both mechanisms of $p\pi$ mesic atom deexcitation leads to satisfactory agreement between the calculated and measured lifetimes of π^- mesons in hydrogen (17).

In addition, the collision processes of the hydrogen mesic atoms with the nuclei Z of other atoms are also accompanied by the processes of meson transfer $p\mu \rightarrow Z\mu$ and $p\pi \rightarrow Z\pi$, which will be discussed below in more detail.

This briefly outlined scheme of the mesoatomic processes in matter had been established by the early sixties after the investigations of Fermi & Teller (8), Wightman (10), Eisenberg & Kessler (11, 12), Leon & Bethe (17), and Gershtein (21). In all of these studies molecular structure was neglected, and in a number of papers (22–25) violations of the Z-law were established as peculiarities of meson capture by atoms of different elements.

However, experiments performed in 1962–1965 convincingly showed that one of the principal causes of the anomalies is molecular structure.

---

[1] The correctness of the deexcitation mechanism described is confirmed by the lifetime measurements for π^- mesons in helium, whose mesic atoms cannot come close to the atomic nuclei because of Coulomb repulsion. As a result, the value of $\tau_{He} = (3.6 \pm 0.7) \times 10^{-10}$ sec (20) exceeds that of τ_H by two orders of magnitude.

ATOMIC CAPTURE OF μ^- MESONS IN CHEMICAL COMPOUNDS

In early studies, the probability of the atomic capture of μ^- mesons in chemical compounds was measured by reaction 3a. Recently, similar measurements have been continued (26). Another method of measuring the atomic capture probabilities $W(Z)$ and $W(Z')$ is the measurement of the relative intensities of mesic X-ray radiation of different elements in chemical compounds. It is essential that the cascade transition stage be close to the beginning of the chain of meso-atomic processes; therefore, the final results are not distorted by the meson transfer processes $p\mu \to Z\mu$ and $p\pi \to Z\pi$.

Integral Intensity of the K Series

The integral intensity $J(Z)$ of the K series of different elements in mixtures of substances and in binary chemical compounds of the type $Z'_k Z_m$ was investigated by Zinov et al (27–31). Because there is no muon transfer in such systems the quantity $J(Z)$ may serve as a measure of the probability $W(Z)$ of atomic capture. The ratio of the capture probabilities $A(Z/Z')$ per atom

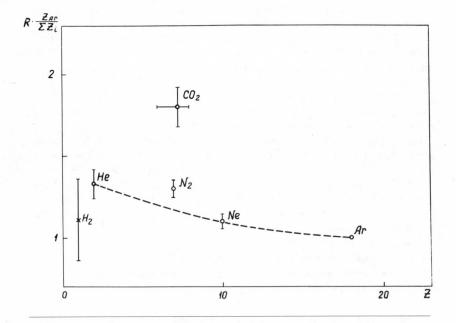

Figure 2 The results of checking the Z-law in gas mixture. R is the ratio of the stopping power of Ar to that of the element with the atomic number Z_i [Zinov et al (29)].

$$A(Z/Z') = kW(Z)/mW(Z') \qquad\qquad 4.$$

shows the degree of deviation from the Z-law, for which $A(Z/Z') = Z/Z'$.

MIXTURES OF INERT GASES   The Z-law was found to be fulfilled satisfactorily in a mixture of inert gases (31) (Figure 2). Small deviations appear to be due to the difference between the ionization potentials, on which the stopping power of elements depends logarithmically.

At the same time, the Z-law is strongly violated in the mixture $Ar + CO_2$, i.e. the measured value of $W(CO_2)$ is larger than the calculated one by about a factor of two (Figure 2).[2]

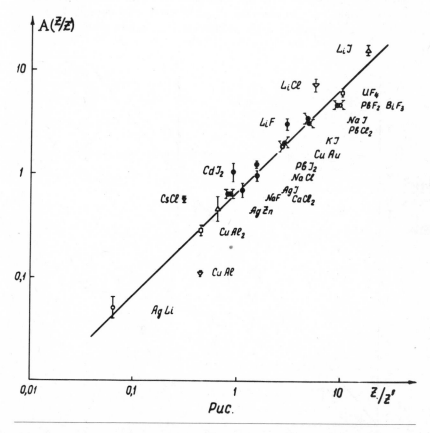

Figure 3   Atomic capture in alloys and halogens of the metals $Z_k Z_m$, in accordance with work of Zinov et al (27). The straight line corresponds to the formula $A(Z/Z') = 0.66\, Z/Z'$.

[2] In particular, this fact provides evidence that in some cases the atomic capture cross section is defined by the final stages of deceleration, at $v < \alpha c$.

ALLOYS AND HALOGEN SALTS   A certain regularity has been established when measuring the quantity $A(Z/Z')$ in metal alloys and chemical compounds of metals with halogens (Figure 3). All the experimental points are grouped around the straight line

$$A(Z/Z') = 0.66 Z/Z' \qquad \qquad 5.$$

This result may be considered as confirmation of the Z-law for mixtures of elements since ionic compounds, in particular halogen salts, may in a sense be thought of as "mechanical mixtures of ions."

OXIDES   The results of measurement of the quantity $A(Z/Z')$ for the normal oxides, i.e. for the oxides $Z_k O_m$ in which the valence of metals coincides with the number of their group in the periodic table, are given in Figure 4. First of all, it is seen that

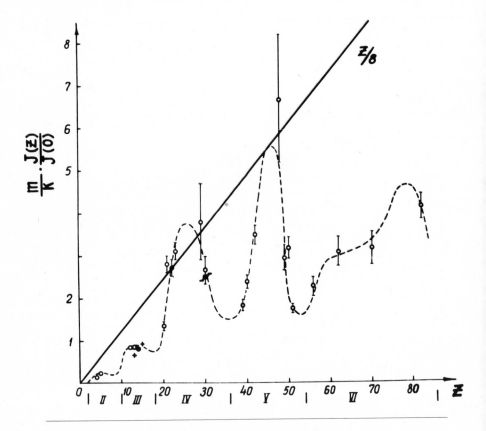

Figure 4   The periodic dependence of the relative probability $A(Z/8)$ of muon capture by metal atoms in oxides $Z_k O_m$ [Zinov et al (27)].

$A(Z/8) < Z/8$, i.e. irrespective of the kind of element forming the oxide, the probability $W(Z)$ of the muon atomic capture into element Z related to the probability $W(O)$ of the capture into oxygen is always smaller than predicted by the Z-law (27).

In addition, the quantity $A(Z/8)$ displays periodic dependence on the Z-value, its minima appearing at the beginning of the periods, i.e. alkali metals. This result provides definite evidence that the type of chemical bond affects the muon atomic capture probability.

This effect is most clearly seen from parallel measurements of the quantity $A(Z/8)$ in different oxides of the same metal, for example,

$$A(MgO) = 0.83 \pm 0.07$$
$$A(MgO_2) = 0.58 \pm 0.03 \quad Z/8 = 1.5$$
$$A(BaO) = 2.27 \pm 0.22$$
$$A(BaO_2) = 3.28 \pm 0.30 \quad Z/8 = 7.0$$

The largest difference (by a factor of two) was found for the pair of oxides

$$\begin{array}{l} A(Sb_2O_3) = 3.48 \pm 0.35 \\ A(Sb_2O_5) = 1.73 \pm 0.09 \end{array} \quad Z/8 = 6.4 \qquad\qquad 6.$$

The Structure of the K and L Series

Much more detailed information on the process of atomic capture of muons can be extracted from the structure of the K and L series, i.e. from measurements of the intensities J_α, J_β, and J_γ of individual components of the K series: K_α (transition 2p-1s), K_β (3p-1s), K_γ (np-1s); and the L series: L_α (3d-2p), L_β (4d-2p), L_γ (nd-2p). The Eisenberg & Kessler calculations (11, 12) predicted a monotonic decrease of the intensity of the highest transitions of the K series with increasing atomic number Z. However, the experimental results do not confirm this dependence: after calcium, where the results of experiments (32, 33) and calculations (11, 12) coincide, the intensity J_γ sharply increases with increasing Z (Figure 5).

In the first experiments of Zinov et al (28), the K-series structure of pure metals was found to differ from that of the same metals as oxides. The spectra of the pairs Ti and TiO_2, V and V_2O_5, and Cr and Cr_2O_3 were studied using an NaI detector. In all three cases the contribution from the K_γ transitions to the total K-series intensity for pure metals was found to be noticeably larger than for the same metals in oxides (Figure 6a).

With the aid of high-resolution Ge(Li) detectors Kessler et al (34) have succeeded in resolving some separated lines of the K_γ and L_γ series. The measured ratios of the intensities

$$R = (J_i/J_\alpha)_{TiO_2}/(J_i/J_\alpha)_{Ti} \qquad\qquad 7.$$

for the separated K_i and L_i lines of titanium are given in Figure 6b. It is immediately seen from the figure that, in going from a metal to its oxide, both the total intensity of the K_γ series and the intensity of each component decreases;

and, the higher the measured transition, $np \rightarrow 1s$, the stronger this decrease. It was also established that the total intensity of all K_ν transitions from the $n \geqq 8$ levels exceeds the predicted values by a factor of two, and the ratio L_α/K_α in titanium is smaller than in TiO_2 by 12%.

Experiments of this kind were continued on a μ^- meson beam by Daniel et al (35) and Tauscher et al (36) and on a π^- meson beam by Grin & Kunselman (37).

In the former studies the ratio of relation 7 was measured for the following pairs: $NaCl/Na_{met}$, Al_2O_3/Al_{met}, CaO/Ca, TiO_2/Ti. The intensity of the K_ν lines in oxides was found to be lower than in pure metals.

The authors of the next study (36) found the relative intensity of the transitions $\alpha(Z) = (5f-3d)/(4f-3d)$ to depend not only on the chemical state of selenium, but also on its physical state. The measured ratio was

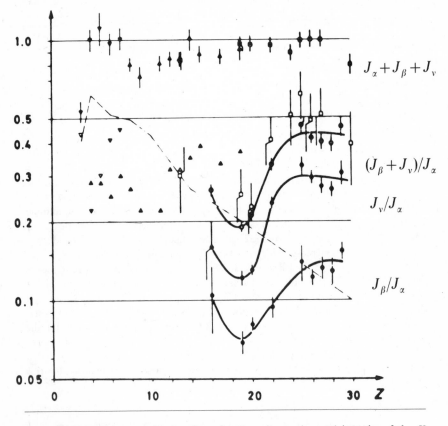

Figure 5 The relative contribution from the K_ν series to the total intensity of the K series, as a function of the element atomic number Z, following Quitman et al (33). The dashed line is the calculation.

$$R = \alpha\,(\text{Se}_{\text{amorphous}})/\alpha\,(\text{Se}_{\text{metallic}}) = 0.74 \pm 0.06 \qquad\qquad 8.$$

The ratios of the intensities of pion X-ray transitions of the type $\alpha(Z_1 Z_2) = (4f-3d)_{Z_1}/(3d-2p)_{Z_2}$ in the compounds $Z_1 Z_2$ and in the equivalent mechanical

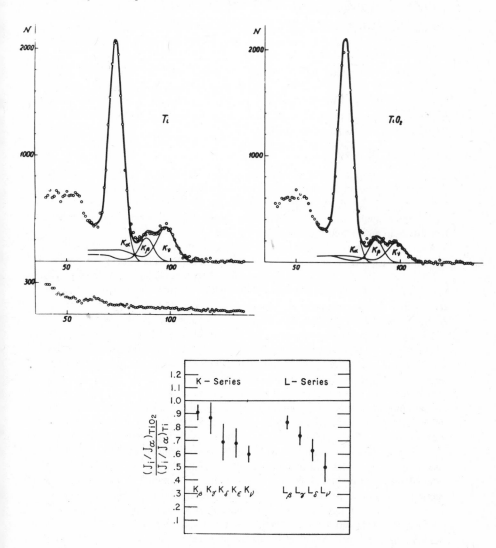

Figure 6 The intensity of the Ti K_v series in the pure metal and the oxide TiO$_2$. (*a–above*) Variation in the total intensity of the K_v series [Zinov et al (28)]; (*b–below*) Variation in the intensities of the separate lines of the K_v series [Kessler et al (34)].

mixtures $Z_1 + Z_2$ have been investigated (36). In particular, the authors studied the pairs $ZnS/(Zn+S)$, $ZnSe/(Zn+Se)$, $FeS/(Fe+S)$, and $FeS_2/(Fe+2S)$.

The most striking difference was observed for the pair $ZnSe/(Zn+Se)$:

$$R(Zn, Se) = \alpha(ZnSe)/\alpha(Zn+Se) = 5.60 \pm 0.35 \qquad 9.$$

Ratios of this type depend on the nature of the chemical bond. For example, the ratios of the transition intensities R (Fe, S) for the pairs $FeS/(Fe+S)$ and $FeS_2/(Fe+2S)$ are different; $R = 2.25 \pm 0.10$ and 1.55 ± 0.10, respectively.

The pair KCl/KCl(solution) is an analogue of the investigated pairs $Z_1 Z_2/(Z_1 + Z_2)$, since the potassium chloride is completely dissociated and equivalent to the mixture of the K^+ and Cl^- ions in the solution. The measured ratio $R = \alpha(KCl)/\alpha(KCl)$ (solution) of the quantities $\alpha = (3d-2p)_K/(3d-2p)_{Cl}$ for the pair KCl/KCl(solution) is $R = 1.40 \pm 0.10$.

Preliminary Discussion

The above-mentioned facts provide convincing evidence for the influence of chemical structure on the processes of atomic capture and the subsequent deexcitation of the resulting mesic atoms and molecules. However, no one has succeeded in bringing them into agreement with the theoretical ideas about cascade transitions in an isolated mesic atom.

Au-Yang & Cohen (38) tried to take into account the chemical bond effect on the atomic capture of mesons. According to their calculations, the observed peculiarities of meson atomic capture in oxides (28) can be explained by the displacement of the electron cloud towards oxygen, which is quantitatively described by the effective charge parameter Z_f [numerically it is equal to the degree of ionicity of the chemical bond multiplied by the number of valence electrons (39)].

Taking into account in this way the change in the effective charges of the atoms forming the molecule and assuming, as before, that the field is centrosymmetric, Au-Yang & Cohen gave a qualitative explanation of the periodic dependence of meson atomic capture in oxides.

However, this approach does not explain the strong dependence of the K series on the kind of chemical compound since the meson cascade transitions must not differ from the cascade transitions in an isolated mesic atom when the centrosymmetric character of the atomic fields in the molecule is conserved.

An attempt has been made (34) to bring into agreement the observed "chemical effects" in the K_v series and the calculations (11, 12) of the meson cascades in an isolated mesic atom. It has been concluded that to obtain this agreement it is necessary to assume essentially different initial distributions of mesons over mesoatomic levels with different angular momenta l, even for elements with close atomic numbers Z (e.g. Ti and Mn). Such an assumption seems to be unnatural.

Thus, all the experimental facts presented here on the study of the atomic capture of negative mesons in chemical compounds force us to admit that the structure of the valence shells of elements appreciably affects the meson atomic capture processes. This effect is especially clearly revealed in the studies of nuclear absorption of π^- mesons in hydrogenous substances (40–45).

NUCLEAR CAPTURE OF π^- MESONS BY PROTONS OF CHEMICALLY BONDED HYDROGEN AND THE MODEL OF LARGE MESIC MOLECULES

The π^- meson charge-exchange reaction on protons occurs in two ways

$$\pi^- + p \rightarrow n + \pi^0 \quad \underset{\longrightarrow\ 2\gamma}{\vphantom{|}} \qquad\qquad 10.$$

$$\pi^- + p \rightarrow n + \gamma \qquad\qquad 10a.$$

with probabilities 0.6 and 0.4, respectively (46), and has some notable features. In liquid hydrogen its probability is close to unity, i.e. each π^- meson stopped in the target induces either reaction 10 or 10a. In addition, the two γ-quanta produced by the decay of a π^0 meson at rest fly in opposite directions. This makes it possible to distinguish reliably reaction 10 from others. When π^- mesons are captured by the nuclei of some other elements, either disintegration of nuclei or emission of nucleons (47) occurs rather than reaction 10. Even in deuterium reaction 10 is suppressed down to $\sim 10^{-3}$ compared with hydrogen (47)[3].

Thus the fact that reaction 10 is observable implies that the π^- meson is absorbed by the hydrogen nucleus and therefore in pure hydrogen its probability can be normalized to unity $[W(\mathrm{H_2}) = 1]$.

Detection of the Effect

In the chemical compound of the type $Z_m \mathrm{H}_n$, reaction 10 is expected to be suppressed as early as at the stage of atomic capture, since some mesons fall into the $Z\pi$ atom levels and make no contribution to reaction 10. In this case the Z-law predicts for the probability

$$W(Z_m \mathrm{H}_n) \approx n/mZ \qquad\qquad 11.$$

However, attempts of Panofsky et al (40) to detect this reaction in LiH and polyethylene $(\mathrm{CH_2})_n$—using facilities available at that time—were a failure. The subsequent studies of Dunaitsev et al (41, 43), Petrukhin & Prokoshkin (42), Chabre et al (44), and Bartlett et al (45) revealed that the probability is suppressed much more strongly, corresponding to the law

$$W(Z_m \mathrm{H}_n) \approx anZ^{-3}/m \qquad\qquad 12.$$

where a is a coefficient of the order of unity.

Figure 7 presents the quantity

$$P = mW(Z_m \mathrm{H}_n)/n \qquad\qquad 13.$$

---

[3] An exception is He[3], in which reaction 10 proceeds with a probability of 0.155 (48). In heavy element nuclei, the charge exchange reaction takes place "in flight" i.e. escaping the stage of mesic atom formation. The probability of such a reaction is, however, rather low, 10^{-5}–10^{-4}. In LiD the probability of charge exchange reaction 10 is equal to $\sim 3 \times 10^{-5}$ (47).

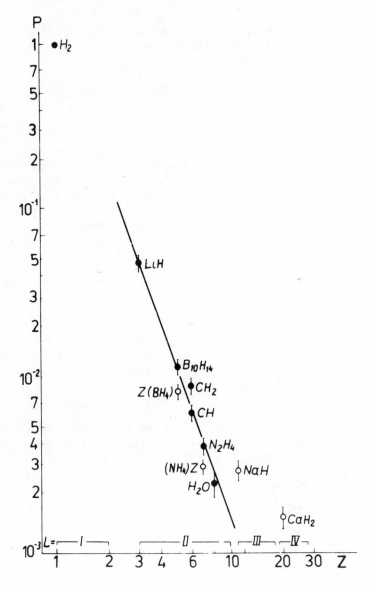

Figure 7 The reduced probability $P = mW(Z_m H_n)/n$ of the charge exchange reaction of π^- mesons on protons in the hydrogenous substances $Z_m H_n$ as a function of the atomic number Z [Krumshtein et al (69)]. The straight line corresponds to the relation $P = aZ^{-3}$ at $a = 1.3$.

Table 1   Hydrides of the second row of the Mendeleev table[a]

Substance	Z	$W, 10^{-3}$	a
LiH	3	35 ± 4	1.26 ± 0.14
$B_{10}H_{14}$	5	12.6 ± 1.4	1.44 ± 0.16
CH_4	6	26.7 ± 0.5	2.40 ± 0.05
N_2H_4	7	5.9 ± 0.7	1.30 ± 0.15
H_2O	8	3.5 ± 0.6	1.12 ± 0.19

[a] This table is compiled on the basis of data from references 1, 69, and 70.

It is seen from Table 1 that the P values for hydrogenous compounds of the elements of the second row of the Periodic Table are well described by the right side of formula 13 with a coefficient $a \approx 1.3$.

Such a strong suppression, e.g. $W(\text{LiH}) \approx 3.5 \ 10^{-2}$ and $W(H_2O) \approx 3.5 \ 10^{-3}$, cannot be explained by π^- meson transfer processes according to the reaction

$$p\pi + Z \rightarrow Z\pi + p \qquad\qquad 14.$$

as was supposed in (40, 49), since in this case probability 12 must depend on the number of collisions of $p\pi$ atoms with Z nuclei, i.e. on the number of nuclei Z per unit volume. Experiment contradicts this conclusion (50). The probability of reaction 10 for ethane C_2H_6 remains unaffected when the ethane density is varied by a factor of 110 (from gas to liquid).

Special experiments aimed at the study of reaction 10 in mixtures of hydrogen with other gases $H_2 + Z$ (51, 52) showed that reaction 14 depends on Z weakly and, therefore, cannot be a cause of dependence 12.

In Figure 8 the ratio of the probabilities $W(C) = W(H_2 + Z)/W(H_2)$ is plotted as a function of the relative concentration $C = n_Z/n_H$ of Z in a gas mixture (n_H and n_Z are the numbers of the atoms of hydrogen and Z contamination per cm^3). It is seen from the figure that even at concentrations $C \sim 1$, i.e. comparable with the heavy atom concentration in hydrogenous substances of the type $Z_m H_n$, the function $W(C)$ equals 0.2–0.1 and is weakly Z-dependent. This result is not associated with the specific features of the gas mixtures since the shape of the curves (Figure 8) is a function of only the C value (51)[4] rather than of the total pressure in the mixture $H_2 + Z$, i.e. the absolute concentrations n_H and n_Z.

These results provide evidence that the mechanism of suppression of reaction 10

[4] The curves in Figure 8 are described by the formula $W(C) = 1/(1 + \Lambda C)$, where $\Lambda = f(Z, C)$ is a function changing slightly within the limits $5 < \Lambda < 10$ for all $Z < 36$ and $C < 1$ (52). In a hydrogenous mixture (50) the transfer reaction $p\pi \rightarrow Z\pi$ is considerably weaker: at $C \sim 1$, $W(C) \approx 0.8$. This picture of transfer is essentially different from that of μ^- mesons, for which the total transition $p\mu \rightarrow Z\mu$ is practically observed already at concentrations as small as $C \sim 10^{-4}$. The indicated difference is due to the long lifetime of μ^- mesons in the K orbit and the strong absorption of π^- mesons by protons even from the high ns-states of the $p\pi$ atom.

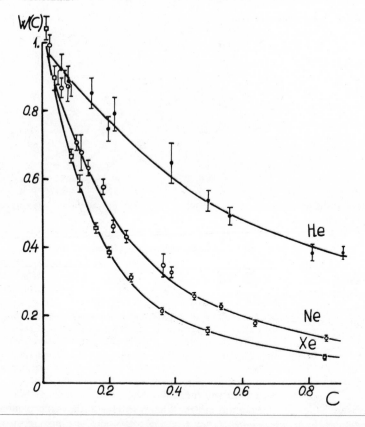

Figure 8   Transfer $p\pi \to Z\pi$ in the gas mixtures $H_2 + Z$ on the basis of work of Petrukhin et al (51, 52).

in the chemical compounds $Z_m H_n$ differs from the transfer mechanism according to reaction 14. Further measurements of the probability for charge-exchange reactions 10 in hydrazine $N_2 H_4$ and in an equivalent mechanical mixture $N_2 + 2H_2$ showed (53) that the ratio of the probabilities is

$$W(N_2 H_4^*)/W(N_2 + 2H_2) \approx 1/30$$

Thus, the results indicate that the explanation for the observed effects should be sought in the particular features of the interaction of mesons with the valence electrons of molecules.

Model of Large Mesic Molecules

The model of large mesic molecules (1, 54–56) was first suggested as the explanation of the features of the process of charge exchange of π^- mesons on the nuclei of chemically bonded hydrogen.

The basic assumption of the model is that in the process of atomic capture a fraction of mesons is captured into highly excited orbits lying near the valence electrons of the molecule. Such an assumption is equivalent to asserting the possibility of the existence of rather stable complexes of the type $Z_m \pi^- H_n$, whose dimensions are tremendous (~ 500 m.a.u.) from the point of view of the meson.

The existence of large mesic molecules seems, at the first glance, unlikely; however, as the simple analysis shows, the idea contains in itself no logical contradictions. It can be easily seen that this hypothesis alone gives the qualitative explanation for many of the observed effects (independence of the probability of reaction 10 on density and state of aggregation, heavy elements contamination, etc). However, more concrete assumptions are required to make quantitative predictions.

PRELIMINARY CONSIDERATIONS   For the sake of simplicity we consider a binary molecule ZH (e.g. LiH) whose level scheme is given in Figure 9. Three types of

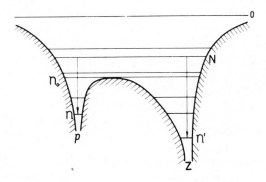

Figure 9 Level scheme for the "Large mesic molecule" $p\pi^- Z$. N denotes mesomolecular levels located in the region of the valence electrons of the molecule ZH; n and n' are the levels of the isolated mesic atoms $p\pi$ and $Z\pi$.

levels are possible in this potential well: the levels n of an isolated hydrogen atom, whose energy E_n is biased by the field of the Z nucleus spaced from the proton by the distance $R \gg 1$

$$E_n \approx -1/2n^2 - Z/R \qquad \qquad 15.$$

the levels n' of an isolated $Z\pi$ atom

$$E_{n'} \approx -1/2(Z/n')^2 - 1/R \qquad \qquad 15a.$$

and, finally, the levels N belonging to the large mesic molecule $Z\pi^- H$ as a whole. The limiting value $n = n_0$ that separates the levels of an isolated mesic atom $p\pi$ from the molecular levels N is determined from the condition $E_{n_0} \leqq U_{max}$, where U_{max} is the height of the potential barrier separating the wells, which is equal to the maximum potential energy of the meson in the field of two fixed nuclei p and Z separated by a distance R:

$$U = -1/r_1 - Z/r_2 \qquad\qquad 16.$$

(r_1 and r_2 are the distances of the meson to the proton and Z nucleus, respectively). The maximum value of U is achieved on the molecule axis at the point $r_1 = R/(1+\sqrt{Z})$, $r_2 = R - r_1$ and equals (54)

$$U_{max} = -(1+\sqrt{Z})^2/R \qquad\qquad 17.$$

We can estimate from this the maximum $n = n_0$ at which level the E_n may be assigned to the hydrogen mesic atom if it is located at the R distance from the Z nucleus in the molecule[5]

$$n_0 = [R/2(1+2\sqrt{Z})]^{\frac{1}{2}} \qquad\qquad 18.$$

for $1 < Z \leq 10$ and $R = 400$–600, $n_0 = 5$–7. For the hydrogen molecule ($Z = 1$, $R = 371$) $n_0 = 8$, which is substantially smaller than Wightman's estimate ($N_0 = 15$) obtained neglecting the effect of the second nucleus in the hydrogen molecule.

On the basis of these arguments and the results of the previous sections we can formulate the main assumptions underlying the model of large mesic molecules.

(a) In the transition of π^- mesons from the state of free motion to the bound state a part of them are captured with a probability W_1 into the mesomolecular orbitals N of the large mesic molecule $Z_m \pi^- H_n$, which lie in the region of the valence electrons of the original molecule $Z_m H_n$. The remainder are captured into the levels n' of the isolated mesic atom $Z\pi$ without falling into the levels n of the mesic atom $p\pi$ at the stage of meson atomic capture.

(b) Later there proceed the radiative and Auger transitions of mesons from the common levels N to separated levels n and n' of the mesic atoms $p\pi$ and $Z\pi$ the probabilities of which are ω_{Nn} and $\omega_{Nn'}$, respectively. The main cause of the suppression of reaction 10 in hydrogenous substances is the low probability $W_2 = \omega_{Nn}/(\omega_{Nn} + \omega_{Nn'})$ of the transition of mesons from the common levels N to the levels n of the isolated mesic atom $p\pi$.

(c) Once the π^- meson reaches one of the lowest levels of the $p\pi$ atom ($n < 7$), it can be transferred into levels of the $Z\pi$ mesic atom according to reaction 14, and therefore captured by a proton only with the probability W_3.

On the basis of these assumptions the total probability of reaction 10 can be represented as the product of the probabilities of all three stages a, b, and c

$$W = aW_1 W_2 W_3 \qquad\qquad 19.$$

where the a coefficient phenomenologically takes into account the features of the chemical bond of hydrogen in the molecules $Z_m H_n$.

We consider in more detail each of these stages.

[5] This simple estimate was subsequently corroborated by precise calculations (57). Since n_0 is weakly dependent on Z, this estimate is practically unaffected by taking into account the screening of the nucleus in heavy atoms.

CAPTURE INTO MESOMOLECULAR LEVELS   No calculations have yet been carried out that give accurately the probability W_1 of meson capture into the levels N of the large mesic molecule. The existing calculations (38, 58–64) are estimates applicable only to an isolated atom. These calculations suggest that low-energy mesons are captured most easily into high mesoatomic levels with quantum numbers $n \geq 15$, the main mechanism of the capture being the Auger ejection of the atomic shell electrons. It is noted in a number of papers that the distribution of captured mesons over levels with different angular momenta l may differ considerably from the statistical $(2l+1)$ distribution (59, 60, 62, 63). In particular, the l-distribution may have a maximum in the region of either the values of $l \sim n/3$ (52) or $l \sim (m_\pi/m_e)^{\frac{1}{3}}$ (63).

At low-energy collisions in hydrogen, adiabatic capture of mesons is most efficient. During this process a meson "replaces" an electron and is captured into one of the mesoatomic levels with an energy approximately equal to the binding energy of the knocked-out electrons. This, in turn, indicates that their orbits are geometrically similar; i.e. the orbit of the captured meson is in the vicinity of the former orbit of the ejected electron.

All of this also remains valid for hydrogenous substances, the only difference being that in this case a valence electron of the molecule $Z_m H_n$ enters the continuum adiabatically. It is also clear that in this case the π^- meson cannot pass into the levels n of the isolated $p\pi$ atom, since in the chemical compound the only electron of the hydrogen atom is consumed to form the chemical bond, and when replaced adiabatically, the meson falls into the common levels N of the system ZH rather than into the n separated levels of the atom $p\pi$.

In case the hypothesis of the geometrical similarity of the meson and electron orbits in adiabatic meson capture is valid, the mesons should mainly be captured into the high-lying levels Nl with small orbital momentum l, since the electrons of the outer shells of the atoms Z forming compounds with hydrogen are, as a rule, in the s-state.

In order to estimate the fraction of mesons captured into the common levels N, we assume that all the electrons of the $Z_m H_n$ molecule are equivalent. Such an assumption is deliberately crude and equivalent to the assumption of the validity of the Fermi-Teller Z-law in the initial stage of meson capture. In this case the probability W_1 of meson capture into the levels N is proportional to the number $2n$ of valence electrons and equal to

$$W_1 = 2n/(n+mZ) \qquad\qquad 20.$$

Later we shall consider the experimental evidence in favor of this formula as well as other possibilities.

CASCADE TRANSITIONS   The mesons captured into the separated levels n' of the $Z\pi$ mesic atom undergo the usual cascade characteristic of isolated mesic atom. The subsequent fate of the mesons falling into the common levels N has some peculiarities.

First of all, the field in which these mesons move differs essentially from a centrally symmetrical one. In such a field, the selection rules of radiative transitions with respect to the angular momentum $\Delta l = \pm 1$, which lengthen the cascade period in an isolated mesic atom, are not so strict. Therefore the probabilities ω_{Nn} and $\omega_{Nn'}$ for transition from the common levels N to the n and n' separated levels increase substantially in this case.

In addition, in describing meson motion in high orbits one has to take into account the screening of the nuclei by the electrons of the inner shells. The estimated values of W_2, taking into account both circumstances, lead to the following result (55)[6]

$$\omega_{Nn'} \approx 10^{10} Z^2 \sec^{-1} \qquad\qquad 21.$$

For the $N \to n'$ transitions to the $n' \geqq 8$ levels Auger transitions prevail over radiative ones (11, 12). Therefore the deexcitation process seems to proceed through two stages, namely the Auger $N \to n''$ transition to the levels $n'' \sim Z$ and the subsequent radiative transitions $n'' \to n'$ in the isolated atom $Z\pi$. The only difference between these two cases is that the population of the n'' levels in the latter case may be considerably different from the statistical $(2l+1)$ distribution. In particular, from the Ns states the transitions to the states $n''p$ are most likely, and the relation $\omega_{Nn'} \sim Z^2$ should be conserved at such step-by-step transitions.[7]

It follows from the expression for the Auger transition probability that the ejection of that electron whose wave function has a maximum overlap with the meson wave function (7, 68) is most likely, i.e. such that their orbits are geometrically similar. Consequently, at the cascade transitions of "valence mesons" the ejection of the valence shell electrons is most probable.

Thus, the probability $\omega_{Nn'}$ for transition of the meson from the common level N to the separated levels n of the $p\pi$ mesic atom is equal to

$$W_2 = \omega_{Nn}/(\omega_{Nn} + \omega_{Nn'}) \approx Z^{-2} \qquad\qquad 22.$$

Apparently these mesons alone induce the subsequent charge exchange reaction 10, since neither of the remaining mesons falling into the n' levels of the $Z\pi$ mesic atom contribute to the reaction.

In principle, the possibility of nuclear capture of π^- mesons by protons directly from the Ns states of the mesic molecule exists, thus escaping the stage of cascade

[6] As is known, $\omega_{Nn'} \sim Z^4$ without taking account of the screening. The resultant dependence $\omega_{Nn'} \sim Z^2$ makes use of the relation $|\psi_N(0)|^2 \sim Z$ for the wave function of the outer atomic electrons in the Ns state moving in the screened field of the nucleus Z (65). This relation is a special case of the Fermi-Segrè formula (66), which has recently been derived anew in the quasiclassical approach (67).

[7] The Auger-transition probability $\omega_{Nn''}^A = \omega_{Nn'} k$. For dipole transitions the conversion coefficient $k = (mc^2/\Delta E)^{7/2}$ depends slightly on Z, since the meson transition energy ΔE between the molecular levels $N \gg Z$ and the atomic levels $n'' \sim Z$ is nearly the same for all the Z atoms. It is also apparent that in the case of the two-stage process ($N \xrightarrow{\text{Auger}} n'' \xrightarrow{\text{Rad.}} n'$) there will be no high transitions $n''p \to 1s$ from the levels $n'' > Z$ in the K_ν series. Recent experiments [Backenstoss et al (30a); Evseev et al and Zinov et al, Private communication] suggest that this reasoning is not senseless.

transitions.[8] The energy E_N of the meson moving in the region of the molecule valence electrons corresponds to the $p\pi$ mesic atom energy levels with quantum numbers $n \sim (m_\pi)^{\frac{1}{2}}$. The rate of nuclear capture from these states, $\Gamma_{capt} \approx 1.6 \times 10^{15}$ $(m_\pi)^{-3/2} \approx 4 \times 10^{11}$ sec^{-1}, still much exceeds the decay rate of the free π^- meson and radiation transition rates ω_{Nn}.

TRANSFER OF π^- MESONS FROM THE $p\pi$ ATOMS TO THE Z NUCLEI   The resultant $p\pi$ atom moves in matter, colliding with the nuclei of other atoms. The transfer rate of the reaction $p\pi \rightarrow Z\pi$ is proportional to the number of nuclei n_Z in volume unit $\gamma_Z = \alpha n_Z$. The deexcitation processes in $p\pi$ mesic atoms and nuclear π^- capture by protons from the high-lying ns-states of the $p\pi$ atom compete with this process. Numerous facts indicate (14–19) that the rate of these processes is proportional to the number of hydrogen nuclei n_H, i.e. $\gamma_H = \beta n_H$. As a consequence, the probability W_3 for a π^- meson to escape the transfer to Z from the $p\pi$ atom moving in matter $Z_m H_n$, is equal to

$$W_3 = \beta n_H/(\beta n_H + \alpha n_Z) = (1 + \lambda C)^{-1} \qquad 23.$$

where $C = n_Z/n_H$. It is noteworthy that the W_3 value depends only on the relative concentration C of the atomic nuclei n_Z and n_H (see also footnote 4).

Recent experiments (65) show that the transfer rate constant λ for the $p\pi$ atoms formed as a result of meson capture in chemical compounds is at least a factor of two smaller than the corresponding constant Λ for the $p\pi$ atoms produced as a result of the mesons stopped in the mixture $H_2 + Z$. Therefore we continue our consideration under the assumption of $\lambda \approx 0$, which will not introduce substantial errors against the background of other uncertainties involved.

Apparently in direct nuclear capture of π^- mesons by protons from the meso-molecular Ns states the transfer stage is absent ($\lambda = 0$).

THE PROBABILITY OF CHARGE EXCHANGE REACTION   The formula for the probability of charge exchange reaction for π^- meson on proton in the chemical compounds $Z_m H_n$, taking into account relations 19, 20, and 22, will assume the following form

$$W(Z_m H_n) = anZ^{-2}/(n + mZ) \qquad 24.$$

One can easily see that the formula obtained describes the empirical dependence 12 satisfactorily.

For complex compounds of the type $Z'_k Z_m H_n$ (e.g. CH_3OH, C_2H_5OH, NH_4Cl etc) the formula is transformed as follows

$$W(Z'_k Z_m H_n) = (avZ^{-2} + a'v'Z'^{-2})/(n + mZ + kZ') \qquad 24a.$$

Here v and v' are the numbers of valence hydrogen bonds with the Z and Z' atoms, respectively.

ACCOUNT FOR THE PECULIARITIES OF CHEMICAL BONDS   In accordance with the spirit of the concepts developed, the meson in the highly excited state N of the system

[8] S. S. Gershtein has called the author's attention to this possibility.

$Z_m \pi^- H_n$ plays the part of a "heavy electron," its orbit being similar to that of the replaced valence electron, and its wave function reflecting the specific features of the chemical bond from which the electron was knocked out. Accordingly the coefficient a in formula 24 takes into account the change in the density of the electron cloud in the vicinity of the proton with the type of chemical bond. Apparently this coefficient should be smaller in ionic compounds than in covalent ones if the hydrogen electronegativity X_H is smaller than the electronegativity X_Z of the Z atom. One may attempt to relate the value of the a coefficient to the degree of bond ionicity σ. One such expression has been given (69)

$$a = 2\sigma + 1.3(1-\sigma) \hspace{4cm} 25.$$

where the first term should be omitted if $X_H < X_Z$ (e.g. in water and acids).

One should not attach crucial significance to expressions of this kind, since uncertainties in determining the W_i probabilities are rather large, and some of these uncertainties associated with the initial atomic capture and the subsequent transfer of the mesons are inevitably included in the a coefficient. Therefore, from now on we shall regard it as a kind of phenomenological parameter, in terms of which it is convenient to carry out an analysis of experimental data and which, in spite of its uncertainty, does reflect the main features of the phenomenon correctly.

Consequences and Verification of Model Assumptions

In addition to the explanation of the empirical dependence $W \sim nZ^{-3}/m$, the described scheme of the processes leads to a number of verifiable consequences.

REACTION 10 INDEPENDENCE OF MATTER DENSITY   Since the main mechanism of suppressing charge exchange reaction 10 operates inside the molecule, the probability of this reaction should not depend on collisions of the molecule with other atoms and, consequently, on the density and aggregate state of the matter. Experiments

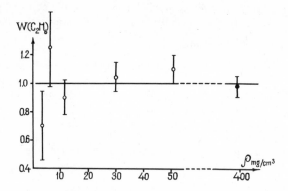

Figure 10   Independence of reaction 10 in ethane of the density of the latter, according to the work of Petrukhin & Prokoshkin (50). ⚡ denotes a gas, ⚡ denotes a liquid.

using ethane (50), in which the density of C_2H_6 was varied in the range of 0.0035 g/cm³ (gas) to 0.390 g/cm³ (liquid) confirm this conclusion (see Figure 10).

REACTION 10 INDEPENDENCE OF HEAVY ELEMENT CONTAMINATIONS   The sharp Z-dependence of the reaction probability is defined by the nuclear charge Z of the atom bonded chemically with hydrogen. Therefore, small mechanical additions of elements even with very high nuclear charges Z' should not essentially change the value of the reaction probability. In fact, addition of iodine salts to methyl alcohol in the proportion $9CH_3OH+NaI$ does not change the probability $W(CH_3OH)$ from that found in pure alcohol (50). Experiments on the gas mixtures CH_4+Z' lead to the same conclusion (70).

CHEMICAL DIFFERENCES   Within the framework of the model described, one should expect different reaction 10 probabilities for substances that consist of the same atoms but differ in the type of chemical bond. A comparison of the values of $P(Z_m H_n) = mW(Z_m H_n)/n$, i.e. the probabilities of reaction 10 per one bond Z—H confirms this conclusion for different combinations of organic substances (see Table 2).

Table 2   Organic compounds with different types of chemical bonding[a]

Substance	W, 10^{-3}	P, 10^{-3}
CH_4	26.7 ± 0.5	6.7 ± 0.1
$(CH_2)_n$	13.7 ± 0.7	6.8 ± 0.3
C_6H_{12}	14.3 ± 0.6	7.1 ± 0.3
C_2H_4	10.0 ± 0.6	5.0 ± 0.3
$(CH)_n$	5.1 ± 0.6	5.1 ± 0.6
C_6H_6	4.5 ± 0.2	4.5 ± 0.2

[a] This table is compiled on the basis of data from references 50, 70, and 71.

In particular, the ratios of the reduced probabilities for organic compounds with different types of bond hybridization are noticeably different from unity:

$$P(CH_4)/P(C_2H_4) = 1.3 \pm 0.1$$
$$P(C_6H_{12})/P(C_6H_6) = 1.6 \pm 0.2$$

At the same time, the P values for substances with different stoichiometric compositions but the same type of chemical bond should not differ. This conclusion is really valid for the combinations of organic substances listed in Table 2, i.e. CH_4 (methane), $(CH_2)_n$ (polyethylene), and C_6H_{12} (cyclohexane), sp^3—hybridization of molecular orbitals; C_2H_4 (ethylene), $(CH)_n$ (polystyrene), and C_6H_6 (benzene), sp^2—hybridization.

The reaction probabilities measured in propyl and isopropyl ether $(C_3H_7)_2O$ do not differ either (71), which provides evidence that isomery does not affect the course of charge exchange reaction 10.

Furthermore, as the bond ionicity increases and, consequently, the density of the electron cloud in the vicinity of the proton becomes lower, the probability of reaction 10 should be suppressed to a larger extent than calculated using formula 24. The measurements carried out in strong acids confirm this conclusion (72).

THE INITIAL MESON CAPTURE   For compounds of the type $Z'_k Z_m H_n$, in which the atom Z' is bonded weakly with hydrogen, e.g. in the ionic compounds $NH_4 Z'$, where $Z' = F$, Cl, Br, I, the contribution from the second term of equation 24a can be neglected. As a result, we obtain

$$4Z^{-2}/W = A + BZ' \qquad\qquad 26.$$

where for NH_4 one should expect the values of $A \sim 10$ and $B \sim 1$. An analysis of the experimental data (Figure 11) yields the values of $A = 6.6 \pm 3.7$ and $B = 1.2 \pm 0.2$. This indicates that in a chemical compound the probabilities of capture of mesons by atoms into the levels of different mesic atoms are nearly proportional to the nuclear charges of the atoms. This fact might be some ground

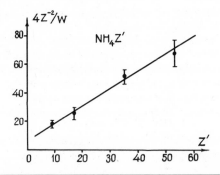

Figure 11   The checking of the atomic capture law in ammonium salts, according to work of Krumshtein et al (69).

for the first assumption of the model concerning the form of W_1. It goes without saying that this conclusion is valid only under the assumption of $W_3 \approx 1$. Otherwise relation 26 is a consequence of the combined action of the two mechanisms, atomic capture and the transfer reaction $p\pi \to Z\pi$. The measurements of the probability of nuclear muon capture in the compounds $NH_4 Z'$ (26) show that the latter assumption is not senseless. However the accuracy of the measurements is still insufficiently high to allow an unambiguous choice of hypothesis.

THE $W_2 \sim Z^{-2}$ DEPENDENCE   The Z-dependence of the second stage of the π^- meson nuclear absorption may be estimated from comparing the probability

$W(CH_4 + Ar)$ in the gas mixture $CH_4 + Ar$ (70) with the probability $W(NH_4Cl)$ (69). In these almost isoelectronic systems the conditions of atomic capture (W_1) and transfer (W_3) may be expected to be the same, since the structure of the electron shell of Cl^- is rather close to that of the noble gas Ar. In this case formula 24 yields the following relation[9]

$$W(CH_4 + Ar)/W(NH_4Cl)$$
$$\approx W_2(CH_4)/(W_2(NH_4^+) \approx (Z_N/Z_C)^2 = 1.36 \qquad 27.$$

The probability ratio is measured to be

$$W(CH_4 + Ar)/W(NH_4Cl)$$
$$= (4.5 \pm 0.4) \times 10^{-3}/(3.2 \pm 0.6) \times 10^{-3} = 1.4 \pm 0.3 \qquad 27a.$$

Unfortunately the accuracy of the measurements is not high enough to draw an unambiguous conclusion.[10]

THE K-SERIES INTENSITY   On the basis of the model described, an increase in the intensity of the K_v series of elements with its increasing valency in chemical compounds (see Figures 5 and 6) is explained as a result of an increase in the number of valence electrons and the corresponding increase in the fraction of mesons captured into the highly excited mesomolecular levels N.

The mentioned correlation is observed in reality. For instance, in metallic chromium $J_v \approx 0.22$ and in chromium oxide $J_v \approx 0.15$, while the ionic compounds (KCl), $J_v \approx 0.07$, which approximately coincides with the intensity of the K series of the noble gases.

Budyashov et al (29) have performed an experiment which seems to model well the mechanism of meson transition from the high-lying molecular states N to the separated levels n' of the mesic atom $Z\pi$. In this experiment the structure of the Ar K series of muons in pure Ar and in the mixture 99.5% $H_2 + 0.5\%$ Ar have been compared. In the latter case virtually all μ^- mesons were captured first to the levels of the mesic atoms $p\mu$ and could yield the K series of $Ar\mu$ only after transfer in the reaction $p\mu + Ar \rightarrow Ar\mu + p$. Figure 12 shows that the structures of the K series are largely different in these two cases: the contribution from the K_v series in pure Ar is $J_v \approx 0.07$, while that in the mixture $Ar + H_2$ is $J_v \approx 0.5$.

A natural explanation of this result is provided by the fact that in the process of

[9] In the latter equality we also assume that $a_N = a_C$, because the type of chemical bond (sp^3—hybridization) is the same in both CH_4 and NH_4^+ (39).

[10] Another possibility of checking the relation $W_2 \sim Z^{-2}$ consists in measuring the reaction 10 probability in deuterated compounds. Similar measurements were recently performed for the set of substances CH_3OH, CD_3OH, and CH_3OD [Goldanskii et al (69a)]. In this case the conditions of capture (W_1) and transfer process (W_3) are practically identical and the Z-dependence of W_2 can be studied in the most direct way. The study of reaction 10 in deuterated compounds is of particular importance because in this way it is possible to isolate any functional group in complex molecules and explore the particular features of its chemical structure independently of the others.

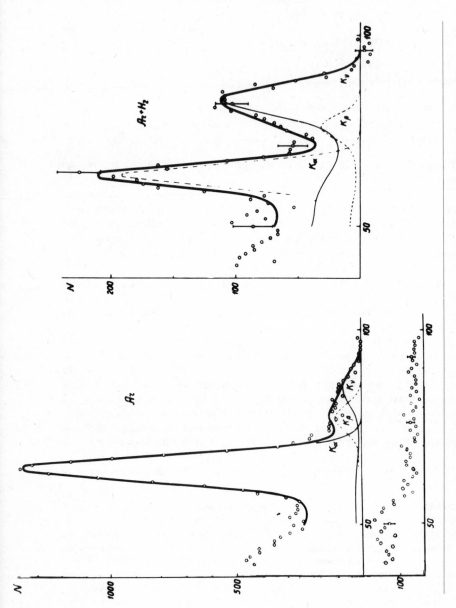

Figure 12   The structure of Ar K series in pure argon and the mixture Ar(0.5% + H$_2$(99.5%). In the second case the contribution from the K_ν series is much more considerable than that in the first one. [Budyashov et al (30)].

collision between the $p\mu$ atom and an Ar nucleus there occurs a highly excited quasimolecule[11] p-μ-Ar, whose cascade transitions are essentially different from those in the isolated mesic atom Arμ. The energy of the pμ atom ground state ($n = 1$) is equal to the energy of the $n' = 18$ level of the mesic atom Arμ; therefore, in the quasimolecule p-μ-Ar the transition from the level $n = 1$ to the level $n' = 1$ corresponds to the rather high transition ($n' = 18$) $\rightarrow$ ($n' = 1$) in Arμ. Because of violated central symmetry in the p-μ-Ar quasimolecule field, the population of the levels $n' < 18$ of the mesic atom Arμ during the cascade transition of the meson will be different from population in the isolated mesic atom Arμ. As a consequence, this should lead to variations in the intensities of the K-series lines of the mesic atom Arμ.

Another possible reason for the strong dependence of the K_v-series intensity on the type of chemical compound is the selective nature of meson capture into mesomolecular orbits with certain values of the angular momentum l. In particular, at adiabatic capture of mesons in hydrogenous substances the anticipated initial ($2l+1$) statistical distribution of mesons is violated due to the increasing contribution from the Nl states with small l which should affect strongly the final result of the cascade in the mesic atom $Z\pi$. In particular, in this case the number of the cascade steps becomes essentially smaller, and thus the information on the chemical bond structure, accumulated by the meson wave function at the atomic capture stage, can be detected even at the final cascade steps.[12]

Systematic Studies

The calculation of meson deceleration and capture, taking into account chemical composition, is a rather laborious problem that is not likely to be solved completely in the near future. Under these conditions, ignoring the existing uncertainties in derivation of formula 24 and determination of the a parameter, one can make a purely phenomenological use of them in an analysis of the data on the measurements of the reaction 10 probability in different substances. Until now about one hundred measurements of this kind have been performed (41–53, 69–73, 75, 79).

ACIDS   Table 3 shows the experimental results for acids (72, 79), which indicate a certain correlation between the value of the a coefficient and acidity. The a coefficient is by definition anomalously small for strong acids, in which practically complete dissociation of the molecules into H^+ ions and acid anions takes place, while for normal acids and acids close to bases, this coefficient amounts to $a = 0.3$–0.5.

With the help of relation 25 one can attempt to relate the degree of dissociation pK_1 of the acid to the degree of ionicity σ of the chemical bond in a molecule (72).

---

[11] Its structure should be analogous to that of the highly excited molecule HCl.

[12] For example, the two-stage transition $Ns \xrightarrow{\text{Auger}} n''p \xrightarrow{\text{Rad.}} 1s$ for the Ns and Nd orbitals is the most probable one, for Np orbitals, three-stage transitions, etc. For ($2l+1$) distribution, the number of transitions is much larger, and the particular features of the initial capture are averaged in the course of cascade.

Table 3   Acids[a]

Acid	$W, 10^{-4}$	a	pK_1
H_2SO_4	<0.4	<0.06	-3.0
HNO_3	<0.6	<0.12	-1.4
$H_2C_2O_4$	<0.1	<0.01	1.3
H_2SeO_3	0.9 ± 0.5	0.17 ± 0.08	2.6
H_3PO_4	1.6 ± 0.4	0.17 ± 0.04	2.1
H_3BO_3	4.2 ± 1.0	0.29 ± 0.05	8.7
$Al(OH)_3$	5.6 ± 1.0	0.48 ± 0.09	9.2

[a] This table is compiled on the basis of data from references 72 and 79.

Oxalic acid $H_2C_2O_4$, in which the general correlation between the quantities σ and pK_1 is violated, is of particular interest here. Krumshtein et al (72) explain this fact as due to association between the molecules $H_2C_2O_4$.

ALKALI   Systematic measurements of the reaction 10 probability W in the alkali $Z'(OH)_q$ have shown (71, 73) that the experimental value of the probability is q times smaller than that calculated by formula 24. This fact may be due to the anomalously high probability of atomic capture of mesons by oxygen. This supposition may be based on the results of work of Zinov et al (27) on measurement of the quantity $A(Z/Z')$ in oxides (see formula 4 and Figure 4), and also on the recent measurements (26) in which the value of $A(OH/Ca) \approx 4$ in $Ca(OH)_2$ has been found to exceed by about an order of magnitude that predicted by the Z-law. Another explanation is suggested in the work of Goldanskii & Bersuker (74).

THE INDUCTIVE EFFECT   In investigating reaction 10 for the set of substances CH_3X, where $X = CH_3S$, I, CN, NO_2, COCl, CH_3CO, etc, Vilgelmova et al (75) observed a correlation between the value of the a coefficient calculated by formula 24, and the induction constants σ_I (76), which characterize the reactivity of the radical CH_3 under the effect of various substituents Z (see Figure 13). Irrespective of the definition of a, the existence of this correlation may prove helpful in evaluating the quantities σ_I for some compounds, since the chemical methods of determining σ_I are rather complicated and sometimes even impracticable.

THE STUDY OF THE TRANSFER STAGE   In works (70, 80) charge exchange reaction 10 has been studied in the mixtures $CH_4 + Z'$, $C_2H_4 + Z'$, and $C_6H_6 + Z'$ with different concentrations $C = n_{Z'}/n_H$ of the Z' atom contamination. In all cases, besides the strong "chemical suppression," a weak additional suppression of the reaction was recorded. Petrukhin et al (70, 80) explain this phenomenon as being conditioned by the transfer that is depicted by the function $W_3 = 1/(1 + \lambda C)$. For the mixture $CH_4 + Ar$ the transfer rate constant λ equals 3.9 ± 0.3, i.e. approximately a factor of two smaller than the corresponding constant $\Lambda = 8.8 \pm 0.9$ in the mixture $H_2 + Ar$ at the same value of $C \approx 0.5$ (52).

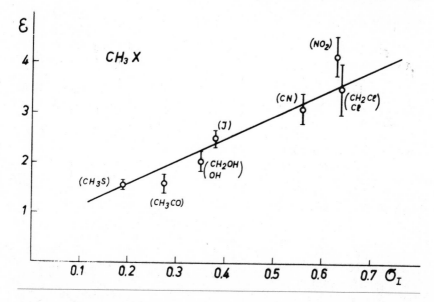

Figure 13 Dependence of the magnitude of the a coefficient calculated by formula (24) and the values of the induction constant σ_I in the compounds CH_3X, in accordance with work of Vilgelmova et al (75). The ordinate axis gives the value of $\varepsilon = a_0/a$, where $a_0 = 2.05 \pm 0.04$. The straight line corresponds to the equation $\varepsilon = \alpha\sigma_I + \beta$ at $\alpha = 4.7 \pm 0.6$, $\beta = 0.6 \pm 0.2$. The substituents X are given in parentheses.

It should, however, be noted that the given values of λ and Λ are calculated assuming the probability of the atomic capture of mesons into mixture components to be proportional to the stopping power of these substances at a meson energy of about 2 MeV. Though this assumption looks natural, it should be checked since in some cases it has proved invalid, e.g. in the mixture $CO_2 + Ar$ (31).

Some Obscurities of the Model

The model of the large mesic molecule combines a rather wide range of phenomena in mesoatomic and mesomolecular physics and explains them from a unique point of view. Formulae 24 and 24a describe semiquantitatively the measured probabilities of reaction 10 in different hydrogenous substances. However, it is necessary to mention some facts that find no explanation in the framework of the model.

CAPTURE OF K^- MESONS AND ANTIPROTONS BY PROTONS   By now the absorption probability for K^- mesons (77) in photoemulsions, and for K^- and $\bar{p}$ (77a, 78) in propane, by the nuclei of chemically bonded hydrogen has been measured. An analysis of the results of these measurements using formula 24 has shown that the a coefficients are respectively equal to

$$a(\pi^-) \approx 2, \quad a(K^-) \approx 5, \quad a(\bar{p}) \approx 11 \qquad\qquad 28.$$

Pawlewicz et al (78) explain this fact as being due to the increase of the transfer rate of mesons from the mesic atoms $p\pi$, pK, and $p\bar{p}$ to the nuclei Z of other atoms in the transition from π^- to K^- and $\bar{p}$. This assumption does not contradict the direct experimental results of Bugg et al (78a) who measured the capture rates of π^-, K^-, and $\bar{p}$ by protons in the mixture of hydrogen with neon.

Another explanation rests on the hypothesis of the preferential filling of the highly excited Ns states at adiabatic capture of mesons in hydrogenous compounds. The probabilities ω_{N_n} of radiative transitions from these states are smaller than the rate of direct nuclear capture of mesons by protons. For mesons with mass m, moving in the region of valence electrons $\Gamma_{capt} = \gamma(m/N)^3 \approx (m)^{3/2}$ (1). Thus, if the direct nuclear capture of π^- mesons by protons occurs with a noticeable but still smaller probability than the cascade transition in the $Z\pi$ atom, then in going from π^- mesons and antiprotons the probability W_2 and, consequently, the a coefficients in formula 24 must increase.

METAL HYDRIDES   For all hydrides of the second row of the Mendeleev Table the a coefficient in formula 24 remains practically constant, the absolute value being no larger than $a \approx 2$ (see Table 1). The formal use of formula 24 in the analysis of the measured reaction 10 probability in the hydrides of the elements of other rows leads, however, to a coefficient values which cannot be considered reasonable [in some cases $a \sim Z/2$ (1)]. This result may possibly be due to the fact that at the stage of the capture of mesons by atoms the valence electrons of the atoms play a more important part than the electrons of the closed atomic shells (see below).

CHARGE EXCHANGE OF π^- MESONS IN ORGANIC COMPOUNDS   According to the definition, the a coefficient in formula 24 reflects the peculiarities of the chemical bond of a hydrogen atom in a molecule and, therefore, should be the same for compounds with identical structure of valence shells, e.g. in the homologous series of saturated hydrocarbons C_mH_n for which $n = 2m+2$. An analysis of the experimental data (80) using formula 24 shows that this requirement is not fulfilled (see Table 4).

Table 4   Saturated organic compounds[a]

Substance	$W, 10^{-2}$	a	α
CH_4	2.67 ± 0.05	2.40 ± 0.05	0.96 ± 0.02
C_2H_6	2.06 ± 0.09	2.22 ± 0.09	0.99 ± 0.04
C_5H_{12}	1.59 ± 0.06	1.99 ± 0.08	0.95 ± 0.04
C_6H_{14}	1.66 ± 0.05	2.14 ± 0.06	1.03 ± 0.03
$C_{12}H_{26}$	1.42 ± 0.06	1.92 ± 0.08	0.94 ± 0.04
$C_{17}H_{36}$	1.40 ± 0.05	1.92 ± 0.06	0.95 ± 0.03
$(CH_2)_n$	1.37 ± 0.07	1.96 ± 0.10	0.99 ± 0.05
C_6H_{12} (cyclohexane)	1.43 ± 0.06	2.04 ± 0.08	1.03 ± 0.04

[a] This table is compiled on the basis of data from references 1, 50, and 70.

Petrukhin et al (80) explain the observed decrease of the a coefficient as being conditioned by the transfer of π^- mesons according to the reaction $p\pi \rightarrow C\pi$. This transfer becomes especially essential at the values of $x = m/n \gtrsim 1$, for instance in cyclic (benzene) and polycyclic compounds (anthracene, etc). With the use of the values of $a = 3.52 \pm 0.17$ and $\lambda = 1.56 \pm 0.14$ in formulae 19 and 23 it is possible to describe satisfactorily all the experimental results available for organic compounds.

However, it should be noted that in the analysis of these experimental data use has been made of a certain hypothesis concerning the form of W_1, particularly of expression 20. In principle, other possibilities may be realized that also deserve discussion here.

In formula 24 the peculiarities of atomic meson capture in chemical compounds are reflected in the product of the factors aW_1, expression 20 for W_1 being based on the assumption that all molecule electrons, both valence and core, are equivalent at the stage of atomic capture. This assumption apparently cannot be made without reservation and is worthy of being discussed in greater detail, especially for organic compounds, where the concept of directed valence (39) makes this hypothesis of the equivalence of all carbon electrons rather implausible. A large set of organic compounds permits a more detailed investigation of this hypothesis.

Figure 14 presents the quantity $Z^{-2}/W = 1/aW_1W_3$ for the saturated compounds

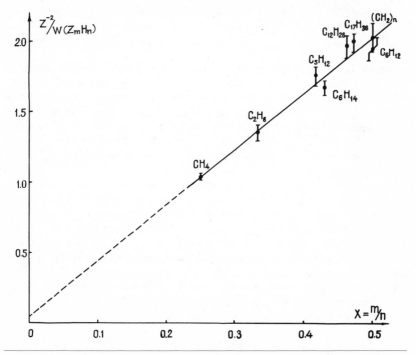

Figure 14 The measured probability W of reaction 10 in the saturated hydrocarbons C_mH_n. The straight line corresponds to the relation $Z^{-2}/W = 4x$.

$C_m H_n$ as a function of the relative carbon concentration $x = m/n$. On the basis of formulae 19 and 23 for the probabilities W_1 and W_3 one should expect to obtain a quadratic relation of the form $Z^{-2}/W = A + Bx + Dx^2$, where $A \approx 1$, $B \approx 6 + \lambda$, $D \approx 6\lambda$. In fact, an analysis of the experimental data gives the following result: $A = 0.04 \pm 0.06$, $B = 4.00 \pm 0.16$, $D \approx 0.15$.

A possible interpretation of the result may be that at absorption of pions in saturated organic compounds, the electrons of hydrogen atoms are not involved in the atomic capture of pions ($A \approx 0$), while transfer $p\pi \to C\pi$ can be neglected ($\lambda \approx 0$)[13] within the existing experimental accuracy. Accordingly one should assume that $W_3 \approx 1$, $aW_1 = \alpha n/vm$, where $v = 4$ is equal to the number of carbon valence electrons. Hence, the probability for reaction 10 in saturated hydrocarbons is expressed through the very simple formula

$$W(C_m H_n) = \alpha n Z^{-2}/vm \qquad\qquad 29.$$

where $\alpha \approx 1$ for the entire homologous series (see Table 4).

Model Modification

Numerous facts, of which a majority are given in the present paper, force us to subject the hypotheses underlying the model of large mesic molecules, and in particular, supposition 20 about the form of the function W_1, to a critical revision. As the previous analysis shows, for saturated organic compounds the relation

$$W_1 = n/mv \qquad\qquad 30.$$

where v is equal to the number of valence electrons of the Z atom, is preferable. This means that the probability of atomic capture of mesons with emission of Auger electrons from the closed $1s^2$ shell of the carbon atom is negligible. Generalization of this result to the elements of other rows leads to the conclusion: The effect of the closed electronic shells of Z atoms on atomic capture of mesons can be neglected; the probability W_1 is expressed by formula 30, where v is equal to the number of electrons in the unfilled shell of the Z atom; the probability $W(Z_m H_n)$ of the charge-exchange reaction 10 is calculated by formula 29.

We have not yet made a detailed quantitative analysis of all of the results of this hypothesis. Here we dwell only upon some of its evident qualitative conclusions:

(a) All the results of measurements of reaction 10 probability in substances $Z_m H_n$ with elements Z from the second row of the Mendeleev Table remain practically unaffected because in this case $v = Z - 2$, and replacement of formula 24 by formula 29, within the available measurement accuracy, is equivalent to some redetermination of the a coefficient.

(b) For hydrides of other rows the dependence $W \sim Z^{-2}$ should be expected rather than $W \sim Z^{-3}$. Preliminary measurements of the probability $W(Z_m H_n)$ in hydrides with large Z do not contradict this conclusion.

---

[13] This assumption is also favored by the equations $P(CH_2) \approx P(CH_4)$ and $P(CH) \approx P(C_2 H_4)$ (Table 2). These equations are satisfied to a good accuracy, although the transfer conditions for these pairs are essentially different.

(c) Formula 29 explains the relations $P(CH_2)_n \approx P(CH_4) \approx P(C_6H_{12})$ and $P(CH)_n \approx P(C_2H_4) \approx P(C_6H_6)$ (see Table 2) in a more natural way than does formula 24.

(d) Sharp deviations from the Z-law (9, 26) are quite natural with the above assumption. In particular, the periodic dependence of the quantity $A(Z/O)$ in oxides $Z_k O_m$ (Figure 4) is a natural consequence of the periodic change of the number of external electrons v with monotonically increasing number Z. Then periodic changes of the K_v-series intensity should also be expected, and the curves of Figure 5 appear to be a fragment of the curve of the type presented in Figure 4.

(e) Finally, the assumption about predominant capture of mesons into N levels lying in the region of the valence electrons gives a simple explanation for the strong difference of atomic capture probabilities in different oxides of the same element, e.g. in Sb_2O_3 and Sb_2O_5 (27).

The subsidiary assumption about predominant population of highly excited Nl states with small l in the process of adiabatic capture enables us to understand also the effect of the chemical bond on the K- and L-series structure (see footnote 12).

In an attempt to estimate the situation as a whole, one has to conclude that the existing experimental accuracies do not allow the analytical expression for the probability $W(Z_m H_n)$ to be established unambiguously, and that the theoretical ideas need further development.

CONCLUSION

The essential influence of the molecular structure of substances on the processes of atomic and nuclear capture of mesons can currently be regarded as a firmly established fact. In principle, this enables one to use mesons to study the electron shells of molecules in problems of both the qualitative and quantitative analysis of substances (53, 75, 76).

Of these two problems, the latter is considerably easier to solve and at present some concrete suggestions and attempts are being made to put the developing possibilities into practice. Some methods of the quantitative and qualitative analysis of samples have been developed using the dependence of the intensity and structure of muon X-ray K series of elements upon their chemical and physical states (81–83). These methods have a number of advantages compared to the common X-ray structure analysis. In particular, the possibility of localizing the muon beam and varying its energy enables one to change the stopping point of a muon and study chemical composition only in the vicinity of this particular point. The design of such an installation at LAMPF, which will be aimed at making a detailed "chemical chart" of a man, is currently under discussion.

In addition, the greater penetrating ability of muon X-rays with respect to the normal X-rays permits an analysis of bulky samples, which is impossible with existing methods.

The unique selectivity of the charge exchange reaction of negative pions with protons and its strong dependence on the chemical state of hydrogen enable one even now to discriminate reliably between chemically bonded and free hydrogen (53). Later on this method can, in principle, be used to investigate the kinetics and catalysis of chemical reactions involving hydrogen, the peculiarities of the hydrogen chemical bond in organic compounds and transition metal hydrides, as well as the structure of hydrogenous substances under critical conditions, e.g. at superlow temperatures, at superhigh pressures, etc.

However, to use the observed effects as a basis of the new method of analysis of substance structure, systematic studies are needed with subsequent comparison of the results with the data obtained by other methods—for instance, the methods of nuclear magnetic resonance, electron paramagnetic resonance, infrared spectroscopy, etc.

The problem of studying the structure of the electron cloud in molecules is substantially more complicated and requires some complex experimental investigations and theoretical calculations. First, it is necessary to separate thoroughly the "physical" effects from the "chemical" ones, i.e. the transfer reaction and atomic capture into inner atomic shells from the effects of the interaction between mesons and the valence electrons of molecules. In this connection, it is appropriate to study simultaneously all the characteristics of X-ray spectra, namely the transition energies and intensities and the residual polarization of the cascade muons. In doing so, the use of Auger electron spectroscopy, which permits determination of the binding energy of Auger electrons to an accuracy of up to 0.1 eV (84), looks rather promising.

In order to succeed in the described investigations which can be united under the title "mesic chemistry," one should increase the existing experimental accuracy which is on the average not higher than 10%. This in turn necessitates theoretical calculations of the process of atomic capture of negative mesons at the present-day level of rigor.

ACKNOWLEDGMENTS
The author expresses his sincere gratitude to S. S. Gershtein for permanent interest in the work and numerous discussions on the problems covered by this paper, to V. S. Evseev, V. I. Petrukhin, V. N. Pokrovsky, Yu. A. Yutlandov, and V. G. Zinov for discussions and communication of some experimental data prior to their publication. Special thanks are due to V. P. Dzhelepov for his encouragement and to V. M. Suvorov for various scientific and technical aid in preparing the manuscript for publication.

Literature Cited

1. Gershtein, S. S., Petrukhin, V. I., Ponomarev, L. I., Prokoshkin, Yu. D. 1969. *Usp. Fiz. Nauk* 97:3–36; 1970. *Sov. Phys. Usp.* 12:1
2. Kim, Y. N. 1971. *Mesic Atoms and Nuclear Structure.* Amsterdam: North-Holland
3. Gershtein, S. S., Ponomarev, L. I. 1973. In *Muon Physics,* ed. V. W. Hughes, C. S. Wu. New York: Academic. In press
4. Burhop, E. H. S. 1969. *High Energy Phys.* 3:109
5. Wu, C. S., Wilets, L. 1969. *Ann. Rev. Nucl. Sci.* 19:527–606
6. Devons, S., Duerdoth, I. 1969. In *Advances in Nuclear Physics,* ed. M. Baranger et al, 2:295–423: New York: Plenum
7. Backenstoss, G. 1970. *Ann. Rev. Nucl. Sci.* 20:467–505

8. Fermi, E., Teller, E. 1947. *Phys. Rev.* 72: 399–408
9. Baijal, J. S., Diaz, J. A., Kaplan, S. N., Pyle, R. V. 1963. *Nuovo Cimento* 30: 711–27
10. Wightman, A. S. 1950. *Phys. Rev.* 77: 521–28
11. Eisenberg, Y., Kessler, D. 1961. *Nuovo Cimento* 19: 1195–1210
12. Eisenberg, Y., Kessler, D. 1963. *Phys. Rev.* 130: 2352–61
13. Fray, G. *Phys. Rev.* 1959. 113: 688–89
14. Fields, T. H., Yodh, G. B., Derrick, M., Fetkovich, J. G. 1960. *Phys. Rev. Lett.* 5: 69–70
15. Bierman, E., Taylor, S., Koller, E. L., Stamer, P., Huetter, T. 1963. *Phys. Lett.* 4: 351–53
16. Derrick, K. et al 1966. *Phys. Rev.* 151: 82–86
17. Leon, H. L., Bethe, H. 1962. *Phys. Rev.* 127: 636–47
18. Day, T. B., Snow, G. A., Sucher, J. 1959. *Phys. Rev. Lett.* 3: 61–64
19. Day, T. B., Snow, G. A., Sucher, J. 1960. *Phys. Rev.* 118: 864–66
20. Block, M. M. et al 1963. *Phys. Rev. Lett.* 11: 301–3
21. Gershtein, S. S. 1962. *Zh. Eksp. Teor. Fiz.* 43: 706–19; 1962. *Sov. Phys. JETP* 16: 501
22. Sens, J. C., Swanson, R. A., Telegdi, V. L, Yovanovitch, D. D. 1958. *Nuovo Cimento* 7: 536–44
23. Lathrop, J. F., Lundy, R. A., Swanson, R. A., Telegdi, V. L., Yovanovitch, D. D. 1960. *Nuovo Cimento* 15: 831–34
24. Eckhause, M., Filippas, T. A., Sutton, R. B., Welsh, R. E., Romanowski, T. A. 1962. *Nuovo Cimento* 24: 666–71
25. Bobrov, V. D. et al 1965. *Zh. Eksp. Teor. Fiz.* 48: 1197–99; 1965. *Sov. Phys. JETP* 21: 798
26. Goldanskii, V. I. et al 1973. *Dokl. Akad. Nauk. SSSR* 211: 60–63
27. Zinov, V. G., Konin, A. D., Mukhin, A. I. 1965. *Yad. Fiz.* 2: 859–67; 1965. *Sov. J. Nucl. Phys.* 2: 613
28. Zinov, V. G., Konin, A. D., Mukhin, A. I., Polyakova, P. V. 1967. *Yad. Fiz.* 5: 591–98; 1968. *Sov. J. Nucl. Phys.* 5: 420 (Preprint JINR P-2039, Dubna, 1965)
29. Budyashov, Yu. G., Ermolov, P. F., Zinov, V. G., Konin, A. D., Mukhin, A. I. 1967. *Yad. Fiz.* 5: 830–33; 1967. *Sov. J. Nucl. Phys.* 5: 529
30. Budyashov, Yu. G., Ermolov, P. F., Zinov, V. G., Konin, A. D., Mukhin, A. I. 1967. *Yad Fiz.* 5: 599–602; 1967. *Sov. J. Nucl. Phys.* 5: 426
30a. Backenstoss, G. et al 1971. *Phys.*

Lett. B 36: 422–25
31. Zinov, V. G., Konin, A. D., Mukhin, A. I. 1964. *Zh. Eksp. Teor. Fiz.* 46: 1919–20; 1964. *Sov. Phys. JETP* 19: 1292
32. Johnson, C. S., Hincks, E. P., Anderson, H. L. 1962. *Phys. Rev.* 125: 2102–11
33. Quitmann, D. et al 1964. *Nucl. Phys.* 51: 609–33
34. Kessler, D. et al 1967. *Phys. Rev. Lett.* 18: 1179–83
35. Daniel, H. et al 1967. *Phys. Lett. B* 26: 281–82
36. Tauscher, L. et al 1968. *Phys. Lett. A* 27: 581–82
37. Grin, G. A., Kunselman, R. *Phys. Lett. B* 31: 116–18
38. Au-Yang, M. Y., Cohen, M. L. 1968. *Phys. Rev.* 174: 468–76
39. Pauling, L. C. 1960. *The Nature of the Chemical Bond and the Structure of Molecules and Crystals.* Ithaca, N.Y.: Cornell Univ. 3rd ed.
40. Panofsky, W. K., Aamodt, R. L., Hadley, J. 1951. *Phys. Rev.* 81: 565–74
41. Dunaitsev, A. F., Petrukhin, V. I., Prokoshkin, Yu. D., Rykalin, V. I. 1962. *Zh. Eksp. Teor. Fiz.* 42: 1680–82; 1962. *Sov. Phys. JETP* 15: 1167
42. Petrukhin, V. I. Prokoshkin, Yu. D. 1963. *Nuovo Cimento* 28: 99–106
43. Dunaitsev, A. F., Petrukhin, V. I., Prokoshkin, Yu. D. 1964. *Nuovo Cimento* 34: 521–28
44. Chabre, M., Depommier, P., Heintze, J., Soergel, V. 1963. *Phys. Lett.* 5: 67–69
45. Bartlett, D., Devons, S., Meyer, S. L., Rosen, J. L. 1964. *Phys. Rev. B* 136: 1452–63
46. Ashkin, J. 1960. *Nuovo Cimento* 16: 490–504
47. Petrukhin, V. I., Prokoshkin, Yu. D. 1964. *Nucl. Phys.* 54: 414–16
48. Zaimidoroga, O. A. et al 1965. *Zh. Eksp. Teor. Fiz.* 48: 1267–78; 1965. *Sov. Phys. JETP* 21: 848
49. Marshak, R. E. 1952. *Meson Physics.* New York: McGraw
50. Petrukhin, V. I., Prokoshkin, Yu. D. 1965. *Dokl. Akad. Nauk. SSSR* 160: 71–72
51. Petrukhin, V. I., Prokoshkin, Yu. D., Suvorov, V. M. 1968. *Zh. Eksp. Teor. Fiz.* 55: 2173–80; 1969. *Sov. Phys. JETP* 28: 1151
52. Petrukhin, V. I., Suvorov, V. M. 1973. Preprint JINR, Dubna
53. Petrukhin, V. I., Ponomarev, L. I., Prokoshkin, Yu. D. 1967. *Khim. Vys. Energ.* 1: 283–85
54. Ponomarev, L. I. 1965. *Yad. Fiz.* 2: 223–31; 1966. *Sov. J. Nucl. Phys.* 2: 160

55. Ponomarev, L. I. 1967. *Yad. Fiz.* 6:389–95; 1968. *Sov. J. Nucl. Phys.* 6:281
56. Ponomarev, L. I., Prokoshkin, Yu. D. 1968. *Comments Nucl. Particle Phys.* 2:176–79
57. Ponomarev, L. I., Puzynina, T. P. 1967. *Zh. Eksp. Teor. Fiz.* 52:1273–82; 1967. *Sov. Phys. JETP* 25:846
58. Rosenberg, R. L. 1949. *Phil. Mag.* 40:759–69
59. Baker, G. A. 1960. *Phys. Rev.* 117:1130–36
60. Martin, A. D. 1963. *Nuovo Cimento* 27:1359–78
61. Condo, G. T. 1964. *Phys. Lett.* 9:65–66
62. Condo, G. T., Hill, R. D., Martin, A. D. 1964. *Phys. Rev. A* 133:1280–94
63. Gershtein, S. S. 1960. *Zh. Eksp. Teor. Fiz.* 39:1170–72; 1961. *Sov. Phys. JETP* 12:815
64. Rook, J. R. 1970. *Nucl. Phys. B* 20:14–22
65. Landau, L. D., Lifshits, E. M. 1963. *Kvantovaja Mekhanika,* 2nd ed. Fizmatgiz
66. Fermi, E., Segrè, E. 1933. *Z. Phys.* 82:729
67. Fröman, N., Fröman, P. O. 1972. *Phys. Rev. A* 6:2064–67
68. Parilis, E. S. 1969. *Auger Effect,* Tashkent, FAN
69. Krumshtein, Z. V., Petrukhin, V. I., Ponomarev, L. I., Prokoshkin, Yu. D. 1968. *Zh. Eksp. Teor. Fiz.* 54:1690–96; 1968. *Sov. Phys. JETP* 27:906
69a. Goldanskii, V. I. et al 1973. JINR Preprint
70. Petrukhin, V. I., Risin, V. E., Samenkova, I. F., Suvorov, V. M. 1973.

71. Krumshtein, Z. V. et al 1972. Preprint JINR P1-6853, Dubna
72. Krumshtein, Z. V., Petrukhin, V. I., Smirnova, L. M., Suvorov, V. M., Yutlandov, I. A. 1970. Preprint JINR P12-5224, Dubna
73. Krumshtein, Z. V., Petrukhin, V. I., Ponomarev, L. I., Prokoshkin, Yu. D. 1968. *Zh. Eksp. Teor. Fiz.* 55:1640–44; 1969. *Sov. Phys. JETP* 28:860
74. Goldanskii, V. I., Bersuker, I. B. 1972. *Dokl. Akad. Nauk. SSSR.* 203:1332–35
75. Vilgelmova, L. et al 1972. Preprint JINR P1-6854, Dubna
76. Hammett, L. P. 1970. *Physical Organic Chemistry.* New York: McGraw
77. Barkas, W. H. et al 1958. *Phys. Rev.* 112:622–23
77a. Vander Velde-Wilquet, C. et al 1972. *Nuovo Cimento Lett.* 5:1099–1103
78. Pawlewicz, W. T. et al 1970. *Phys. Rev. D* 2:2538–44
78a. Bugg, W. M. et al 1972. *Phys. Rev. D* 5:2142–45
79. Petrukhin, V. I. 1971. *Rep. IV Int. Conf. High Energy Phys. Nucl. Structure, Dubna,* 41–43
80. Petrukhin, V. I., Risin, V. E., Suvorov, V. M. 1973. Preprint JINR, Dubna
81. Knight, J. D., Schillaci, M. E., Naumann, R. A. 1971. *Rep. IV Int. Conf. High Energy Phys. Nucl. Structure, Dubna*
82. Zinov, V. G., Konin, A. D., Mukhin, A. I. 1972. Preprint JINR P14-6407, Dubna
83. *The Meeting on Muons in Solid State Physics, 1971, Bürgenstock, Switzerland*
84. Hollander, J. M., Shirley, D. A. 1970. *Ann. Rev. Nucl. Sci.* 20:435–66

SOME RELATED ARTICLES APPEARING
IN OTHER ANNUAL REVIEWS

From the *Annual Review of Astronomy and Astrophysics,* Volume 12 (1974)

Equation of State at Ultra High Densities, V. Canuto
Nucleo-Cosmochronology, D. N. Schramm

From the *Annual Review of Biophysics and Bioengineering,* Volume 3 (1974)

^{13}C-NMR, H. Sternlicht

From *The Excitement and Fascination of Science* (A special publication)

The Organization of Science, R. W. Gerard
Thirty Years of Atomic Chemistry, W. F. Libby
Fifty Years of Physical Chemistry in the California Institute of Technology, L. C.
  Pauling

REPRINTS

The conspicuous number aligned in the margin with the title of each article in this volume is a key for use in ordering reprints.

Available reprints are priced at the uniform rate of $1 each postpaid. Payment must accompany orders less than $10. A discount of 20% will be given on orders of 20 or more. For orders of 200 or more, any Annual Reviews article will be specially printed.

The sale of reprints of articles published in the Reviews has been expanded in the belief that reprints as individual copies, as sets covering stated topics, and in quantity for classroom use will have a special appeal to students and teachers.

AUTHOR INDEX

CUMULATIVE INDEXES

CONTRIBUTING AUTHORS VOLUMES 14-23

443

CHAPTER TITLES VOLUMES 14-23